FLUORESCENCE MICROSCOPY OF LIVING CELLS IN CULTURE

PART B

METHODS IN CELL BIOLOGY, VOLUME 30

Series Editor

LESLIE WILSON

Department of Biological Sciences
University of California, Santa Barbara
Santa Barbara, California

Prepared under the Auspices of the American Society for Cell Biology

FLUORESCENCE MICROSCOPY OF LIVING CELLS IN CULTURE

PART B
Quantitative Fluorescence Microscopy — Imaging and Spectroscopy

METHODS IN CELL BIOLOGY, VOLUME 30

Edited by

D. LANSING TAYLOR
DEPARTMENT OF BIOLOGICAL SCIENCES
CENTER FOR FLUORESCENCE RESEARCH IN BIOMEDICAL SCIENCES
CARNEGIE-MELLON UNIVERSITY
PITTSBURGH, PENNSYLVANIA

YU-LI WANG
CELL BIOLOGY GROUP
WORCESTER FOUNDATION FOR EXPERIMENTAL BIOLOGY
SHREWSBURY, MASSACHUSETTS

ACADEMIC PRESS, INC.
Harcourt Brace Jovanovich, Publishers
San Diego New York Boston
London Sydney Tokyo Toronto

Front cover photograph: Multiple spectral parameter mapping image created by the simultaneous display of three individual images; fluorescence image of tubulin in a single, well-spread, relatively nonmotile 3T3 fibroblast. (See Chapter 17, Figure 3 in Fluorescence Microscopy of Living Cells in Culture, Part B.)

This book is printed on acid-free paper. ∞

Academic Press, Inc.
San Diego, California 92101

United Kingdom Edition published by
Academic Press Limited
24-28 Oval Road, London NW1 7DX

Library of Congress Catalog Card Number: 64-14220

ISBN 0-12-684755-X (alk. paper)

Printed in the United States of America
90 91 92 93 9 8 7 6 5 4 3 2 1

CONTENTS

Contents of *Fluorescence Microscopy of Living Cells in Culture. Part A. Fluorescent Analogs, Labeling Cells, and Basic Microscopy* (Volume 29 of Methods in Cell Biology)

CONTRIBUTORS

Numbers in parentheses indicate the pages on which the authors' contributions begin.

DAVID A. AGARD, Howard Hughes Medical Institute and Department of Biochemistry, University of California, San Francisco, San Francisco, California 94143 (353)

DONNA J. ARNDT-JOVIN, Department of Molecular Biology, Max Planck Institute for Biophysical Chemistry, D-3400 Göttingen, Federal Republic of Germany (417)

DANIEL AXELROD, Biophysics Research Division and Department of Physics, University of Michigan, Ann Arbor, Michigan 48109 (245, 333, 399)

G. J. BRAKENHOFF, Department of Electron Microscopy and Molecular Cytology, University of Amsterdam, 1018 TV Amsterdam, The Netherlands (379)

G. R. BRIGHT, Department of Biological Sciences, Center for Fluorescence Research in Biomedical Sciences, Carnegie Mellon University, Pittsburgh, Pennsylvania 15213 (157, 449)

JOSEPH BRYAN, Departments of Medicine and Cell Biology, Baylor College of Medicine, Houston, Texas 77030 (47)

P. CONRAD, Department of Biological Sciences, Center for Fluorescence Research in the Biomedical Sciences, Carnegie Mellon University, Pittsburgh, Pennsylvania 15213 (449)

R. DEBIASIO, Department of Biological Sciences, Center for Fluorescence Research in the Biomedical Sciences, Carnegie Mellon University, Pittsburgh, Pennsylvania 15213 (449)

ELLIOT L. ELSON, Department of Biological Chemistry, Washington University School of Medicine, St. Louis, Missouri 63110 (307)

L. ERNST, Department of Biological Sciences, Center for Fluorescence Research in the Biomedical Sciences, Carnegie Mellon University, Pittsburgh, Pennsylvania 15213 (449)

G. W. FISHER, Department of Biological Sciences, Center for Fluorescence Research in the Biomedical Sciences, Carnegie Mellon University, Pittsburgh, Pennsylvania 15213 (157)

DAVID GROSS, Department of Biochemistry and Program in Molecular and Cellular Biology, University of Massachusetts, Amherst, Massachusetts 01003 (193)

EDWARD H. HELLEN, Biophysics Research Division and Department of Physics, University of Michigan, Ann Arbor, Michigan 48109 (399)

BRIAN HERMAN, Laboratories for Cell Biology, Department of Cell Biology and Anatomy, University of North Carolina School of Medicine, Chapel Hill, North Carolina 27599 (219)

YASUSHI HIRAOKA, Howard Hughes Medical Institute and Department of Biochemistry, University of California, San Francisco, San Francisco, California 94143 (353)

SHINYA INOUÉ, Marine Biological Laboratory, Woods Hole, Massachusetts 02543 (85)

ŽELJKO JERIČEVIČ, Rugjer Boskovic Institute, Zagreb, Croatia, Yugoslavia (47)

THOMAS M. JOVIN, Department of Molecular Biology, Max Planck Institute for Biophysical Chemistry, D-3400 Göttingen, Federal Republic of Germany (417)

LESLIE M. LOEW, Department of Physiology, University of Connecticut Health Center, Farmington, Connecticut 06032 (193)

N. NANNINGA, Department of Electron Microscopy and Molecular Cytology, University of Amsterdam, 1018 TV Amsterdam, The Netherlands (379)

M. NEDERLOF, Department of Biological Sciences, Center for Fluorescence Research in the Biomedical Sciences, Carnegie Mellon University, Pittsburgh, Pennsylvania 15213 (449)

HONG QIAN, Department of Biological Chemistry, Washington University School of Medicine, St. Louis, Missouri 63110 (307)

J. ROGOWSKA, Department of Biological Sciences, Center for Fluorescence Research in the Biomedical Sciences, Carnegie Mellon University, Pittsburgh, Pennsylvania 15213 (157)

K. RYAN, Department of Biological Sciences, Center for Fluorescence Research in the Biomedical Sciences, Carnegie Mellon University, Pittsburgh, Pennsylvania 15213 (449)

JOHN W. SEDAT, Howard Hughes Medical Institute and Department of Biochemistry, and Biophysics, University of California, San Francisco, San Francisco, California 94143 (353)

PETER SHAW, Howard Hughes Medical Institute and Department of Biochemistry, University of California, San Francisco, San Francisco, California 94143 (353)

JESSE E. SISKEN, Department of Microbiology and Immunology, College of Medicine, University of Kentucky, Lexington, Kentucky 40536 (113)

LOUIS C. SMITH, Department of Medicine, Baylor College of Medicine, Houston, Texas 77030 (47)

D. TAYLOR, Department of Biological Sciences, Center for Fluorescence Research in the Biomedical Sciences, Carnegie Mellon University, Pittsburgh, Pennsylvania 15213 (449, 157)

ROGER Y. TSIEN, Department of Physiology-Anatomy, University of California, Berkeley, California 94720 (127)

H. T. M. VAN DER VOORT, Department of Electron Microscopy and Molecular Cytology, University of Amsterdam, 1018 TV Amsterdam, The Netherlands (379)

E. A. VAN SPRONSEN, Department of Electron Microscopy and Molecular Cytology, University of Amsterdam, 1018 TV Amsterdam, The Netherlands (379)

A. WAGGONER, Department of Biological Sciences, Center for Fluorescence Research in the Biomedical Sciences, Carnegie Mellon University, Pittsburgh, Pennsylvania 15213 (449)

BRENT WIESE, Departments of Medicine and Cell Biology, Baylor College of Medicine, Houston, Texas 77030 (47)

DAVID E. WOLF, Worcester Foundation for Experimental Biology, Shrewsbury, Massachusetts 01545 (271)

IAN T. YOUNG, Department of Applied Physics, Delft University of Technology, Delft, The Netherlands (1)

PREFACE

Fluorescence techniques are uniquely suitable for probing living cells because of their sensitivity and specificity. Since fluorescence from a single cell can be detected with a microscope both as an image and as a photometric signal, fluorescence microscopy has great potential for qualitative and quantitative studies on the structure and function of cells. However, owing to previous technical limitations, fluorescence has been used primarily for staining fixed cells for many years. It has not been until recently that the true power of the techniques has evolved for use with single living cells. The most important advances that have made this possible include the development of (1) probes for specific structures or environmental parameters; (2) methods for delivering fluorescent probes into living cells; (3) methods for detecting weak fluorescence signals from living cells; and (4) methods for acquiring, processing, and analyzing fluorescence signals with microscopes.

The primary purpose of this and the accompanying volume is to provide readers with detailed descriptions of methods in these four areas. While techniques for fluorescence spectroscopy in solution are described in various sources, there has been no convenient source for the methods specifically applied to living cells. Even with an extensive literature search, one often finds crucial technical details, including instrumentation, sample handling, and precautions, left out in many research articles. It is our hope that these volumes will provide enough detail to make the new developments approachable by most investigators. Although some biological perspectives are provided in many chapters, the main emphasis of the volumes is practical laboratory methods; the job of biological reasoning and experimental design is left to individual investigators. The books are thus targeted primarily at experienced cell biologists who wish to apply modern fluorescence techniques. However, they should also be of great interest to biochemists and molecular biologists who attempt to correlate results in test tubes with activities in living cells. In addition, many chapters should be valuable to those specializing in instrumentation, including microscopy, electronic imaging, and digital image processing.

The two volumes represent a collective effort of many investigators. The chapters were assembled by specific areas which, in our view, were important or held great promise in the future. We then invited those researchers with extensive experience in the particular area to make contributions. There was a certain degree of subjectiveness in choosing the topics. On the one hand, we have included topics crucial to, but not specific for, fluorescence microscopy of living cells, including microscopy cell culture, microinjection, microscopy photometry, and low light level imaging. On the other hand, we decided to sacrifice several useful topics that were either not in a mature stage of development or where we were unable to obtain a commitment from an authority.

The first volume (Part A) deals with the preparation, delivery, and detection of fluorescent probes. The first half is focused on the preparation of specific structural probes, including fluorescent analogs that can be utilized by living cells in structural assembly, fluorescent molecules that bind to specific cellular components, and probes that can be used to label particular cellular compartments. There are special challenges in the preparation of each class of probes, including proteins, small peptides, heterocyclic compounds, lipids, and polysaccharides. Subsequent chapters discuss factors that determine the destination of probes and methods for delivering probes to specific sites in living cells. The second half of the first volume discusses the detection of fluorescent probes in living cells, including issues related to sample physiology (microscopy cell culture), optics (basic fluorescence microscopy), and signal detection (electronic photometry and imaging, immunoelectron microscopic detection of fluorophores). The last few chapters introduce modern techniques in image detection and provide a continuity to quantitative analytical methods covered in Part B.

The second volume (Part B) explores a combination of the theoretical and technical issues related to the quantitation of fluorescence signals in the living cell with a light microscope. The first section explores the engineering principles required in the characterization of the performance of an imaging system. The use of system validation procedures and quantitative fluorescent standards are explored in detail. The remainder of Part B is devoted to specific applications and optical methods. A mix of theoretical and practical issues is discussed, including the measurement of membrane potential, ionic concentrations, tracer diffusion coefficients, total internal reflection, fluorescence polarization, and three-dimensional reconstruction. Thus, the two-volume set defines a technical continuum from organic chemistry, through biochemistry, cell biology, physics, and engineering, to computer science. The present status of the field reflects the occurrence of a revolution in cell biological research.

We would like to thank all contributing authors for providing us with their extensive experience in various areas. Most of them have worked closely with us in planning their chapters and minimizing overlaps, then submitting excellent manuscripts in a timely fashion and answering questions which arose during editing.

D. LANSING TAYLOR
YU-LI WANG

Chapter 1

Image Fidelity: Characterizing the Imaging Transfer Function

IAN T. YOUNG

Department of Applied Physics
Delft University of Technology
Delft, The Netherlands

I. Introduction

Fluorescence microscopy is today one of the most significant tools for the examination of cells and cellular constituents. Fluorescent probes and fluorescently marked immunological probes offer us the ability to visualize and quantify basic structures within the cell. The ability to perform quantitative measurements on fluorescence images, however, can be severely limited by the same instrument that provides us with the images—the microscope itself. When a fluorescence image is converted to a digital image for subsequent computer processing, the effect of the scanning instrument as well as the quantitization process can further compound the problem. It is the purpose of this chapter to study the various limitations and distortions inherent and introduced in quantitative fluorescence microscopy and to describe ways to compensate and/or eliminate them.

A. The Reality of Distortion

To begin, it is important to realize that distortion in fluorescence images—or, for that matter, any image—is unavoidable. Even if electrooptical sensors were linear and introduced no noise and microscope lenses had no geometrical aberrations, no chromatic aberrations, and no glare, images observed through a microscope and recorded through a sensor would still contain distortion. At the most basic level this is caused by the finite size of microscopes, their lenses, and, most importantly, their apertures. It is not necessary for us at this time to go through the theory that describes this result. Suffice it to say that the diffraction limits of light optics (dictated by the wave nature of light) do not permit us to produce arbitrarily sharp images.

This phenomenon is illustrated in Fig. 1. The cytoskeletal actin molecules in a fibroblast have been labeled with a fluorescently tagged

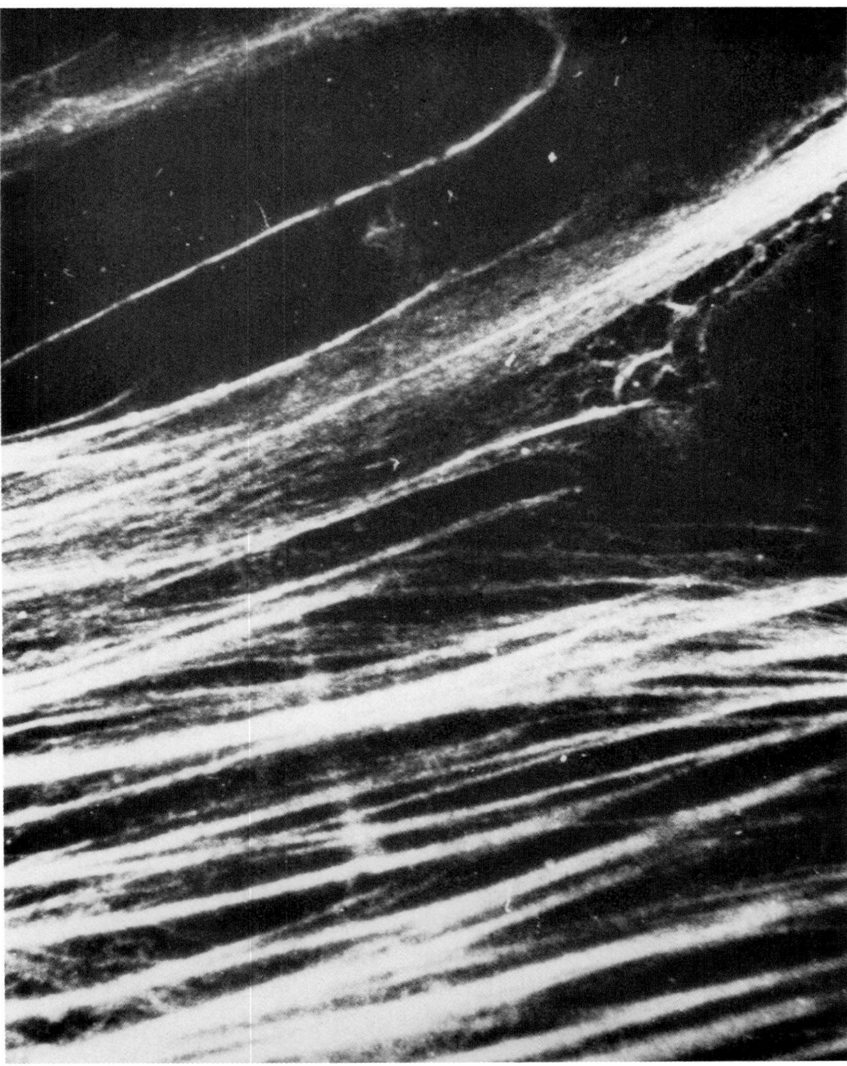

FIG. 1. Actin molecules in a fibroblast stained with a fluorescently tagged antibody.

antibody. The actin filaments, while they have lengths that can be measured in micrometers, have diameters (widths) that are measured in nanometers—two orders of magnitude smaller than the wavelength of visible light.

A single labeled actin filament cannot be resolved. What we see in the image is a distortion of the light distribution along the length of the filament. However useful though the resulting image may be, it remains a distorted version of reality.

B. Models of a Fluorescence Imaging System

To understand the origin and the nature of distortions in a quantitative fluorescence microscope system, it is essential that we understand the various components that form such a system and how they work together to produce a digital image in a computer memory. We begin with a

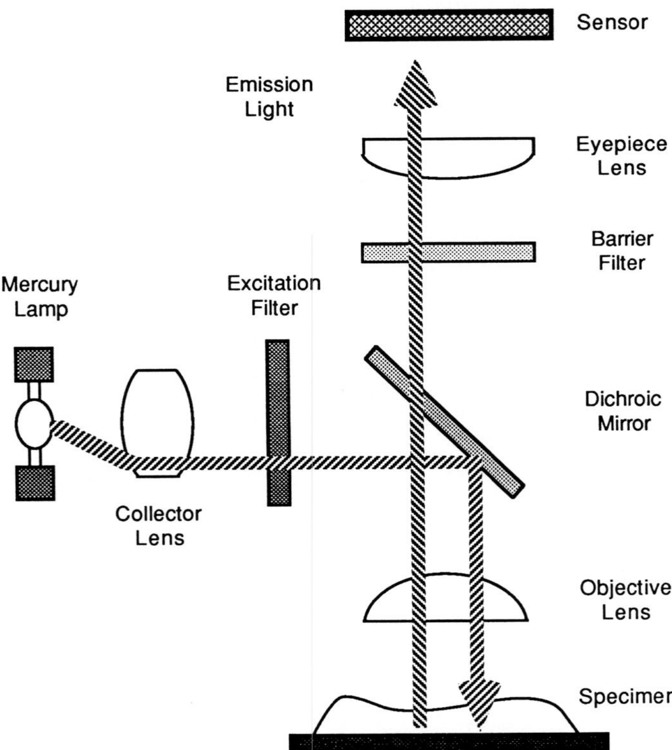

FIG. 2. Schematic diagram of a fluorescence imaging system.

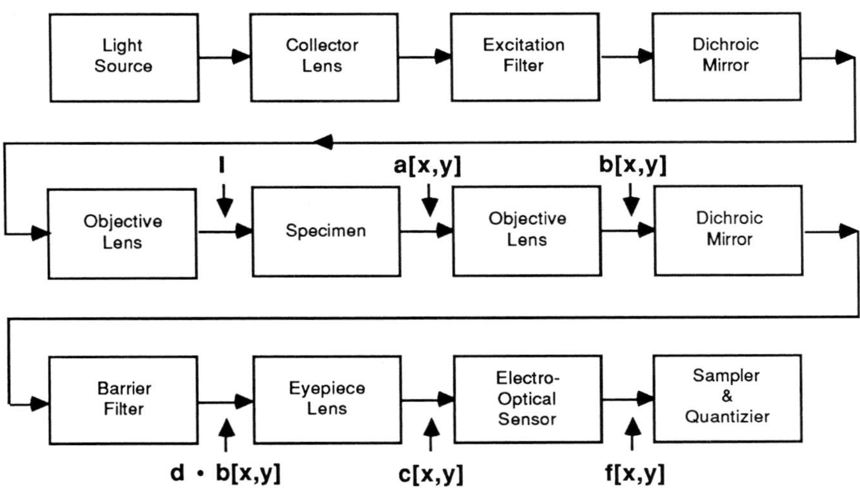

FIG. 3. System diagram of a fluorescence imaging system. The illumination I produces $a[x, y]$ from the specimen. Various lenses and filters interact to produce the final image $f[x, y]$.

schematic model of such a system as shown in Fig. 2. This model represents an epiillumination fluorescence imaging system [after Ploem, 1967]. Unless we indicate otherwise, we will always be referring to this epiillumination model.

The model does not directly indicate how the various distortions are introduced nor how they can be described. It does, however, offer us a starting point for building an analytical description of the imaging process. This description, by its very nature, requires the use of a mathematical formalism, that is, a set of equations to describe the various components and their interrelationships.

While the model shown in Fig. 2 may indicate the physical layout of a quantitative fluorescence microscope system, it is not the best choice for representing the "flow of data" in such a system. The model shown in Fig. 3 is more appropriate and is usually referred to as a system diagram.

Each of the components is now a subsystem and the "flow of data" is more easily represented and described. Based upon this model, we are now in a position to introduce a major assumption concerning many of these subsystems. Specifically, we assume that the (optical) imaging components of this system form a linear, shift-invariant system. To understand the importance of this assumption as well as its consequences, it is necessary to define carefully each of these terms.

II. The Concept of a Linear, Shift-Invariant (LSI) System

A. What Is Linearity?

Informally, the concept of linearity is quite straightforward. Let us start with a distribution of fluorescently labeled objects that under a certain illumination produce a distribution of light $a[x, y]$. Through an imaging system this yields a distribution of light $b[x, y]$. Let us now alter the fluorescent dyes (fluorochromes) or the illumination such that $a[x, y]$ becomes $2a[x, y]$, that is, twice as much light is produced by the objects. If the imaging system is linear, then $b[x, y]$ will become $2b[x, y]$. In general, if $a[x, y] \rightarrow b[x, y]$ and the system is linear, then $\eta a[x, y] \rightarrow \eta b[x, y]$. (The symbol "$\rightarrow$" is to be read as "will give under the imaging operation.") In words, we say that, if the input image ($a[x, y]$) is multiplied by a scale factor (η) and the system is linear, then the resulting output image ($b[x, y]$) will be multiplied by the same scale factor. This, however, is only half of the definition of linearity. The remainder is as follows.

Consider now that a given combination of illumination and fluorochrome produce $a_1[x, y]$ which yields image $b_1[x, y]$. A second combination produces $a_2[x, y]$, which in turn yields image $b_2[x, y]$. We now construct a combination of illumination and fluorochromes such that fluorescently labeled objects produce a distribution of light $a_1[x, y] + a_2[x, y]$. If the imaging system is linear, then the resulting distribution of light will be $b_1[x, y] + b_2[x, y]$. If the input image is the sum of two images ($a_1[x, y]$, $a_2[x, y]$) and the system is linear, then the resulting output image will be the sum of the two output images ($b_1[x, y]$, $b_2[x, y]$).

These two conditions can be summarized in a single statement. Let $a_1[x, y] \rightarrow b_1[x, y]$, $a_2[x, y] \rightarrow b_2[x, y]$, and η_1 and η_2 be scale factors. If the imaging system is linear, then

$$\eta_1 a_1[x, y] + \eta_2 a_2[x, y] \rightarrow \eta_1 b_1[x, y] + \eta_2 b_2[x, y] \tag{1}$$

As can be seen from the previous discussion, this relation summarizes the necessary and sufficient conditions for a system to be considered linear.

One of the "by-products" of this definition should be clear: If we let η_1 and η_2 be zero in Eq. (1), then we have the result that, for a linear system, zero in gives zero out ($0 \rightarrow 0$). While this might seem like a trivial observation at this point, it will have important consequences when we come to the problem of shading correction in fluorescence imagery.

B. What Is Shift Invariance?

We are used to the idea that, if a cell is observed in the upper left portion of a microscope field of view or the lower right portion of that field of view, the result will be the same. We expect that the image produced by a microscope will be *invariant* to shifts. While this is the goal of all microscope designs, it is a goal that is only approached, never reached. Using the terminology introduced in the previous section, let us consider the input distribution of light $a[x, y]$ and the output distribution of light $b[x, y]$. Shift invariance means that if the input image is shifted from $a[x, y]$ to $a[x - x_0, y - y_0]$ then the output image will be shifted from $b[x, y]$ to $b[x - x_0, y - y_0]$.

C. Is a Fluorescence Imaging System LSI?

While the two potential properties *linearity* and *shift invariance* are interesting, they would not be worth pursuing if they did not represent a reasonable description of the imaging process in a fluorescence microscope. In the context of mathematical definitions no quantitative fluorescence microscope system will be LSI. To an excellent *approximation*, however, the system will be LSI and further, the insight that we gain by using this assumption will help us in the evaluation of the effects generated by nonlinearities as well as spatial variance.

D. Fluorescence Image as a Superposition Result—Convolution

Let us now develop a basic result that follows from our LSI assumption for a quantitative fluorescence microscope system. We start by defining a basic test object that has the property that it has a spatial position, a finite total brightness, but no spatial extent. This test object is called a unit impulse $\delta[x, y]$ and we may think of it as a pinpoint of light on a black background. Mathematically this impulse function has the properties that

(i) $$\delta[x, y] = 0 \qquad \text{unless} \quad x = y = 0 \tag{2a}$$

(ii) $$\int_{-\infty}^{+\infty} \int_{-\infty}^{+\infty} \delta[x, y] \, dx \, dy = 1 \tag{2b}$$

Equation (2a) states that the position of the impulse is ($x = 0$, $y = 0$) but that it has no spatial extent. Equation (2b) states that the total brightness of the impulse is one. It is possible to show that *any* distribution of fluorescent light produced by illumination and fluorochromes can be represented by a weighted sum of these impulses. Formally, any distribution can be represented by:

$$a[x,y] \approx \sum_{u=-\infty}^{+\infty} \sum_{v=-\infty}^{+\infty} a[u,v] \cdots \delta[x-u,y-v]\,\Delta u\,\Delta v \qquad \text{(3a)}$$

This is a complicated equation but it says, essentially, that a collection of unit impulses $\{\delta[x,y]\}$ at different positions with weighting coefficients $\{a[u,\ v]\}$ can produce an arbitrary image $a[x,y]$. It is beyond the scope of this chapter to prove this statement. For further details the reader is referred to Oppenheim *et al.*, 1982. The standard form of this statement is to allow the sum of Eq. (3a) to pass to an integral giving

$$a[x,y] = \int_{-\infty}^{+\infty}\int_{-\infty}^{+\infty} a[u,v]\cdot\delta[x-u,y-v]\,du\,dv \qquad \text{(3b)}$$

The interpretation of Eq. (3b) is the same as that of Eq. (3a). We are now in a position to develop the consequences of this last equation as well as the assumption that the imaging system is LSI.

Consider a single term in Eq. (3a), $(a[u,v]\,\Delta u\,\Delta v)\cdot\delta[x-u,y-v]$. Let us name the output image that results from a single input impulse $\delta[x,y]$ as $h[x,y]$. As a result of the LSI assumption we have the following:

(i) $\delta[x,y] \rightarrow h[x,y]$ (*by definition*)
(ii) $(a[u,v]\,\Delta u\,\Delta v)\,\delta[x,y] \rightarrow (a[u,v]\Delta u\,\Delta v)\cdot h[x,y]$ (*by linearity*)
(iii) $\delta[x-y,\ y-v] \rightarrow h[x-u,\ y-v]$ (by *shift invariance*)
(iv) $(a[u,v]\,\Delta u\,\Delta v)\cdot\delta[x-u,\ y-v] \rightarrow (a[u,v]\,\Delta u\,\Delta v)\cdot h[x-u,\ y-v]$ (*by LSI*)
(v) $\Sigma\Sigma\, a[u,v]\cdot\delta[x-u,\ y-v]\Delta u\,\Delta v \rightarrow \Sigma\Sigma\, a[u,v]\cdot h[x-u,\ y-v]\,\Delta u\,\Delta y$ (*by linearity*)
(vi) $\int\int a[u,v]\cdot\delta[x-u,\ y-v]\,du\,dv \rightarrow \int\int a[u,v]\cdot h[x-u,\ y-v]\,du\,dv$ (*by linearity*)

The last line of this proof is a central result. In Eq. (3b) we indicated that an arbitrary input image can be represented as a collection of weighted impulses. We also stated earlier that *by definition* if the input image was $a[x,y]$ then the resulting output image was $b[x,y]$. In line (vi) above we show that the image $b[x,y]$ can be computed through knowledge of $h[x,y]$ and by application of the *convolution* equation:

$$b[x,y] = \int_{-\infty}^{+\infty}\int_{-\infty}^{+\infty} a[u,v]\cdot h[x-u,y-v]\,du\,dv \qquad \text{(4)}$$

This important result, sometimes written as $b[x, y] = a[x, y] * h[x, y]$, says that an output image formed by an LSI system can be described as the result of the convolution between $a[x, y]$ and the image $h[x, y]$ formed by a single input impulse $\delta[x, y]$. The image $h[x, y]$ contains all of the information necessary to describe the imaging system because, if we know $h[x, y]$, we can then compute the output $b[x, y]$ for any other $a[x, y]$. The response $h[x, y]$ to a single input impulse $\delta[x, y]$—a single *point* of light—is referred to in optics as the *point spread function* (PSF) and in system theory as the impulse response.

In summary, each arbitrary input image can be thought of as a weighted collection of impulses; each weighted impulse generates a weighted point spread function; the sum of the weighted point spread functions is the resulting output image.

III. Characterizing LSI Systems with Sinusoids

A. Sinusoids in/Sinusoids out

We are almost in a position to characterize the image fidelity of a quantitative fluorescence microscope system. What we will show in this section is that the key input image to observe, as it passes through a LSI optical system, is a sinusoidal signal. There are two reasons:

1. If the input signal to an LSI system is a sinusoid with frequency ω, then the output signal will also be a sinusoid *with precisely the same frequency* ω. The amplitude of the sinusoid may change, the phase of the sinusoid may change, but the frequency will be the same.
2. It is possible to represent virtually any input image as a weighted sum of sinusoids.

Using the property given in Eq. (1) together with the two statements above, it is possible for us to describe how sinusoidal terms in the input image will be altered as they pass through an LSI optical system.

B. The Complex Sinusoid and Convolution

There are a number of ways to represent a sinusoid. In this chapter we shall use the complex exponential form described by Euler's relation:

$$e^{j\omega x} = \cos(\omega x) + j\sin(\omega x) \qquad j = \sqrt{-1} \tag{5}$$

Using this formulation we can prove the first statement above. Consider a sinusoidal input image:

$$a[x, y] = \exp[j(\omega_x x + \omega_y y)] \tag{6}$$

Note that $a[x, y]$ has two distinct sinusoidal frequencies, ω_x and ω_y, one in the x direction and one in the y direction. We can now derive the result that the output image $b[x, y]$ has the same basic character as the input image and that only the complex *amplitude* of the sinusoidal term will be affected by the LSI system. Using Eq. (6) in Eq. (4) gives

$$b[x, y] = e^{j(\omega_x x + \omega_y y)} \left(\int_{-\infty}^{+\infty} \int_{-\infty}^{+\infty} h[u, v] \exp[-j(\omega_x u + \omega_y v)]\, du\, dv \right) \tag{7}$$

The first term in Eq. (7) is the sinusoidal term with the same frequencies, ω_x and ω_y, as in the input image. The term in parentheses represents the change in amplitude caused by the LSI system. This new amplitude will, of course, be dependent upon the specific values of ω_x and ω_y as well as the form of the PSF, $h[x, y]$. This dependency is usually summarized by:

$$b[x, y] = H(\omega_x, \omega_y) \exp[j(\omega_x x + \omega_y)] \tag{8a}$$

where

$$H(\omega_x, \omega_y) = \int_{-\infty}^{+\infty} \int_{-\infty}^{+\infty} h[x, y] \exp[-j(\omega_x x + \omega_y y)]\, dx\, dy \tag{8b}$$

In Eq. (8b) the variables x and y are dummy spatial variables of integration.

In Fig. 4 we see the effect of this phenomenon through the application of a PSF to four test images, each with a different value of ω_x. The effect of $h[x, y]$ in Fig. 4 is to change the complex amplitudes of the sinusoids from the initial values of one to the values shown. Test patterns, such as those used in Fig. 4, are sometimes referred to as sinusoidal gratings.

C. Description of an Image in Terms of Complex Sinusoids—the Fourier Representation

We stated previously that any input image could be represented as a weighted sum of sinusoids. This statement is essential if we are to use the results of Eq. (7) to describe how an image propagates through an LSI system. The foundation for this statement lies in the results of the nineteenth century French mathematician/physicist Jean Baptiste Joseph Fourier. In his work on the diffusion of heat, he showed how a very wide variety of physical signals, including those concerning us in this chapter,

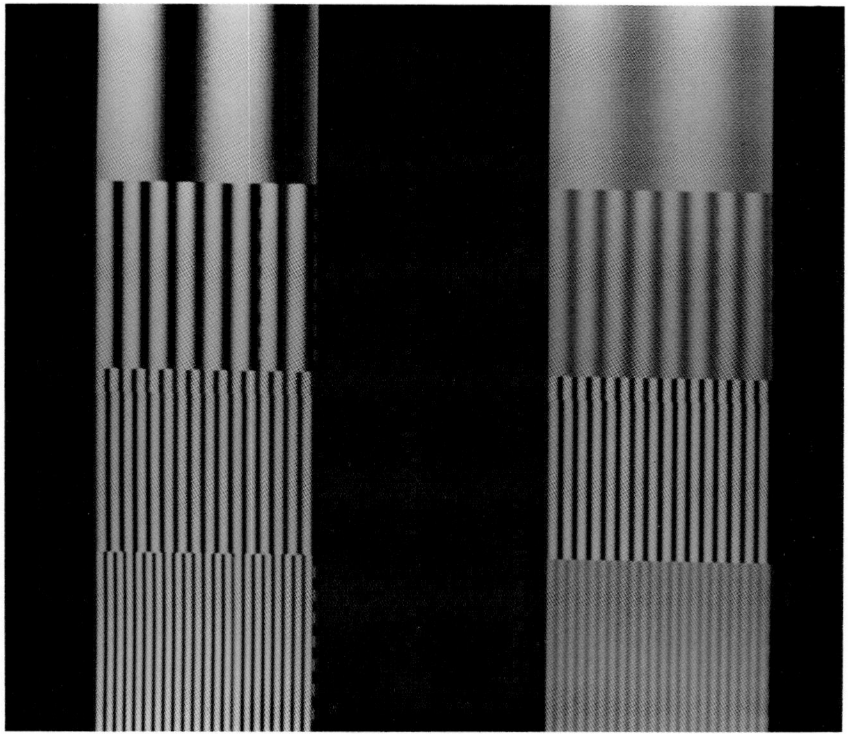

FIG. 4. Four sinusoidal test images of the same input amplitude are shown. Each has been filtered by the same LSI linear filter. This filter has been chosen such that it enhances the middle frequencies and suppresses low and high frequenices.

could be represented by weighted sums or weighted integrals of sinusoidal functions. Deriving the Fourier representation is not within the scope of this chapter; we must instead accept the result. A clear (albeit nonrigorous) derivation is given in Chapter 4 of Oppenheim *et al.* (1983).

In Eq. (3) we saw that that *any* distribution of fluorescent light produced by illumination and fluorochromes could be represented by a weighted sum of impulses. It is also true that any distribution of light can be represented by:

$$a[x, y] \approx \left(\frac{1}{2\pi}\right)^2 \sum_{m=-\infty}^{+\infty} \sum_{n=-\infty}^{+\infty} A(m\,\Delta\omega_x, n\Delta\omega_y) \times \exp[j(m\,\Delta\omega_x x + n\,\Delta\omega_y y)]\,\Delta\omega_x\,\Delta\omega_y \tag{9}$$

Equation (9) says that each complex sinusoid, represented by $e^{j(\quad)}$, has an amplitude given by $\{A(\quad)\,\Delta\omega_x\,\Delta\omega_y\}$. The sum of all these weighted terms, together with a scale factor, gives an almost exact representation of the input image $a[x,y]$. If we allow the sum to pass to an integral by letting $\Delta\omega_x \rightarrow 0$ and $\Delta\omega_y \rightarrow 0$, then we have the equation that gives an exact representation for $a[x, y]$. At the same time we give the equation that tells us how to compute the amplitudes $A(\quad)$ for each of the complex sinusoids.

$$a[x,y] = \left(\frac{1}{2\pi}\right)^2 \int_{-\infty}^{+\infty}\int_{-\infty}^{+\infty} A(\omega_x,\omega_y)\exp[j(\omega_x x + \omega_y y)]\,d\omega_x\,d\omega_y \tag{10a}$$

$$A(\omega_x,\omega_y) = \int_{-\infty}^{+\infty}\int_{-\infty}^{+\infty} a[x,y]\exp[-j(\omega_x x + \omega_y y)]\,dx\,dy \tag{10b}$$

Careful comparison of Eqs. (10b) and (8b) will show that they are identical. The complex amplitudes of $a[x,y]$ are given by $A(\omega_x, \omega_y)$ and the complex amplitudes of $h[x,y]$ are given by $H(\omega_x, \omega_y)$. The relations in Eq. (10) are known as *Fourier transform* pairs. $A(\omega_x, \omega_y)$ and $H(\omega_x, \omega_y)$ are the Fourier transforms of $a[x,y]$ and $h[x,y]$, respectively. Likewise, $a[x,y]$ and $h[x,y]$ are the *inverse* Fourier transforms of $A(\omega_x, \omega y)$ and $H(\omega_x, \omega_y)$, respectively. Fourier transform pairs are frequently written as

$$A(\omega_x, \omega_y) = \mathcal{F}\{a[x,y]\} \qquad \text{(Forward transform; space} \rightarrow \text{frequency)}$$

$$a[x,y] = \mathcal{F}^{-1}\{A(\omega_x, \omega_y)\} \qquad \text{(Backward transform: frequency} \rightarrow \text{space)}$$

The ability to build up an image out of more and more terms in a Fourier representation, as suggested by Eq. (9), is illustrated in Fig. 5.

Finally, we are in a position to relate the convolution equation [Eq. (4)] to the Fourier description of an input image and the Fourier description of the PSF (point spread function). The key question is If an input image $a[x,y]$ is convolved with a PSF $h[x,y]$ to produce an output image $b[x,y]$, what is the relationship between their respective Fourier transforms? We can answer this question by considering

$$B(\omega_x, \omega_y) = \mathcal{F}\{b[x,y]\} \tag{11a}$$

Using the convolution relation from Eq. (4), this is

$$B(\omega_x, \omega_y) = \mathcal{F}\{a[x,y] * h[x,y]\} \tag{11b}$$

By manipulating the Fourier transform we can show that this becomes

$$B(\omega_x, \omega_y) = A(\omega_x, \omega_y)H(\omega_x, \omega_y) \tag{12}$$

Equation (12) illustrates the central role played by $H(\omega_x, \omega_y)$. The output image has a Fourier spectrum that is directly related to both the

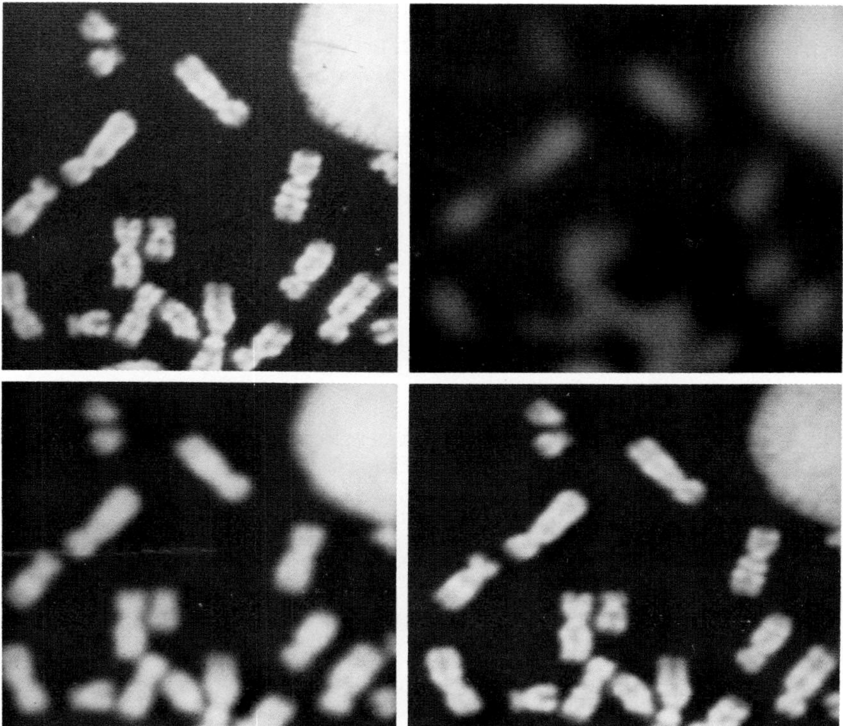

FIG. 5. The original image is shown in (a). In (b)–(d) we see the effect of including an increasing number of terms in the Fourier representation as defined in Eq. (10a).

input image spectrum and the Fourier transform of the PSF. The relationship is a multiplicative one. For any frequency (ω_x, ω_y) where $H(\omega_x, \omega_y) = 1$, the output complex amplitude will be identical to the input amplitude. For any frequency (ω_x, ω_y) where $H(\omega_x, \omega_y) = 0$, the output amplitude at that frequency will be zero. For this reason, $H(\omega_x, \omega_y)$ plays the role of a *filter*, determining which terms will be amplified ($|H| > 1$), which terms will be attenuated ($|H| < 1$), and which terms will be unchanged ($|H| = 1$). The common name for $H(\omega_x, \omega_y)$ is the optical transfer function (OTF); the absolute value of the OTF, $|H(\omega_x, \omega_y)|$, is referred to as the modulation transfer function (MTF). As an image is "transferred" from input to output, its Fourier spectrum is multiplied by the OTF. Just as the PSF, $h[x, y]$, is a complete description of an LSI optical system, so is the OTF, $H(\omega_x, \omega_y)$, a complete description. This follows from the fact that OTF = $\mathscr{F}\{\text{PSF}\}$ and PSF = $\mathscr{F}^{-1}\{\text{OTF}\}$.

The images that we are considering, $a[x, y]$, $h[x, y]$, etc., represent the distribution of light *intensity*, I, over a certain region. Normally, we consider light as an electric field, E, propagating in a certain direction with a certain magnitude and phase. For two reasons, however, we will continue to develop our description of quantitative fluorescence microscopes in terms of I instead of E. First, we assume that our sources, while they may be near-monochromatic, will be spatially incoherent. Mathematically, this means that the intensity (actually the irradiance) $I = \langle \mathrm{Re}\{E\} \rangle = \frac{1}{2}E \cdot E^*$ will be independent of phase. Second, our image sensors, indicated in Fig. 2, are not sensitive to the complex electric field but rather to the intensity distribution I. Our convolutions thus become the convolutions of intensity distributions.

IV. Characterizing the Fluorescence Imaging System

A. Model of a Fluorescence Imaging System (Reprise)

Based upon the Fourier models developed in the previous section, we are now prepared to describe each of the components in the system diagram of Fig. 3. It is important to understand that a number of assumptions will be used to develop this description: (1) that components are linear, (2) that components are shift invariant, and (3) that the descriptions hold *per wavelength* of light, either excitation or emission. In an important sense our descriptions will be monochromatic. All of these descriptions form an ideal model of a fluorescence imaging system. To the extent that a real system deviates from this ideal, our results will be progressively less applicable.

The first four subsystems in Fig. 3—light source, collector lens, excitation filter, and dichroic mirror—can be reasonably combined into a single subsystem whose function is to produce a certain distribution of excitation light at the back focal plane of the epiillumination objective lens. Let us term this illumination $I_0[x, y, \lambda_{ex}]$. When the illumination system is properly aligned, the distribution of excitation light on the specimen will be given by $I_{ex}[x, y, \lambda_{ex}] = I_{ex}[\lambda_{ex}]$, an illumination that is constant (uniform) in x and y but definitely a function of the excitation wavelength(s) λ_{ex}.

If the illumination is not constant, this can contribute to the phenomenon known as shading. The amount of shading can be assessed by Fourier techniques by measuring the deviation from a single frequency term at $\omega_x = \omega_y = 0$ (Young, 1970) or by nonlinear, morphological techniques (Dorst, 1985). Both techniques, when properly applied, will have the

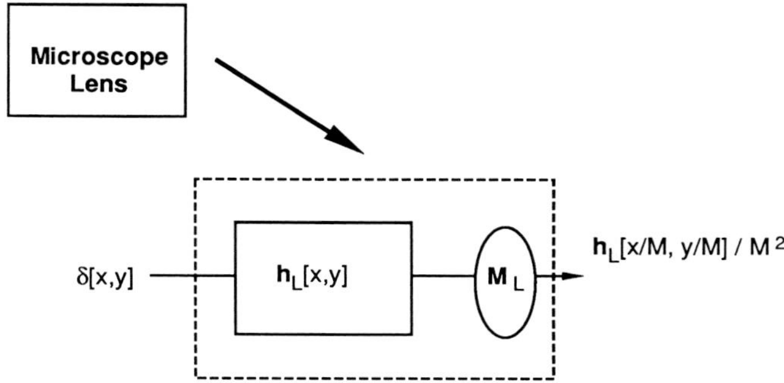

FIG. 6. System diagram of a microscope lens showing the role of the PSF and the magnification.

advantage that they will assess not only the shading caused by nonuniform illumination but also the shading introduced by nonuniform sensitivity and offset in the electrooptical sensor. Any correction technique subsequently applied, linear or nonlinear, will then have the advantage of correcting simultaneously for all sources of nonuniformity.

Assuming, however, that the illumination is uniform, the interaction with the fluorochrome in the specimen (using a low-concentration model) will produce a distribution of light $a[x, y, \lambda_{em}]$ given by:

$$\begin{aligned} a[x, y, \lambda_{em}] &= I_{em}[x, y, \lambda_{em}] \\ &= I_{ex}[\lambda_{ex}] \cdot \kappa[x, y, \lambda_{ex}, \lambda_{em}] \end{aligned} \tag{13}$$

where $\kappa[x, y, \lambda_{ex}, \lambda_{em}]$ represents the term describing the interaction between excitation light and the fluorochrome to produce the emission light $I_{em}[x, y, \lambda_{em}]$. The term $\kappa[x, y, \lambda_{ex}, \lambda_{em}]$ will be a function of the absorption cross section as well as the quantum yield of the fluorochrome (taking into account the excitation and emission spectra). The wavelengths associated with $I_{em}[\ \ ]$ will, of course, be longer than those associated with $I_{ex}[\ \ ]$.

This image will then be altered by the objective lens of the microscope. The effect of a lens, in system terms, is to alter the original light distribution through the effect of the PSF and to magnify (scale) the image. In Fig. 6 we see these represented in a system diagram.

In this figure M_L represents the magnification associated with the lens. For an ordinary thin lens M_L can be negative. While this is an important consideration in the design and analysis of optical instruments such as binoculars, it is not really relevant to quantitative fluorescence microscopy. For this reason, we shall treat the magnification as always being positive. The order of the operations in Fig. 6, first the PSF and then the scaling, is crucial. The scaling of a spatial image from coordinates $[x, y]$ to $[x,/M, y/M]$ cannot be described as an LSI operation. Further, the two operations convolution and scaling are not commutative. The term $1/M^2$ represents the scale factor required to guarantee that the total intensity exiting the lens equals the total intensity entering the lens, independent of magnification.

Starting with the objective lens in Fig. 3 we have for the effect of the PSF:

$$b[x, y, \lambda_{\text{em}}] = I_{\text{em}}[x, y, \lambda_{\text{em}}] * h_{\text{o}}[x, y] \tag{14}$$

We have "hidden" here the possibility that the PSF of the objective lens will be a function of λ_{em}, that is, we assume an apochromatic lens over the wavelengths associated with the emission image. The image resulting from the objective lens will thus be a distorted version of the actual spatial distribution of light $I_{\text{em}}[x, y, \lambda_{\text{em}}]$, distorted by the PSF. The magnification component will then give $b[x,/M_{\text{o}}, y/M_{\text{o}}, \lambda_{\text{em}}]/M_{\text{o}}^2$. After the eyepiece lens, however, the image will once again be altered by the PSF of that lens, $h_{\text{e}}[x, y]$, to

$$c[x, y, \lambda_{\text{em}}] = \left\{ d(\lambda_{\text{em}}) b\left[\frac{x}{M_{\text{o}}}, \frac{y}{M_{\text{o}}}, \lambda_{\text{em}}\right] \frac{1}{M_{\text{o}}^2} \right\} * h_{\text{e}}[x, y] \tag{15}$$

which, after the magnification M_{e}, becomes $c[x/M_{\text{e}}, y/M_{\text{e}}, \lambda_{\text{em}}]/M_{\text{e}}^2$.

The penultimate element in the system chain of Fig. 3 is the electrooptical sensor. This sensor will, of course, have a light-sensitive surface and the possibility to integrate the image signal over time to collect more emission photons and thereby improve the signal-to-noise ratio (SNR). If the sensor is composed of spatially discrete photosensitive elements, as, for example, in a charge-coupled device (CCD) array, then the image sampling may occur as well. We shall not consider the sampling and quantizing aspects of this problem in detail. The final image $f[x, y]$ produced by the sensor will then be a function of the spectral sensitivity of the sensor $S(\lambda)$ as well as its PSF, $h_{\text{s}}[x, y]$:

$$f[x, y] = \frac{1}{M_{\text{e}}^2} \left(\int_{\lambda_1}^{\lambda_2} c\left[\frac{x}{M_{\text{e}}}, \frac{y}{M_{\text{e}}}, \lambda_{\text{em}}\right] S(\lambda)\, d\lambda \right) * h_{\text{s}}[x, y] \tag{16}$$

where integration occurs over the spectral range of the emission light $\lambda_1 \leq \lambda_{\text{em}} \leq \lambda_2$. As with the previous two lenses, the magnification must be considered. This yields as a final result $f[x/M_{\text{s}}, y/M_{\text{s}}]/M_{\text{s}}^2$.

The system diagram in Fig. 3, when modified to reflect the various components that are now described by models, is shown in Fig. 7.

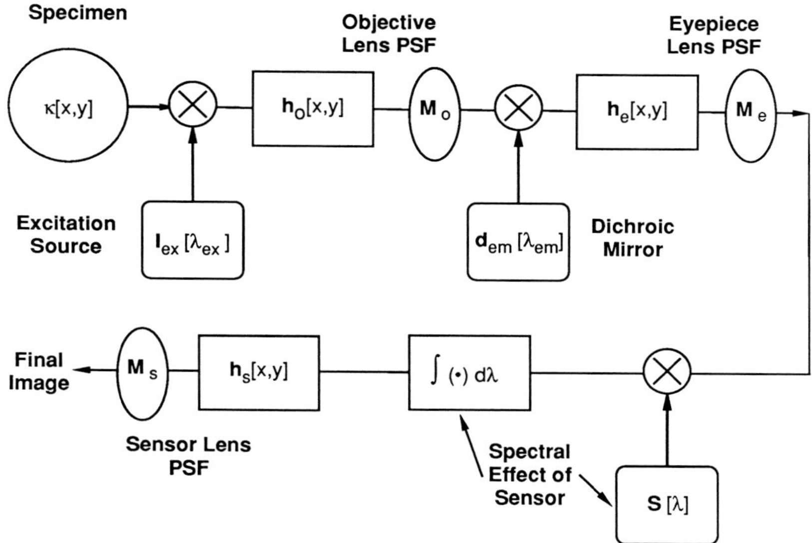

FIG. 7. System diagram of a fluorescence imaging system based on the various mathematical models developed in the text.

B. Role of Microscope "Parameters"

It is clear from Fig. 7 that the terms involving spatial coordinates are the three point spread functions $h_o[x, y,]$, $h_e[x, y]$, and $h_s[x, y]$ and the magnification (scaling) that occurs with each lens. The total PSF of the fluorescence imaging system will therefore be related to each of these terms.

The next step involves the specification of the different PSFs. We assume that each system has aberration-free lenses that have circular apertures. Under these assumptions, each PSF is known to be of the form $h[x, y] = h[r, \phi] = h[r]$. That is, each PSF is circularly symmetric and thus depends only upon the radial distance r from the center of the PSF. The specific form of $h[r]$ is given by the Fraunhofer far-field diffraction pattern for an incoherent source (O'Neill, 1963; Martin, 1966; Goodman, 1968; and Castleman, 1979):

$$h[r] = \left(2\frac{J_1(ar)}{ar}\right)^2 \tag{17}$$

where J_1 is a Bessel function of the first kind and a is determined by the parameters of the lens. It is beyond the scope of this chapter to derive Eq. (17). The reader is instead referred to Chapter 8 of Born and Wolf

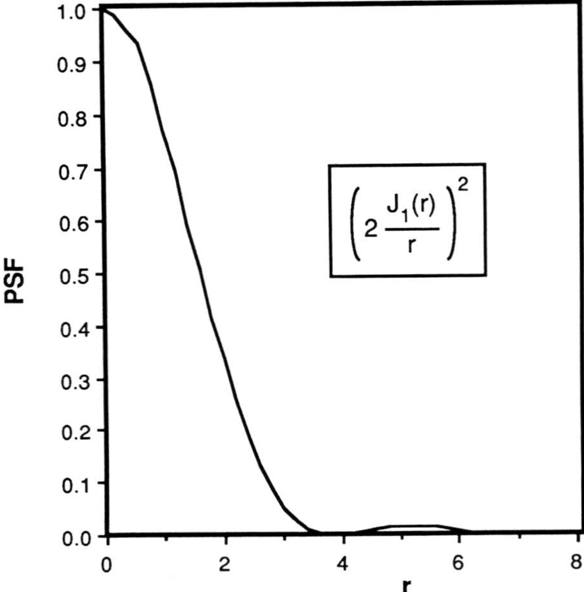

FIG. 8. $h[r]$ Normalized point spread function (PSF) corresponding to an aberration-free lens with a circular aperture [$a = 1$ in Eq. (17)].

(1980). For each lens we have

$$a = \frac{(2\pi NA)}{\lambda} \tag{18}$$

where λ is the wavelength of light and NA the numerical aperture of the lense being studied. A graph of this PSF is given in Fig. 8.

Viewed on a screen, this PSF appears as a series of concentric rings called the Airy pattern. The center position r_o of the first dark ring occurs at the position where $J_1(ar) = 0$. This value of r is associated with the *Rayleigh criterion* for the optical resolution of a lens:

$$ar_o = 1.220\pi \Rightarrow r_o = \frac{(0.61\lambda)}{\mathrm{NA}} \tag{19}$$

In this formula, we see clearly that the PSF of a lens is related to both the wavelength of light passing through the lens and the numerical aperture of the lens itself. For green light ($\lambda \approx 0.5$ μm) and the NA associated with a high magnification, oil-immersion lens (NA ≈ 1.3), r_o is approximately $\lambda/2 = 0.25$ μm. The NA represents the light collection ability of a lens. It is a function of the acceptance angle of the lens and the index of refraction of the medium through which the light must travel to reach the lens. For lenses which work in air NA $\leqslant 1$; for lenses which work in oil or water NA ≤ 1.4 (see Taylor and Salmon, Volume 29, this series).

C. The Fundamental Lens Transfer Function $H(\omega_r)$—The OTF

In Eqs (8b) and (12), we indicated that the OTF of a lens could be calculated from the PSF using the Fourier transform and that, further, this description was extraordinarily useful in characterizing the way other images would propagate through a given lens. We are now in a position to use that approach. In Eq. (17) we have the PSF in polar coordinates instead of Cartesian coordinates; that is, we have $h[r]$ instead of $h[x,y]$. The Fourier transform in polar coordinates is known as the *Hankel transform* and is most appropriate in the situation where we assume lenses with circular symmetry. The relevant equations are (Papoulis, 1968)

Forward equation: space → radial frequency

$$H(\omega_r) = \int_0^\infty rh[r]J_0(\omega_r r)\, dr \tag{20a}$$

Back equation: radial frequency → space

$$h[r] = \int_0^\infty \omega_r H(\omega_r) J_0(r\omega_r)\, d\omega_r \tag{20b}$$

where J_0 is the Bessel function of the zeroth kind. Applying Eq. (20a) to Eq. (17) gives the OTF of an abberration-free (microscope) lens:

$$H(\omega_r) = \begin{cases} \left[\dfrac{4}{\pi a^2}\left[\cos^{-1}\left(\dfrac{\omega_r}{2a}\right) - \dfrac{\omega_r}{2a}\sqrt{1-\left(\dfrac{\omega_r}{2a}\right)^2}\right]\right] & \omega_r \leq 2a \\ \left[\,0\,\right] & \omega_r > 2a \end{cases} \tag{21}$$

This important OTF (normalized to one) is shown in Fig. 9.

The OTF is always positive, and hence qualifies as its own MTF, but more importantly it is strictly bandlimited to a cutoff frequency:

$$\omega_c = 2a = \frac{(4\pi \mathrm{NA})}{\lambda} \tag{22}$$

Below this frequency, ω_c, the spectrum of an image is multiplied [as in Eq. (12)] by a decreasing but nonzero amplitude factor. For all frequencies in the input image above ω_c, however, the multiplication factor is exactly zero. For green light ($\lambda \approx 0.5\ \mu$m) and the NA associated with a high

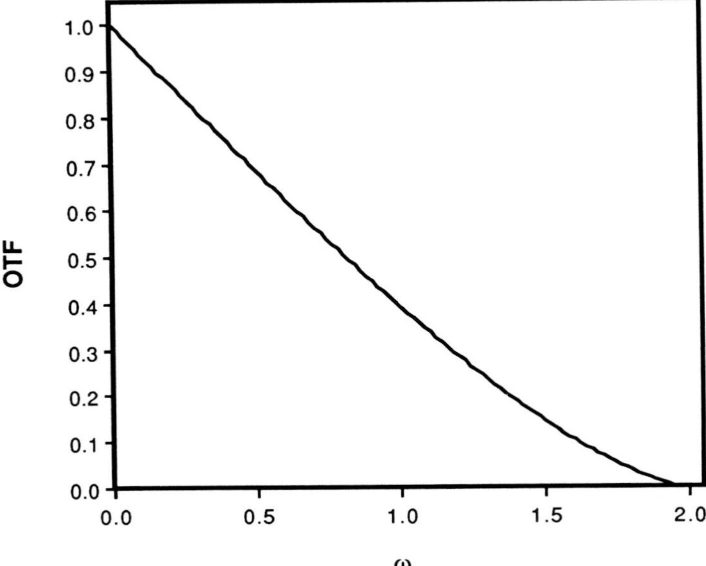

FIG. 9. $H(\omega_r)$, Normalized optical transfer function (OTF) corresponding to an aberration-free lens with a circular aperture [$a = 1$ in Eq. (21)].

magnification, oil-immersion lens (NA $\approx$ 1.3), the cutoff frequency is given by $\omega_c \approx 10.4\pi$ radians/micrometer or $f_c = 5.2$ cycles/micrometer. We see in Eq. (22) that, as the numerical aperture of a lens increases or the wavelength of light used decreases, the bandwidth of the optical system increases.

D. The Effect of Magnification

In Eq. (14) we see that, for the objective lens, the relationship between the input and output images is given by $b[x,y] = I_{em}[x,y] * h_o[x,y]$. (We have dropped the dependence on λ_{em} to simplify the notation.) In Fourier terms this becomes $B(\omega_x, \omega_y) = \ddot{I}(\omega_x, \omega_y)\, H_o(\omega_x, \omega_y)$ following the use of Eq. (12) and where $\ddot{I} = \mathcal{F}\{I\}$. To account for the magnification, it is necessary for us to scale $b[x,y]$ by M_o. Scaling in the spatial domain translates easily to scaling in the frequency domain. If $B(\omega_x, \omega_y)$ corresponds to $b[x,y]$, then

$$\frac{1}{M_o^2}\, b\left[\frac{x}{M_o}, \frac{y}{M_o}\right] \leftrightarrow B(M_o\omega_x, M_o\omega_y) \tag{23}$$

Using the result in Fig. 6 we have that, after magnification of the image, the spectrum will be given by:

$$B(M_o\omega_x, M_o\omega_y) = \ddot{I}(M_o\omega_x, M_o\omega_y)\cdot H_o(M_o\omega_x, M_o\omega_y) \tag{24}$$

and it is this spectrum that serves as input to the eyepiece lens. The output of the eyepiece lens will therefore be given by:

$$C(\omega_x, \omega_y) = \ddot{I}(M_o\omega_x, M_o\omega_y)\cdot H_o(M_o\omega_x, M_o\omega_y)\cdot H_e(\omega_x, \omega_y) \tag{25}$$

and, after the magnification associated with the eyepiece, this will be

$$C(M_e\omega_x, M_e\omega_y) = \ddot{I}(M_eM_o\omega_x, M_eM_o\omega_y)\cdot H_o(M_eM_o\omega_x, M_eM_o\omega_y) \cdot H_e(M_e\omega_x, M_e\omega_y) \tag{26}$$

At this point, it is useful to look at what the effect of the scaling factors is on the individual terms. The eyepiece lens will have a lower magnification than the objective lens and a lower NA as well. We might expect, therefore, that the total spectrum would be dominated by the poor bandwidth of the eyepiece lens. This is not the case, however, as the input image to the eyepiece lens is M_o times larger than the original image. A line in the original specimen that was 1 μm in diameter will now be M_o μm in diameter. The effect this has on both the spatial extent of an object as well as its spectral extent is illustrated in Fig. 10.

Figure 10 shows that the effect of the magnification is to make each subsequent lens element have an *effectively* broader spectrum.

E. The Total OTF

The total of all the system components can then be written as

$$F(\omega_x, \omega_y) = \ddot{I}(M_1\omega_x, M_1\omega_y)\cdot H_o(M_1\omega_x, M_1\omega_y)\cdot H_e(M_2\omega_x, M_2\omega_y) \cdot H_s(M_s\omega_x, M_s\omega_y) \tag{27}$$

where $M_1 = M_oM_eM_s$ and $M_2 = M_eM_s$. The net effect is that the sensor lens will have little if any effect on the overall OTF and that the dominant term will be, as expected, the objective lens.

F. Other Possible Effects

Our model is a simplified one. We assume that lenses are LSI, aberration-free, apochromatic, and can be fully described by talking only about the coordinates x and y, apochromatic, that is, forgetting the z-axis of a microscope completely. We have assumed that the illumination is a plane wave with perpendicular incidence on the specimen and that it has no

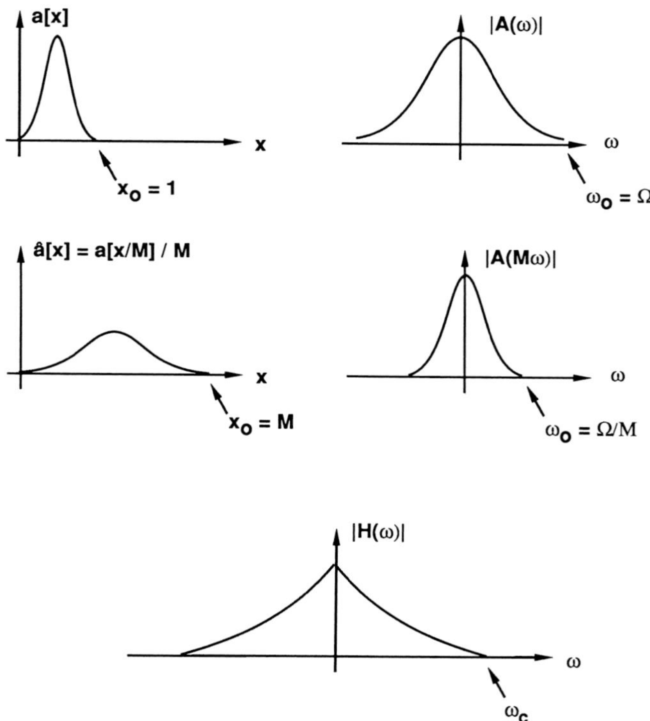

FIG. 10. The line $a[x]$, with width 1 has spectrum $A(\omega)$ with width Ω. After magnification by $M > 1$, the width increases to M and the spectral width *decreases* to Ω/M. If the magnified line $\hat{a}[x]$ is put through a system with spectrum $H(\omega)$, then the filter will appear equivalently wider than it would have for the unmagnified $a[x]$.

particular polarization that might interact with either the specimen or the lens to produce polarization effects (wanted or unwanted) in the final image. We have also assumed that light emitted from a specimen will pass directly to the lens. All of these assumptions are, to a greater or lesser extent, justifiable. Two of the most important—the absence of glare and the absence of scatter—are worth discussing in a little more detail.

Glare is defined in the broadest sense as unwanted light that reaches the sensor. It is caused by the reflections of light that occur at lens surfaces (including internal surfaces in complex lenses), dust, scattering in the lens mounts, and the reflections that may occur between the specimen and the lens. In Chapter 12 of Piller (1977) an excellent discussion of glare is given as well as suggested procedures for its suppression. The potential effect that glare has on accurate measurement of fluorescent light intensities is illustrated in Fig. 11a.

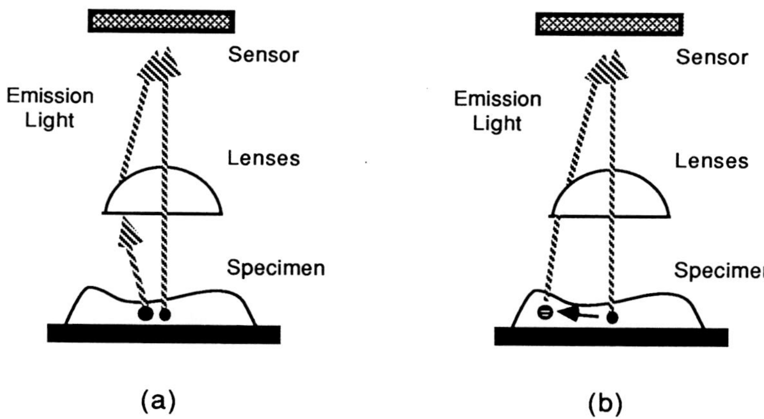

FIG. 11. Glare (a) and Scatter (b) in lens systems. (a) Fluorescence emission from one point within the specimen goes "directly" to coordinate point on the sensor. From another point within the specimen, light is emitted at an angle. This ray is bent by the lens and is imaged at the same coordinate point as the first ray. This gives a false idea as to the amount of light emanating from the first point. (b) Again the light from a point is imaged to a coordinate point on the sensor. Another ray from the same isotropically emitting point scatters from a second point and is imaged at the same coordinate point on the sensor surface. This gives, once again, a false idea as to the amount of light emanating from the first point. We expect that only a constant fraction of light originating from each point will reach the sensor.

Scatter can be thought of as the effect produced by emission light that does not make its way directly to the sensor from a small volume of fluorochrome but instead reaches the sensor only after reflecting off another structure within the specimen. The emitted light is *scattered* by structures within the specimen and thus correspondence between where the light is sensed and where it is emitted is confounded. This effect is illustrated in Fig. 11b. One of the few ways of dealing with the scattering problem, particularly in thick specimens, is through the use of confocal optics (see Brakenhoff *et al.*, Chapter 14, this volume). We shall not discuss this topic here; the reader is instead referred to Section VIII,B.

V. The Contrast Modulation Transfer Function (CMTF)

A. The Definition

In the previous section we emphasized the use of (complex) sinusoidal test images to characterize an LSI imaging system. Based upon the developments in Eqs. (7) and (12), there can be no doubt that this is the

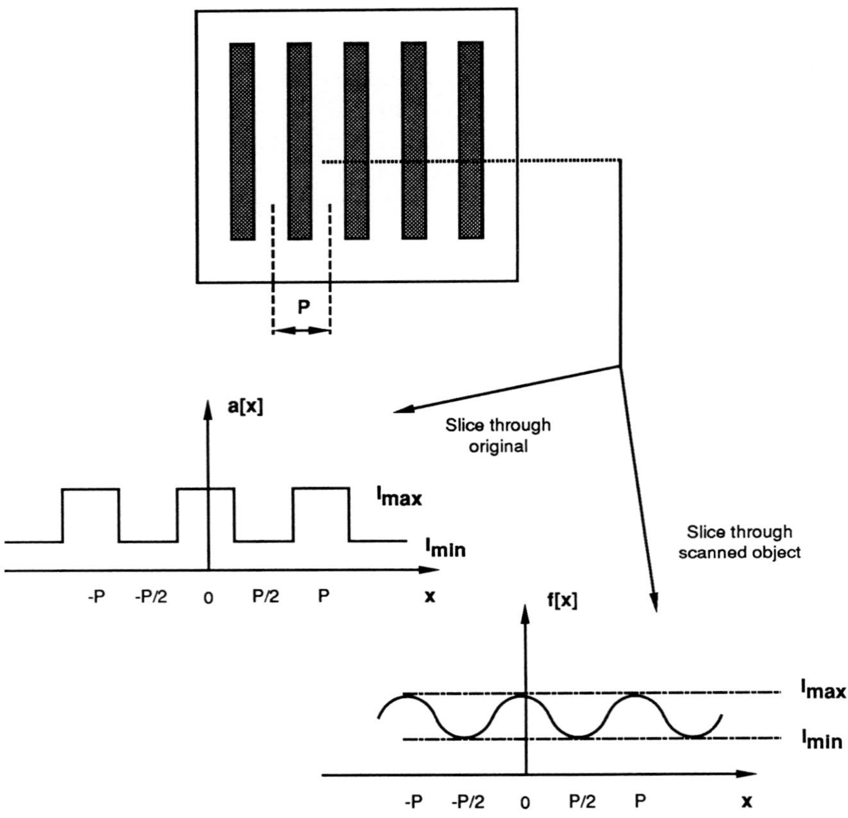

FIG. 12. The test pattern consists of a bar chart with a periodic distance P between the bars. In the original image, $a[x, y]$, a slice through the pattern yields $a[x]$, a square wave pattern. After the pattern has passed through an optical system, the pattern will be convolved with the total PSF. This yields a new image, $f[x, y]$. A slice through this output image yields $f[x]$ a smoothed version of $a[x]$. The contrast is defined in terms of measures made at $x = 0$ and $x = P/2$ for different values of P (see text).

proper approach. Unfortunately, such test images are difficult to prepare. A more common test image is one based upon bars of illumination separated by bars of no illumination. As we shall show in Section VI, this test pattern can be used indirectly to assess the OTF of a system. We must begin, however, by defining a suitable measure on such a pattern.

It is important to realize that, although the input pattern will be a rectangular bar chart (as illustrated, for example, in Fig. 12), the output pattern will *not* be a bar chart. While Eq. (7) indicates that a sinusoid in will give a sinusoid out (with a different amplitude), it is not true that a bar pattern in will give a bar pattern out (with a different amplitude). What will in general happen, is that the *contrast* of the output pattern will decrease as the size and spacing of the input bars becomes smaller. In a sense, as the frequency of the input bars increases, the output amplitude will decrease. This observation is in agreement with the form of the OTF shown in Fig. 9.

The standard definition for contrast was developed for fringe patterns by the American physicist A. A. Michelson. He defined the visibility (ν) of a fringe as

$$\nu = \frac{I_{\max} - I_{\min}}{I_{\max} + I_{\min}} \tag{28a}$$

where $I_{\max}$ and $I_{\min}$ are as indicated in Fig. 12. We define the contrast as equal to the visibility, $C = \nu$. If the distance between the bars and the width of the bars is given by the spacing s we can then write:

$$C(s) = \frac{I_{\max}(s) - I_{\min}(s)}{I_{\max}(s) + I_{\min}(s)} \tag{28b}$$

In this form, the contrast is a clear function of the spacing, $C(s)$. As $I_{\max} \geq I_{\min} \geq 0$, we have that $C(s) \geq 0$ and thus it qualifies as a modulation transfer function. We refer to $C(s)$, therefore, as the *contrast modulation transfer function,* CMTF. The spacing is usually expressed in line-pairs per millimeter. At 500 line-pairs/mm. we would be describing a test pattern with alternating light and dark bars where the width of the bars was 1 μm.

B. The Measurement

Experimental determination of CMTF is relatively straightforward. Bar chart test patterns on a microscope slide are available from a number of commercial firms (for example, Opto-Line Corporation, Massachusetts). After placing the test pattern on top of a fluorescent standard such as uranyl glass, a fluorescent image can be scanned and digitized. Digital measurement of the contrast in the recorded image, at various positions corresponding to various bar spacings (see Fig. 13), will then give a characterization of CMTF for the *complete* system as shown in Fig. 3.

It is important to remember that such measurements must be made in a variety of directions. While the lenses in the complete system may be circularly symmetric, the scanning and digitizing system will not be so. The curve shown in Fig. 13c represents only one direction in the two-dimensional image. By rotating the test pattern it is possible to examine other directions. At the very least, two orthogonal directions should be measured, one parallel to the x-scanning and the other parallel to the y-scanning.

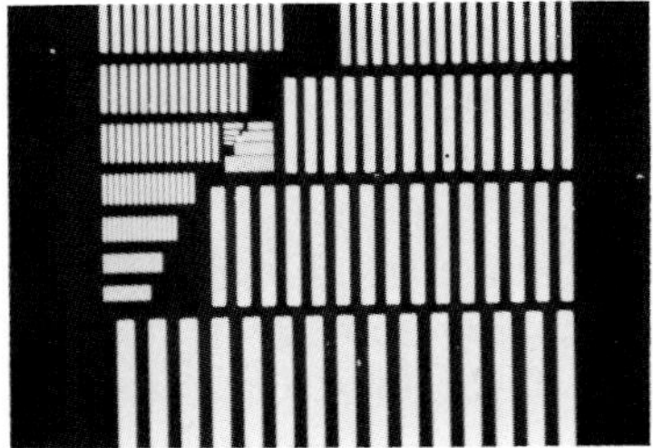

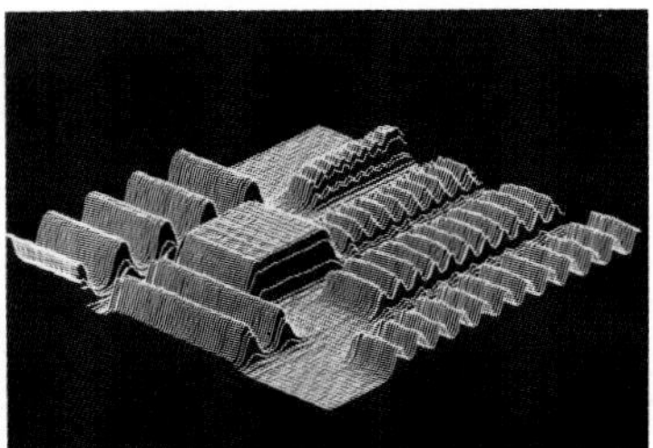

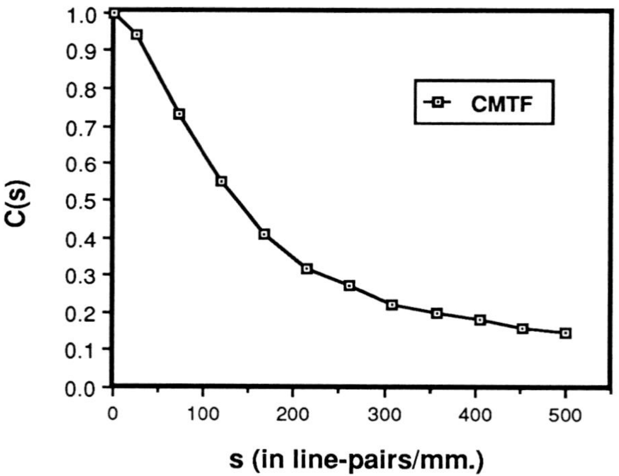

FIG. 13. (a) A two-dimensional test pattern of bars. The bars start at a spacing of $P_0 = 2\ \mu\text{m}$. Subsequent sets of bars increase in spacing by $P_{n+1} = P_n \times 10^{0.1} = 1.2589 P_n$. (b) After scanning by a quantitative microscope system the bars are degraded. (c) The CMTF can be derived from the decrease in contrast in the *output* bars as a function of P_n.

VI. Relating the CMTF and the OTF

A. The Theory

In the previous section we indicated that it is possible to relate the CMTF and the OTF. In this section we would like to show how this can be done so that the OTF can be derived from the measured CMTF. This derivation is based upon the observation that the bar chart pattern itself can be represented as a sum of weighted sinusoids, that is, a Fourier representation. We will show that the CMTF is related to the OTF through a simple matrix algebra equation and thus inversion of the matrix will give the OTF in terms of the CMTF.

The square wave test pattern shown in Fig. 12 can be considered as the input signal. We assume that the duty cycle of the square wave is 50%, the width of the black bars equals the width of the white bar. For this one-dimensional, periodic test image $a[x]$, the Fourier representation is given by:

$$a[x] = a_0 = \sum_{n>0} a_n \cos\left(\frac{2\pi nx}{P}\right) \tag{29a}$$

where P is the period of the bars and with

$$a_n = \begin{cases} (I_{\max} + I_{\min}/2) & n = 0 \\ (I_{\max} - I_{\min})[\sin(n\pi/2)/(n\pi/2) & n > 0 \text{ and odd} \end{cases} \tag{29b}$$

This input signal is now passed through our total optical system [Eq. (27)] with (one dimensional) transfer function $H_T(\omega)$ to give an output image $f[x]$. Based upon Eq. (8a) the output is, therefore,

$$f[x] = H_T(\omega = 0)a_0 + \sum_{n \text{ odd}} H_T\left(\omega = \frac{2\pi n}{P}\right) a_n \cos\left(\frac{2\pi nx}{P}\right) \tag{30}$$

We assume that $H_T(\omega = 0) = 1$, that is, no light is lost or gained in the optical components. We then measure the intensity in the output image at two points $x = 0$ (a brightness maximum) and $x = P/2$ (a brightness minimum), to give $f[x = 0]$ and $f[x = P/2]$. These can be written as

$$f[x = 0] = \frac{I_{\max} + I_{\min}}{2}\left(1 + \sum_n 2C_0 H_T\left(\frac{2\pi n}{P}\right)\frac{\sin(n\pi/2)}{(n\pi/2)}\right) \tag{31a}$$

$$f[x = P/2] = \frac{I_{\max} + I_{\min}}{2}\left(1 + \sum_n 2C_0 H_T\left(\frac{2\pi n}{P}\right)\frac{\sin(n\pi/2)}{(n\pi/2)}\cos(\pi n)\right) \tag{31b}$$

where the sums are once again over positive, odd values of n and where C_0 is the contrast in the original (input) image using the definition in Eq. (28a). The spacing of the bars is given by $s = P$ so that we can evaluate $C(s = P)$ in the output image as

$$C(s = P) = \frac{f[0] - f[P/2]}{f[0] + f[P/2]} \tag{32}$$

The term in the denominator of Eq. (32) can be shown to be

$$f[0] + f[P/2] = I_{\max} + I_{\min} \tag{33}$$

and Eq. (32) becomes

$$C(s = P) = \frac{4C_0}{\pi} \sum_{n\ \mathrm{odd}} H_{\mathrm{T}}\left(\frac{2\pi n}{P}\right) \frac{\sin(n\pi/2)}{n} \tag{34}$$

From Eq. (34) we see that, for a given spacing $s = P$, the CMTF $C(s = P)$ is a linear combination of the Fourier MTF evaluated at a specific set of frequencies $\omega_n = 2\pi n/P$. These frequencies are precisely the Fourier frequencies contained in the bar test pattern.

B. Practical Implementation

If measurements are made for different bar spacings ($s = P$, $s = P/3$, $s = P/5, s = P/7, \ldots$), then absorbing $4C_0/\pi$ in H_{T} leads to the following set of simultaneous equations:

$$\begin{aligned}
C(s = P) &= H_{\mathrm{T}}\left(\frac{2\pi}{P}\right) - \frac{1}{3} H_{\mathrm{T}}\left(\frac{2\pi 3}{P}\right) + \frac{1}{5} H_{\mathrm{T}}\left(\frac{2\pi 5}{P}\right) - \frac{1}{7} H_{\mathrm{T}}\left(\frac{2\pi 7}{P}\right) + \frac{1}{9} H_{\mathrm{T}}\left(\frac{2\pi 9}{P}\right) + \cdots \\
C(s = P/3) &= 1H_{\mathrm{T}}\left(\frac{2\pi 3}{P}\right) - \frac{1}{3} H_{\mathrm{T}}\left(\frac{2\pi 9}{P}\right) + \cdots \\
C(s = P/5) &= 1H_{\mathrm{T}}\left(\frac{2\pi 5}{P}\right) + \cdots \\
C(s = P/7) &= 1H_{\mathrm{T}}\left(\frac{2\pi 7}{P}\right) + \cdots
\end{aligned} \tag{35a}$$

These equations can be rewritten in matrix form as

$$\begin{bmatrix} C(P) \\ C(P/3) \\ C(P/5) \\ C(P/7) \\ C(P/9) \\ \cdot \\ \cdot \\ \cdot \end{bmatrix} = \begin{bmatrix} 1 & -1/3 & +1/5 & -1/7 & +1/9 & \cdots & \cdots \\ & 1 & 0 & 0 & -1/3 & \cdots & \cdots \\ & & 1 & 0 & 0 & \cdots & \cdots \\ & & & 1 & 0 & \cdots & \cdots \\ & & & & 1 & \cdots & \cdots \\ & & & & & \cdots & \cdots \\ & & & & & \cdots & \cdots \\ & & & & & \cdots & \cdots \end{bmatrix} \begin{bmatrix} H(q) \\ H(3q) \\ H(5q) \\ H(7q) \\ H(9q) \\ \cdot \\ \cdot \\ \cdot \end{bmatrix} \tag{35b}$$

where $q = 2\pi/\mathrm{P}$. The experimental contrast measurements give us the various values of $C(s = P)$, $C(s = P/3)$, $C(s = P/5)$, $C(s = P/7)$, etc. and these can used to solve for the values of $H_{\mathrm{T}}(\omega)$ at the frequencies

$\omega_n = 2\pi n/P$ for n odd. If we denote the three matrices in Eq. (35b) by C, M, and H_T, then the solution for the OTF will be given by $H_T = CM^{-1}$ where M is the matrix of numerical coefficients. These coefficients are constant, that is, they are not a function of the experimental data and M^{-1} need only be computed once. In this manner, we can find values of the OTF based upon values of the CMTF.

The solution of Eq. (35b) for the samples of the OTF can be implemented easily on a programmable calculator or a personal computer. A "cookbook" solution has been published in the literature and is simple to use (Limansky, 1968). The number of terms that have to be solved for in Eq. 35 is generally quite reasonable; the matrix is typically on the order of 20×20. This is because the twentieth term corresponds to the forty-first harmonic frequency ($n = 41$). If the basic period P is equivalent to 100 line-pairs/mm, then the forty-first harmonic will correspond to 4.1 cycles/μm, in other words, frequencies close to or beyond the cutoff frequency $f_c = \omega_c/2\pi$ of the lens system.

An example of the types of curves that may result from measurement of the CMTF and the OTF using this approach is shown in Fig. 14.

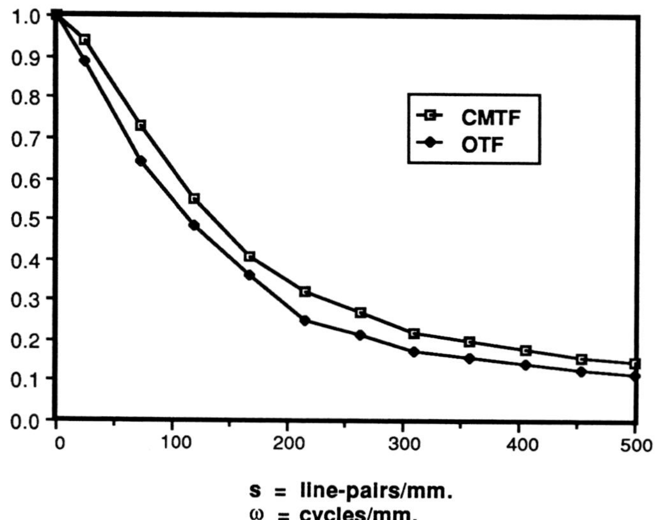

FIG. 14. Using the techniques developed in the text, the contrast modulation transfer function (CMTF) and the optical transfer function (OTF) have been measured for a quantitative microscope system. For the CMTF, the measurements were made over the interval $0 \leq s \leq 500$ line-pairs/mm. For the OTF, the measurements were made over the interval $0 \leq \omega \leq 500$ cycles/mm. (After Young (1982).

VII. Implications

A. The Resolving of Fine Detail

It should be obvious at this point that the finite bandwidth of our entire quantitative microscope prevents us from resolving arbitrarily small detail in a specimen. At a practical level, the Rayleigh criterion [Eq. (19)] limits our ability to distinguish between two fluorescing points in a specimen if the distance separating them is less than $r_o \approx 0.5$ μm. This is illustrated in Fig. 15.

It is also reasonable to ask if the Rayleigh criterion is a suitable criterion for characterizing the resolution of a microscope (or any other optical) system. As we have seen in preceding sections, the total transfer $H_T(\omega_x, \omega_y)$ is the complete characterization of the system. Yet the Rayleigh criterion is one number that attempts to tell the whole story concerning resolution. This is clearly an impossible task. If the total optical system is circularly symmetric, then r_o is related to the cutoff frequency through $\omega_c = 2.44\pi/r_o$; this can be derived from Eqs. (19) and (22). Within this cutoff frequency, for frequencies $0 \leqslant \omega_r = (\omega_x{}^2 + \omega_y{}^2)^{1/2} \leqslant \omega_c$, however, the Rayleigh criterion says nothing about the shape of the

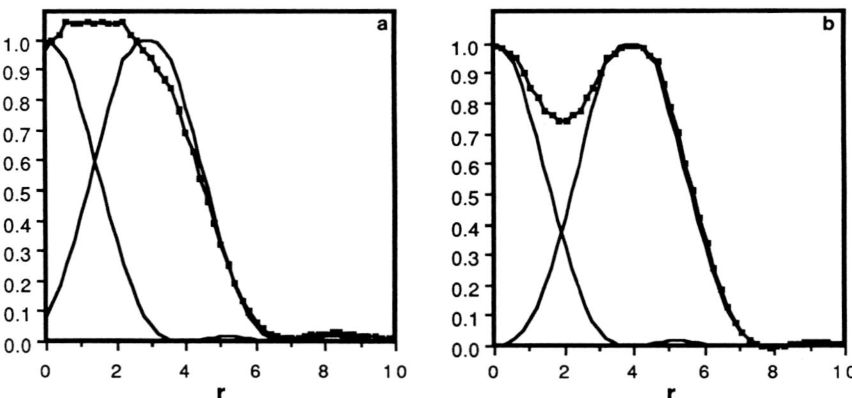

FIG. 15. Two point sources separated by a distance D (in a, $D = 2.8$; b, $D = 3.8$; c,$D = 4.8$) are imaged through an optical system with a PSF as shown in Fig. 8. Because of linearity and shift invariance, this gives rise to an image that is the sum of two impulse responses: $h[r] + h[r - D]$. The individual contributions ($h[r]$, $h[r - D]$) are drawn with thin lines in the above diagrams; the sum is drawn with a thick line. For $D \geq 3.8$ (the Rayleigh distance), there is no problem in seeing the two PSFs as separate. For $D < 3.8$, it becomes impossible to distinguish between two narrow sources or one slightly broader source. Note that the curves are drawn for $a = 1$ in Eqs. (17) and (18).

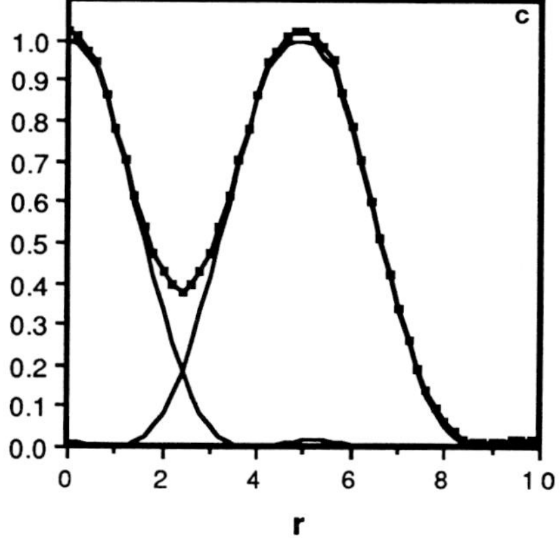

FIG. 15. (*continued*)

spectrum. For an aberration-free, in-focus image the OTF will be as shown in Fig. 9. If the image is not in focus or the lenses contain aberrations, then the amplitude of the transfer function may differ significantly from that in Eq. (21) and illustrated in Fig. 9. This is illustrated in Fig. 16.

While above the cutoff frequency there is no distinction between the idealized model and reality, below the cutoff frequency there is not only a difference in the amplitude of the output spectrum but also in its phase.

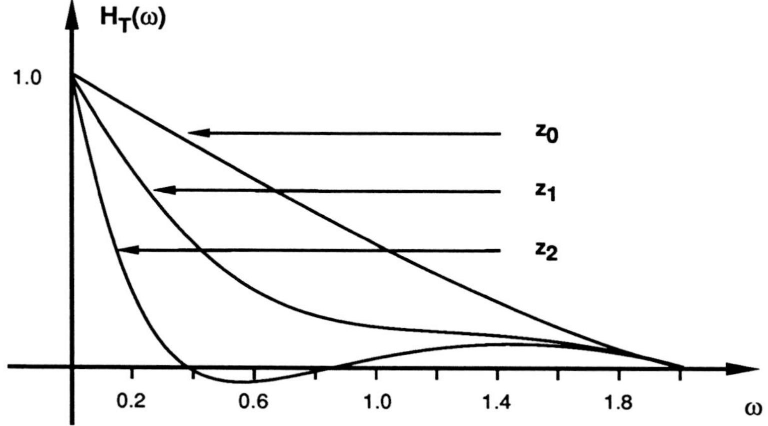

FIG. 16. For an in-focus optical system ($z_0 = 0$) the OTF is the same as that in Fig. 9. For increasing amounts of defocus $z_2 > z > z_0$ the OTF deviates more and more from the form given in Eq. (21). All three of the curves, however, have the same cutoff frequency.

This potential phase change or *phase reversal* is not an academic problem. As shown in Fig. 17, the reversal of phase caused by defocusing can lead to serious artifacts in a digitized image.

B. Visualizing Below-Resolution Strutures

We are now in a position to answer the question of how it is possible to visualize structures in a fluorescence microscope whose dimensions are smaller than r_o (see Inoué, Chapter 3, this volume). How can we visualize actin filaments with a diameter of 5.5 nm and how can we hope to make measurements on such an image? If we start by considering an object consisting of a single point with a total intensity of I_o, then this can be modeled by $a[x,y] = I_o\ \delta[x,y]$. By our LSI hypothesis, the image we will see in the microscpe is $I_o h_T[x,y]$, the scaled PSF of the entire optical system. While $\delta[x,y]$ has no extent (to speak of), the PSF may have an extent of a few $r_o(\approx 1.0\ \mu m)$. This extent is observable. Further, it is possible for us to measure the total fluorescence intensity I_o. If we define the total intensity I_T in the final, output image as

$$I_T = \int_{-\infty}^{+\infty}\int_{-\infty}^{+\infty} f[x,y]\ dx\ dy \tag{36}$$

then, using the result that $f[x,y] = I_o h_T[x,y]$, we have

$$\begin{aligned} I_T &= \int_{-\infty}^{+\infty}\int_{-\infty}^{+\infty} I_o h_T[x,y]\ dx\ dy \\ &= I_0 \int_{-\infty}^{+\infty}\int_{-\infty}^{+\infty} h_T[x,y]\ dx\ dy \end{aligned} \tag{37}$$

If

$$\int_{-\infty}^{+\infty}\int_{-\infty}^{+\infty} h_T[x,y]\ dx\ dy = 1 \tag{38}$$

which is equivalent to $H_T(\omega_x = 0,\ \omega_y = 0) = 1$ [see Eq. (8b)], then $I_T = I_o$, a correct measurement. Equation (38) is equivalent to saying that photons are neither lost nor gained in the optical system. Should the "gain" of the optical system deviate from one, then a simple one-time measurement of the gain factor can provide a suitable correction term.

If we are dealing with a fluorescently labeled filament (such as actin), then an appropriate model for the fluorescence is given by the line $a[x,y] = I_L\delta[x]$, where, without loss of generality, we have assumed that the line is along the y-axis. Using the convolution in Eq. (4) this

(a)

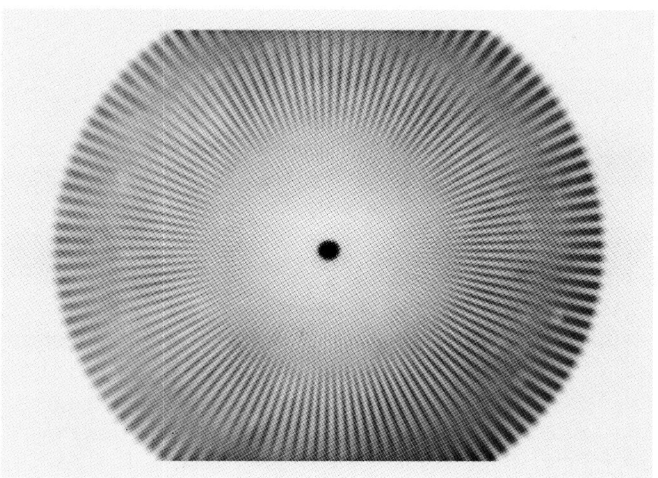

(b)

FIG. 17. (a) Original image; (b) a defocussed version of the original. Note the phase reversal rings caused by the negative-going amplitudes in the OTF.

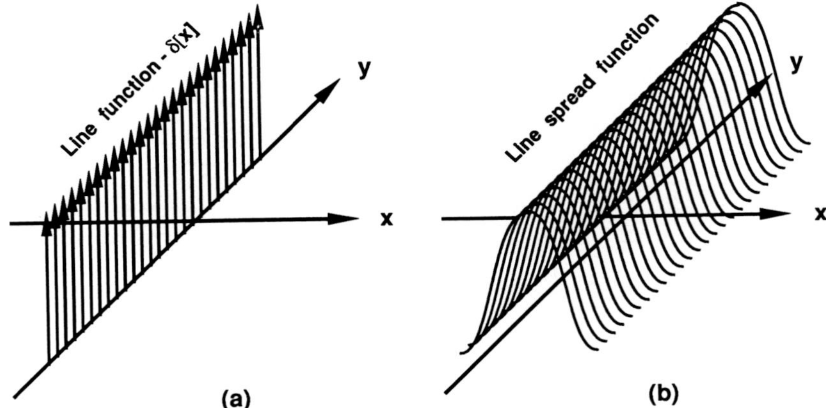

FIG. 18. (a) Original line image; (b) line spread function generated by $h_T[x, y]$.

becomes

$$f[x,y] = \int_{-\infty}^{+\infty} \int_{-\infty}^{+\infty} I_L \, \delta[u] h_T[x-u, y-v] \, du \, dv \tag{39a}$$

$$= I_L \int_{-\infty}^{+\infty} h_T[x, y-v] \, dv$$

With a simple change of variables this becomes

$$f[x,y] = I_L \int_{-\infty}^{+\infty} h_T[x,y] \, dy \tag{39b}$$

This effect is illustrated in Fig. 18. For each value of y, the projection of the PSF upon the x-axis is repeated. Thus, what we observe is not the line of impulses defined by $I_L \, \delta[x]$ but rather the line spread function which is related to the PSF by Eq. (39b). Fluorescently labeled actin filaments, with their diameter of approximately 0.01λ, are the equivalent of this line image. What we see through our imaging system is this line image distorted by the line spread function.

C. Sampling the Image for Analysis

One of the advantages of dealing with an image that has been passed through a system with a finite bandwidth is that it is now possible to sample the image with a finite frequency and obtain a collection of numbers that completely represent the presampled continuous image. We quote without proof the Nyquist sampling theorem: A signal limited to bandwidth ω_c can be completely represented by the collection of samples obtained by periodic sampling at a frequency higher than $2\omega_c$ (= Nyquist frequency). That is, the sampling frequency must satisfy $\omega_s > 2\omega_c$. Notice that this is a strict inequality. We must be careful to choose the correct value for ω_c. For the final images $f[x, y]$ that are almost always bandlimited by the objective lens, we have

$$\omega_s > \frac{8\pi \mathrm{NA}}{\lambda} = \frac{4.88\pi}{r_o} \tag{40}$$

If our original image $a[x, y]$ itself has a bandwidth considerably less that than given by the objective lens, then a lower sampling frequency will be possible. Measurement of the bandwidth of $a[x, y]$ is not easy, however, and thus Eq. (40) based on $f[x, y]$ should be used as a guideline.

The phrase *completely represented* in the sampling theorem must be carefully interpreted. It means that it is possible to reconstruct the original image without any error at all. This reconstruction occurs by means of ideal low-pass filtering when the reconstruction process is viewed in the Fourier frequency domain. When the same process is viewed in the spatial domain, it can be considered to be an interpolation for the values of the image between the sample points; values that were lost during sampling. The interpolation is performed with the aid of sinc( ) functions as given (for the one-dimensional case) by Eq. (41):

$$\hat{a}[x] = \frac{2\pi}{\omega_s} \sum_{n=-\infty}^{+\infty} a\left[x = \frac{2\pi n}{\omega_s}\right] \mathrm{sinc}\left(\frac{\omega_s x}{2} - \pi n\right) \tag{41}$$

where $\sin(q) = \sin(q)/q$. The term $2\pi/\omega_s$ is simply the sampling distance in the x and y directions. A thorough proof and discussion of the sampling theorem is given in Chapter 8 of Oppenheim *et al.* (1983). In that chapter, it is also shown that the actual form of an individual sampling pulse is not important. Any pulse $p_s[x, y]$ whose spatial extent is less than $2\pi/\omega_s$ can be the basis of the sampling system. The only difference in the reconstruction formula [Eq. (41)], when $p_s[x, y]$ is used for sampling instead of $\delta[x, y]$, will be a simple gain factor. The sampling as well as the reconstruction of a bandlimited (one-dimensional) image is illustrated in Fig. 19.

It is interesting to consider whether the analysis of sampling given above tells the whole story with respect to quantitative fluorescence image analysis. Several points are worth making in this regard.

1. A signal that is band limited *must* have infinite spatial extent. That is, it must extend over the interval $(-\infty \leq x, y \leq +\infty)$. It is beyond the scope of this chapter to prove this well-known result. Clearly, our images are

always of finite extent and thus they can not be band limited. The error introduced by our approximation, however, is negligible.

2. While the samples we propose to process contain, in principal, all of the information necessary to reconstruct them, our reconstruction techniques do not have ideal low-pass filters available. Such filters are, in fact, unrealizable. The most common techniques for reconstruction are rather coarse compromises with this fact and, instead, make use of interpolation

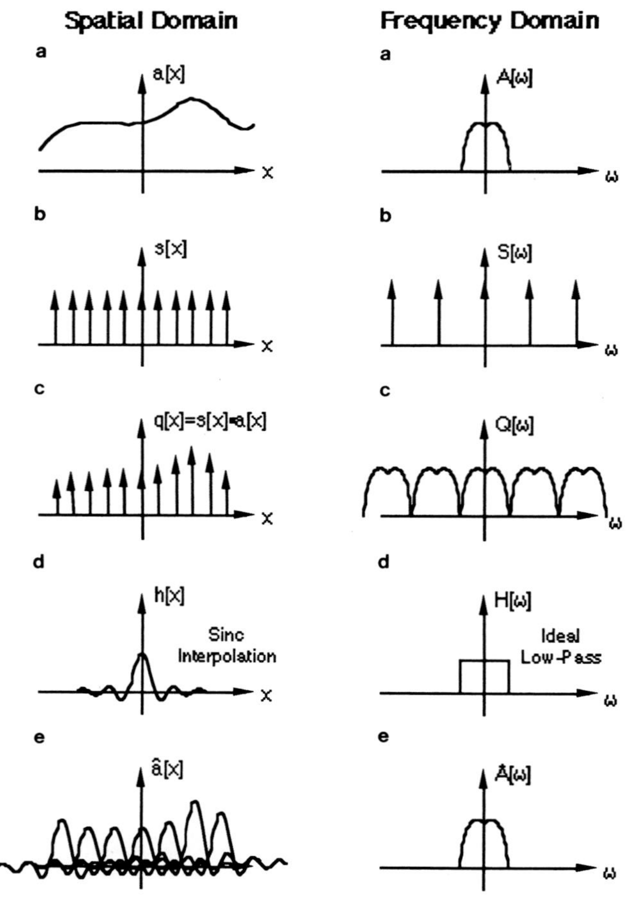

FIG. 19. Space domain representation: (a) original image, (b) sampling pulses, (c) sampled image, (d) impulse response of reconstruction filter, (e) result of interpolation. Frequency domain representation: (a) original spectrum, (b) sampling spectrum, (c) spectrum of sampled image, (d) frequency response of reconstruction filter, (e) result of filtering.

techniques that tend to cause artifacts. Several examples of such interpolation techniques are shown in Fig. 20.

3. One goal of image analysis is not simply the reconstruction of images but specifically the collection of measurements from the images. To this extent the sampling problem is much more serious. Accurate measurements require sampling frequencies far in excess of that predicted by the Nyquist frequency $2\omega_c$. There are several ways of looking at this result. Although the interpolation formula [Eq. (41)] states that all of the data needed for accurate reconstruction and therefore accurate measurement are contained in the samples, it does not say how to extract the measurements. We are forced to truncate the infinite sum in Eq. (41) to a finite sum and to approximate the sinc( ) functions which have infinite extent by terms with finite extent. These two compromises lead to inaccuracies in the

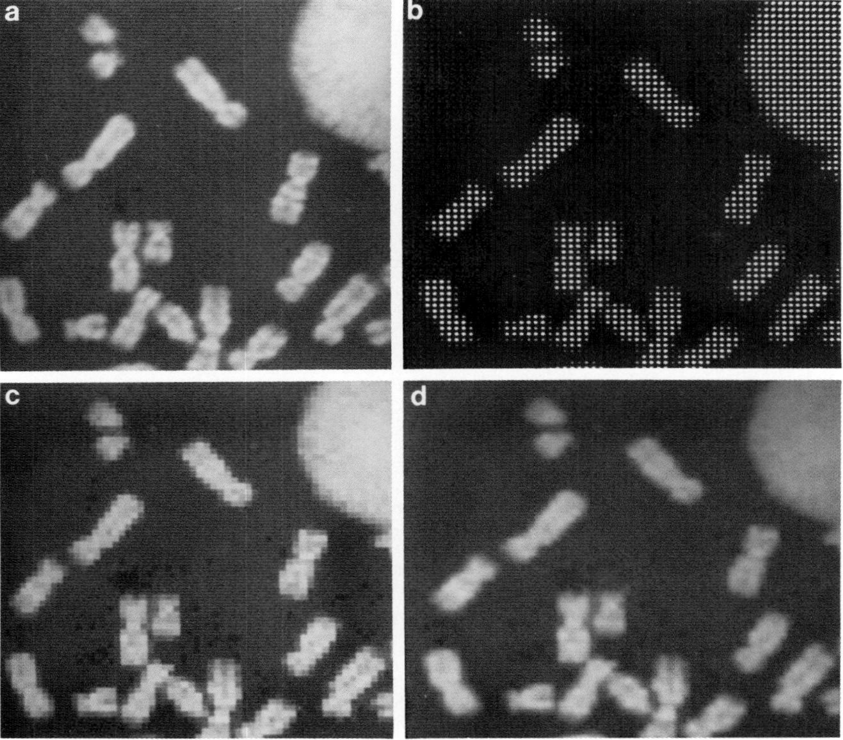

FIG. 20. (a) original image, (b) sampled image, (c) image reconstructed using zero-order hold (ZOH) technique, (d) image reconstructed using first-order hold (FOH) technique.

estimation of "analog" measures based upon "digital" data. The solution is either very complex interpolation formulas or *oversampling,* sampling in excess of the Nyquist frequency.

As a "rule-of-thumb," the measurement of area requires sampling densities on the order of 50 samples per average cell diameter (Young, 1988). The measurement of length requires sampling densities on the order of 25 samples per unit "analog" length (Dorst and Smeulders, 1986). The measurement of certain texture features requires sampling densities on the order of 150 samples per average cell diameter (Young *et al.*, 1986)! Notice that each of these specifications is independent of bandwidth considerations. They arise, instead, from considerations of the quantization error introduced by the spatial sampling process. Translating such specifications back to the domain of microscope images, where cells, cell structures, and chromosomes may have dimensions on the order of micrometers, means that sampling densities should be on the order of 15 to 30 samples per micrometer; a sampling density that is two to four times the value predicted by the Nyquist sampling theorem (see Harms and Aus, 1984).

VIII. The Analysis of the z-Axis

A. Classical Depth-of-Field

At an earlier point in this chapter, we mentioned that our discussion of quantitative microscopy involved a number of assumptions, one of which was to neglect the effect of the z-axis. We have essentially considered specimens to be two-dimensional instead of three-dimensional objects. Only in Fig. 16 did we refer to an issue that concerns the z-axis, focusing. In fact, our specimens are decidedly three-dimensional. The very nature of the imaging process guarantees that the value of intensity assigned to a given $f[x_0, y_0]$ will be a function of the values along the z-axis in the neighborhood of (x_0, y_0). This phenomenon is illustrated in Fig. 21.

In Born and Wolf (Section 8.8.2, 1980), the authors show that the form of the illumination along the z-axis for $(x = 0, y = 0)$ is given by

$$I[x = 0, y = 0, z] = I_0\left(\frac{\sin(u/4)}{u/4}\right)^2 \qquad \text{where} \quad \frac{u}{4} = \frac{\pi z \mathrm{NA}^2}{2\lambda} \tag{42}$$

This distribution of excitation light is the weighting function shown in Fig. 21 and the "cone" indicated in the figure is the geometrical shadow associated with the three-dimensional illumination pattern. Associated with the shadow and the light distribution in Eq. (42) is an equivalent

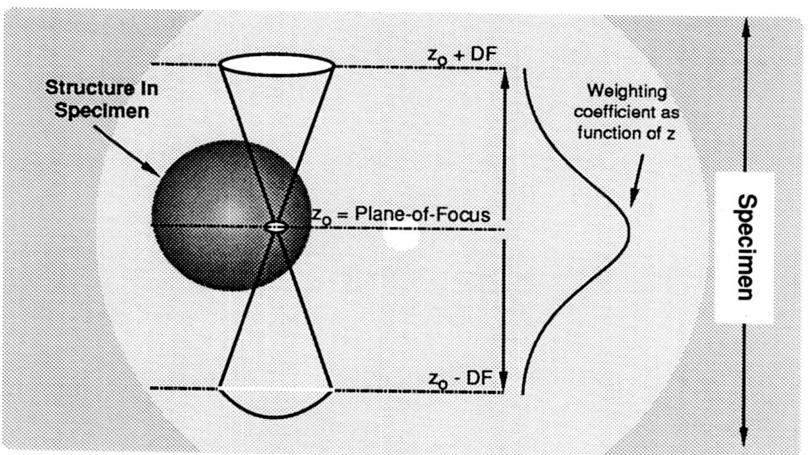

FIG. 21. In the thick specimen shown, the regions above and below the plane of focus ($z = z_0$) contribute to the measurements made at z_0. While the spatial extent of the region increases as one moves away from z_0, the weight of these contributions decreases with distance from z_0. Significant contributions to the measurements made at z_0 come from the interval $z_0 \pm$ DF.

depth of focus, **DF**. Piller (1977) gives a formula for **DF** in micrometers:

$$\mathrm{DF} = \frac{1000}{7\mathrm{NA}M_1} + \frac{\lambda}{2\mathrm{NA}^2} \tag{43}$$

where λ is the wavelength of illumination in micrometers, M_1 is the total magnification as defined in Eq. (27), and NA is the numerical aperture of the objective lens. The resulting **DF** is expressed in micrometers. Using typical values ($\lambda = 0.5\ \mu$m, $M_1 = 1000\times$, and NA = 1.3) we have **DF** = 0.26 μm, or about half a wavelength. In a system configuration where the total magnification M_1 is 40 $\times$, and the NA might be 0.4, the depth of focus changes significantly to **DF** = 10.49 μm. This second situation might occur if we were using a CCD array sensor. If the physical size of a CCD photoelement were 16 μm, then 40 $\times$ magnification would make the sampling grid equal to 0.4 μm. This sampling interval might be appropriate for the screening of slides for the presence of fluorescently tagged cells. Figure 22 shows **DF** for various values of NA and M_1.

B. Confocal Depth of Field

One of the interesting new developments in quantitative microscopy is the confocal microscope. By inserting an aperture on the excitation side

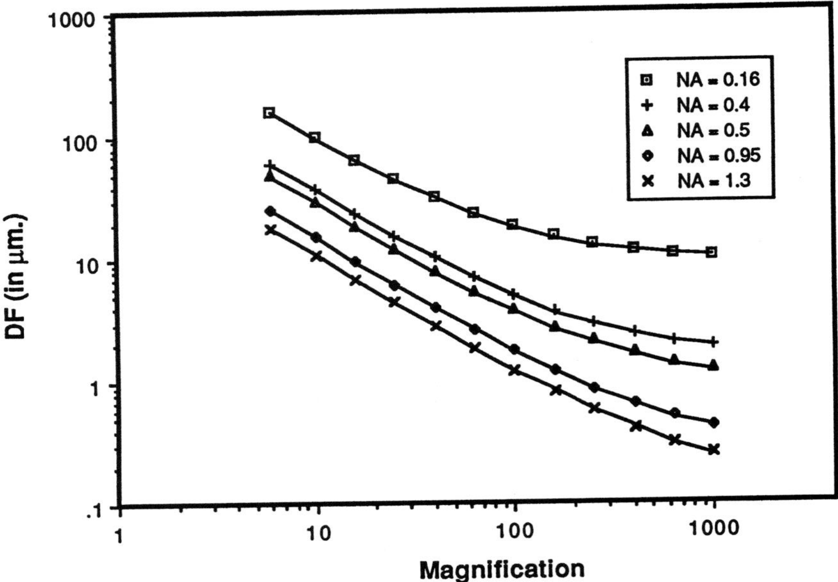

FIG. 22. The depth of focus, DF, is shown here for various values of the numerical aperture, NA, and the total magnification M_1.

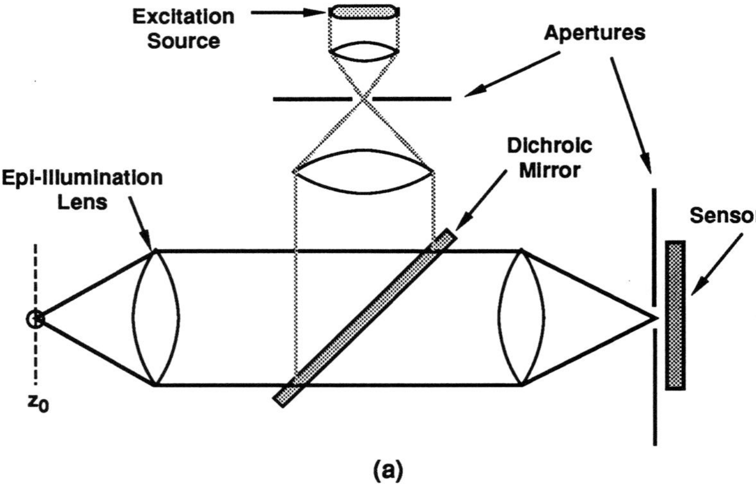

FIG. 23. (a) The ray geometry of a confocal epiillumination fluorescence microscope. The pinhole formed by each circular aperture leads to a PSF proportional to $J_1^4(r)/r^4$ (Brakenhoff *et al.*, 1979). This should be compared to the conventional microscope PSF given in Eq. (17). (b) The aperture on the emission side blocks rays coming from the out-of-focus plane z_1 but permits rays from the in-focus plane z_0 to reach the sensor. This decreases the depth of field and gives better resolution along the z-axis.

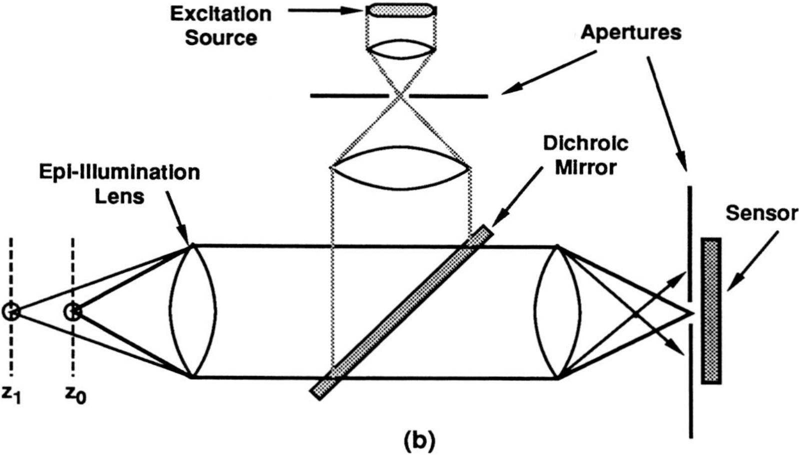

FIG. 23. (*continued*)

and the emission side of the objective lens, it is possible to (1) partially blocks rays of light emanating from above and below the desired plane of focus and (2) increase the bandwidth of the PSF. This configuration increases the resolution of the imaging sysem in the z-direction as well as the x-y plane. It offers, in addition, the ability to acquire three-dimensional images $a[x, y, z]$ by stepping the z-axis through a number of positions, acquiring a series of two-dimensional images, and putting these together as tomographic sections (Brakenhoff *et al.,* 1979, 1985). Figure 23 shows a ray-tracing analysis of a confocal microscope configuration and Fig. 24 shows an example of a specimen recorded with and without the confocal technique (see also Chapter 14 by Brakenhoff *et al.,* this volume).

From Fig. 23 we see that the price we have to pay for the increased three-dimensional resolution is the efficiency of light transfer in the fluorescence microscope. Since only a single point in the specimen is illuminated at any instant, the time required to build up a fluorescent image with a sufficient SNR increases significantly.

IX. Summary

A. The Utility of the System Approach

In this chapter, we have presented the basic ideas that govern the system approach to describing and analyzing quantitative imaging systems for fluorescence microscopy. Throughout the chapter we have tried to indicate

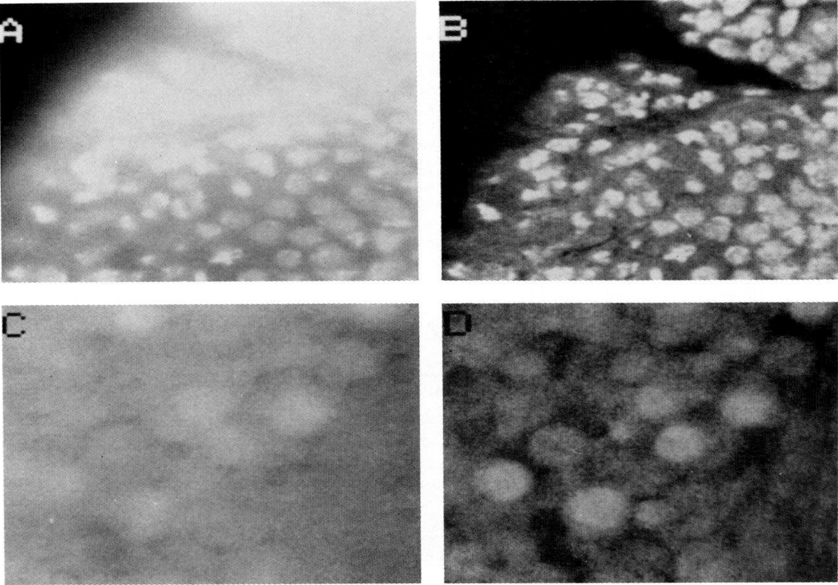

FIG. 24. Confocal and nonconfocal images from a confocal laser scanning microscope (adaption of Zeiss model LSM-44). Whole mount *Drosophila melanogaster* embryos double stained for DNA with 50 μM mithramycin and affinity purified anti-topoisomerase II with rhodamine-labeled goat anti-rabbit IgG. Excitation with an argon ion external laser. Objective: 63 × oil immersion, 1.25 NA. A and B, DNA staining: nonconfocal and confocal modes, respectively, excitation 457 nm, emission bandpass 520–535 nm. Optical slice through the top surface of an embryo, total field, 110 μm. C and D, Rhodamine immunofluorescence: nonconfocal and confocal, respectively, excitation 528 nm, emission above 550 nm. Mid-plane slice through the embryo, total field, 60 μm. Images are averages of 16 2-second scans. No further image processing was performed. Data from Jovin, Arndt-Jovin, Robert-Nicoud, *Abt. Mol. Biol.*, Max Planck Inst. Biophys. Chem, Goettingen, FRG.

the assumptions that are being used and what influence, if any, small deviations from these assumptions can have.

Our use of the linear, shift-invariant (LSI) model to describe the complete system has given us an elegant tool, the optical transfer function (OTF), to describe the imaging system's resolution capabilities. It has also enabled us to understand how an object much smaller than a wavelength of light can be visualized. Using simple test patterns we have also seen how it is possible to measure the contrast modulation transfer function (CMTF) and, from that, derive the OTF. Further, we have seen how microscope parameters such as numerical aperture, wavelength, and magnification combine to describe system performance.

B. Other Issues

Many issues have not been addressed in this chapter. These include the potential influence of noise on the total image quality, the ability to remove noise and distortion through the use of modern filtering techniques, and the ability to resolve detail and restore images at frequencies beyond the optical cutoff frequency ω_c. These topics are quite complex and space does not permit a presentation of sufficient depth. These topics are introduced in Papoulis (1968) and Dudgeon and Mersereau (1984).

A problem of equal importance is the validation of the assumptions of *linearity* and *shift invariance.* As these properties have been crucial to our development, it is highly appropriate that we be able to test a quantitative fluorescence imaging system for them. Testing, however, is not easy. At the very least, the complete imaging system (including the sensor) should be tested for one aspect of the linearity condition. Following the linearity definition we know that, if $a[x, y]$ is the input image and $b[x, y]$ the corresponding output image, then $\eta a[x, y] \rightarrow \eta b[x, y]$. By recording the digital output brightness as a function of *known* input brightness, we can test for this property. In particular, we can test to see if zero light in gives zero value out, a necessary condition for linearity [see Eq. (1)].

Using a feedback-controlled light-emitting diode (LED), for example, it is possible to generate known quantities of light in circular apertures ranging in diameter from 5 to 50 μm. By controlling the LED from the same digital computer that is used for the image acquisition, it is possible to calibrate the brightness input/output characteristics of a complete system over four orders of brightness magnitude and thereby to validate the scaling property of a linear system (Young, 1983).

The testing of the other aspects of linearity as well as the shift-invariance of the system is very complicated. Asymmetries in the optics, the previously mentioned scatter and glare, and distortion in sensors will all contribute to a departure from the hypothesis of shift invariance. While space does not permit a thorough discussion of this topic, certain simple procedures are available to assess the extent of *shift variance* in the imaging system. In particular, one can use the technique of repetitive scanning of a single object in a number of positions in the microscope field. These images can then be compared either directly or parametrically. Direct comparison can be based upon an error measure between the various images (translated to the center of the *digitized* field). If all the translated images are the same—a shift-invariant system—then the error measure will be zero. As the difference between the images increases—a reflection of the shift variance—the error measure will increase.

A parametric comparison might also be termed an operational comparison. Since we propose to use our imaging system to acquire measurements from images, we can compare the measurements derived from the same object when it is scanned at different positions in the microscope field. If the system is shift variant, then the measurements may differ. The area of an object might change, for example, while its total brightness remains the same. Through this parametric approach we can determine if the influence of the shift variance is sufficiently small to permit us to ignore its effects on the desired measures.

C. The Distortion of Reality

We began this chapter by describing how every acquired fluorescence image is a compromise between the actual distribution of light and the constraints of physics and engineering that prevent us from observing it. This distortion is inherent in the observation process. We can reduce it, minimize it, or ignore it, but we cannot make it go away. At the very best, the data we obtain through quantitative fluorescence microscopy will be a distortion of reality. We have tried to present in this chapter the tools for understanding that distortion.

Acknowledgments

We would like to acknowledge the support of NATO (Grant 85/0324) and The Netherlands foundation for Medical Research, MEDIGON (Grant 900-538-016) in this project. We would also like to thank Dr. Donna Arndt-Jovin, Dr. Tom Jovin, Prof. Mack Fulwyler, Prof. Brain Mayall, Dr. Piet Verbeek, Prof. Frans Groen, and Dr. Bob Duin for their advice and insight over the years.

References

Born, M., and Wolf, E. (1980). *"Principles of Optics,"* 6th Ed. Pergamon, Oxford.

Brakenhoff, G. J., Blom. P., and Barends, P. (1979). *J. Microsc.* **117,** 219–232.

Brakenhoff, G. J., Van der Voort, H. T. M., Van Spronsen, E. A., Linnemans, W. A. M., and Nanninga, N. (1985). *Nature (London)* **317,** 748–749.

Castleman, K. R. (1979). *"Digital Image Processing."* Prentice-Hall, New York.

Dorst, L. (1985). *In* "Optics in Engineering Measurement," SPIE–599, 155–159.

Dorst. L., and Smeulders, A. W. M. (1986). *In* "Pattern Recognition in Practice II" (E. S. Gelsema and L. N. Kanal, eds.), pp. 73–80. Elsevier, Amsterdam.

Dudgeon, D. E., and Mersereau, R. M. (1984). *"Multidimensional Digital Signal Processing."* Prentice-Hall, New York.

Goodman, J. W. (1968). *Introduction to Fourier Optics."* McGraw-Hill, New York.

Harms, H., and Aus, H. M. (1984). *Cytometry* **5,** 228–235.

Limansky. I. (1968). *Electron. Eng.* **June,** 50–55.

Martin, L. C. (1966). "The Theory of the Microscope." Elsevier, New York.

O'Neill, E. L. (1963). *Introduction to Statistical Optics."* Addison-Wesley, Reading, Massachusetts.
Oppenheim, A. V., Willsky, A. S., and Young, I. T. (1983). *"Systems and Signals."* Prentice-Hall, New York.
Papoulis, A. (1968). *"Systems and Transforms with Applications in Optics."* McGraw-Hill, New York.
Piller, H. (1977). *"Microscope Photometry."* Springer-Verlag, Berlin.
Ploem, J. S. (1967). *Z. Wiss. Mikrosk.* **68,** 129–1143.
Young, I. T. (1970). *In* "Automated Cell Identification and Cell Sorting" (G. L. Wied and G. F. Bahr, eds.), pp. 187–194. Academic Press, New York.
Young, I. T. (1983). *In* "Optics in Engineering Measurement," SPIE–397, 326–335.
Young, I. T., Balasubramanian, Dunbar, D. L., Peverini, R. L., and Bishop, R. P. (1982). *IEEE Trans. Biomed. Eng.* **BME-29,** 70–82.
Young, I. T., (1988). "Sampling Density and Quantitative Microscopy," *Anal. Quant Cytol. Histol.* (in press).
Young, I. T., Verbeek, P. W., and Mayall, B. H. (1986). *Cytometry* **7,** 467–474.

Chapter 2

Validation of an Imaging System: Steps to Evaluate and Validate a Microscope Imaging System for Quantitative Studies

ŽELJKO JERIČEVIĆ,[1] BRENT WIESE, JOSEPH BRYAN AND LOUIS C. SMITH

Departments of Medicine and Cell Biology
Baylor College of Medicine
Houston, Texas 77030

The ability to obtain and store microscope images in a digital format, to perform numerical operations to obtain ratio images, difference images, etc., and to display these results in pseudocolor has generated great excitement and interest in digital imaging microscopy. In some applications, this technology has been used only as an expensive method of qualitative documentation. If the power of digital imaging is to increase our understanding of the dynamics of cellular phenomena, careful attention must be given to a systematic evaluation of each component and

[1] Permanent address: Rugjer Bošković Institute, Zagreb, Croatia, Yugoslavia.

procedure in image acquisition, image processing, and data analysis. We envision a typical imaging system as a two-dimensional array containing about 250,000 fluorimeters, each with unique and independent properties that must be characterized before their data, or pixel values, can be used with confidence. It is common knowledge that all fluorimeters are different and that comparison of results from two instruments requires some form of standardization and normalization. The problem of making an erroneous comparison is acute with digital imaging systems. Simple numerical operations will create invalid data when the spatial errors between images are greater than one pixel, in other words, when values from separate fluorimeters are compared. For common imaging systems, pixel separations are from 0.2 to 0.5 μm, depending on the objective lens.

In this chapter, we discuss some of the sources of error that can severely limit attempts to obtain quantitative information by digital imaging fluorescence microscopy. Conceptually, the sources of error can be divided into two groups: (1) those associated with the experimental apparatus used for image acquisition: linearity, geometric distortion, modulation transfer function, and the sampling frequency and (2) those associated with image processing: shading correction, geometric correction, and image registration. There is, in most instances, no simple experimental way to avoid certain sources of error associated with image acquisition. Consequently, these errors must be removed by image processing, without the introduction of different errors by the processing operations. The primary emphasis of this chapter is to describe the utilization of statistical criteria to evaluate the performance of image processing algorithms.

I. Shading Correction

Shading correction and background subtraction are basic procedures in image processing (Castleman, 1979). These two operations are the first steps in a sequence of more complex operations, such as filtering, geometric corrections, registration, segmentation, and the application of various transforms. This initial operation must be performed correctly, or the introduction of noise will severely compromise all subsequent processing.

A shading correction is a point operation used to remove camera- and light source-induced, space-dependent, photometric nonlinearities, i.e., an operation used to correct intensity distortion of an image obtained from a uniform field (Castleman, 1979; Benson *et al.,* 1985). For a fluorescent image, the information needed for a shading correction is the dark current image, D, acquired with the camera shutter closed, and a shading correction mask, S, which is the image obtained from a uniformly fluorescent field. The image processing operation for a shading correction can be represented as:

$$C_{x,y} = \frac{I_{x,y} - D_{x,y}}{S_{x,y} - D_{x,y}} M \tag{1}$$

where C is the corrected image, I is the original uncorrected image, M is the mean value of the shading mask, equal to

$$M = \sum_{x=0}^{X-1} \sum_{y=0}^{Y-1} (S_{x,y} - D_{x,y})/XY$$

and x, y are the space coordinates in integer numbers of pixels, ranging from 0 to $X - 1$ for x and from 0 to $Y - 1$ for y, where X and Y are the image size in number of pixels.

Because the dark current and shading correction mask are both experimentally acquired images, they each contain noise which is randomly distributed in image space. This additional noise is introduced into the corrected image by the operation described by Eq. (1) and will increase the existing noise in the original image. Moreover, the noise in the original image will be amplified further if the intensity values in the shading correction mask are low. Although the noise is independent of the pixel intensity value, amplification of noise occurs for the following reason. In the case of a low-intensity shading mask, the random fluctuations are relatively large compared to the magnitude of the intensity values of the shading mask. The disruption of image quality by noise can be obvious. Visual inspection of such an image shows a "salt and pepper" effect (Castleman, 1979) which makes this type of processed images unacceptable. It should be emphasized that the procedure described by Eq. (1) always transfers additional noise from the dark current and shading correction masks to the corrected image, although the added noise may not be obvious by visual inspection if the processing used a shading correction mask of higher intensity values where the noise contribution is small.

Using the general formula for error propagation, Eq. (2), it is possible to estimate the impact of noise in the I, S, and D images transferred to the corrected image C.

$$\Delta C = \frac{\partial C}{\partial I} \Delta I + \frac{\partial C}{\partial S} \Delta S + \frac{\partial C}{\partial D} \Delta D \tag{2}$$

Equation (2) is based on the calculation of the incremental change in function C of the three variables I, S, and D, using only the linear members of a Taylor expansion. The increments of the independent variables

$(\Delta I, \Delta S, \Delta D)$ are estimated using a low pass filter. The partial derivatives are developed from Eq. (1). Summation of Eq. (2) over the image space represents the propagation of noise from images I, S, and D into image C. To avoid the possibility that the errors might cancel each other, absolute or squared values should be used in the summation.

The partial derivatives of the function $C(I, S, D)$ are

$$\frac{\partial C}{\partial I} = \frac{M}{S - D} \tag{3}$$

$$\frac{\partial C}{\partial S} = \frac{I - D}{S - D}\left(1 - \frac{M}{S - D}\right) \tag{4}$$

$$\frac{\partial C}{\partial D} = \frac{1}{S - D}\left(M\frac{I - S}{S - D} + D - I\right) \tag{5}$$

In all equations, the propagation of errors is inversely proportional to the difference $(S - D)$. The consequence of this relationship is that the quality of the final image increases with higher values of the shading mask and lower values of the dark current mask. Equation (3) shows that noise will be enhanced for pixels for which the intensity value is smaller than the mean value of the mask and reduced for the pixels for which the intensity value is greater than the mean value of the mask. We do not know *a priori* the errors in image I, which represents our sample. There is no simple way to estimate these errors because the original image can, and very often does, contain genuine high frequency information. Therefore, the errors described by the first term in Eq. (2) cannot be corrected at the level of shading correction. By contrast, filtering in the frequency domain is able to correct for the propagation of these errors.

Equations (4) and (5) are more interesting in this context. Because of the nature of images S and D, their errors can be easily estimated by means of a low pass filter, since they should have no high frequency content except for noise. Subtraction of the low pass-filtered image from the original image gives a reasonable noise approximation. Moreover, it is possible, not only to estimate the errors in image C, but also to reduce the propagation of errors described by the second and third terms of Eq. (2).

Since the shading correction always needs to be performed, a procedure to remove this additional noise has been devised. We apply a low pass filter, for example, a simple convolution with a square pulse, to the dark current image and to the shading correction mask. It is well known that the low pass filter reduces contrast in the image by averaging and actually degrades its information content (Pratt, 1978; Niblack, 1986). However, the dark current image and the shading correction mask are, by design, images with low information content. In principle, both should be smooth images with slowly changing intensity values. The application of a low pass filter to these images does not reduce their information content, but does remove noise. Incorporation of a low pass filter into Eq. (1) gives

$$C_{x,y} = \frac{I_{x,y} - L[D_{x,y}]}{L[S_{x,y}] - L[D_{x,y}]} \text{Ml} \tag{6}$$

where $L[\ \ ]$ is the low pass operator applied on the image, and Ml, the mean value, is

$$\text{Ml} = \sum_{x=0}^{X-1} \sum_{y=0}^{Y-1} (L[S_{x,y}] - L[D_{x,y}])/(XY)$$

The use of Eq. (6) avoids degradation of images by the introduction of additional noise described by the second and third terms of Eq. (3). Specific camera defects or black spots in the image represent a special case for low pass filtering. Images containing such defects must be processed with a special mask for filtering because the defects are always at the same place in the image. Otherwise, when the pixel values representing camera defects are averaged by the low pass filter, the actual values of the neighboring pixels are inaccurate after processing.

To illustrate these points, two images were subjected to shading correction procedures using Eqs. (1) and (6) with both a high intensity and a low intensity shading mask. These results demonstrate that significantly more noise amplification occurs when the low intensity shading mask is used. The noise propagation, calculated with Eqs. (2)–(5), is summarized in Tables I and II. Fourier spectra of the dark current (Fig. 1A–D) and shading mask (Fig. 1E–H) before and after the low pass filter were calculated to illustrate the noise contained in the experimental images. In such spectra, the noise is easily visible as high frequency components (Fig. 1B, D, F, H). In these isometric projections of the Fourier spectra magnitude, the frequency 0,0 is located in the center of the figures, while the high frequencies are at the edges. The figures are scaled to illustrate the reduction of noise in the spectra of the dark current and the shading correction mask after the application of a low pass filter. Without low pass filtering, the noise would be transferred into the sample image during the shading correction operation.

Figure 2 shows the fluorescent image of several cells with stress fibers (Fig. 2A) before the shading correction operation. Figure 2B is the same image after shading correction using Eq. (1) and a high intensity shading mask, but without low pass filtering of the dark current and the shading masks. The noise amplification is not obvious from a comparison of Fig. 2

TABLE I

SHADING CORRECTION ERROR VALUES FOR FIG. 2[a]

	Input	Shading mask	Dark current mask	Total
High shading mask				
Absolute value	Not calculated	1.886823*e* + 04	1.618027*e* + 05	1.806709*e* + 05
Squared value	Not calculated	5.363348*e* + 03	2.077932*e* + 05	2.131565*e* + 05
Low shading mask				
Absolute value	Not calculated	8.938782*e* + 04	1.557450*e* + 05	2.451328*e* + 05
Squared value	Not calculated	9.783795*e* + 04	2.250738*e* + 05	3.229117*e* + 05

$$\frac{\partial C}{\partial I}\Delta I + \frac{\partial C}{\partial S}\Delta S + \frac{\partial C}{\partial D}\Delta D = \Delta C$$

[a] The input is the first term, the shading mask second, the dark current mask is third term, and the total is the result of Eq. (2), either the absolute or squared values summed over the whole image space.

TABLE II

SHADING CORRECTION ERROR VALUES FOR FIG. 3[a]

	Input	Shading Mask	Dark Current Mask	Total
High Shading Mask				
Absolute value	Not calculated	2.858549*e* + 04	1.644215*e* + 05	1.930070*e* + 05
Squared value	Not calculated	3.935454*e* + 04	2.117202*e* + 05	2.510747*e* + 05
High shading mask and low-pass of input image				
Absolute value	3.039179*e* + 05	2.853600*e* + 04	1.644463*e* + 05	4.969002*e* + 05
Squared value	7.303529*e* + 05	4.203650*e* + 04	2.117818*e* + 05	9.841712*e* + 05
Low shading mask				
Absolute value	Not calculated	7.796852*e* + 04	1.602354*e* + 05	2.382039*e* + 05
Squared value	Not calculated	8.633126*e* + 04	2.376475*e* + 05	3.239787*e* + 05
Low shading mask and low-pass				
Absolute value	3.041657*e* + 05	7.768623*e* + 05	1.602112*e* + 05	5.420631*e* + 05
Squared value	7.410328*e* + 05	8.841699*e* + 04	2.372533*e* + 05	1.066703*e* + 06

$$\frac{\partial C}{\partial I}\Delta I + \frac{\partial C}{\partial S}\Delta S + \frac{\partial C}{\partial D}\Delta D = \Delta C$$

[a] The input is the first term, the shading mask is the second, the dark current mask is the third term and the total is the result of Eq. (2), either the absolute or squared values summed over the whole image space.

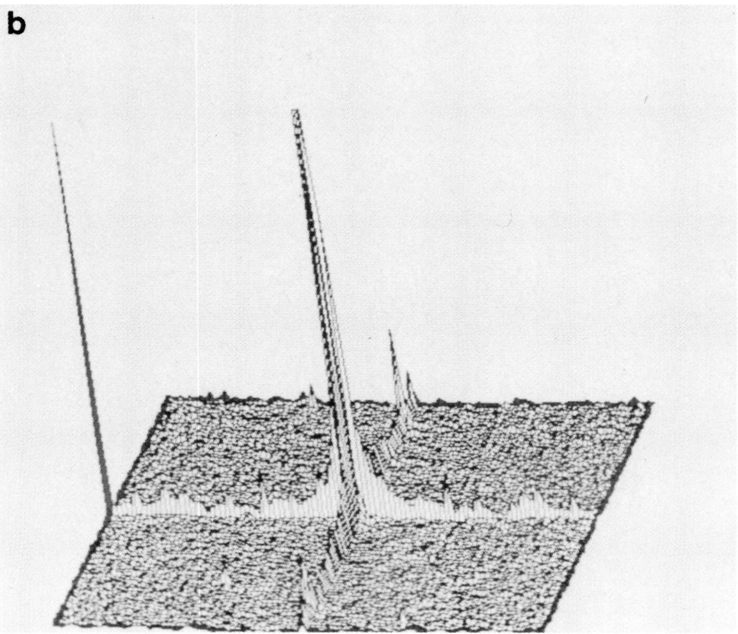

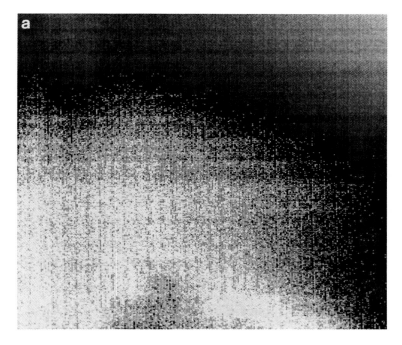

FIG. 1. (A) Dark current; (B) the magnitude of the Fourier spectrum of the dark current mask; (C) dark current mask after a low pass filter; (D) the magnitude of the Fourier spectrum of the dark current mask after a low pass filter; (E) high intensity shading correction mask; (F) magnitude of the Fourier spectrum of the shading correction mask; (G) high intensity shading correction mask after a low-pass filter; (H) the magnitude of the Fourier spectrum of the shading correction mask after a low-pass filter.

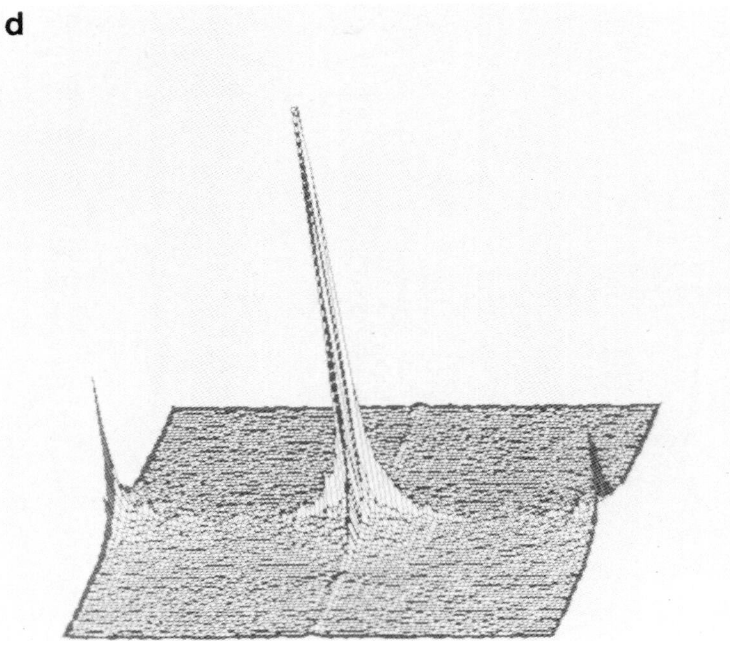

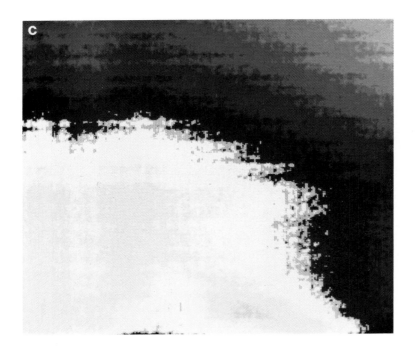

FIG. 1. (*continued*)

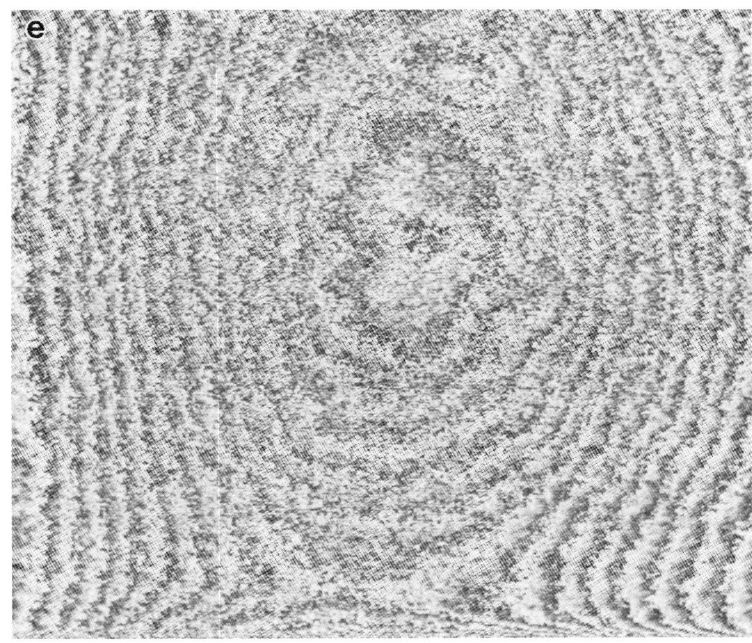

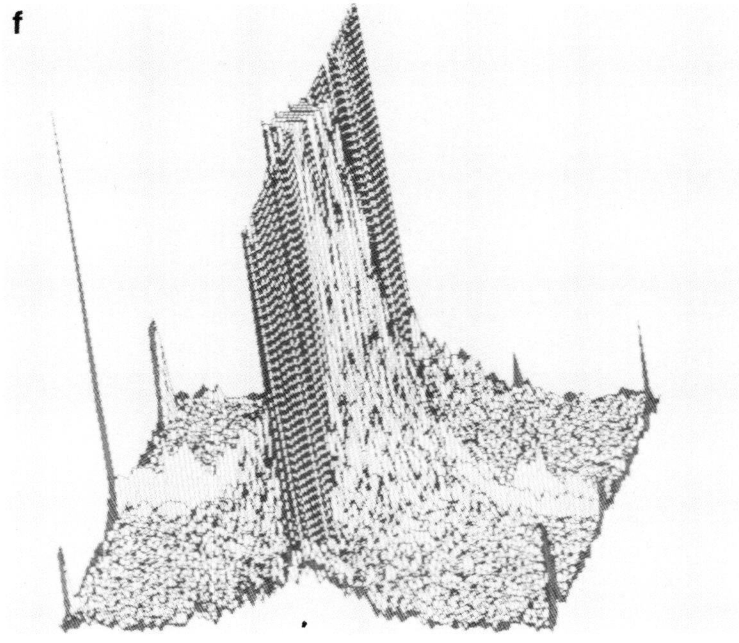

FIG. 1. (*continued*)

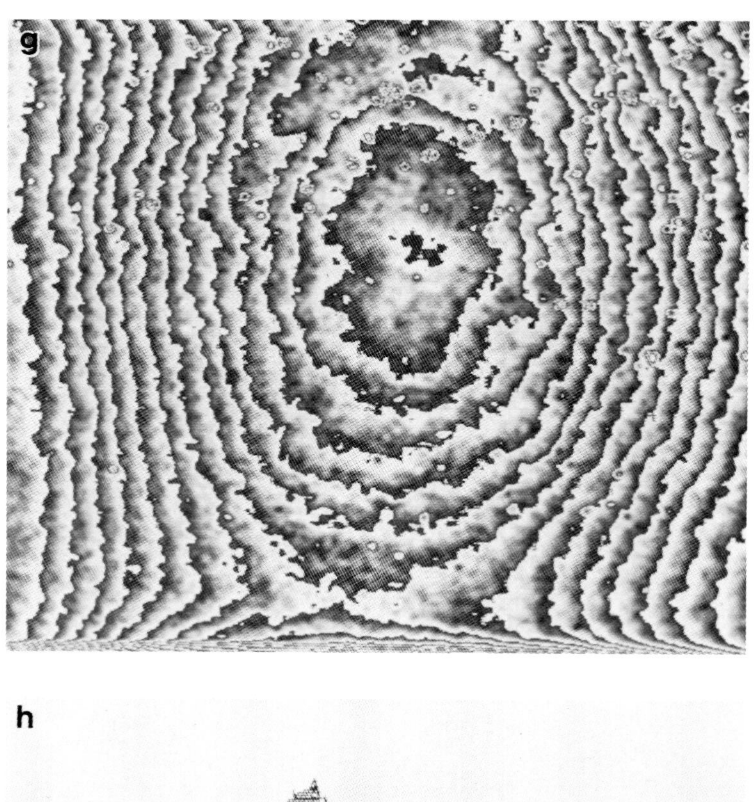

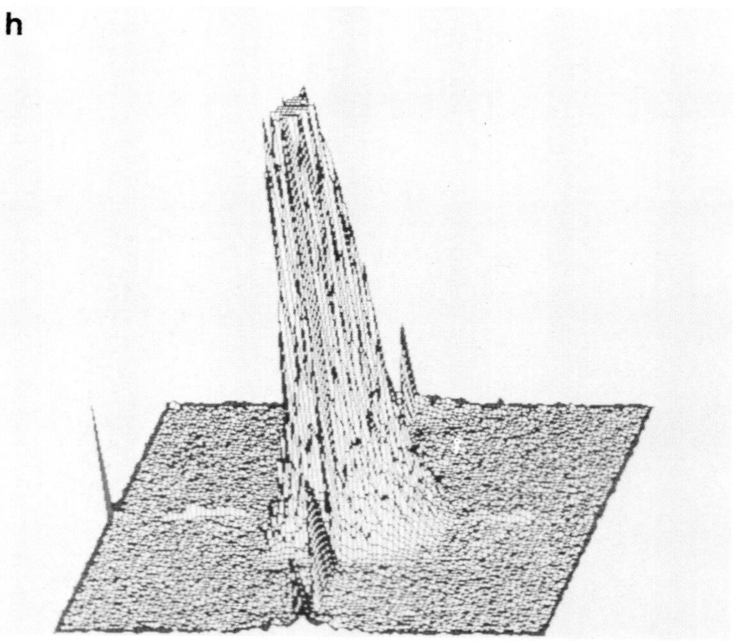

FIG. 1. (*continued*)

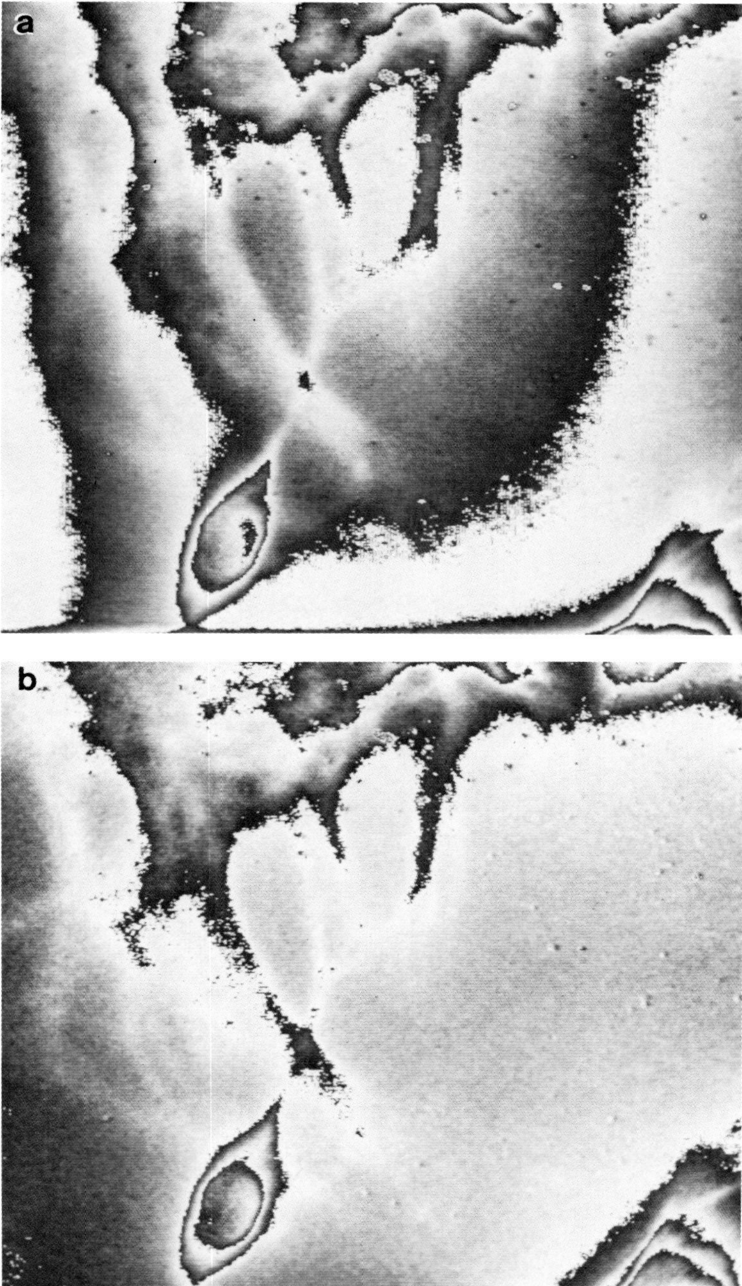

FIG. 2. Cells with stress fibers. (A) Image before shading correction. (B) Image after shading correction without low pass filtering of the masks and with a high shading correction mask. (C) Sample after shading correction without low pass filtering of the masks and with a low shading correction mask. (D) Sample after shading correction with low pass filtering of the masks and a low shading correction mask.

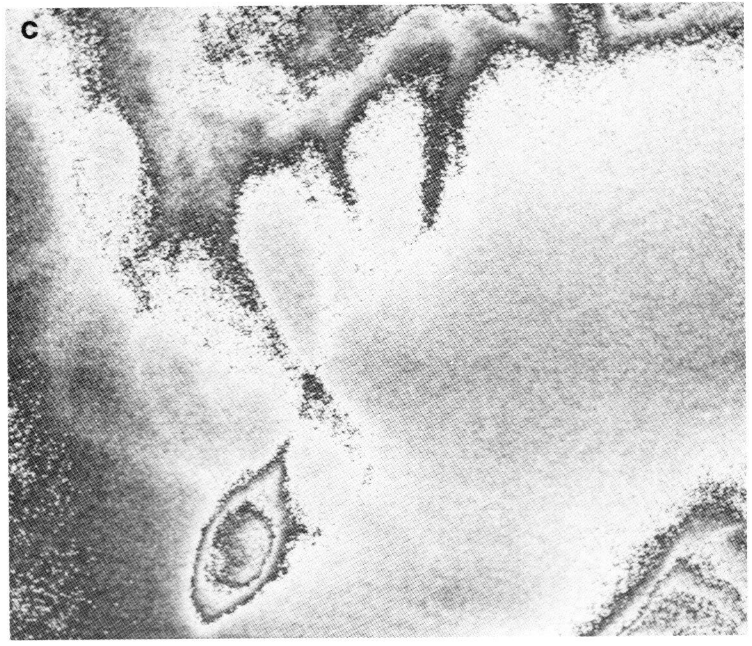

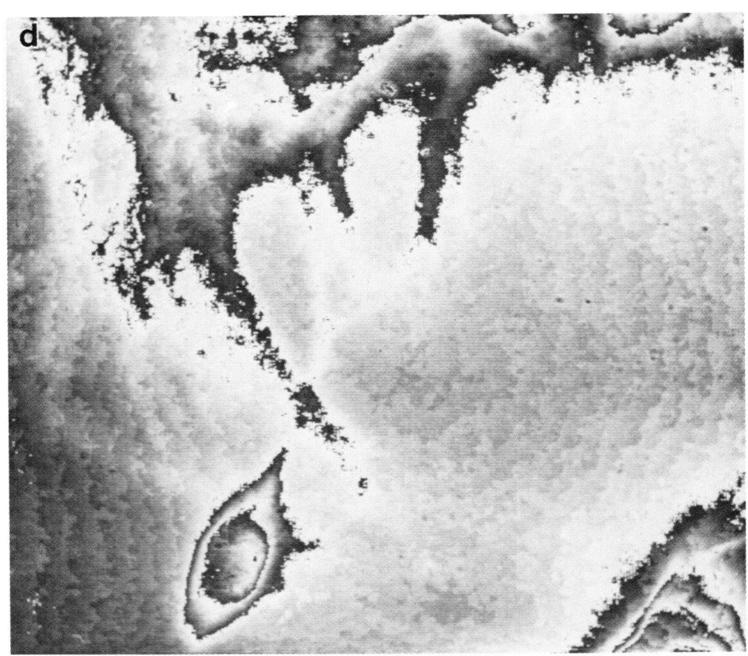

FIG. 2. *(continued)*

images with low information content. In principle, both should be smooth images with slowly changing intensity values. The application of a low pass filter to these images does not reduce their information content, but does remove noise. Incorporation of a low pass filter into Eq. (1) gives

$$C_{x,y} = \frac{I_{x,y} - L[D_{x,y}]}{L[S_{x,y}] - L[D_{x,y}]}\,\mathrm{Ml} \tag{6}$$

where $L[\ ]$ is the low pass operator applied on the image, and Ml, the mean value, is

$$\mathrm{Ml} = \sum_{x=0}^{X-1}\sum_{y=0}^{Y-1}(L[S_{x,y}] - L[D_{x,y}])/(XY)$$

The use of Eq. (6) avoids degradation of images by the introduction of additional noise described by the second and third terms of Eq. (3). Specific camera defects or black spots in the image represent a special case for low pass filtering. Images containing such defects must be processed with a special mask for filtering because the defects are always at the same place in the image. Otherwise, when the pixel values representing camera defects are averaged by the low pass filter, the actual values of the neighboring pixels are inaccurate after processing.

To illustrate these points, two images were subjected to shading correction procedures using Eqs. (1) and (6) with both a high intensity and a low intensity shading mask. These results demonstrate that significantly more noise amplification occurs when the low intensity shading mask is used. The noise propagation, calculated with Eqs. (2)–(5), is summarized in Tables I and II. Fourier spectra of the dark current (Fig. 1A–D) and shading mask (Fig. 1E–H) before and after the low pass filter were calculated to illustrate the noise contained in the experimental images. In such spectra, the noise is easily visible as high frequency components (Fig. 1B, D, F, H). In these isometric projections of the Fourier spectra magnitude, the frequency 0,0 is located in the center of the figures, while the high frequencies are at the edges. The figures are scaled to illustrate the reduction of noise in the spectra of the dark current and the shading correction mask after the application of a low pass filter. Without low pass filtering, the noise would be transferred into the sample image during the shading correction operation.

Figure 2 shows the fluorescent image of several cells with stress fibers (Fig. 2A) before the shading correction operation. Figure 2B is the same image after shading correction using Eq. (1) and a high intensity shading mask, but without low pass filtering of the dark current and the shading masks. The noise amplification is not obvious from a comparison of Fig. 2

TABLE I

SHADING CORRECTION ERROR VALUES FOR FIG. 2[a]

	Input	Shading mask	Dark current mask	Total
High shading mask				
Absolute value	Not calculated	1.886823*e* + 04	1.618027*e* + 05	1.806709*e* + 05
Squared value	Not calculated	5.363348*e* + 03	2.077932*e* + 05	2.131565*e* + 05
Low shading mask				
Absolute value	Not calculated	8.938782*e* + 04	1.557450*e* + 05	2.451328*e* + 05
Squared value	Not calculated	9.783795*e* + 04	2.250738*e* + 05	3.229117*e* + 05

$$\frac{\partial C}{\partial I}\,\Delta I + \frac{\partial C}{\partial S}\,\Delta S + \frac{\partial C}{\partial D}\,\Delta D = \Delta C$$

[a] The input is the first term, the shading mask second, the dark current mask is third term, and the total is the result of Eq. (2), either the absolute or squared values summed over the whole image space.

TABLE II

SHADING CORRECTION ERROR VALUES FOR FIG. 3[a]

	Input	Shading Mask	Dark Current Mask	Total
High Shading Mask				
Absolute value	Not calculated	2.858549*e* + 04	1.644215*e* + 05	1.930070*e* + 05
Squared value	Not calculated	3.935454*e* + 04	2.117202*e* + 05	2.510747*e* + 05
High shading mask and low-pass of input image				
Absolute value	3.039179*e* + 05	2.853600*e* + 04	1.644463*e* + 05	4.969002*e* + 05
Squared value	7.303529*e* + 05	4.203650*e* + 04	2.117818*e* + 05	9.841712*e* + 05
Low shading mask				
Absolute value	Not calculated	7.796852*e* + 04	1.602354*e* + 05	2.382039*e* + 05
Squared value	Not calculated	8.633126*e* + 04	2.376475*e* + 05	3.239787*e* + 05
Low shading mask and low-pass				
Absolute value	3.041657*e* + 05	7.768623*e* + 05	1.602112*e* + 05	5.420631*e* + 05
Squared value	7.410328*e* + 05	8.841699*e* + 04	2.372533*e* + 05	1.066703*e* + 06

$$\frac{\partial C}{\partial I}\,\Delta I + \frac{\partial C}{\partial S}\,\Delta S + \frac{\partial C}{\partial D}\,\Delta D = \Delta C$$

[a] The input is the first term, the shading mask is the second, the dark current mask is the third term and the total is the result of Eq. (2), either the absolute or squared values summed over the whole image space.

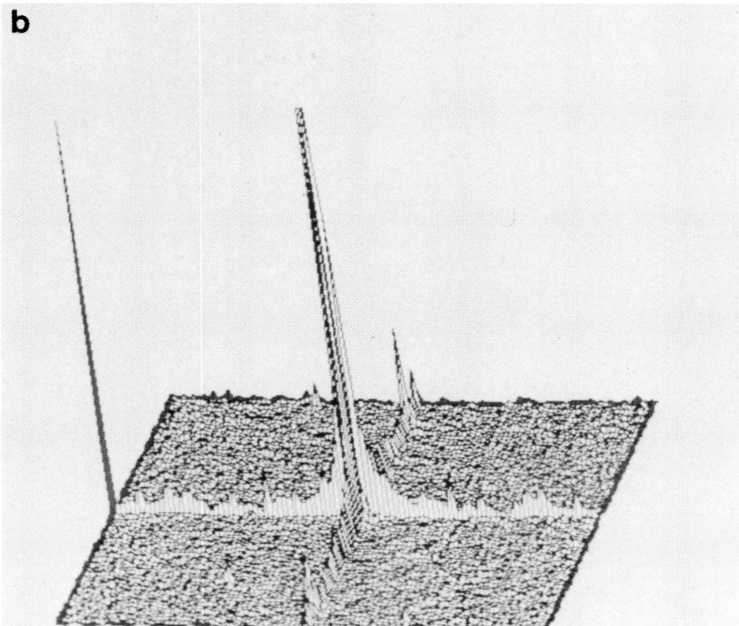

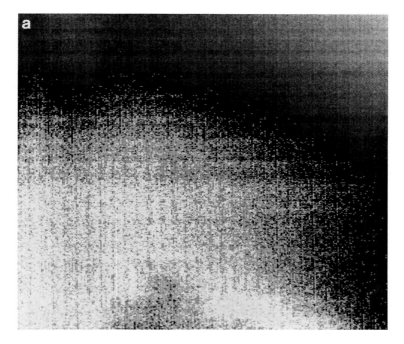

FIG. 1. (A) Dark current; (B) the magnitude of the Fourier spectrum of the dark current mask; (C) dark current mask after a low pass filter; (D) the magnitude of the Fourier spectrum of the dark current mask after a low pass filter; (E) high intensity shading correction mask; (F) magnitude of the Fourier spectrum of the shading correction mask; (G) high intensity shading correction mask after a low-pass filter; (H) the magnitude of the Fourier spectrum of the shading correction mask after a low-pass filter.

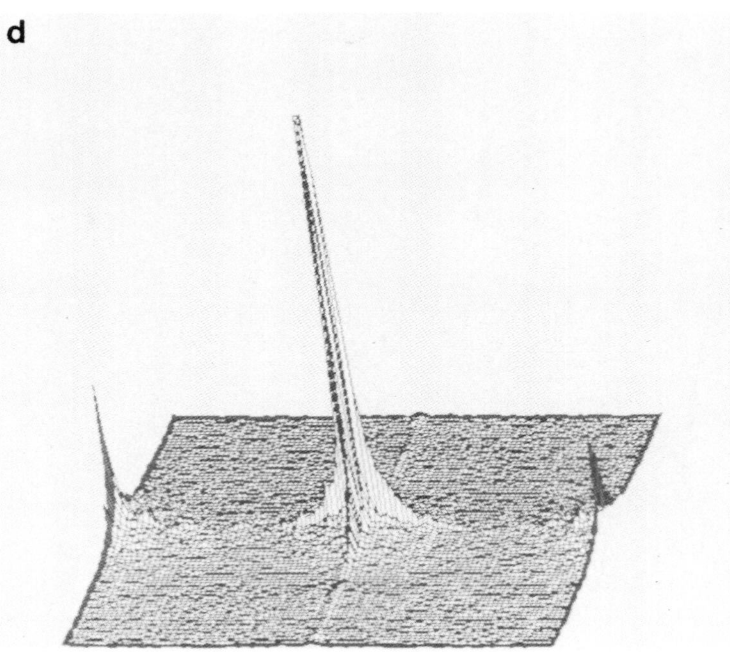

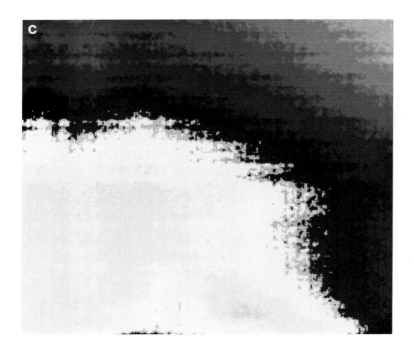

FIG. 1. (*continued*)

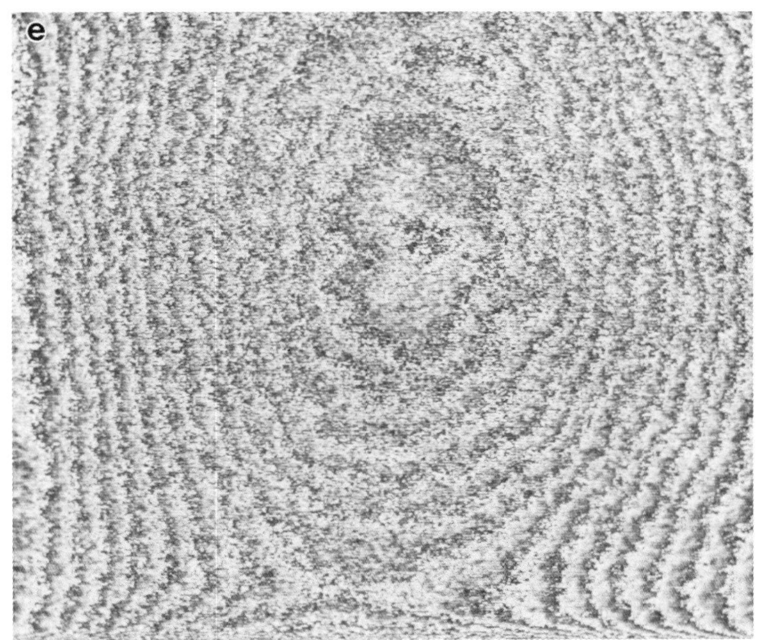

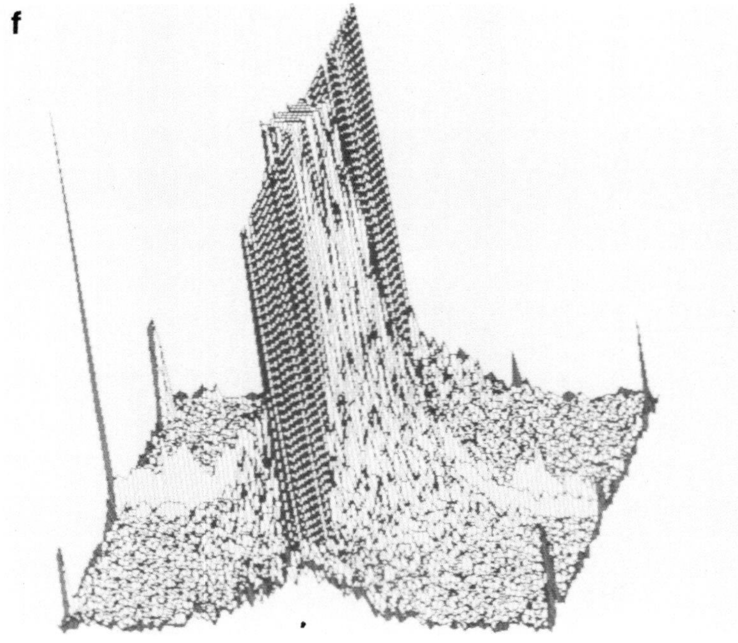

FIG. 1. (*continued*)

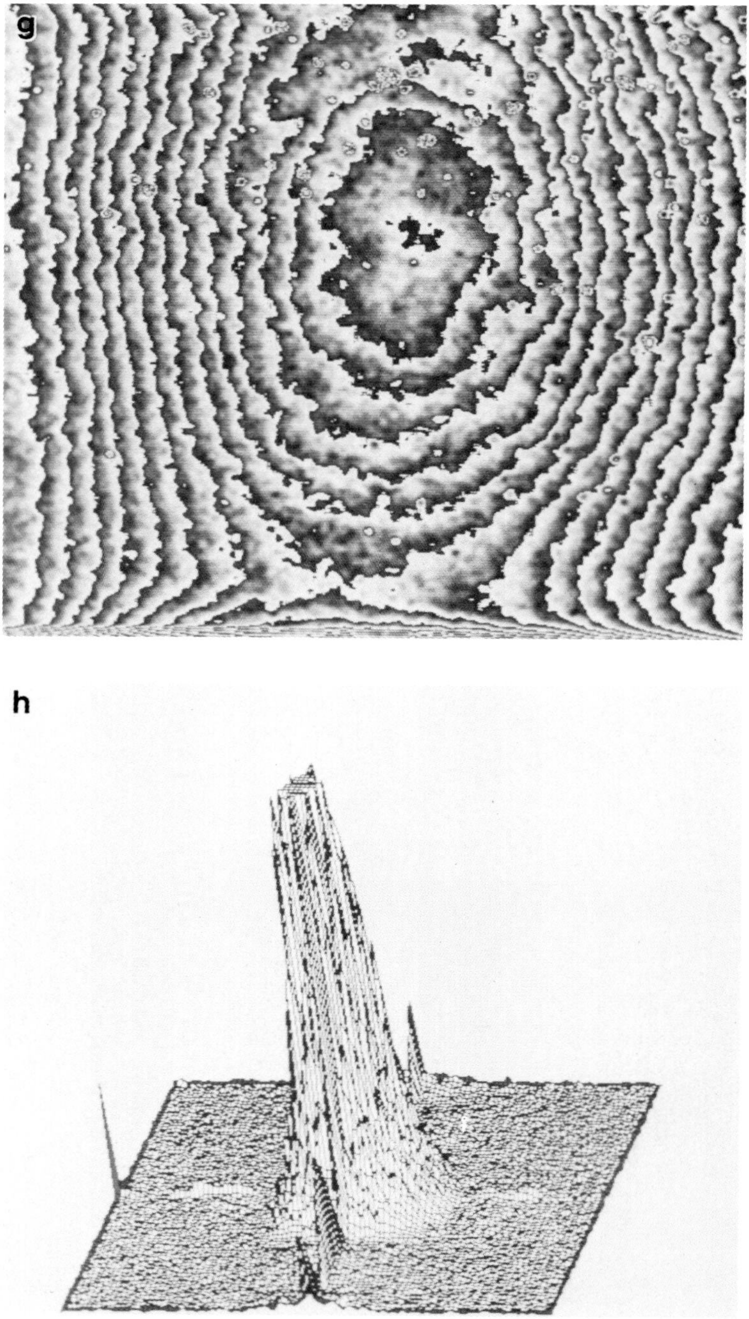

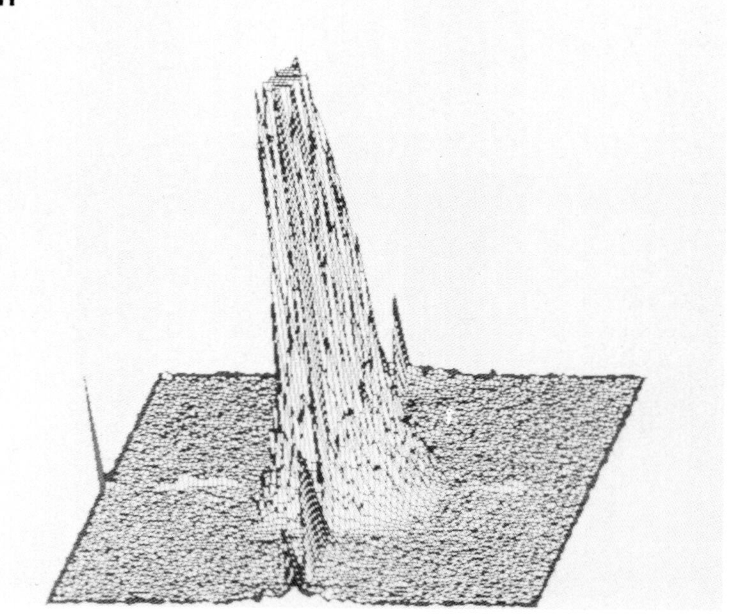

FIG. 1. (*continued*)

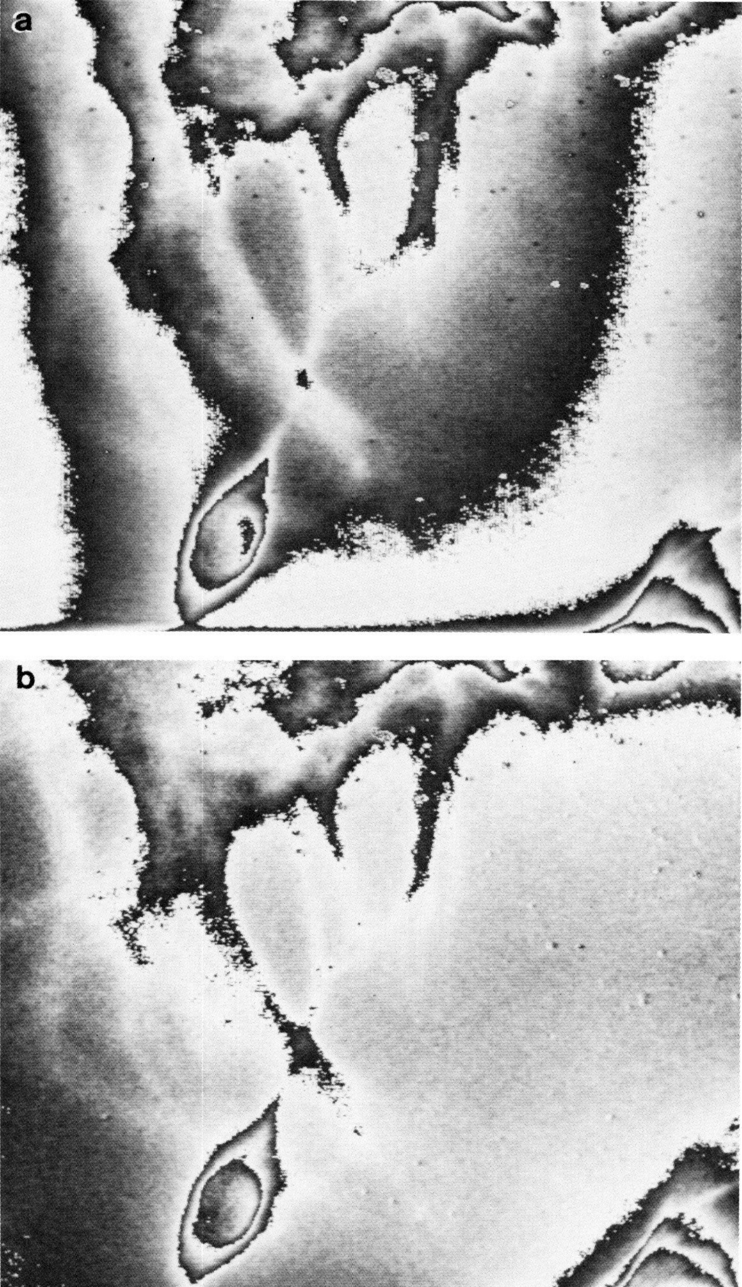

FIG. 2. Cells with stress fibers. (A) Image before shading correction. (B) Image after shading correction without low pass filtering of the masks and with a high shading correction mask. (C) Sample after shading correction without low pass filtering of the masks and with a low shading correction mask. (D) Sample after shading correction with low pass filtering of the masks and a low shading correction mask.

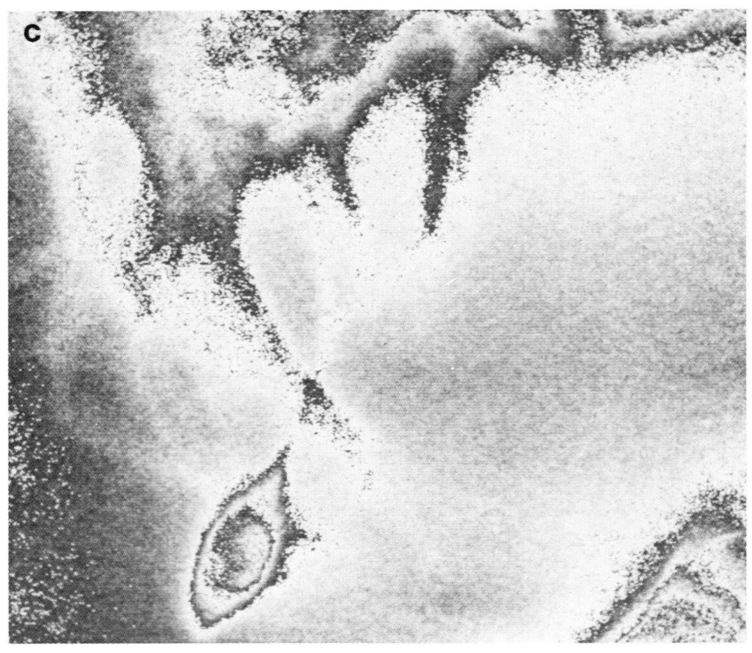

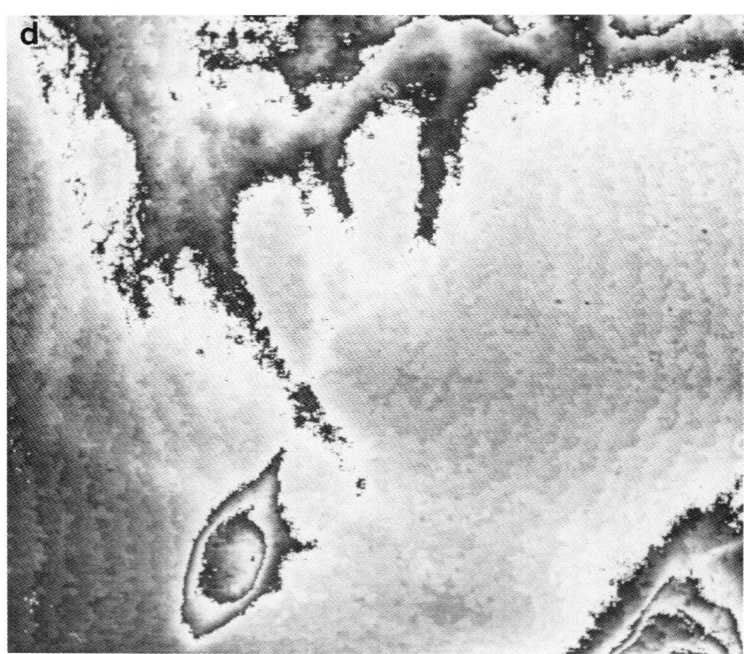

FIG. 2. *(continued)*

A and B because the original level of noise is too low. As discussed earlier, if the low-intensity shading mask is used, the noise is disportionately amplified, as illustrated by comparison of Fig. 2 with Fig. 2B. However, application of a low pass filter to the masks reduces the noise in the corrected image can be reduced to the level observed in images corrected with a high intensity shading mask (Figs. 2B and 2D). The improvement comes from the use of Eq. 6, which does not propagate noise from the masks to the corrected image.

Our system has a noise level of about 2–3%, as reported earlier by Benson *et al.* (1985). With this low level, it is not possible to observe noise directly in an image, unless flat field images containing no objects are used. In these featureless images, the noise can be identified visually by simple manipulation of the image display. Figure 3 illustrates image noise using a uniformly fluorescent uranium oxide glass. The flat field image before

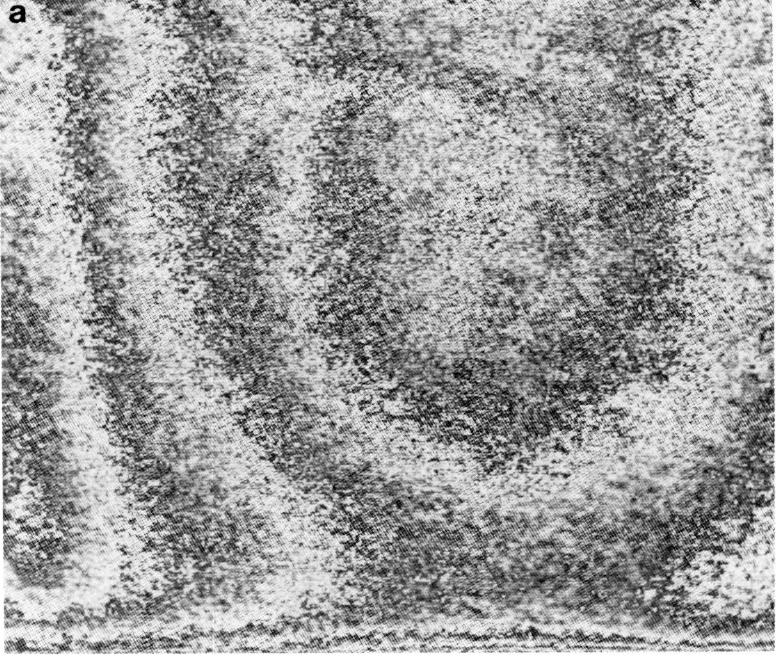

FIG. 3. Uranium oxide glass as an example of a uniform image. (A) Image before shading correction. (B) Image after shading correction without low pass filtering of the masks. (C) Image after shading correction with low pass filtering of the masks. (D) Sample after shading correction with low pass filtering of the masks and of the original image.

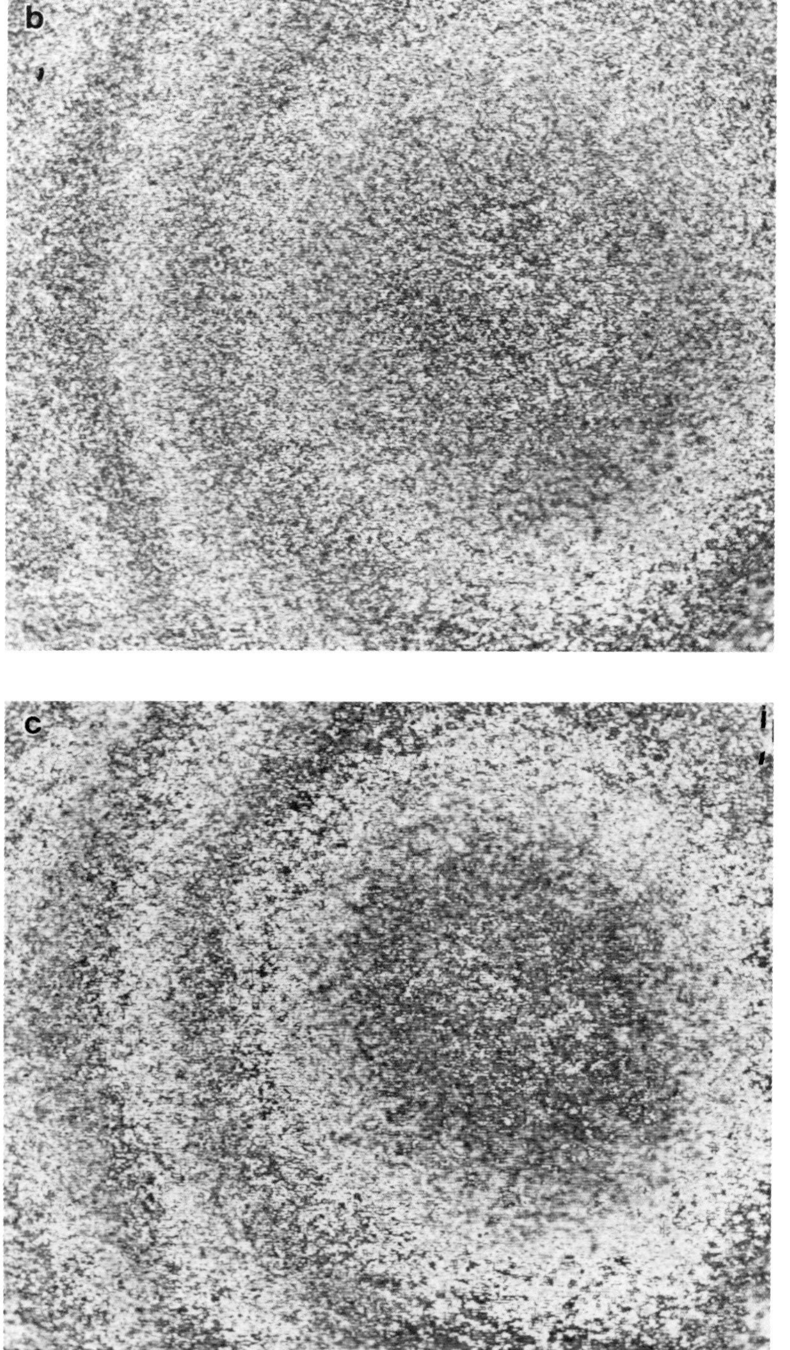

FIG. 3. *(continued)*

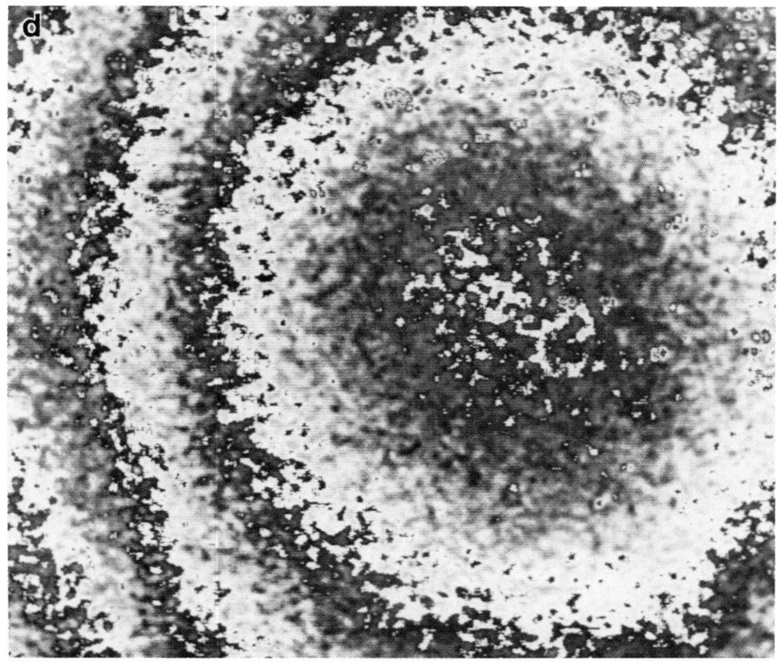

FIG. 3. *(continued)*

correction is shown in Fig. 3A. The image corrected with an unfiltered mask is shown in Fig. 3B. The gray scale, defined by the system look-up tables, were set so that it is possible to see that Fig. 3B is more noisy than Fig. 3A. Application of a low pass filter to the mask used for correction reduces the noise level in the processed image to approximately the same level as original image; compare Fig. 3A with Fig. 3C. Further improvement can be achieved by filtering the original image, as can be seen by comparing Fig. 3D with the other images in Fig. 3.

Nonlinear intensity values can arise from both the camera and the light source. The procedure described above does not distinguish between these two sources and correction is done for both errors at the same time, although their separate contributions are unknown. This lack of information can limit severely any attempt to achieve objective quantification for comparison of different images or various regions of the same image. The usual procedure is to assume that the illumination is uniform. Our experience has shown that this is not a reasonable assumption for microscopy systems using only a mercury arc or a xenon lamp. It is possible, however, to separate the contributions from the light source from those of the camera by measuring the light intensity using a photobleaching procedure. The extent to which the incident light is uniformly distributed in the object plane can be determined by photobleaching a uniform, featureless film of a fluorescent compound. We have used FITC-IgG mixed in agar, spread on a microscope slide, and sealed under a coverslip. The space-dependent photobleaching process for this specimen can be described by the following equation:

$$I_{x,y,t} = b_{x,y} + a_{x,y} \exp(-k_{x,y}t) \tag{7}$$

where I is a dependent variable, the intensity of the fluorescent signal; t is an independent variable, the elapsed time; x and y are spatial coordinates of the pixel; and a, b, and k are parameters of the model. Parameter a is the intensity of the fluorescent signal at the initial time, parameter b is the intensity of the fluorescent signal at infinite time, and k is the rate constant for the photobleaching reaction occurring at a particular pixel. A rate constant is calculated for every pixel from a time series of images. The procedure generates a rate constant map of the uniformly distributed fluorophore, since the photobleaching rate constant (k) is proportional to the amount of light incident on the sample and the concentration of the fluorescent dye. For a sample that is spatially uniform, spatial differences in the rate constant map are produced only by spatial heterogeneity of the excitation light. In the linear range of camera response, intensity distortion in the individual images due to the camera will not affect the rate constants calculated from 20 or more images since the distortion will affect every image in the time series in the same way. Figure 4 illustrates the difference between the shading correction mask, which is a combination of the spatial heterogeneity of the incident light and the camera response to a signal, and the photobleaching rate constant map, which represents only the nonuniformity of incident light. The experimental procedure has been described by Benson *et al.* (1985) and the numerical procedure by Jeričević *et al.* (1987). The problem of nonuniform excitation light has been addressed by Spring and Lowy (1988 Volume 29 this series).

II. Linearity

It is obviously important to establish the linear intensity range of the detector. If image acquisition is done within this range, linear system theory can be used to facilitate image comparison. Although a correction for linearity can be done after or during sampling (Pratt, 1978), subsequent

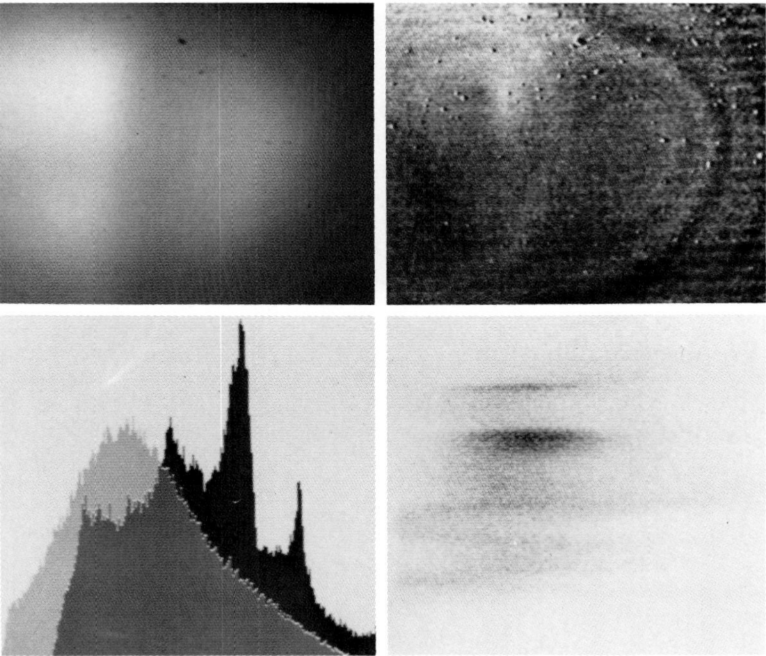

FIG. 4. Comparison of shading mask (A) and photobleaching rate constant map (B) for the same field. (C) One-dimensional histograms; dark is the image in A, light gray is the image in B, with overlap in between. (D) Two-dimensional histogram with intensities from image B at abscissa (x) and intensities from image A at ordinate (y). The range of intensities is from 0 to 255.

processing and the attendant noise introduction are avoided if an experiment is conducted so that measurements are done within the linear range of camera response. The system must be calibrated periodically, usually about once a month. An example of a procedure has been described by Benson *et al.* (1985). Inoué (1986) has provided a more detailed discussion of detector responses and linearity.

III. Geometric Correction and Registration

The geometric correction and registration of digital images (Castleman, 1979) are basic procedures in image processing, usually done after shading correction but preceding any comparison of image data or image reconstruction. The failure to perform these operations with sufficient accuracy will make subsequent image comparison inaccurate and probably misleading. Distortions in different parts of an image arise from the nonlinear dependence of pixel size on their individual spatial coordinates. These distortions are caused by the system optics and by the photoactive surface of the camera (Castleman, 1979; Pratt, 1978; Niblack, 1986). A geometric correction is a multistep numerical procedure for removing such distortions. The correction procedures can also be used to warp images to any desired geometry.

A. Geometric Correction

Geometric correction consists of three principal steps: (1) specification of the distortion; (2) a mapping procedure; and (3) an interpolation procedure.

1. Specification

Specification of the distortion can be based on a physical model of distortion. In that case, the contribution(s) from every component of the imaging system must be known. Since this information is extremely difficult to obtain, methods based on a mathematical description of the distortion, which do not require an understanding of its physical causes, are preferred. The specification of the distortion requires an empirical identification of one set of image features present in both the distorted and in the ideal, undistorted image. Those features can be related mathematically to give a numerical description of the distortion. The numerical description is achieved by fitting the data with a polynomial equation of an arbitrary degree. A polynomial function is chosen because it is linear in its parameters and the least squares fitting of a polynomial is then a linear problem. Such a descriptive method does not require an understanding of the physical causes of the distortion, but does require that the control points be obtained under the same conditions as those used for experimental image acquisition. This method is described in detail by Jeričević *et al.* (1988a) and will be outlined briefly in this chapter.

2. Mapping

Two mapping strategies can be considered: input-to-output and output-to-input. With the output-to-input mapping method, the arguments for the correction polynomial functions are the integer coordinates of the output image and the function values are the real coordinates in the distorted input image. This method is recommended because it does not leave gaps in the image and higher-order interpolation schemes do not represent conceptual difficulties. For example, interpolation of the intensity values can be done using the real coordinates in the input image and the interpolated value written as the pixel value at the integer coordinates for the output image. Because all of the pixel values in the input image are known before the calculation, interpolation from any number of neighboring pixels can be done without the implementation problems encountered in the input-to-output scheme. An additional advantage of the output-to-input method is that an output image can be constructed by calculating pixel coordinates in regular order, usually by column or by row, which simplifies the calculations and can be used to vectorize the algorithm and thereby increase speed.

3. Interpolation

The simplest interpolation scheme is the nearest-neighbor interpolation where the real coordinates are rounded to the closest integer value. Although this procedure is the most rapid, it can produce aliasing if done repeatedly. A better procedure is a bilinear scheme in which the four neighboring points are used, as described in detail by Castleman (1979).

4. Implementation

If the distortion of the image can be described by a polynomial equation, the following equations can be written for every pair of pixel coordinates:

$$x = f(i, j) \tag{8}$$

$$y = g(i, j) \tag{9}$$

where the function $f(i, j)$ and $g(i, j)$ are polynomials of arbitrary degree in two variables, i and j, which are the integer coordinates of the pixels in the output image. The real coordinates in the input image are x and y.

A general and efficient way to evaluate a polynomial is by the Horner nesting scheme (Press *et al.*, 1986). To use this method, the members of a multidimensional polynomial must be presorted appropriately with respect to the order of the powers of the independent variables. For a two-dimensional polynomial, the members are sorted as in a two-dimensional array, with the indices representing the powers of the variables. An example for a third degree polynomial is presented in Eq. (10).

$$\begin{aligned} f(i, j) = {} & c_0 i^0 j^0 + c_1 i^1 j^0 + c_2 i^2 j^0 + c_3 i^3 j^0 \\ & + c_4 i^0 j^1 + c_5 i^1 j^1 + c_6 i^2 j^1 \\ & + c_7 i^0 j^2 + c_8 i^1 j^2 \\ & + c_9 i^0 j^3 \end{aligned} \tag{10}$$

With this particular arrangement, each row on the right-hand side of Eq. (10) can be evaluated separately by Horner's algorithm for powers of i. The result of this evaluation is then stored in the appropriate element of a temporary array. After all rows are evaluated with respect to values of i, the resulting temporary variables are evaluated by Horner's scheme with respect to powers of j. The method is computationally efficient and completely flexible with respect to polynomial degree. The number of coefficients for a two-dimensional polynomial of degree k can be calculated from the formula $n = [(k + 1)(k + 2)]/2$.

To calculate the polynomial coefficients, control points and the least squares method (Pratt, 1978; Niblack, 1986) are used. The control points from the distorted image are used to construct the **x** and **y** vectors. The corresponding points from the ideal image are used to construct $f(i, j)$ and $g(i, j)$ from Eqs. (8) and (9). In this context, the right-hand side of Eq. (10) is the product of one of the m rows in the matrix A with the column vector of n unknown coefficients **c**. Written in matrix notation, Eqs. (8) and (9) for m control points and n coefficients of the correction polynomials become Eqs. (11) and (12), respectively.

$$\mathbf{x}(m) = \mathrm{A}(m, n)\mathbf{c}(n) \tag{11}$$

$$\mathbf{y}(m) = \mathrm{A}(m, n)\mathbf{d}(n) \tag{12}$$

Equations (11) and (12) both must be solved in order to make the geometric correction. However, the problem is the same for both equations and only Eq. (11) will be discussed.

If the degree for the polynomial is the same for x and y, the Vandermonde maxtrix A (Press *et al.*, 1986) is the same for both equations, which simplifies the calculation of vector $\mathbf{d}(n)$ after Eq. (11) has been solved. The unknown vector of polynomial coefficients $c(n)$ cannot be calculated by inversion directly from matrix Eq. (11). The system of equations is overdetermined and since $m > n$, matrix A is not square and cannot be inverted to calculate vector c. The structure of matrix A is shown in the right side of Eq. (10), which is a product of a single row of matrix A with the vector **c**.

The usual procedure recommended in the digital image processing literature (Castleman, 1979; Pratt, 1978; Niblack, 1986) is to use Eq. (13):

$$\mathbf{c}(n) = [\mathrm{A}^{\mathrm{T}}(n, m)\mathrm{A}(m, n)]^{-1}\mathrm{A}^{\mathrm{T}}(n, m)\mathbf{x}(m) \tag{13}$$

where T and $[\ \ ]^{-1}$ denotes the transpose and inverse matricies, respectively. Equation (13) can be found in most numerical analysis textbooks as the general solution for the linear least squares problem. Authors of more specialized monographs (Press *et al.*, 1986; Lawson and Hanson, 1974; Golub and Van Loan, 1983) observe that this is the least acceptable numerical approach to the problem.

An alternate computational approach can be used to solve Eq. (11) without inverting the A matrix. For example, Lawson and Hanson (1974) have made extensive use of the singular value decomposition method, which is, in most cases, the best choice. We have chosen to use orthonormal decomposition, as shown in Eq. (14), because it allows us to use back-substitution in the next step, which reduces the overall number of computations to a minimum. Matricies $\mathrm{Q}(m,n)$ and $\mathrm{U}(n,n)$ are an orthonormal and an upper triangular matrix, respectively, calculated from matrix A by a modified Gram–Schmidt procedure.

$$\mathbf{x}(m) = \mathrm{Q}(m,n)\mathrm{U}(n,n)\mathbf{c}(n) \tag{14}$$

From the property of an orthonormal matrix that its inverse is equal to its transpose, we can develop Eq. (15) from Eq. (14).

$$\mathrm{Q}^{\mathrm{T}}(n,\ m)\mathbf{x}(m) = \mathrm{U}(n,\ n)\mathbf{c}(n) \tag{15}$$

After multiplication of the left-hand side of the maxtrix Eq. (15), the resultant column vector is used in the calculation of vector $\mathbf{c}(n)$ by back substitution, because $\mathrm{U}(n,\ n)$ is an upper triangular matrix. By calculating the vectors $\mathbf{c}(n)$ and $\mathbf{d}(n)$, the specification of the distortion is completed and the mapping operations of the geometric correction can now be done. In the mapping procedure, the polynomials for the x and y coordinates must be evaluated for every pixel. For computational efficiency, when higher degree polynomials are used, the evaluation should be implemented by a Horner's scheme, as previously described for Eq. (10).

In practical terms, control points were selected on the image of a reference object, a rectangular grid with 10-μm-square openings. The reference grid is a commercially available product from Applied Image Inc., Rochester, NY 14609, and was assumed to be perfectly regular and symmetrical. The image of this grid, shown in Fig. 5 and obtained using the imaging system described by Benson *et al.* (1985), clearly is not rectangular. On this distorted image, we superimpose an artificial regular net, which can be identified as the sharper white lines in Fig. 5, particularly at the upper and lower regions of the image. Figure 5 also shows the non-uniform variation in the spatial relationship of the reference grid to the grid image in the uncorrected image. The control points were sampled all

FIG. 5. Uncorrected image of a 10 μm grid with the superimposed net showing control points connected by white lines to the appropriate experimental points.

over the image to ensure proper correction. The sampling included the least distorted part of the image, usually the center, as well as the more distorted edges. By statistical criteria, the appropriate number of control points should be no less than 10 for each coefficient of the polynomial equation. The parameters of the correction polynomial were calculated using the version of the least squares method described above.

We have evaluated the use of correction polynomials of various degrees to remove distortion from the microscopic image of a test grid. Visual inspection of the corrected image using a second degree polynomial shows that the result is not satisfactory (Fig. 6). By contrast, third degree or higher polynomials do not show any visible distortion (Figs. 7 and 8). The grid in the corrected image coincides with the artificial net with no apparent deviation, and it is not possible by visual inspection to decide which degree of the polynomial is adequate. More objective criteria are necessary; the deviations of the polynomial correction functions from the

Fig. 6. Image of a grid and the superimposed net, corrected using a second degree polynomial.

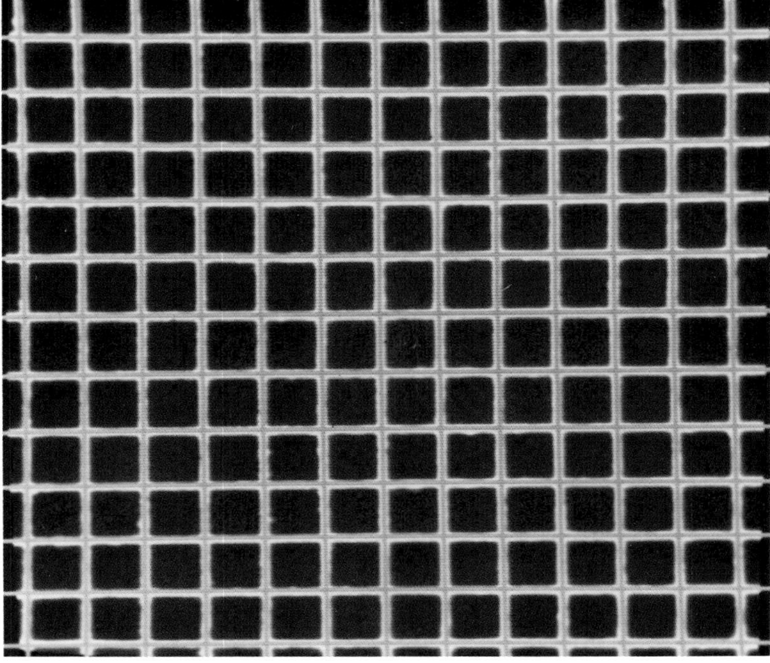

Fig. 7. Image of a grid and the superimposed net, corrected by a third degree polynomial.

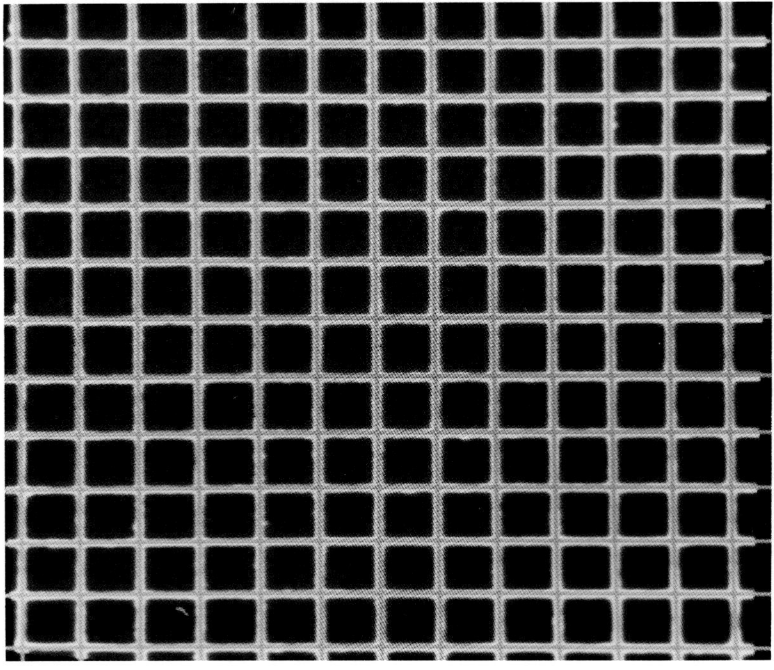

FIG. 8. Image of a grid and the superimposed net, corrected by a fourth degree polynomial for the x direction and a fifth degree polynomial for the y direction, which gave the best correction using single floating point precision.

control points, as represented by the least squares sums, are informative in this respect. The dependency of the logarithm of the least squares sum as a function of the degree of the fitting polynomial is shown in Fig. 9.

Several considerations are important. The best correction is not necessarily achieved by using a higher degree of polynomial. A limit arises from the numerical difficulties of decomposing a Vandermonde matrix A constructed from too many basis functions (Press *et al.*, 1986). As the degree of the polynomial increases, the basis functions become indistinguishable from the collected data points. Sixteen is the highest degree of a two-independent-variables polynomial equation which can be used with 169 points. Sixteen- and seventeen-degree polynomials have 153 and 171 coefficients, respectively. We estimate that the best fit is obtained with a fourth degree polynomial for the x direction and a fifth degree polynomial for the y direction, using the single precision floating point calculations. If a better correction is desirable, more control points must be sampled and/or

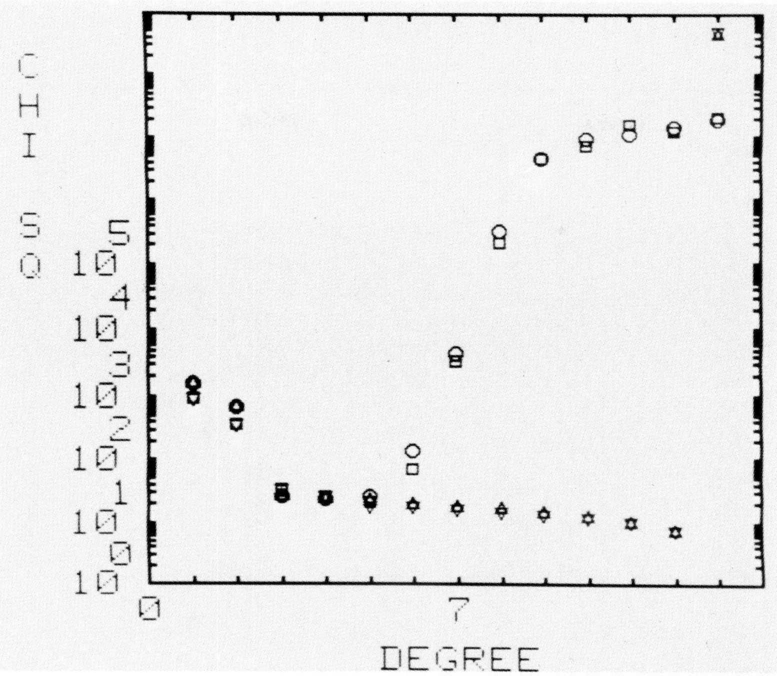

FIG. 9. Dependence of the least squares sums on the degree of the polynomial used for correction. (○) The least squares sum values for the x correction polynomial and (□) the y correction polynomial using the single precision floating point calculations. (△) The least squares sum values for the x correction polynomial and (▽) the y correction polynomial using the double precision floating point calculations. For these calculations, a set of 169 tie points was used.

a higher precision floating point representation must be used. Both single (32 bit) and double (64 bit) floating point versions of the geometric correction program have been tested, as summarized in Tables III and IV and in Fig. 9. The use of double precision gives a greater range of stability for the algorithm, but has a higher computation cost.

The uncorrected image and the corrected images obtained using increasing powers of the polynomial correction function have been compared with the image obtained from the best possible correction using single precision calculations. The improvement of the correction is shown in the two-dimensional histograms displayed in Fig. 10. The convergence of the two-dimensional histograms toward a straight line is obvious as higher degrees of polynomials are used. Correlation coefficients were calculated by comparing images with the best corrected image (Table V). The

TABLE III

SINGLE PRECISION FLOATING POINT COMPUTATIONS[a]

		x		y	
k	n	$\Sigma\delta^2$	$\left(\frac{\Sigma\delta^2}{m-n}\right)^{1/2}$	$\Sigma\delta^2$	$\left(\frac{\Sigma\delta^2}{m-n}\right)^{1/2}$
1	3	1,600.154	3.104752	949.6494	2.391815
2	6	692.8916	2.061763	367.9990	1.502552
3	10	26.69348	0.4097359	33.44294	0.4586207
4	15	23.59717	0.3914440	26.08590	0.4115689
5	21	26.28296	0.4214114	20.87351	0.3755492
6	28	137.2864	0.9867433	69.24120	0.7007655
7	36	4,527.786	5.834681	3,449.493	5.092744
8	45	329,033.1	51.51206	220,082.0	42.12903
9	55	4,690,562	202.8430	4,589,365	200.6429
10	66	9,666,167	306.3434	7,130,038	263.1039
11	78	10,973,620	347.2596	15,803,070	416.7255
12	91	14,273,800	427.7820	12,305,480	397.1934
13	105	18,352,050	535.4912	20,376,690	564.2568

[a] $m = 169$.

TABLE IV

DOUBLE PRECISION FLOATING POINT COMPUTATIONS[a]

		x		y	
k	n	$\Sigma\delta^2$	$\left(\frac{\Sigma\delta^2}{m-n}\right)^{1/2}$	$\Sigma\delta^2$	$\left(\frac{\Sigma\delta^2}{m-n}\right)^{1/2}$
1	3	1600.161	3.104758	949.6500	2.391816
2	6	692.8914	2.061763	367.9976	1.502549
3	10	26.68360	0.4096601	33.42724	0.4585130
4	15	23.45397	0.3902544	25.89163	0.4100335
5	21	22.24394	0.3876814	18.40962	0.3526887
6	28	18.87526	0.3658785	17.86615	0.3559639
7	36	17.41972	0.3619052	16.17111	0.3486937
8	45	16.43037	0.3640096	14.80305	0.3455133
9	55	13.79058	0.3478074	13.42582	0.3431768
10	66	12.01326	0.3415166	11.86761	0.3394399
11	78	10.01740	0.3317850	10.51546	0.3399331
12	91	7.336028	0.3066784	7.390851	0.3078222
13	105	438,043,200	2,616.185	517,423,100	2,843.367

[a] $m = 169$.

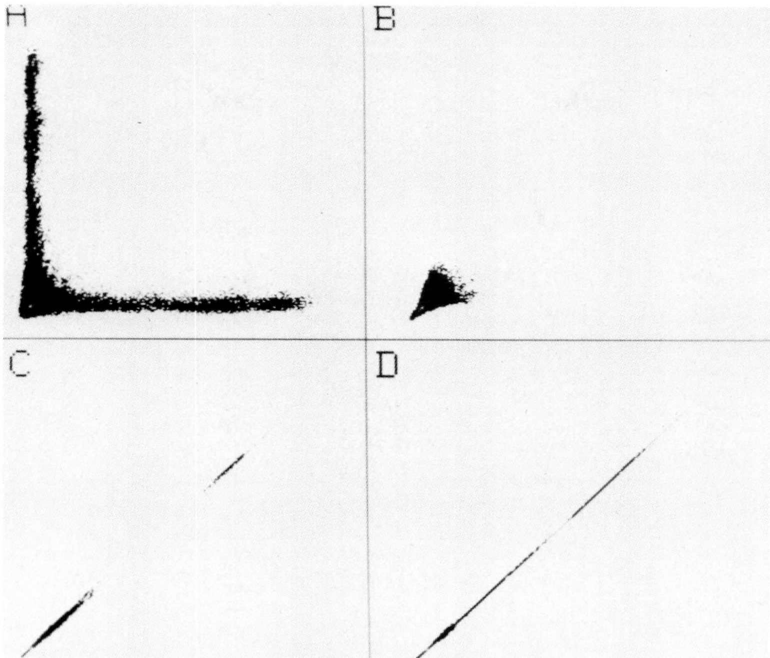

FIG. 10. Two-dimensional histograms of the image in Fig. 8 versus the indicated image. (A) uncorrected, (B) second degree correction, (C) third degree correction, (D) fourth degree correction.

TABLE V

CORRELATION OF VARIOUS CORRECTION DEGREES

	Uncorrected	Second	Third	Fourth
Correlation coefficient	0.255	0.842	0.994	0.997
Correlation index	0.000	0.827	0.994	0.997

greatest improvement in the correction is achieved between second and third degree polynomials. Subsequently, the improvement in the least squares sum reaches a weak minimum. The computational effort, however, increases considerably with the degree of the polynomial. The number of floating point additions and multiplications in the evaluation of third

FIG. 11. Image of the difference between the image corrected by the second degree polynomial and the best correction.

and fifth degree polynomials for correction of a 512 × 512 pixel image is 789,504 and 1,318,400, respectively. While it is reasonable to reduce the number of operations, this should not be done at the expense of corrupting the data or compromising data interpretation. Therefore, we set the extent of processing empirically to meet the objectives of a particular experiment. For example, the same degrees of correction would not be necessary for characterizing cell motility and for characterizing resonance energy transfer. The first measurements are at a cellular level, while the second are at a molecular level. These have different requirements for accuracy and precision.

The difference between a second degree correction and the best correction is shown in Fig. 11 and between a third degree correction and the best correction in Fig. 12. The values of the mean and standard deviation for the difference images are also useful as a characterization of the correction operations. The mean pixel intensities of difference images obtained by subtracting the fifth degree polynomial corrected image from the second,

FIG. 12. Image of the difference between the image corrected by the third degree polynomial and the best correction.

third and fourth degree polynomial corrected images were 0.313 ± 29.956, 0.0624 ± 6.0803, and 0.0591 ± 4.2086, respectively.

An important feature to note in Fig. 11 is the systematic pattern of the differences and the strongly pronounced grid pattern for the second degree polynomial correction. There are only a few areas close to the corners and one in the center which exhibit a random-like distribution of differences. Systematic differences in most of the image area are due to the inability of the second degree polynomial to achieve a correction of the initial grid image (Figs. 5 and 6). The third degree correction exhibits a much better random distribution of differences, with only a minimal grid pattern (Fig. 12). The improvement achieved with a third degree polynomial is obvious. It gives an essentially flat field with small evenly distributed differences between the corrected image and the reference image. From the analysis of the least squares sum, the two-dimensional histograms, the distribution of the differences, and the number of numerical operations, our choice for this particular grid image would be to correct the distorted image with a third degree polynomial. A slightly better correction is possible, but is not justified by the computation expense.

This geometric correction procedure has been applied to the study of various biological problems, including cell motility and intracellular Ca^{2+} (Weir *et al.,* 1987). Because cells are moving throughout the microscope field and pixels are different in size at different parts of the image, a geometric correction of each image is necessary before the length of a cell path can be measured accurately. We have also observed, when comparing images sampled at the different wavelengths, that the magnification and distortion of images are wavelength dependent. Therefore, comparison on a pixel by pixel basis, as with studies using image ratios (Bright *et al.,* 1987), may produce artifacts unless a geometric correction operation is performed in such a way that the corrected images are magnified to the same degree and the distortion associated with image acquisition has been removed.

B. Registration of Images

After geometric correction, a comparison of different parts of the same image or between different images by numerical operations is possible, provided, in the latter case, there is proper alignment of the images. With fluorescence microscopes that employ filter cubes for wavelength selection, image misregistration arises primarily from translational shifts of the image as a result of the mechanical movement of optical filters. Processing for the geometric correction also produces a slight translation of the corrected image with respect to the original image. The first coefficient in the polynomial for geometric correction can be interpreted as translational movement [Eq. (10), although that particular coefficient is smaller than 1.0 if the artificial net and the grid image are aligned properly. For this reason, the construction of the artificial net should begin from the center of the image where distortion is the least (Figs. 5–8). Another source of error is rotation of the image if the grid used for the geometric correction and the samples are not positioned identically along the optical image axis of the microscope. This error could be tolerated if the comparison of interest involves only a comparison of different areas within the same image. Comparisons of multiwavelength images of the same specimen would also be valid if data sets were all rotated in the same way. A more extensive discussion of rotational errors, particularly those that occur in electron microscopy, is available (Frank, 1980).

For translational registration, the scheme proposed by Pratt (1978) is adequate. A comparison area or window is specified such that some

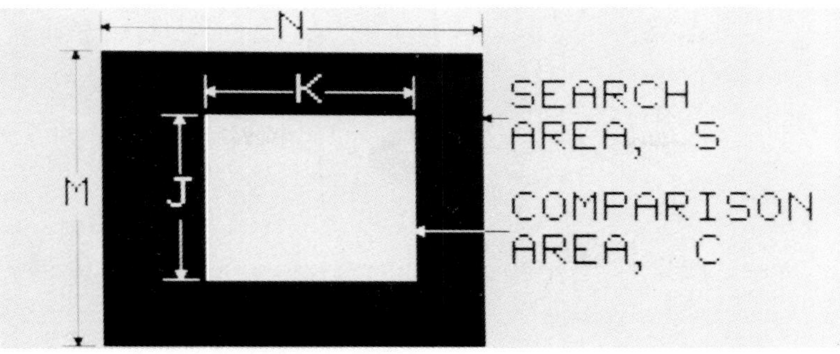

FIG. 13. Registration by calculation of the map of the matching function. The comparison area C is moved inside the search area S and, for every position, the matching function is calculated.

characteristic feature found in both images is located inside the window. In the second image, a search area is designated and the smaller comparison window is moved inside the search window (Fig. 13). For every position of the comparison window in the search area, a matching function is calculated and stored as a registration map, which can be displayed as an image or as an isometric projection.

We have chosen to use the correlation index (Akhnazarova and Kafarov, 1982) as the matching function because of its high discriminative power. The correlation index (r) is defined by the following equation.

$$r = (1 - \Sigma(I_2 - I_1)^2/\Sigma(I_1 - \bar{I}_1)^2)^{1/2} \qquad (16)$$

where I_1 and I_2 are the pixel values from the first and second image, respectively, and $\bar{I}_1$ is the mean pixel value of the comparison area in the first image. The sums in Eq. (16) are calculated over two dimensions for every overlap position of the comparison and search domains. The resulting values of the correlation index are stored as a two-dimensional map, illustrated in Fig. 15. The value of the correlation index occurs in a closed interval between 0 and 1. Consequently, for comparative purposes, it is not necessary to take the square root and extend the computational time. This procedure is described in detail by Jeričević *et al.* (1988b).

The discriminative power of the correlation index is generally superior to that of the correlation coefficient for linear regression and other functions described by Pratt (1978) and Groen *et al.* (1985). The fluorescent particles, shown as Fig. 14, were registered by using the correlation index (Fig. 15A) and the correlation coefficient (Fig. 15B). The center of the

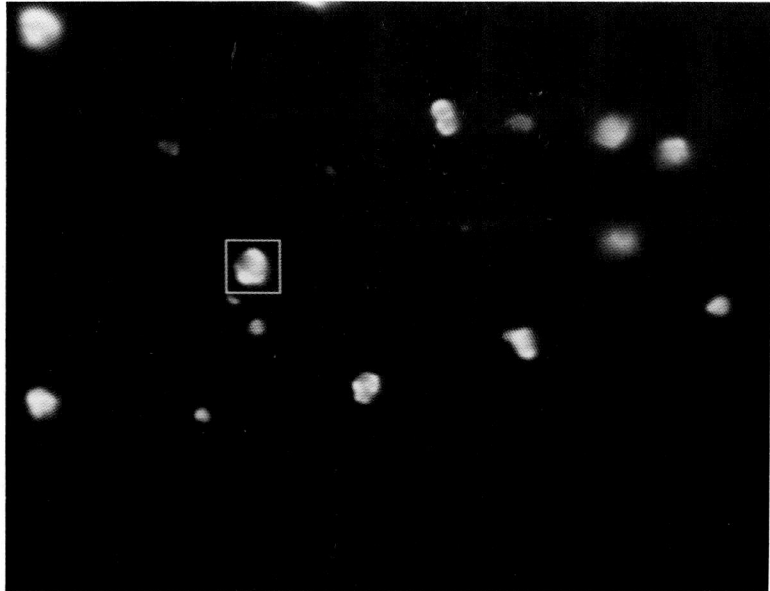

FIG. 14. Fluorescent particles were used as an example for the registration of an image with itself. The comparison area is designated by the square.

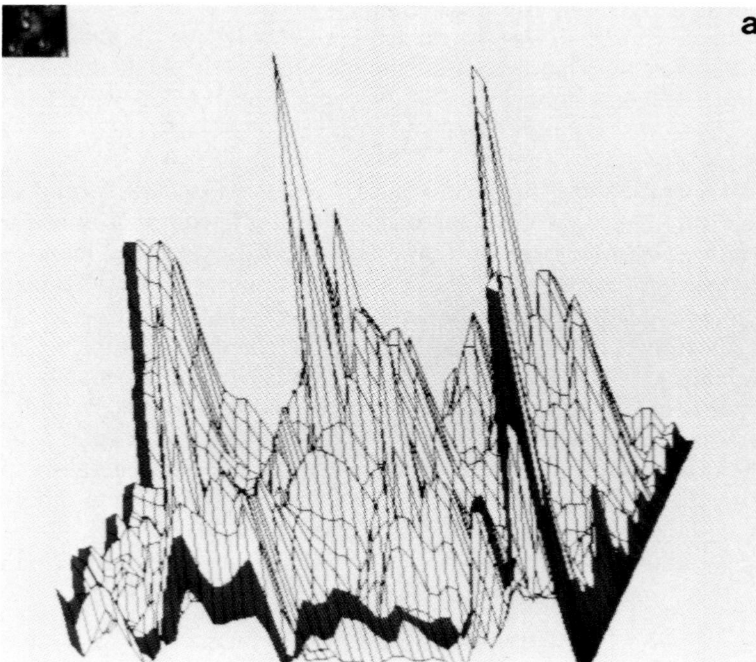

FIG. 15. Isometric projections. (A) Registration of image in Fig. 14 using the correlation coefficient as a matching function. (B) Registration of image in Fig. 14 using the correlation index as a matching function. The map is displayed as an image in the upper left corner of the

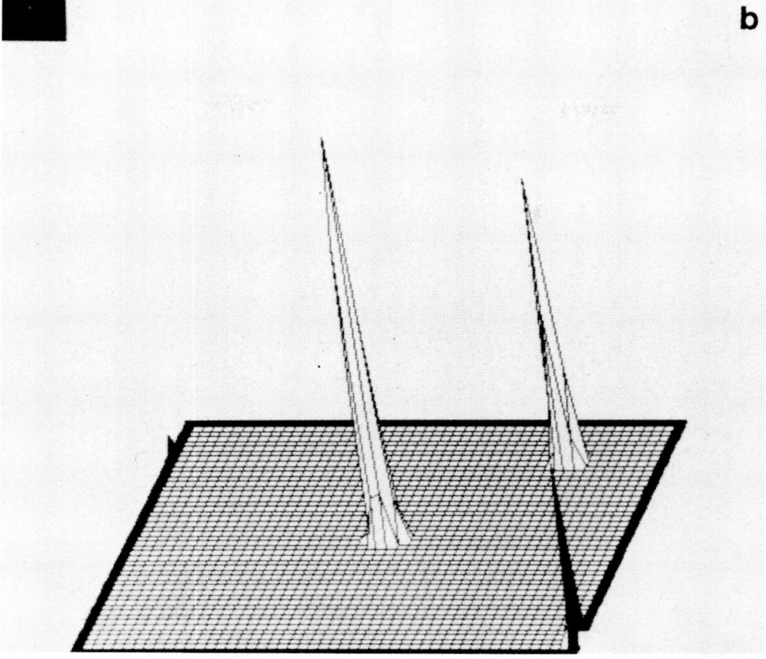

FIG. 15. (*continued*)

isometric projection of the registration maps corresponds to a 0, 0 x, y pixel shift between images. It is clear that the correlation index with a single, well-defined peak provides a better criteria for registration than the correlation coefficient, which gives multiple peaks with comparable values. Even highly repetitive grid structures were successfully registered by the correlation index algorithm (Fig. 16). Attempts to achieve registration of these images with the correlation coefficient method failed.

IV. Modulation Transfer

The modulation transfer function (MTF) is a quantitative description of an imaging system. To measure the MTF, the relationship of the object to its image is determined with special optical targets, usually opaque lines with various spatial frequencies (Inoué, 1986). For computational purposes, a related function, the point spread function (PSF), must be

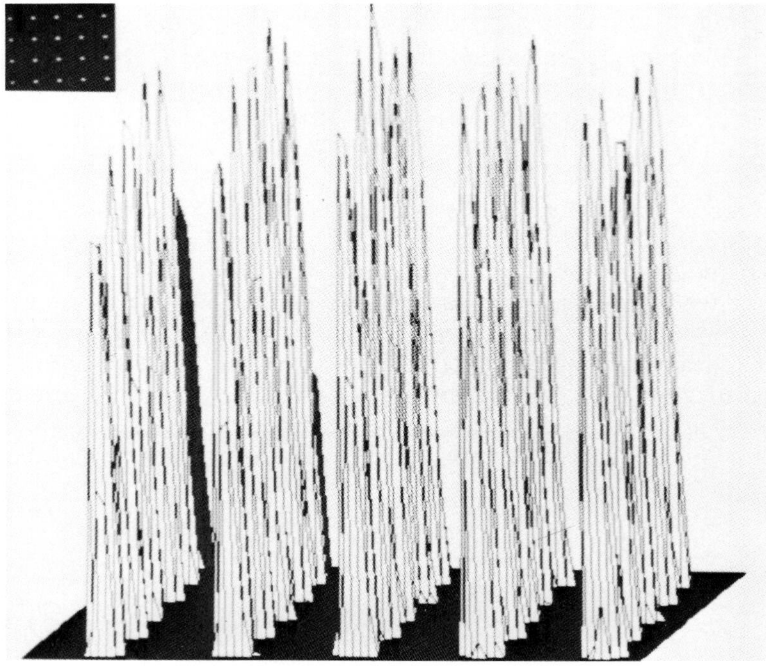

FIG. 16. Isometric projection of the registration map of the grid image, using the correlation index as a matching function. The map is displayed as an image in the upper left corner of the figure.

determined. MTF and PSF are Fourier transforms of each other. For the detailed exposition of the Fourier theory in optics, the mathematically inclined reader should see Steward (1987), Goodman (1968), Gaskill (1978), and related chapters in Castleman (1979), Inoué (1986), and Weaver (1983) (see also Ian T. Young, Chapter 1, this volume).

The PSF of a microscope is the image of a point source smaller than the resolution of the lens. In the nomenclature of linear system theory, the PSF is the impulse response of the system. Because of the nonideality of microscope, the impulse or delta function will be blurred into a finite-sized distribution of intensities or responses. In this context, any arbitrary image acquired with the system is the convolution of the ideal image with the PSF. The ideal image can be recovered by deconvolution of the actual image with its PSF. One of the most impressive examples of such a procedure in fluorescence microscopy has been provided by Agard and associates, who developed a method of three-dimensional reconstruction

(Agard, 1984). The evaluation of the PSF for a microscope can be done by a theoretical calculation using microscope parameters. However, the microscope is only one part of an imaging system and additional blurring of the signal can be caused by other elements of the system such as the camera. Therefore, it is usually necessary to determine the PSF experimentally. The image of a fluorescent bead which is smaller than the resolution of the microscope can be used as an appropriate approximation of the PSF. It should be noted that such an experimentally obtained image contains noise that must be removed by filtering before the PSF is used for deconvolution. This procedure will give the impulse response of the imaging system as a whole, not just the microscope. Because of the associative and comutative properties of convolution, it is appropriate to do deconvolution using an empirically sampled PSF. If there is geometric distortion in an image, it must be corrected first, because distortion implies that the PSF is space dependent, which would further complicate deconvolution.

V. Sampling Frequency

Conversion of an analog signal to a digital form makes it possible to perform operations with a general purpose computer using algorithms readily available from the linear systems theory. The process of acquiring discrete data imposes certain limits on such data. One of the most important limitations of digital image processing is the frequency of sampling. The sampling procedure can be envisioned as the selection of only a limited number of equidistant points from a continuous image. A sampling theorem states that at least two samples per cycle of the highest or Nyquist frequency must be obtained (Castleman, 1979). If this sampling theorem is ignored, the result is aliasing, where high frequency components are folded over in the spectra and represented as lower frequency components. The most familiar example of aliasing, due to improper sampling in the time rather than the spatial domain, is found in movies in which wheels seem to be spinning in the wrong direction. The sampling theorem describes a theoretical limitation that cannot be achieved in practice. Because of imperfect instruments and distorted, noisy signals, the limit will often be greater by a factor of two or three. A resolution target is the experimental device which provides information about the overall resolution of the imaging system. Our imaging system contains an inverted microscope which has about twofold lower resolution than does an upright microscope, based on a comparison of the same resolution target and the same camera. The sampling factor should be estimated for every system, based on the microscope resolution, the modulation transfer function, the signal-to-noise ratio, and analytical objective of the particular experiment. Inoué (1986) provides a detailed discussion of resolution targets. It is clear that it is not possible to obtain, by image processing, data which were not present in the original image. Unlike the shading or geometric corrections, which can be done after image acquisition, the sampling strategy has to be designed properly before data acquisition.

VI. Summary

The processing methods described in this article are general and can be used with any computer with adequate software without reference to a specific experimental apparatus. To achieve appropriately corrected images, the operations must be performed correctly and in a specific order to extract meaningful information from digital images. The operations are subtraction of the dark current, a shading correction after low pass filtering of the shading mask, a geometric correction with a reference grid obtained at the longest wavelength of interest, and finally registration. Since the digital data sets are too large for manual evaluation, statistical criteria must be used to define the analytical capabilities and error limits of the system before the more interesting cellular studies can be undertaken.

Acknowledgments

This work was supported by Robert A. Welch Foundation Q-343, HL-15648, HL-23741, HL-26973, HL-33750, GM-26091, and GM-17468.

References

Agard, D. A. (1984). Annu. Rev. Biophys. Bioeng. **13,** 191–219.

Akhnazarova, S., and Kafarov, V. (1982). "Experiment Optimization in Chemistry and Chemical Engineering," p. 312. Mir Publ., Moscow.

Benson, D. M., Plant, A. L., Bryan, J., Gotto, A. M., Jr., and Smith, L. C. (1985). *J. Cell Biol.* **100,** 1309–1323.

Bright, G. R., Fisher, G. W., Rogowska, J., and Taylor, D. L. (1987). *J. Cell Biol.* **104,** 1019–1033.

Castleman, K. R. (1979). "Digital Image Processing," p. 479. Prentice-Hall, New York.

Frank, J. (1980). In "Computer Processing of Electron Microscope Images" (P. W. Hawks, ed.), pp. 187–222. Springer-Verlag, New York.

Gaskill, J. D. (1978). "Linear Systems, Fourier Transforms and Optics," p. 554. Wiley, New York.

Golub, G. H., and Van Loan, C. F. (1983). "Matrix Computations," p. 476. Johns Hopkins Univ. Press, Baltimore.

Goodman, J. W. (1968). "Introduction to Fourier Optics," p. 287. McGraw-Hill, New York.

Groen, F. C. A., Young, I. T., and Ligthart, G. (1985). *Cytometry* **6,** 81–91.

Inoué, S. (1986). "Video Microscopy," p. 584. Plenum, New York.

Jeričević, Ž., Benson, D. M., Bryan, J., and Smith, L. C. (1987). Anal. Chem. **59,** 658–662.

Jeričević, Ž., Benson, D. M., Bryan, J., and Smith, L. C. (1988a). *J. Microsc.* **149,** 233–245.

Jeričević, Ž., Wiese, B., Rice, L., Bryan, J., and Smith, L. C. (1988b). *SPIE Proc.* **909,** 328–335.

Lawson, C. L., and Hanson, R. J. (1974). "Solving Least Squares Problems," p. 340. Prentice-Hall, New York.

Niblack, W. (1986). "An Introduction to Digital Image Processing," p. 215. Prentice-Hall, New York.

Pratt, W. K. (1978). "Digital Image Processing," p. 750. Wiley, New York.

Press, W. H., Flannery, B. P., Tenklsky, S. A., and Vetterling, W. T. (1986). "Numerical Recipes," p. 818. Cambridge Univ. Press, London.

Steward, E. G. (1987). "Fourier Optics: An Introduction," 2nd Ed., p. 269. Wiley, New York.

Weaver, H. J. (1983). "Applications of Discreet and Continuous Fourier Analysis," p. 375. Wiley, New York.

Weir, W. G., Cannell, M. B., Berlin, J. R., Marban, E., and Lederer, W. J. (1987). *Science* **235,** 325–328.

Chapter 3

Imaging of Unresolved Objects, Superresolution, and Precision of Distance Measurement with Video Microscopy

SHINYA INOUÉ

Marine Biological Laboratory
Woods Hole, Massachusetts 02543

The basics of microscope image formation, point and line spread functions, and modulation transfer functions (MTF) are covered in the two previous chapters (see also Castleman, 1979; Hecht, 1987; and Inoué, 1986). In this chapter, we will examine the utility of video microscopy to visualize, resolve, and measure widths and distances to precisions that are conventionally considered to be below the limit of resolution of the light microscope.

I. Visualizing Objects Narrower than the Resolution Limit of the Light Microscope

To start with, we must clearly distinguish between resolving an image and visualizing an object.

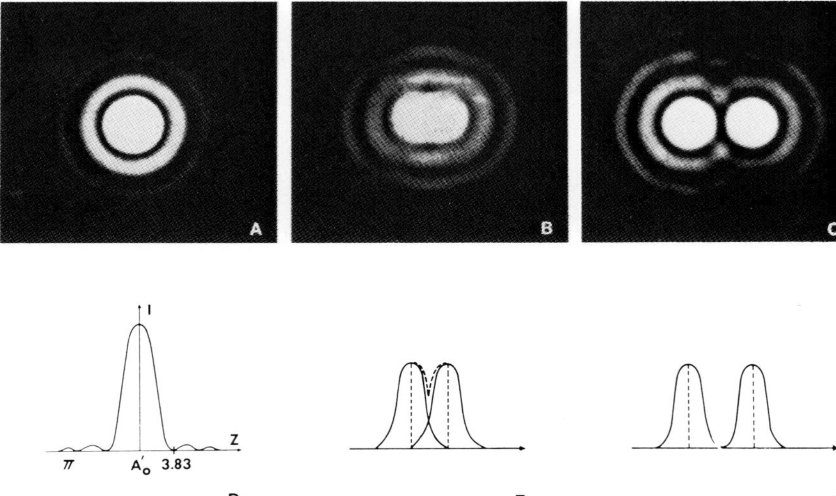

FIG. 1. Airy disk and intensity distribution in the diffraction image. (A) Photograph of Airy disk; (D) intensity distribution. (B, C, E, F) Images of neighboring Airy disks (B, C), and their corresponding intensity distributions (E, F), separated by different distances. In C (and F), where the distance between the disk centers is larger than their radii, the disks (and intensity peaks) are clearly separate, or resolved. In B, the center-to-center distance equals the radius, the disks are just barely distinguishable, and the images are just resolved (the Rayleigh Criterion). [The curves in E and F are schematic only. Had they been drawn exactly as in D, the sum of the two curves in E (dash line) would have dipped in the middle by only 26.5% from their peak.] If the disk centers are closer together than their radii, their intensity distributions merge into a single peak (D), and they are said not to be resolved. (After Françon, 1961.)

In an image-forming system, resolution is generally expressed as a measure of the ability to separate images of two neighboring object points (Born and Wolf, 1980; Sect. 8.b.2; Françon, 1961). In the case of a light microscope, producing diffraction-limited images, the resolution limit is defined as the minimum distance between two self-luminous or incoherently illuminated objects or structures whose diffraction images[1] can visually be distinguished as coming from two points. When the diffraction images of the two points overlap to an extent that they can no longer be distinguished from that of an individual object, the two are said not to be resolved or that the distance is less than the limit of resolution (Fig. 1). The

[1] In this chapters, I use the term diffraction pattern to refer to the pattern produced at the objective lens back aperture by the spatial periodicity of the specimen, and diffraction image to refer to the diffraction limited image (such as the Airy disk) produced in the image plane by the objective lens.

minimum distance (d) between such objects or structures visually resolvable with a light microscope is commonly given by:

$$d = \frac{1.22\lambda_0}{NA_{obj} + NA_{cond}} \tag{1}$$

where λ_0 is the vacuum wavelength of the light used, NA_{obj} is the numerical aperture of the objective lens, and NA_{cond} is the actual working NA of the condenser set by its immersion medium and iris diaphragm setting (Fig. 2). At this separation, the diffraction images of the two

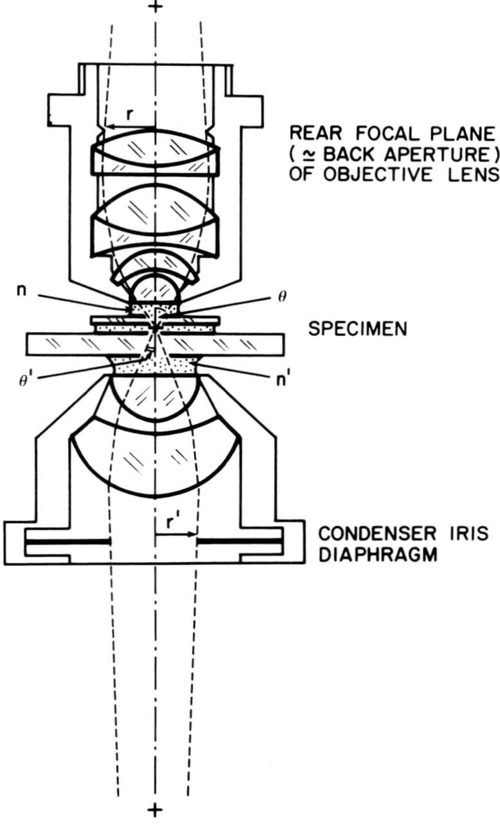

FIG. 2. The working numerical aperture of the condenser, NA_{cond}, is $n' \sin \theta'$, where n' is the refractive index of the medium between condenser and specimen, and θ' is the angle shown. NA_{cond} is proportional to r', the radius of the condenser iris opening. Thus, changing the condenser iris setting changes NA_{cond}. Analogous considerations for the objective lens are also illustrated. (From Inoué, 1986.)

incoherently illuminated objects overlap with a ~26.5% depression that signals the twoness of the object (Rayleigh criterion).[2]

Naturally the Rayleigh criterion assumes that the objective lens is capable of, and is in fact being used with, the correct immersion medium (including the medium bathing the specimen), tube length, coverslip thickness, ocular type, projection distance, and wavelength, etc., that provide a diffraction limited image free from measurable aberrations.[3]

While the lateral resolution in the plane of focus is defined as above [or in some cases using the Sparrow criterion where the central minimum just disappears (Sparrow, 1916)], we can nevertheless visualize isolated objects or structures whose widths are far below the limit of resolution.

Forming an image of an object whose diameter or width is far below the resolution limit of the microscope is not a new event. Dark-field microscopy has been used for decades to visualize the dynamic behavior of smoke and other colloidal particles, bacterial flagella, etc. However, those images could not readily be recorded.

In the past few years the light microscope with a variety of contrast modes has come to be used for visualizing and studying the behavior of unresolved thin objects. The major gain has come from the ability of analog and digital video devices to dramatically boost the image contrast and speed of image acquisition, and to subtract or filter out unwanted optical noise, average out random electronic and photon statistical noise, and to record the image on media that allow immediate playback (e.g., Allen *et al.,* 1981a,b; Inoué, 1981, 1986, 1987a).

Importantly for microscopy, the contrast boosting ability of video allows the use of the best corrected (Plan Apochromatic) objective lenses not only for bright field and fluorescence, but even for polarized light and differential interference contrast (DIC) microscopy. In the past, in these latter contrast modes, such lenses could not be used well because of the poor extinction due to the crystalline elements they contained.

With the contrast boosted electronically, the condenser can also be used at an NA much higher than was previously possible. Thus, the diffraction image becomes narrower and the limit of resolution is improved [Eq. (1)]. At the same time, more light becomes available to the sensor, bringing it

[2] Strictly speaking, we are here talking about periodic or complex objects (rather than just two points) whose separation is d. The case of solely distinguishing two points or lines from a single one can be considered a special case, as discussed under the section on superresolution, since in this case we have *a priori* knowledge or assumption about the specimen, i.e., that it is either one or two.

[3] For through-focal photographs of diffraction images free of aberrations and in the presence of various aberrations, see, e.g., Cagnet *et al.* (1962).

into a range where its S/N ratio is more favorable (e.g., see Fig. 7–16 in Inoué, 1986). In addition, as discussed in Section V, the depth of field decreases dramatically with the rise of system NA, so that axial resolution is improved and the influence of out-of-focus objects is reduced.

With these improvements brought about by video, objects (especially filaments) whose width are far below the resolution limit are now routinely visualized by fluorescence, DIC, dark field, and rectified polarized light microscopy. The main reason such thin objects could not be visualized readily without video enhancement was not because their diameters were below the resolution limit of the microscope but because their diffraction images have low contrast. By boosting the contrast (and using conditions that give greater lateral and axial resolution) with video, and in the case of dark-field and fluorescence images with help from the greater light-detecting sensitivity of the intensifier tubes, one sees such dynamic events in real time as: 65-nm-diameter sperm acrosomal process growing (Tilney and Inoué,1982); 25-nm-diameter microtubules assembling and disassembling, gliding or moving particles (Allen *et al.,* 1985; Schnapp *et al.,* 1985; Koonce and Schliwa, 1986); 10-nm-diameter actin filaments sliding (Yanagida *et al.,* 1984; Toyoshima *et al.,* 1987); and lipid bilayers in emulsions extending, deforming, and fusing with each other (Kachar *et al.,* 1984).

II. Diffraction Patterns of Very Narrow Objects

We shall next consider the widths of the images formed by structures whose width approximates or is narrower than the conventional limit of resolution.

For a given optical system, the point or line spread function describes the distribution of intensity in the image of an infinitely narrow point or line (see Young, Chapter 1, this volume). For a well-corrected, diffraction-limited system, the function would essentially be an Airy disk (Fig. 1) for absorbing or self-luminous points (which includes in-focus images in fluorescence and dark-field microscopy). [For DIC images, see Galbraith (1982) and Galbraith and Sanderson (1980); for rectified and nonrectified polarized light images, see Inoué (1986).] What is the width of the diffraction image formed by a single object whose width lies between the infinitely thin and approximately an Airy disk diameter?

Given the proper optical corrections (and appropriate video sampling), these unresolved objects form a diffraction image, expanded to the size of an Airy disk or larger as determined by the microscope optics and size of the object. In detail, the diffraction images vary in peak intensity as well as in width. As the object becomes smaller, the peak intensity decreases

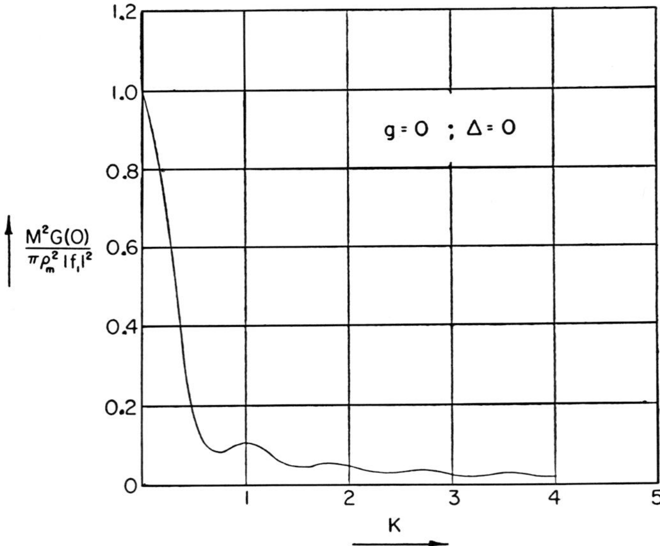

FIG. 3. The central energy density in the images of opaque particles as a function of K the radius of the particles (cf Fig. 6). (From Smith, 1960. See original article for further details.)

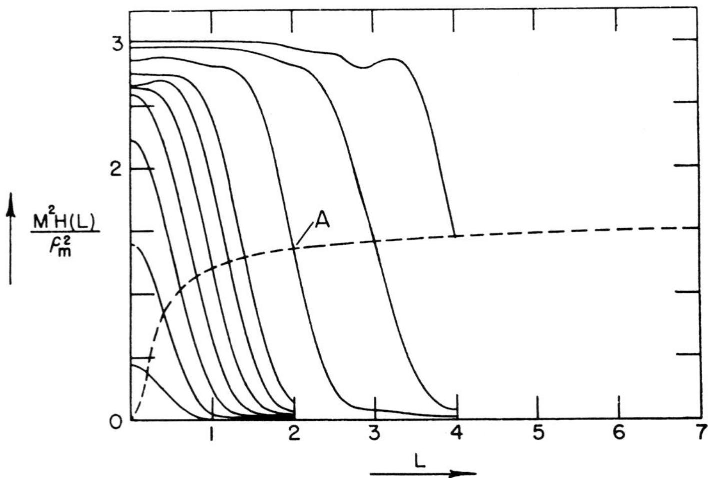

FIG. 4. The solid curves show the distribution of irradiance in the image formed by an Airy-type objective when the source is a self-radiant disk of a specified radius. The abscissa of the intersection of the broken curve with a solid curve indicates the radius in Airy units of the corresponding source disk. For example, the distribution of irradiance in the image of a circular disk 2 Airy units in radius is given by the solid curve passing through the intersection A. (From Smith and Osterberg, 1961. See original article for further details.)

nonmonotonically as shown in Fig. 3. (The curve shown in Fig. 3 is for an absorbing object; cf. Fig. 6.)

The in-focus diffraction image for self-luminous (which includes fluorescence), thin, circular disks of 0.2 to 2.0 Airy units in radius, calculated by Smith and Osterberg (1961), is reproduced in Fig. 4. As shown, the radius of the source disk, for disk diameters of 3 to ~0.5 Airy units, can be approximated by the radius of the diffraction image at half height. For objects with radii below 0.5 Airy units, the half-peak diameter of the diffraction image departs drastically from the diameter of the source.

Figure 5 shows the energy distribution in the diffraction image for a circular hole in an opaque screen, and Fig. 6 the energy distribution for an opaque particle (Smith, 1960). The in-focus diffraction image is again smaller for narrower objects, although the two are not related linearly.

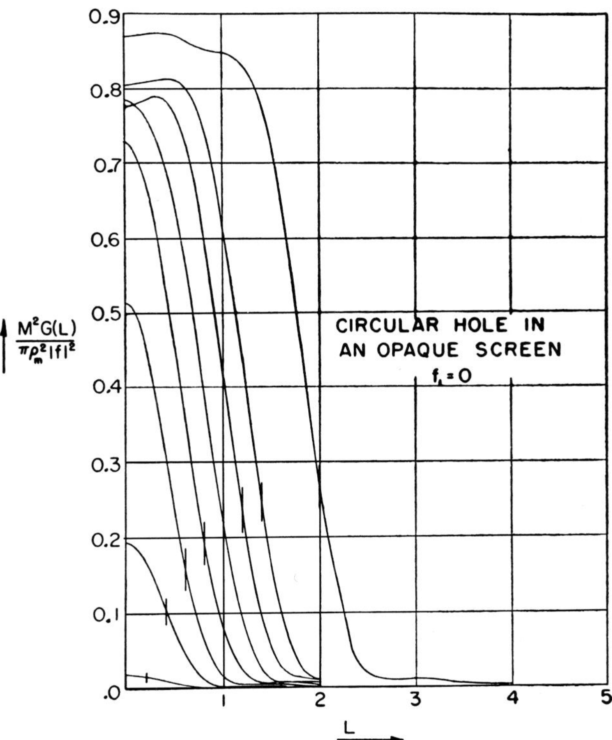

FIG. 5. Diffraction images of circular holes in an opaque screen. The short vertical line on each curve indicates the radius of the corresponding image according to geometrical construction. (From Smith, 1960. For further details see original article.)

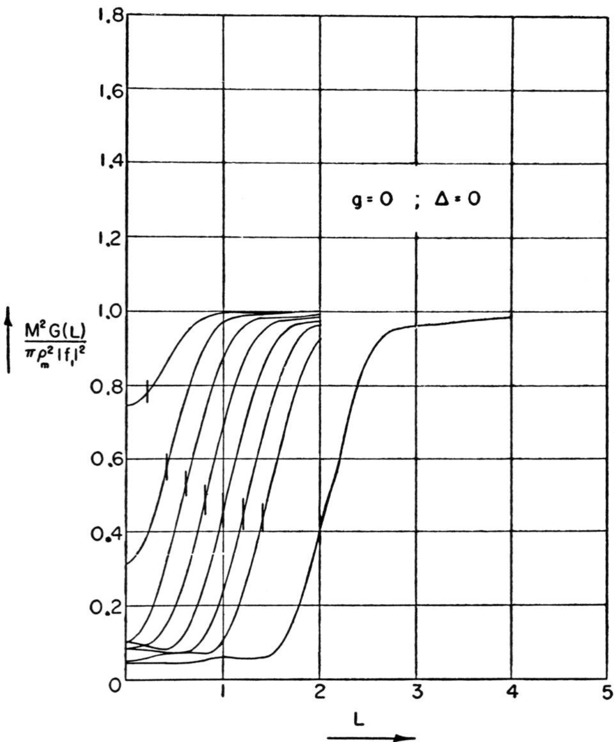

Fig. 6. Diffraction images of opaque particles. The short vertical line on each curve indicates the radius of the corresponding geometrical image. (From Smith, 1960. See original article for further details.)

The width of the diffraction image at half height for self-luminous and small opaque objects thus approximates the geometrical width of the object down to approximately 0.5 Airy disk radius. This relation does not hold for small transmitting apertures nor for particles which only differ in refractive index from their surroundings (Fig. 7).

Another question then arises. In fluorescence, or DIC, or polarized light microscopy, where especially with the aid of video we can visually discern objects as thin as individual microtubules, can we tell, from the diffraction image observed in the microscope, how many individual filaments make up the diffraction image? To paraphrase, when a single microtubule well-separated from others is observed in video-enhanced DIC contrast, its image appears "inflated to" the width of a typical

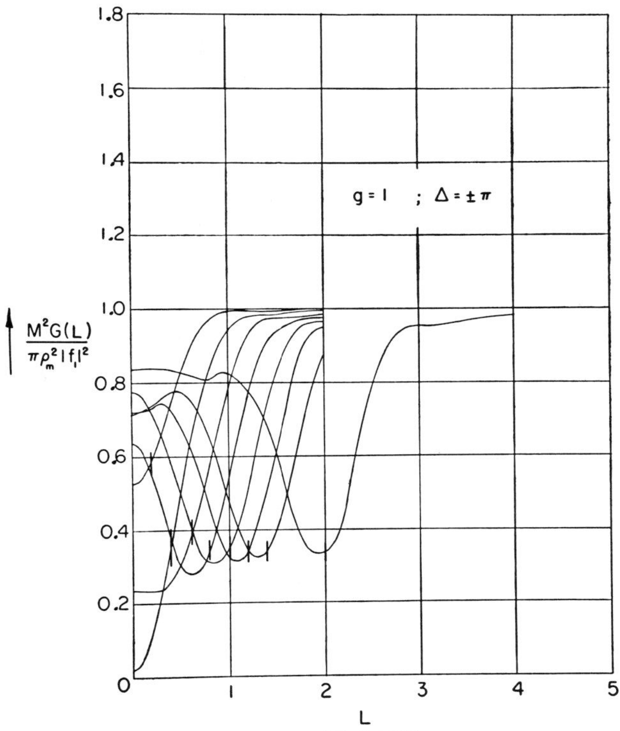

FIG. 7. Diffraction images of particles which differ from the surrounding only by having a different refractive index. The short vertical line on each curve indicates the radius of the corresponding geometrical image. (From Smith, 1960. See original article for further details.)

diffraction image (Fig. 8). What happens to the diffraction image when more microtubules are packed in a width (and depth) less than ~0.5 Airy unit?

One can clearly tell where two microtubules overlap (Fig. 9), and in DIC microscopy, preliminary observation suggests that the area of the diffraction pattern is roughly proportional to the number of microtubules, i.e., the mass of the material contained in the "unresolved" fine thread. Similar relations are expected for fluorescence images and for rectified polarized light microscopy.

In polarization microscopy we have noticed a rather striking difference in the width of images of individual microtubules which were additively

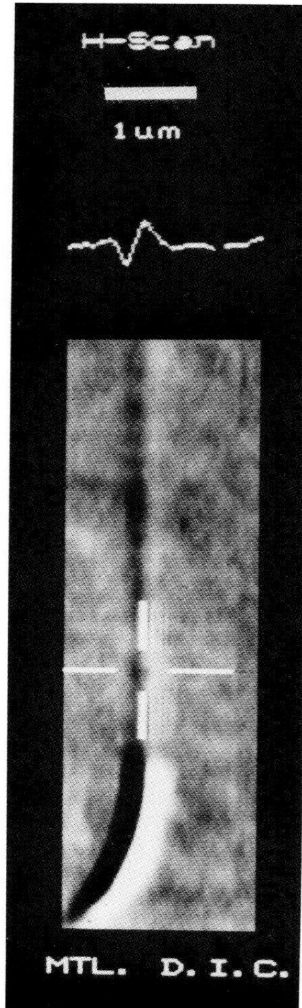

Fig. 8. Intensity distribution in DIC diffraction image of a single MAPs-free microtubule grown on purified axoneme (curved, high-contrast object at bottom). Each point in the line trace (generated by the Image-1/AT processor) represents the intensity across a ten-pixel long, one-pixel wide slit. 100/1.40 Plan Apo objective lens with rectified DIC condenser (Nikon). Magnification, 11,500.

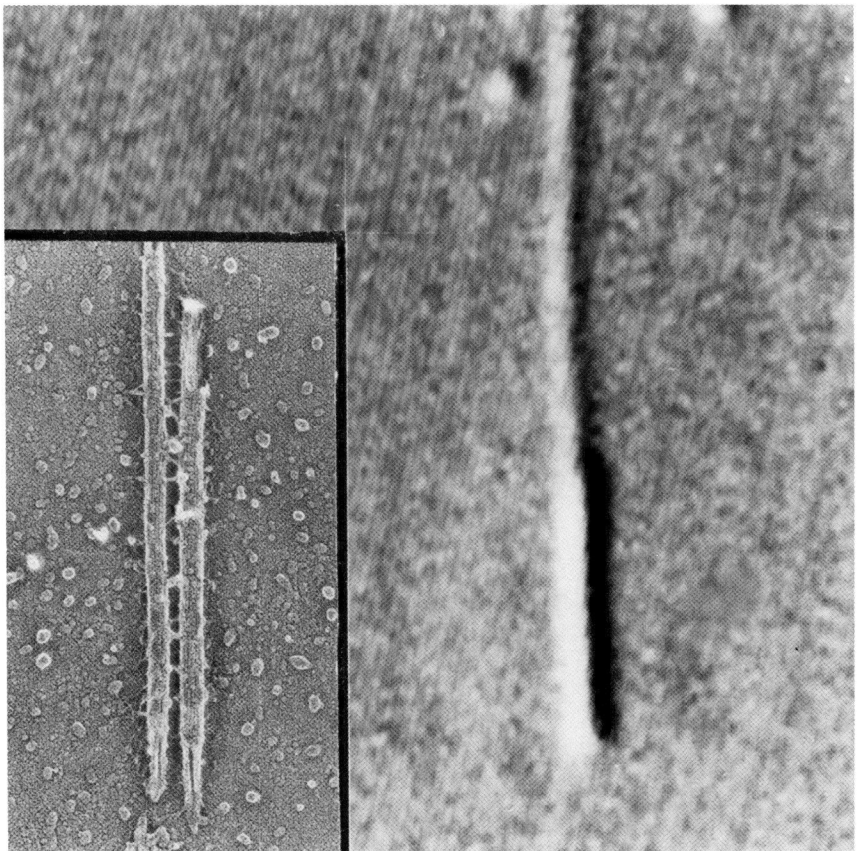

FIG. 9. DIC diffraction image of single and overlapping microtubules. Magnification: approximately 10,000×. Inset shows electron micrograph of overlap region. Magnification: Approximately 100,000×. (Courtesy of B. Schnapp, Marine Biological Laboratory, and National Institutes of Health.)

and subtractively compensated. In the additive compensation the diffraction image of the microtubule (or an axoneme), which is brighter than the gray background, appeared rather broad and with unclearly defined edges (Fig. 10). In the subtractive compensation, the dark image of the microtubule appeared considerably thinner and with rather sharply defined edges. It is not yet clear whether this asymmetric behavior upon compensation reflects our sensory response, an intrinsic physical optical effect, or the

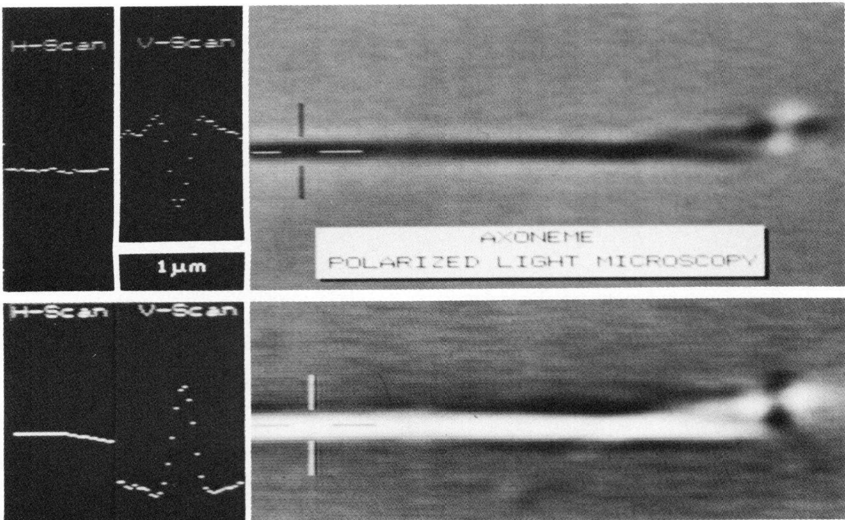

FIG. 10. Rectified polarized light image of axoneme in additive (bottom) and subtractive (top) compensation. Each point in the line trace represents the intensity across a one-pixel wide slit. 100/1.35 NA Plan Apochromatic objective lens with rectified condenser (Nikon). Magnification, 17,000.

property of the instrument.[4] Be that as it may, the birefringence of individual microtubules in the subtractive compensation does yield a narrow-appearing diffraction image that makes them stand out distinctly in video-enhanced, rectified polarized light microscopy.

III. Superresolution

Under some special conditions discussed below, one can obtain "superresolution" and in fact resolve objects that are closer together than the Rayleigh criterion.

[4]The high NA Plan Apochromatic lenses used to obtain these images do not provide a perfectly uniform aperture function; the rectifier provides an imperfect match so that there exists some residual rotation at the lens surfaces, and the lens also exhibits detectable degrees of birefringence with radial and lateral axes. In addition, we would expect the image to also be affected by "edge birefringence" since the specimen is not embedded in a matched index medium (see Inoué, 1959, 1986 p. 498; Takenaka and Rikukawa, 1974).

One case of superresolution is predicted when the field of view is not infinitely wide (the case of ordinary microscopy), but is confined to a very small area (e.g., Cox *et al.*, 1982; McCutchen, 1967; Fellgett and Linfoot, 1955; Toraldo, 1955). An example of this type of superresolution occurs with confocal scanning microscopy, where the field at any instant is limited to an Airy diffraction pattern of the entrance pinhole formed by the condenser lens. In this case, a superresolution by a factor of two is predicted by Cox *et al.* (1982) and Cox and Sheppard (1986) for fluorescent objects, and a 1.4-fold improvement in resolution is described by Brakenhoff *et al.* (1979) and White *et al.* (1987). The theoretical bases of this type of superresolution is briefly as follows.

From the classical work of Toraldo (1955), expanded on by Harris (1964) and based on information sampling theory, one can show that the overlapping diffraction images produced by two narrow self-luminous objects, lying in a limited microscope field, can be deconvoluted to establish their separations, e.g., down to two-tenths of the Rayleigh criterion distance (Figs. 11 and 12). As Harris explains, the Fourier components of a diffraction pattern can be uniquely defined beyond the physical limit of the objective lens aperture, provided the object lies in a limited narrow field. In other words, given precise enough information about the diffracting waves that do pass the objective lens aperture, the coefficients for the Fourier components which would fall outside of the aperture, i.e., those that are due to spacings which are too fine to be "resolved" by the lens at the specified wavelength, can nevertheless be uniquely approximated theoretically for objects in a small limited field.

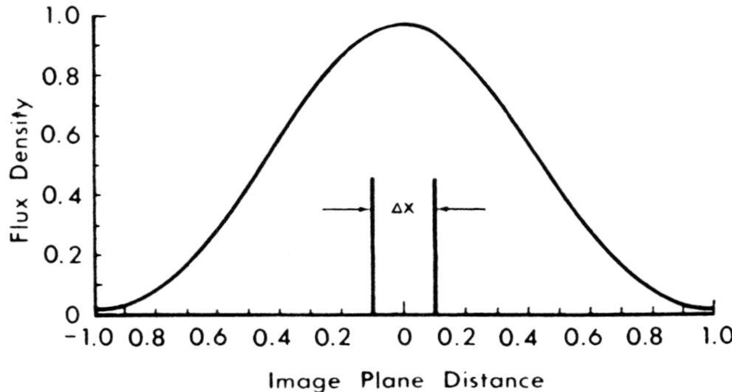

FIG. 11. The composite image formed by two monochromatic mutually incoherent point sources. The vertical lines show the position of the two points. The image plane distance has been normalized to the Rayleigh criterion distance, so that Δx is 0.2 of the Rayleigh criterion separation. (From Harris, 1964.)

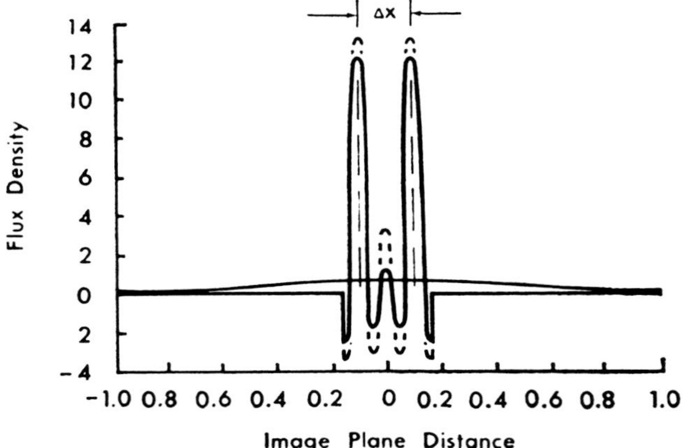

FIG. 12. The restoration of the image of Fig. 11 is shown. The dashed curve is the representation of the two point sources, which is achieved by using the first four terms of a Fourier series over the interval ±0.15. The similar solid curve is the restoration accomplished in the illustrative example. The broad, solid curve is the original image of Fig. 11 replotted to the scale of the image. (From Harris, 1964.)

Thus, the theorem predicts that close study of a diffraction image can reveal structures below the limit of resolution, provided the field is narrowly confined and the noïse level is appropriately low [see Cox and Sheppard (1986) for a mathematical treatment of superresolution that includes the noise factor and the two polarization vectors].

Superresolution could also be achieved if the central peak of the diffraction image (point or line spread functions) could be made narrower than the width of the standard Airy disk. This can be achieved to a certain extent by modifying the aperture function of the objective lens. For example, as shown in Fig. 13, in the presence of a central obstruction in the objective lens, more of the diffracted energy appears in the first and higher order fringes, but the zero order fringe becomes narrower (e.g., Born and Wolf, 1980). Thus, while extraneous fringes would obscure the image of a complex structure, the distance between two equally bright points or lines isolated and located near the center of a clean field could be measured with somewhat improved resolution using an objective lens with central obstruction (Boivin and Boivin, 1980).

In some cases, the situation can be further improved by applying a graded series of absorbing rings toward the outer limits of the objective lens aperture. Such "apodizing" treatment, which reduces the energy distributed into the outer diffraction rings (at some cost to the narrowness

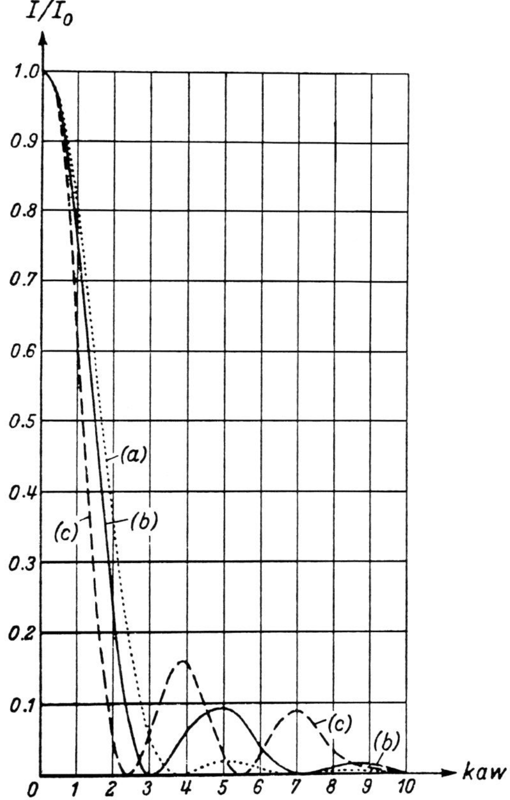

FIG. 13. Illustrating the effect of central obstruction on the resolution. Normalized intensity curves for Fraunhofer patterns (diffraction images) of (a) circular aperture, (b) annular aperture with $\varepsilon = \frac{1}{2}$, and (c) annular aperture with $\varepsilon \rightarrow 1$. ε is the ratio of the inner to outer radius of the aperture. (From Born and Wolf, 1980.)

of the central disk diameter), can help to bring out the weak diffraction image of an object lying close to a much brighter object. The dim image of the less bright object would otherwise be masked by the bright first or second order diffraction rings of the brighter object (see, e.g., Hecht and Zajac, 1979).

Another condition giving rise to superresolution occurs when two adjoining structures give rise to diffraction patterns that are reversed in phase. Since I have never formally reported on this condition, which I had formulated and experimentally verified in 1957, I will expand on this subject here.

Consider the double slit diffraction experiment in Fig. 14. Two narrow slits (S_2, S_3), separated by a distance (d), are illuminated by a monochromatic source (of wavelength λ) through slit S_1. The diffraction images of slits S_2 and S_3 are formed by lens (L) whose aperture and λ determine the width of each of the slit images. As d is decreased, the two diffraction images overlap and eventually become unresolvable once their separation is reduced beyond the Rayleigh (or Sparrow) criterion.

We now add the following polarizing components to the experimental setup as shown in Fig. 15. These are a polarizer (P) immediately preceding or following S_1; two birefringent crystals (C_1 and C_2) of equal retardations placed immediately preceding or following S_2 and S_3, respectively, with their slow axes at 90° to each other and lying at azimuths of 45° to the polarizer azimuth PP′ (see lower half of Fig. 15); and a third crystal, or compensator, (C_3) followed by an analyzer (A) with their axes as indi-

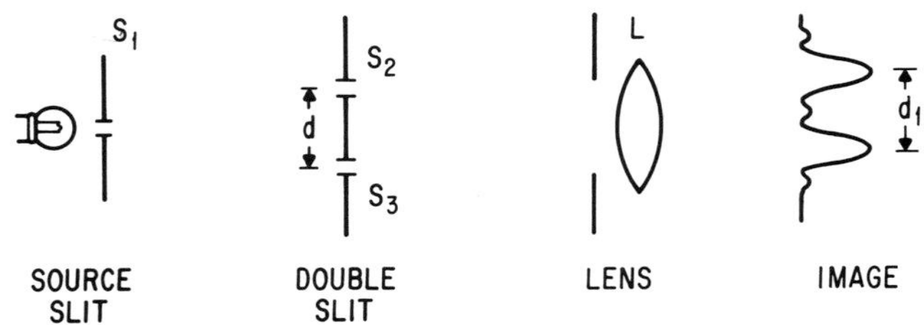

FIG. 14. Diffraction images formed by double slit. See text.

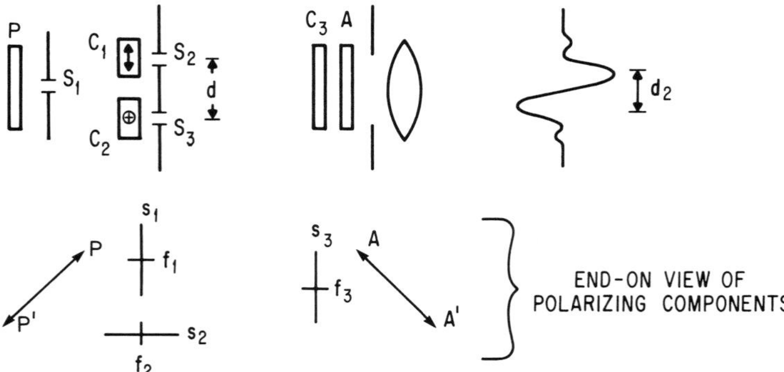

FIG. 15. Double slit experiment coupled with phase reversal of the slit images. Figures in the lower half show the azimuth orientation of the polarizing components used to achieve the phase reversal. See text for further details.

cated. When these polarizing components are added, the phases of the diffraction images of S_2 and S_3 are reversed by 180°. Thus the diffraction images become a bright and dark pair on a gray background. As d is reduced below the Rayleigh criterion, they continue to appear as a bright–dark pair but with reduced amplitude, rather than merging into a single unresolved peak. Depending on the orientation of the slow axes S_1 and S_2 relative to S_3, the bright peak appears to one side or the other of the dark peak. Thus the two slits (microscopic objects), separated by distances considerably below the classical resolution limit, produce a white/black or black/white diffraction image on a gray background. In practice, slit separations of one-fifth the Rayleigh criterion produced images that were readily discernible by visual observation in the model experiment.

This experiment shows that the duality and relative positions of two objects can be discerned at separations considerably below the classical resolution limit if the phases of the diffraction patterns are inverted. Likewise, in scanning light microscopy (or other field scanning imaging systems) it should be possible to improve image resolution and precision of position determination by scanning through an aperture that would produce a similar diffraction pattern. Such a pattern can be generated, for example, with half masks placed at the condenser aperture plane that shift the phases of two parts of a coherent illuminating beam by 180° relative to each other.[5]

To summarize this section, the conventional limit of resolution, affected by the objective as well as condenser lens NA, provides a reasonable criterion for defining the smallest distances resolvable in a complex structure. However, when *a priori* knowledge regarding the structure is available, including the fact that it is present in a limited field (as in confocal microscopy) or that the phases of the diffraction images are inverted, the classical limit need not apply. In fact, the diffraction pattern of an isolated object can contain structural information that extends way below the conventional limit of resolution.

IV. Positional Information and Lateral Setting Accuracy

Another problem in light microscopy related to image resolution (but not identical with it) is the precision and accuracy by which we can determine the position of a small object. In addition to defining its spatial

[5] In this case, the beam need only arise from a coherent source and (unlike the case of the experiment discussed above) would not specifically need to be polarized.

coordinates, we can use (relative) position measurements to derive distances, velocities, etc.

The precision (or setting accuracy) for determining the position of an object point, lying in the object plane, reduces to the question "how well can we determine the location of a diffraction image (such as the center of an Airy disk) in the image plane?" In other words, how good a lateral setting accuracy can we achieve? From our earlier discussions it should not be surprising that we can, in fact, determine the center of the Airy disk with extremely high precision.

When two more or less similar points or lines are spaced very close together, the problem of measuring their separation reduces to that of resolution discussed in the previous sections. But if they are well-separated, the centers of corresponding points of the diffraction images can be closely defined (if necessary, by taking symmetric parts, unique points, or the whole envelope of the diffraction images into account, Fig. 16). Thus, one can rather readily measure distances between well-separated points or lines to precisions of the order of one-tenth of the resolution limit of the microscope or better. The measurement is limited by the precision of distance calibration rather than by the size of the diffraction image. Françon (1961) in fact describes a lateral setting accuracy of one-hundredth of the Airy diameter. Likewise, with video enhancement, one can see a small step, of the order of one-tenth of an Airy radius, in the diffraction image of a filament or an edge with an otherwise smooth contour (Inoué and Inoué, unpublished observation).

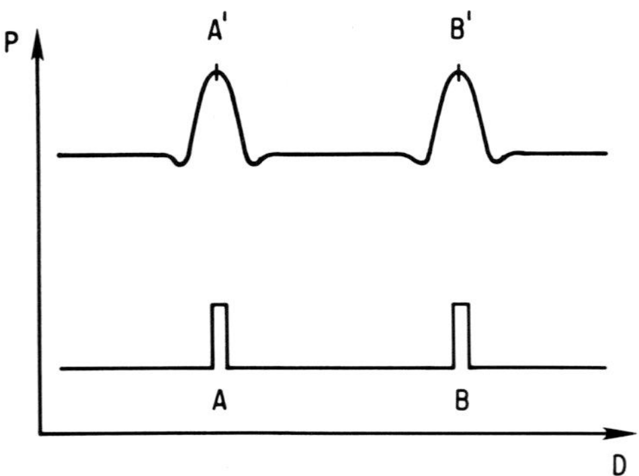

Fig. 16. The distance (D) between two unresolved narrow lines or points (A and B) can be determined with great accuracy (down to one-hundredth or less of the Airy radius) from the separation of the centroids of the diffraction images (A′ and B′) if they are adequately separated.

Denk *et al.* describe a case of very high precision relative position determination using the light microscope. These authors measured the vibrational amplitudes of auditory hair bundles with a sensitivity for the power spectrum density approaching precision of a few picometers/$\sqrt{\text{Hz}}$. In their approach, the laser-illuminated phase image of the edge of a hair bundle is projected by the microscope onto a quadrant detector photodiode. Outputs of the quadrants are individually amplified and compared to determine the rapid, few-nanometers displacement of the edge image (W. Denk and W. W. Webb, 1987, and personal communication).

Likewise, Kamimura (1987) describes the measurement of namometer size displacements of 1-μm-diameter polystyrene beads, by photoelectrically determining the displacement of two halves of the microscope image.

Gelles *et al.* (1987) have determined the successive positions of subresolution diameter (kinesin-coated) beads gliding on microtubules down to step sizes of 4 nm. They used video-enhanced DIC microscopy to obtain diffraction images of the gliding bead and determined their centroid by checking the cross-correlation between the bead's image and the previously digitized image of a bead. Thus, by using the whole diffraction image of the unresolved (150-nm diameter) bead, these authors have shown that the light microscope image can directly reveal nanometer-range quantal molecular steps by which particles are transported.

A case which is somewhat more complex than measuring the separation of two points or lines is the measurement of absolute distances between two edges. In general, the edges can be represented by a change in fluorescence, absorbance, reflectance, optical path difference, birefringence, etc., in the microscopic object. These parameters (P) are expressed in the lower parts of Fig. 17.

In contrast to measuring the separation of two points or lines (Figs. 1 and 16), the problem is no longer that of measuring a simple translational distance of identical or similar diffraction images. Instead, we need to interpret the positions of the edges from the diffraction images. Even when the distance between the edges is considerably larger than the Airy radius, we cannot readily determine the exact distance without a detailed knowledge of the relationship between the object's geometrical edge and the shape of the diffraction image. The curves from Smith and Osterberg discussed earlier provide some clues, but further data are needed for different types of objects imaged with alternate contrast modes.

To summarize this section, the position of optically isolated small points or lines can be determined with very high precision—down to distances of the order of a hundredth of the Airy radius. The distance between two separate edges is, however, much more difficult to measure accurately.

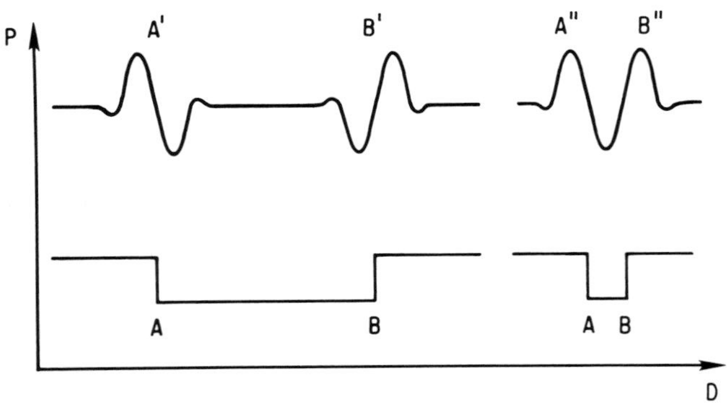

FIG. 17. The distance (D) between two edges is considerably more difficult to measure accurately than between two narrow lines or edges (Fig. 16). Each edge produces an asymmetrical diffraction pattern (A′, B′) from which the location of the true geometrical edges (A, B) must be inferred.

V. Ultrathin Optical Sectioning and Axial Setting Accuracy

Video enhancement has also improved our ability to obtain very thin optical sections, or shallow depth of field, with the light microscope. This means that the axial resolution and axial setting accuracy (both measured along the optical axis of the microscope) are also improved.

While there is no single agreed-upon formula that defines these parameters, Françon (1961; also see Inoué, 1986) expresses the axial setting accuracy (2ξ) as

$$2\xi = \frac{\lambda}{4n\sin^2(u/2)} \tag{2}$$

where λ is the wavelength, n the refractive index of the immersion medium, and u the half angle of the cone of light captured by the objective lens. What is important to note is that the axial setting acurracy, and the related axial resolution and thinness of the depth of field, rises inversely, essentially with the second power of the numerical aperture. From Eq. (2) Françon would predict that, in the ideal case, we should get an axial setting accuracy of 0.29 μm with a 1.4 N A oil immersion lens.

We have in fact obtained optical sections that are 0.1–0.15 μm thick with rectified polarized light, just under 0.2 μm thick in phase contrast microscopy, and approximately 0.2 μm thick with DIC (Fig. 18)! These

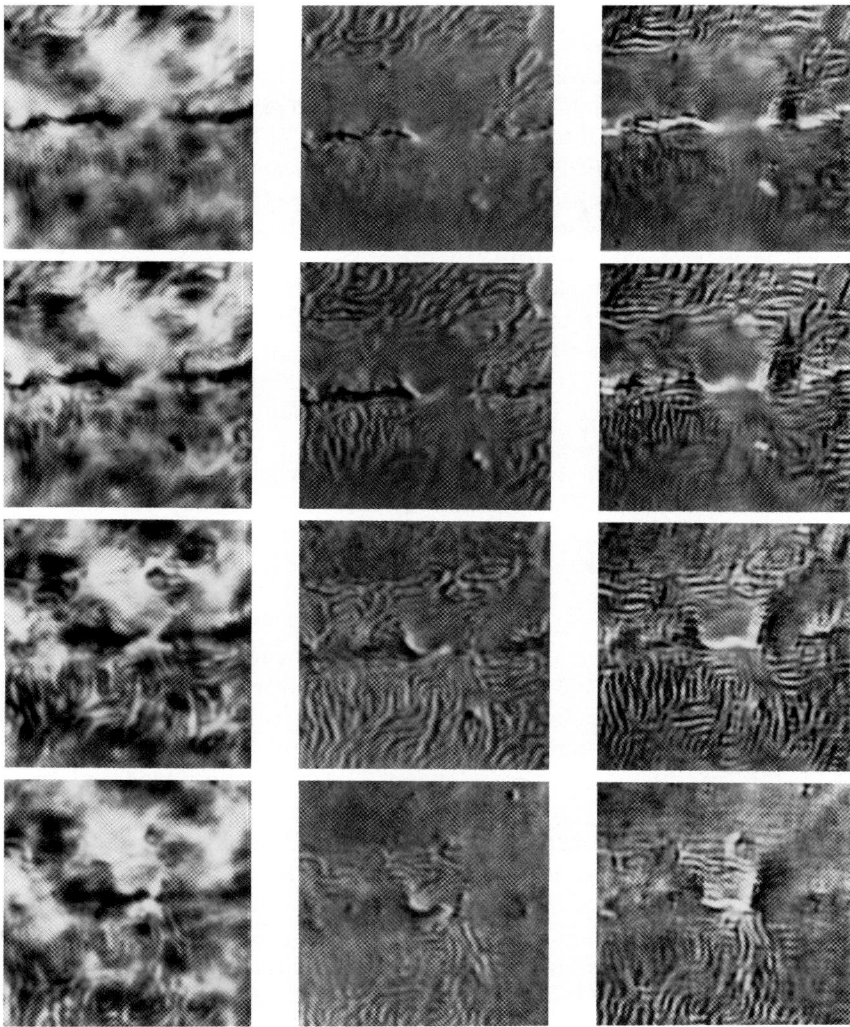

FIG. 18. Shown are 0.22 μm step, through-focal optical sections of buccal epithelial cell surface in phase contrast. The same ×100/1.4 NA Plan Apo objective lens (Nikon) was used for all images. Left column: At this stepping distance in phase contrast, the successive optical sections are too far spaced for the image details to be contiguous. In other words, the depth of field is less than 0.2 μm. Middle column: Rectified DIC condenser and Nomarski prism. Details in the consecutive optical sections now just overlap each other. The depth of field is around 0.20–0.25 μm. Right column: Rectified polarized light microscopy. As with phase contrast, the successive optical sections do not overlap, but unlike the phase image, there is much less interference from out-of-focus structures. Scale bar (lower left): 10 μm.

ultrathin sections were achieved by the use of: (1) a new generation of well-corrected, high NA objective lenses (Nikon new series 100/1.4 and 60/1.4, and Zeiss Axiophot 63/1.4, Plan Apochromats); (2) a high NA condenser, fully and uniformly illuminated with high intensity monochromatic (546 nm) light through a light scrambler (Ellis, 1985; Inoué, 1986, Figs. III-21, III-22); (3) optical contrast enhancement and image correction with a condenser rectifier for polarization and DIC optics, together with improved lens coating (Nikon, Inc.); (4) analog enhancement with black level and gain controls on the video camera (DAGE-MTI 65M Newvicon); and (5) further digital enhancement (Universal Imaging Corporation Image-I). Optical sectioning, through-focal sectioning, recording, and analysis were also aided by (6) a solid, optical bench microscope with precision stage (Inoué, 1986); (7) stepper motor drive and timed controller for the fine focus; and (8) a 450-TV-line-resolution laser disk recorder (Panasonic OMDR T2-2021 FBC).

We do not yet have a theory to explain how we obtain such ultrathin optical sections, since the observations exceed the setting accuracy calculated according to Eq. (2) (which one would expect to be considerably smaller than the thickness of the optical section).

One might think that our observation is related to a smaller Airy disk radius and a consequent shrinking of the axial elongation of the three-dimensional diffraction pattern. As discussed earlier (Section III), the radius of the Airy central disk is reduced in a lens with central obstruction. In a high extinction polarizing or DIC system, the back aperture of the objective lens is not completely uniformly extinguished even after rectification; birefringence in the lens elements and ellipticity introduced at the lens surfaces conspire to reduce the extinction at higher NA. Thus to some extent, the aperture function is similar to a lens with a partially absorbing central stop that could reduce the radius of the Airy disk as discussed earlier.

The trouble with the explanation is that the three-dimensional diffraction pattern is supposed to elongate axially, rather than shrink, as the Airy radius is decreased by the introduction of central obstruction (Linfoot and Wolf, 1953). Therefore, the axial resolution and setting accuracy should decrease, and the thickness of the optical sections should in fact increase.

Whatever the theoretical foundation, the ultrathin optical sections now attained with video microscopy allows us to better discriminate objects which are part of a complex three-dimensional structure. For example, the dynamic behavior of individual microtubules amidst a large number of other microtubules can be distinguished in the optical section of living cells viewed with high extinction polarized light microscopy (Cassimeris *et al.*, 1988; Inoué, 1988).

We do not yet have equivalent data for fluorescence microscopy where, as mentioned before, we can consider the light waves arising from adjacent specimen points as being totally incoherent. The point is that, if anything, the resolution obtained on incoherently illuminated objects should be greater than for coherently or partially coherently illuminated objects (Hopkins and Barham, 1950; Born and Wolf, 1980). However, studies using the laser scanning confocal microscope with a 1.4 NA. optical system show that the optical system for Epifluorescence is approximately 0.7 μm thick (White *et al.*, 1987; B. W. Amos, personal communication; see also Brakenhoff *et al.*, 1979).

With fluorescence microscopy, light contribution from out-of-focus objects may affect the in-focus image more severely than with contrast-producing methods based on phase differences. The comparison has, however, not yet been clearly established, especially for diffraction-limited aberration-free optics.

Be that as it may, a number of new developments have made it possible to isolate clean optical sections which are especially helpful for fluorescence microscopy. These include mathematical deconvolution from serial optical sections of fluorescence video images (see Agard *et al.*, Chapter 13, this volume) and confocal microscopy (see Chapter 14 by Brakenhoff *et al.*, this volume).

Whatever means are used to obtain optical sections, serial optical sections provide the basis for three-dimensional and stereoscopic reconstruction. In addition to the chapters mentioned above, articles relating to this topic can be found in Somlyo (1986; see also Inoué *et al.*, 1985).

VI. Other Factors Affecting the Choice of Pixel Dimensions, Calibration, and Choice of Magnification

In contrast to bright-field, dark-field, fluorescence, and phase-contrast microscopy, or conventional photography, in which the apertures should be radially symmetric and resolution is independent of spacing direction, resolution in video is direction dependent. Thus, in conventional raster scan video, the horizontal resolution is generally different from, and independent of, the vertical resolution, which is limited by the number of active video scan lines.

The number of active scan lines, which are those that are visible on an underscanned monitor (i.e., the number of total scan lines per frame, e.g., 525, minus those hidden in the vertical blanking period, e.g., 42) for standard 525/60 video used in the United States and Japan today is around

480 lines. For 625/50 video used in the British Commonwealth and many parts of Europe and South America it is around 580 lines.[6] The limiting vertical resolution is affected by the relative positions of the specimen structure and the scan lines, and is usually taken to be 0.7 (the Kell factor) times the number of active scan lines (see Inoué, 1986).

In video the horizontal resolution is defined as the number of black plus white lines that can be resolved horizontally at the center of the screen for a distance equal to the height of the screen. Many monochrome video cameras provide horizontal limiting resolutions of 700 to 800 lines or greater, but video tape recorders tend to limit the monochrome horizontal resolution to between 320 and 450 lines.

It is important to note that these definitions of TV resolution in fact give only half of the line pair resolution ordinarily used for defining microscopic and photographic resolution (Inoué, 1986). Also, the resolution is a function of image contrast and brightness so that the limiting resolution may or may not be attainable, depending on the nature of the image.

Horizontal resolution is ultimately limited by the bandwidth used to convey the video signal, 10 MHz being required for every 800 black plus white lines to be resolved. Thus, video deals with rather high frequency electronic signals with the attendant problems of radio frequency and other electromagnetic interference, signal distortions, etc. The response of the video system to square waves is particularly important, since improper response can introduce ringing or reverse contrast streaking. The boundaries and dimensions of objects in the video image can then be distorted. For critical measurements, one needs to test for (or at the least be aware of) the system's square wave response in addition to its higher frequency modulation transfer function (see Young, Chapter 1, this volume; also Inoué, 1986). Additionally, one needs to consider the lag, blooming, and burn characteristics of the video camera as well as geometrical distortions in the camera and monitor. While these and other factors affect the quality of the video image and its resolution, we cannot adequately discuss these important topics within the limited space of this chapter. The reader is referred to my text *Video Microscopy* (Inoué, 1986) for a fuller treatment of these subjects.

With modern charge-coupled device (CCD) cameras there is generally less lag and, in some cases, less blooming so that the dimensions of moving objects or bright areas can be determined more accurately than with many

[6] 525/60 and 625/50 each refer to the number of scan lines per frame and the number of fields per second. In the conventional 2 : 1 interlace system, the alternate scan lines in two fields are interlaced to give a full frame. The 525/60 (2 : 1 interlace) video format thus utilizes 30 frames a second, with a frame made up of two interlaced fields, each with 262.5 scan lines. Thus, there are 60 fields every second.

vidicon or intensifier cameras. Some CCD cameras provide pixel arrays as large as 1000 × 1000 or greater, although they tend to require scan rates that are slower than video rates. On the other hand, one can achieve very much improved S/N ratio, dynamic range, and reduced geometric distortion with a CCD sensor (see Aikens *et al.*, 1988).

CCD cameras and digital image processors that operate at video rates currently tend to provide pixel arrays that number around 512 horizontally and 480 vertically. Barring pixel smearing, or "blooming," that was prevalent in earlier CCD cameras, we should then expect a horizontal resolution of about 250 line pairs and a vertical resolution of about 240 line pairs. These numbers also correspond to the maximum spatial frequencies for the width and height of the picture which would be free from aliasing. When aliasing is present, spatial frequencies greater than these figures would be represented in the output pictures by a number of lines proportionately smaller than those present in the input, and with their locations shifted in phase.

The ratio of the horizontal to vertical dimensions, known as the aspect ratio, is commonly 4 : 3 for standard broadcast compatible video. Depending on the digital image processor, different fractions of the active horizontal scan are digitized into the horizontal pixel numbers, some taking the whole active scan line and digitizing this into 640 or 512 pixels. Since the vertical scan is generally represented as 512 pixels for the full scan (or 480 for the active scan lines), each pixel in the digitized picture may or may not correspond to a unit square area in the original image. Thus, in some digitizers the pixel represents a rectangular rather than square area of the incoming image. Therefore, proper precaution must be exercised when one calibrates geometrical parameters using pixel numbers or (if the picture is rotated digitally) during image processing. In general, for digital quantitation of image dimensions, it is safer to use image processors providing square pixels.

The exact frame width and pixel dimensions in the video picture are calibrated by using stage micrometers. Since the aspect ratio of the scanning pattern in the cameras (as well as monitor) is adjustable and can vary, and the pixels in solid-state cameras and digital image processors may or may not have an exact square array, it is prudent to calibrate the image separately for the horizontal and vertical video directions.

Taking into account the limited MTF and resolution of analog video devices and the digital nature of the CCD camera and digital image processors, how large should we make the microscope image in order to gain the full advantage of video microscopy?

We will initially assume that the image brightness is nonlimiting and that we wish to match the video resolution approximately with that of the microscope.

In dealing with an analog video system and standard microscope image, we need to make the closest line pairs that are just resolvable with the microscope visible in the video picture. That means that the microscope image should be magnified adequately onto the video camera face plate so that the contrast of the video picture (which depends on the line spacing seen by the video camera according to its MTF characteristics) is adequate for detection. Given adequate image brightness, an average camera MTF dictates that we need an image approximately 200 times the objective NA to be projected onto the face plate of a 1-in. camera tube. For a 100/1.4 objective lens, one would need an approximately 2.5 to 3 × ocular magnification onto the camera face plate. Given a 280 × optical magnification, if the image were digitized to 512 × 480 pixels, each pixel would represent horizontal and vertical dimensions of 89 and 70 nm, respectively, at the specimen plane.

If on the other hand, we wish to examine the diffraction image itself, we need a greater magnification by the microscope. For example, to clearly view the behavior of individual molecular filaments, we find that the microscope image needs to be magnified so that the diameter of the diffraction image occupies some 8 to 10 pixels for fluorescence and polarized light imaging and 20 or so pixels for DIC. These are also the minimum number of pixels across the width of a diffraction image that are needed to quite closely define the shape of the Airy pattern within the first minimum (Figs. 8 and 10). Most such studies are carried out with the aid of a digital image processor in order to raise the contrast of the diffraction image obtained by use of full condenser aperture illumination with well-corrected high NA objectives, and to subtract away the interfering optical noise which concurrently becomes prominent. To make the diffraction image this large, the microscope magnification is raised to about 400 times the objective NA on the target of a 1-in. camera tube. For a 100/1.4 objective lens this would require an ocular projection of 5 to 6× since 1.4 × 400/100 = 5.6. This corresponds to a video frame width of about 20 μm, and pixel dimensions of about 40 nm × 30 nm measured at the specimen plane. (see also Castleman, 1987; Gelles *et al.*, 1987).

While it is desirable, and often necessary, to have this large a microscope magnification, often there is a conflict with available light levels. The luminance (more properly the radiance at the wavelength involved) of the image is often too low, especially in fluorescence microscopy, to produce a detectable frame-rate video signal even with an intensified camera. As we use cameras with greater sensitivity, the noise level rises until eventually we reach the level of (random) photon statistics.

In order to reduce the noise and gain an intelligible image, it is often necessary to integrate and reduce the random picture noise. This is achieved at the cost of time resolution. For static objects the loss of time resolution may not matter, but for objects that are moving or changing dynamically, this imposes significant constraint. Besides increasing the illumination and signal strength (which may not be practical), one is then obliged to give up spatial resolution by reducing the microscope (ocular) magnification. As the image radiance on the camera face plate is improved (by inverse square of the magnification) by dropping the ocular magnification, the camera will hopefully reach its operating range. If not, one needs to switch to a more sensitive camera with its attendant noise.

Whatever the final combination of camera type, signal integration, and microscope magnification, one ends up having to choose between spatial and temporal resolution and the amount of S/N one is able to live with (see, e.g., Inoué, 1986). The smaller the signal source, and the more dynamic the image, the greater becomes the constraint. It is remarkable that video in fact allows us to obtain images as good as those that are already obtainable (for a recent summary, see Inoué, 1987).

In general, we give up field size in order to minimize the video or processor resolution from limiting the fidelity of our microscope image. While the loss of field size is acceptable for a number of applications, it can seriously interfere with observations of living cells requiring high resolution coupled with an overview of the cell behavior.

As high definition TV with its higher resolution and improved S/N becomes practical, we should benefit in microscopy by the larger number of pixels available per video frame. As the video resolution is improved, initial magnification by the microscope can be reduced so that there would be a significant gain in light level. Or, when light is not a limiting factor, we can gain an improved field size without running into limitations in fidelity governed by the pixel dimensions of the video system.

Acknowledgments

Bob Knudson, Ted Inoué, and Dan Green made possible the through-focal series in Fig. 18 by designing the fine-focus stepper system and controller programs. Dr. Bruce Schnapp of MBL and NIH provided the original of Fig. 9. Linda and Bob Golder carried out the artwork for the original illustrations. I am grateful to these individuals and the authors of the figure sources, and for support by NIH grant R37 GM31617-07 and NSF grant DCB 8518672.

References

Aikens, R. S., Agard, D. A., and Sedat, J. W. (1988). *In* "Methods in Cell Biology" (Y. L. Wang and D. L. Taylor, eds.), Vol 29. Academic Press, San Diego.

Allen, R. D., Travis, J. L., Allen, N. S., and Yilmaz, H. (1981a). *Cell Motil.* **1,** 275–289.

Allen, R. D., Allen, N. S., and Travis, J. L. (1981b). *Cell Motil.* **1,** 291–302.

Allen, R. D., Weiss, D. G., Hayden, J. H., Brown, D. T., Fujiwake, H., and Simpson, M. (1985). *J. Cell Biol.* **100,** 1736–1752.
Boivin, R., and Boivin, A. (1980). *Opt. Acta* **27,** 587–610.
Born, M., and Wolf, E. (1980). "Principles of Optics" 6th Ed. Pergamon, Oxford. Brakenhoff, G. J., Blom, P., and Barends, P. (1979). *J. Microsc.* **117,** 219–232.
Cagnet, M., Françon, M., and Thrierr, J. C. (1962). "Atlas of Optical Phenomena." Springer–Verlag, Berlin.
Cassimeris, L., Inoué, S., and Salmon, E. D. (1988). *Cell Motil. Cytoskel.* **10,** 1–12.
Castleman, K. R. (1979). "Digital Image Processing." Prentice–Hall, New York.
Castleman, K. R. (1987). *Appl. Opt.* **26,** 3338–3342.
Cox, I. J., and Sheppard, C. J. R. (1986). *J. Opt. Soc. Am.* **3,** 1152–1158.
Cox, I. J., Sheppard, C. J. R., and Wilson, T. (1982). *Optik* **60,** 391–396.
Denk, W., and Webb, W. W. (1987). *Bull. Am. Phys. Soc.* **32,** 645.
Ellis, G. W. (1985). *J. Cell Biol.* **101,** 83a.
Fellgett, P. B., and Linfoot, E. H. (1955). *Proc. R. Soc. London Ser. A* **247,** 369–407.
Françon, M. (1961). "Progress in Microscopy." Row, Peterson, Evanston, Ilinois.
Galbraith, W. (1982). *Microsc. Acta* **85,** 233–254.
Galbraith, W., and Sanderson, R. J. (1980). *Microsc. Acta* **83,** 395–402.
Gelles, J., Schnapp, B. J., and Sheetz, M. P. (1987). *Nature (London)* **331,** 450–453.
Harris, J. L. (1964). *J. Opt. Soc. Am.* **54,** 931–936.
Hecht, E. (1987). "Optics." Addison–Wesley, Reading, Massachusetts.
Hopkins, H. H., and Barham, P. M. (1950). *Proc. Phys. Soc.* **63,** 737–744.
Inoué, S. (1959). *J. Opt. Soc. Am.* **49,** 508.
Inoué, S. (1981). *J. Cell Biol.* **89,** 346–356.
Inoué, S (1986). "Video Microscopy." Plenum, New York.
Inoué, S. (1987). *Appl. Opt.* **26,** 3219–3225.
Inoué, S. (1988). *Zool. Sci.* **5,** 529–538.
Inoué, S., Molè-Bajer, J., and Bajer, A. S. (1985). *In* "Microtubules and Microtubule Inhibitors" (M. De Brabander and J. De Mey, eds.), pp. 269–276. Elsevier, Amsterdam.
Kachar, B., Evans, D. F., and Ninham, B. W. (1984). *J. Colloid Interface Sci.* **100,** 287–301.
Kamimura, S. (1987). *Appl. Opt.* **26,** 3425–3427.
Koonce, M. P., and Schliwa, M. (1986). *J. Cell Biol.* **103,** 605–612.
Linfoot, E. H., and Wolf, E. (1953). *Proc. Phys. Soc.* **66,** 145–149.
McCutchen, C. W. (1967). *J. Opt. Soc. Am.* **57,** 1190–1192.
Schnapp, B. J., Vale, R. D., Sheetz, M. P., and Reese, T. S. (1985). *Cell* **40,** 455–462.
Smith, L. W. (1960). *J. Opt. Soc. Am.* **50,** 369–374.
Smith, L. W., and Osterberg, H. (1961). *J. Opt. Soc. Am.* **51,** 412–414.
Somlyo, A., ed. (1986). *Ann. N. Y. Acad. Sci.* **483,** 387–456.
Sparrow, C. M. (1916). *Astrophys. J.* **44,** 76–86.
Takenaka, H., and Rikukawa, K. (1974). *Japan J. Appl. Phys.* **14,** (Suppl), 429–433.
Tilney, L. G., and Inoué, S. (1982). *J. Cell Biol.* **93,** 820–827.
Toraldo di Francia, G. (1955). *J. Opt. Soc. Am.* **45,** 497–501.
Toyoshima, Y., Krone, S., McNally, E., Niebling, K., Toyoshima, C., and Spudich, J. A. (1987). *Nature (London)* **328,** 536–539.
White, J. G., Amos, W. B., and Fordham, M. (1987). *J. Cell Biol.* **105,** 41–48.
Yanagida, T., Nakase, M., Nishiyama, K., and Oosawa, F. (1984). *Nature (London)* **307,** 58–60.

Chapter 4

Fluorescent Standards

JESSE E. SISKEN

Department of Microbiology and Immunology
College of Medicine
University of Kentucky
Lexington, Kentucky 40536

I. Introduction

In quantitative fluorescence microscopy, the aim is to measure and interpret the light output from an excited fluorophore. To do this in a meaningful way, one has to know the limits of reliability of the optical and detection systems, that they are operating at optimal levels, and the conditions employed are reproducible. In addition, one needs to be able to interpret the fluorescence signal in terms of the phenomenon or reaction under study. Thus, as has long been recognized, standards are required for two different purposes (e.g., Ploem, 1970; Ruch, 1973; Sernetz and Thaer, 1973; Haaijman and Van Dalen, 1974). The first is for the evaluation of the quality and capabilities of optical and detection systems, and for their standardization, correction, and adjustment. The second is to serve as calibration references so that fluorescence output can be interpreted in terms of the underlying phenomena of interest.

Early on, the main standards available for fluorescence microscopy were phosphor crystal, uranyl glass, and solutions of fluorochromes (see Ploem, 1970, 1975). Since then, a number of new standards have been developed which are of considerable use. The purpose of this chapter is to describe some of the standards that are available and how they may be used. They will be discussed in the two categories indicated above. We will refer to them as system standards and analytical standards.

II. System Standards

In this section, we will describe four kinds of materials which are available for use as system standards. We will indicate their characteristics, discuss how they may be used, and point out their advantages and disadvantages.

A. Standard Cells

Historically, several types of cells have been used as standards for quantitative adsorption and fluorescence measurements. These include polymorphonuclear leukocytes, lymphocytes, sperm, and nucleated erythrocytes (RBSs). However, only the RBCs, which are autofluorescent when fixed in glutaraldehyde, are of general relevance to the aims of this chapter. They have been used in flow cytometry for some time, earlier as standards for both fluorescence and light scatter, though now, with the availability of fluorescent beads (see below), they are used mainly for standardization of light scatter measurements. Their preparation is simple and, as described for chick RBCs by Herzenberg and Herzenberg (1978), involves fixation of PBS-rinsed RBCs in 1% glutaraldehyde in PBS for 48 hours at room temperature. During this time, autofluorescence increases and then stabilizes. The suspension is then washed and resuspended in saline in which the cells are stable for weeks if refrigerated. Aliquots can be frozen and, upon thawing, can be washed and resuspended at a desired concentration.

In the microscope one sees that the nucleus is brightly fluorescent compared to the cytoplasm which also displays significant fluorescence (Fig. 1). Total fluorescence of the cell, however, is relatively constant. In

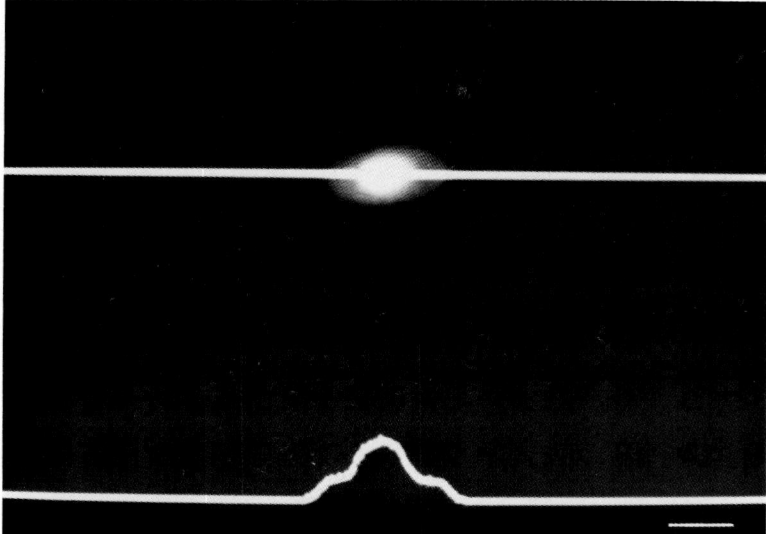

FIG. 1. Autofluorescence of a glutaraldehyde-fixed chick red blood cell with a scan of intensity values across the cell. The horizontal line in the top portion shows the position along which the intensity scan was done. The image was obtained by averaging 128 frames and subtracting an image of the background obtained in the same way. No shading correction was done and no thresholding or other operations were carried out on the image. Bar = 5 μm.

an earlier study with a relatively simple imaging system, we (Barrows *et al.*, 1984) obtained coefficients of variation (CVs; standard deviation × 100/mean) for fluorescence intensity per cell of about 18%, which is similar to that obtained with such cells by flow cytometry (Becton Dickinson Instruction Manual). Thus, one of their possible uses is for the determination of the reproducibility of a system. Differences in CVs between expected and actual will indicate how reproducible the system really is.

Other uses of these cells derive from their physical characteristics. They are fairly large, about 11 μm in their longest diameter, ellipsoidal, and relatively flat. As a result, more of their volume falls within the depth of focus of the objective lens compared to spherical beads of similar cross-sectional area. This means that they yield relatively less out-of-focus signal and have fewer potential problems with autoabsorption. Since there is a considerable difference in signal intensity between nucleus and cytoplasm, they are useful for adjusting excitation light intensity and for setting video camera controls in order to be certain signals are within the dynamic range of the system. For visual or qualitative work, they are useful for setting display monitor controls and could serve for adjusting phase and interference contrast systems. They are inexpensive and easy to prepare. Their main disadvantage is that the variance of their fluorescence intensities is high compared to fluorescent beads.

B. Manufactured Beads

Probably the first researchers to use beads for fluorescence were Haaijman and Van Dalen (1974), who coupled fluorescein isothiocyanate (FITC) and tetramethylrhodamine isothiocyanate (TRITC) to Sephadex G-25 beads. Since then, many kinds of fluorescently labeled beads have been synthesized and a number of companies now supply them (Table I). For the most part, their manufacture has been aimed at the flow cytometry market, where they are routinely used, among other things, for the alignment of wavelengths and light scatter. However, many of these beads are of use in fluorescence microscopy and some are now being prepared especially for the microscopist (see below).

1. Types Available and Their Properties

As indicated above, there are many kinds available. They are nominally homogeneous and spherical but this is not always the case. For example, some have been sold in which every bead has a single brightly fluorescent pimple on it, while others may have a small nonstaining region. The potential problem with such beads is that the amount of fluorescence measured per bead may be a function of its orientation in the field, thus contributing to a larger CV. Investigators should carefully examine all preparations of beads received to determine that they are acceptable. They are available in diameters from submicrometer to a millimeter or more with stated CVs of 1–5% for beads in the size range of interest to microscopists.

TABLE I

Some companies That Market Fluorescent Beads

Becton Dickinson, P.O. Box 7375, Mountain View, California 94039
Covalent Technology Corp., 3941 Research Park Drive, Box 1868, Ann Arbor, Michigan 48108
Coulter Electronics, P.O. Box 2145, Hialeah, Florida 33012
Flow Cytometry Standards Corp., P.O. Box 12621, Research Triangle Park, North Carolina 27709
Polysciences, Inc., 400 Valley Road, Warrington, Pennsylvania 18976
Seragen Diagnostics Inc., P.O. Box 1210, Indianapolis, Indiana 46206

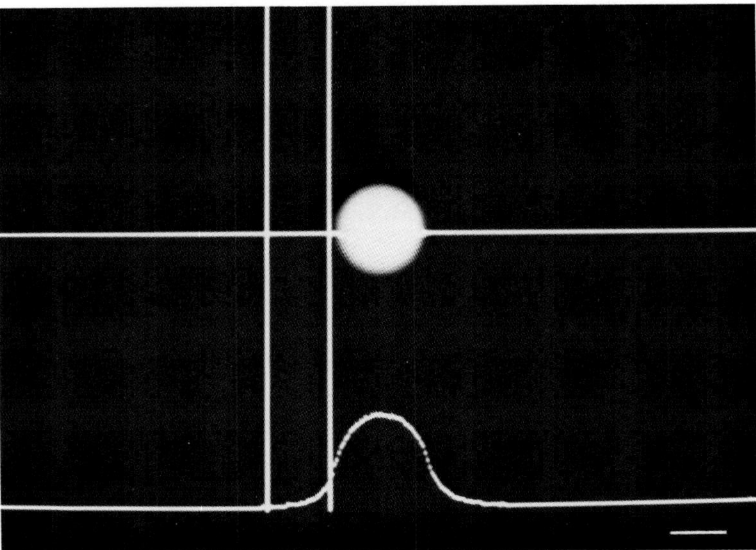

FIG 2. Fluorescence of 9.77 μm bead (Fluorosphere, Coulter Electronics) at best focus. The line at the bottom shows fluorescence intensity along the scan line. The vertical line nearest the bead marks its edge. Note that fluorescence can be detected in the region between the two vertical lines. Other details are as in Fig. 1.

If beads in suspension are to be used for the standardization of video imaging systems, one needs to select an appropriate size range because the smaller the size, the greater the Brownian motion and thus the blurring of any images acquired. Drift may also be a problem. In our experience, unless one takes measures to keep the beads immobile, bead of less than 5 μm in diameter are not of much use. Beads of 10 μm are quite immobile, but have the disadvantage that much of their volume will be out of the depth of field of most objectives, causing considerable out-of-focus signal and, therefore a halo around the bead. Figures 2 and 3 are images of a bead 9.77 μm in diameter (Fluorosphere, Coulter Electronics). Note that a scan across the bead (Fig. 2) or under the bead (Fig. 3) shows that some fluorescence can be detected outside the main image of the bead. At least some of this is due to the portion of the bead which is out of the plain of focus of the objective.

In most cases, beads are supplied as suspensions in water or buffer which may contain proteins such as bovine serum albumin (BSA) or gelatin and a preservative such as azide. One might want to dilute the suspensions for use and recommendations for diluents are available from suppliers. One

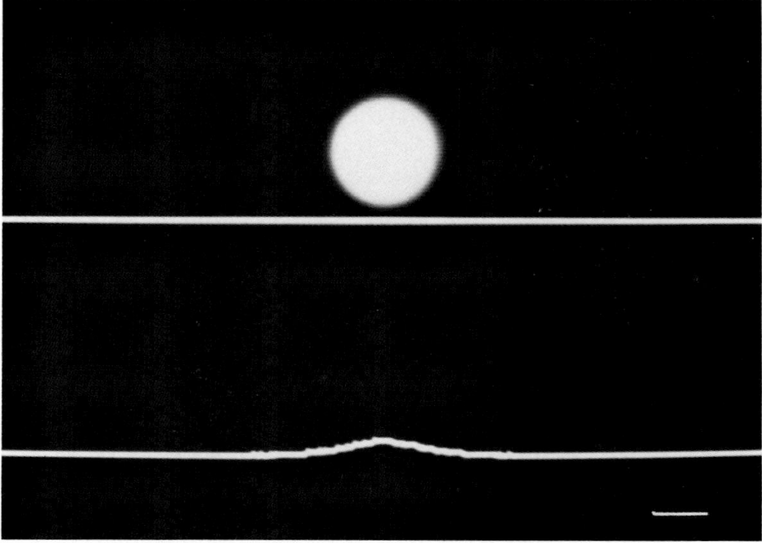

FIG. 3. The same image as Fig. 2 except, in this case, a scan was done below the main image of the bead and through the region of the fluorescent halo. Other details are as in Fig. 1.

company (Flow Cytometry Standards Corporation, Research Triangle Park, NC) supplies beads already fixed to slides.

The beads available are made from a variety of polymers. They can be obtained "plain," i.e., with no reactive groups available on them, or may contain various reactive groups to which one can attach selected dyes or other molecules such as antigens or antibodies. They are available with various fluorescent dyes bound to them spanning a range of excitation and emission wavelengths. They can also be obtained with multiple dyes attached to the same beads, which would appear to be useful to those interested in multiparameter work. Others can be obtained with graded amounts of dye attached to them, which makes them at least potentially useful for linearity testing and for analytical calibration (see below). In some cases the dye molecules are bound so that only the surface is fluorescent while in others, the dyes seem relatively homogeneously distributed throughout the bead. Textural details can be seen in some if excitation intensities are kept low.

In general, these beads are relatively constant with respect to their fluorescence intensities. Each lot is available with a nominal CV which is usually in the 1–5% range. In addition, as discussed below, one recent innovation is to supply the beads with attached fluorochromes which are sensitive to their environment, specifically with calcium-sensitive dyes attached.

The photostability of such beads is a function of the dye molecules attached and these appear to be as photolabile on beads as they are in or on cells. For example, those containing fluorescein dyes undergo fairly rapid photobleaching, making them relatively unsuitable as system standards. Some, however, are made with dyes that are much less photolabile (see for example, Smith *et al.,* 1986).

2. Usage

In this section, we will deal with the uses of standards for evaluation and checkout of optical and signal detection systems and for setting system parameters. This discussion will be brief since Chapter 2 by Jeričević *et al.,* this volume, deals in depth with this topic.

Perhaps the first thing one ought to do when putting a new system into operation is to use bead standards to determine the quality of each objective and the optical system as a whole. One way to do this is to digitize the image of a bead at best focus and above and below focus and examine pixel intensities across the diameter of the bead. In an ideal system at perfect focus, i.e., where the specimen is of uniform thickness and entirely within the focal plane of the objective, a nearly square wave configuration of the scan should be obtained, with all pixels outside the image being at zero level and a flat plateau at the maximum if the sample is of uniform intensity.

What is actually observed is only an approximation of this since, among other things, the beads are spherical, the standard is thicker than the depth of focus of the objective, there are imperfections in the optical system, movement of the object may occur while the image is being digitized, and focus may not be optimal. In addition, any asymmetries in the above- or below-focus images will indicate problems in the optical system.

With respect to the performance of a system, some variables of concern are the stability of the light source and detector system (video camera or PMT) and the ability to achieve exact focus each time, especially when one must work rapidly to avoid photobleaching with low-intensity images. The question then is how well can one reproduce measurements of fluorescence intensity using beads of known variance. One way of doing this is to perform rapid, multiple measurements of the same bead with and without refocussing. Another is by successively measuring a number of different beads in the same place in the field, thus avoiding any contribution of potential shading errors. In both cases, variances should approximate the nominal variances of the standards. Excessive variances in any of the above measures would indicate problems. In addition, one can evaluate shading errors and their correction by measuring beads in different locations in the field. Smith *et al.* (1986) used beads to determine excitation intensity and illumination uniformity as a function of lamp position and optical properties of individual lenses (see Jeričević *et al.*, Chapter 2, this volume).

Another use of beads is to check the linearity of a system. This can be done in at least two ways. One is to measure the fluorescence of a population of beads at different excitation intensities by inserting neutral density filters of known absorptions into the light path. This should yield a plot of percent transmission versus intensity which is linear. A second way is to measure a series of beads to which different numbers of fluorochrome molecules are attached. Such beads are now commercially available. Again, a plot of fluorescence intensity versus mean dye content should be linear so long as fluorescence intensities are maintained within the dynamic range of the detector.

Finally, beads can be useful where problems exist in day-to-day or hour-to-hour variations in excitation intensities and/or camera output. By measuring the fluorescence of standard beads under standard conditions each time a system is used and periodically during use, one can evaluate the temporal stability of a system and make suitable adjustments if this should be a problem. The beads can also be used to normalize data collected at different times or to compensate for changes in conditions such as may result from an aging light source or the installation of a new one. Thus, bead standards can be used to satisfy the needs cited many years ago by Ploem (1970).

C. Standard Solutions

As indicated earlier, solutions of fluorochromes have been used for the standardization of microscopes and detector systems and as reference standards for many years. In 1970, Ploem described in detail several ways of using fluorescein diacetate-containing solutions for such purposes. These included the measurement of fluorescence of microdroplets embedded in resin and of solutions contained in either a blood cell counting chamber or a microcuvette. He reported that the reproducibility was satisfactory with the first two but not with the latter, possibly due to differences in the internal dimensions of the chamber. From his experiences with these, he proposed a standardization system by which experimental data could be compared each day to the fluorescence of a standard size microdroplet containing a specific concentration of fluorophore at a specific pH. This would allow the fluorescence of all specimens to be expressed in microdroplet units. This was a very reasonable idea which today could be implemented as suggested above with fluorescent beads.

Another way to use standard solutions on the microscope stage is with capillary tubes. Sernetz and Thaer (1973) showed that quartz glass capillaries mounted in glycerine could be used for this purpose. They demonstrated the utility of this method for such ends as examining the relationship between fluorescence intensity and concentration of dye, for determining limits of detectability, and for the study of emission spectra of dyes in pure solutions versus those located intracellularly. In my own laboratory, we have used an ordinary hemocytometer filled with various concentrations of fluorescein to test the linearity of a fluorescence imaging system, the details of which are shown in Fig. 4 (Barrows *et al.*, 1984). For

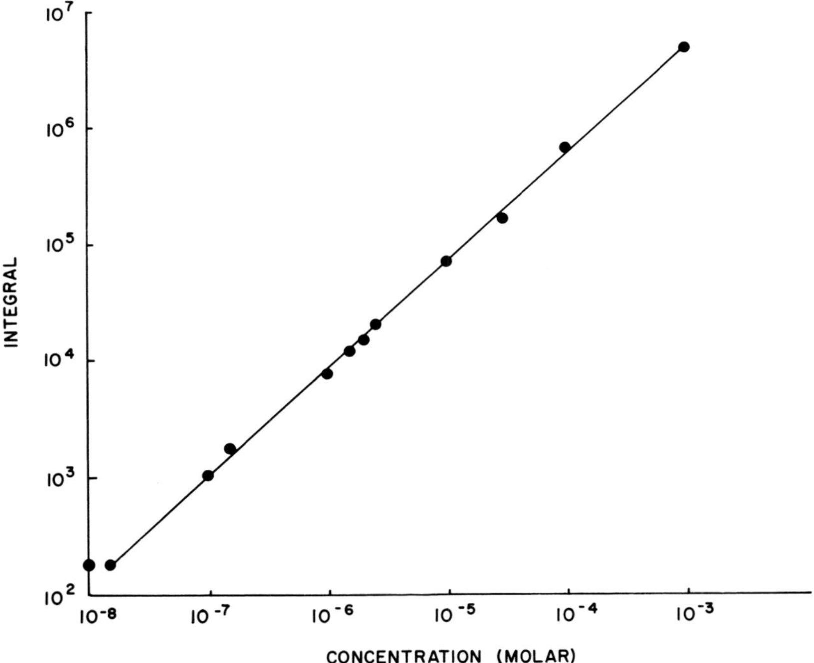

FIG. 4. Fluorescence intensity of various concentrations of fluorescein as used to indicate the linearity of an imaging system. Measurements were made in a small region in the center of the field with no background subtraction or shading correction. Neutral density filters were used where necessary to keep intensity values within the dynamic range of the system and intensities were accordingly corrected based upon the known transmittances of the filters. (Reproduced with the permission of The Histochemical Society, Inc. and the Elsevier Science Publishing Co., Inc.)

this purpose, we disabled camera autocontrols and gamma corrector and used neutral density filters in the excitation pathway to keep fluorescence levels within the dynamic range of our SIT camera. We were careful to use only those portions of the mirrored surface of the hemocytometer which contained no grid lines in the optical path, and focus was midway between the mirrored surface of the hemocytometer and the bottom surface of the cover glass.

A further possibility is to use flattened glass capillaries (Vitro Dynamics, Rockaway, NJ) (as used, e.g., by Tanasugarn *et al.*) which are optically superior to round capillaries. In our experience, 0.1 mm × 2 mm capillaries seem appropriate for these purposes. They fill easily by capillary action, are not subject to significant drying from the open ends, which could be sealed if necessary, and are easily attached to glass slides using double-sided adhesive tape. Solutions can, in fact, be used in flat capillaries for a number of the same purposes as beads but are probably less convenient. Perhaps the most important use of solutions is as described in later sections.

D. Solid Standards

Several types of solid standards have been used which provide a stable, reproducible level of fluorescence and are relatively insensitive to photobleaching by the excitation light. Their use for system adjustment and checkout, therefore, is similar to that of solutions. Ploem (1970) described the use of fluorescent phosphor crystals and uranyl glass for such purposes. More recently, the use of Starna polymer blocks (Starna Cells Inc., Atascadero, CA), of which there are a number available, has been described (Benson *et al.*, 1985; T. D. Coates, personal communication). Benson *et al.*, (1985) utilized such a block to generate a flat "mask" image which they used for shading corrections of images.

III. Analytical Calibration Standards

A different use for standards is for the calibration and interpretation of fluorescence data. That is, given a valid signal, what does it mean in terms of quantity of intracellular probe or reaction product. One way to calibrate such signals is to compare the signal to one or more appropriate standards. Most such calibrations are done with standard solutions and some of these will be discussed first. However, standard beads are also of potential use and will be mentioned in this context as well.

A. Solutions

As stated earlier, solutions can be used as models for intracellular events. Thus, Sernetz and Thaer (1973) studied the spectral characteristics of fluorescein in albumin-containing solutions in order to understand the interactions of the dye in the intracellular environment. A more recent example is the use of microdroplets of benzo[*a*]pyrene in mineral oil as a model for intracellular photobleching (Benson *et al.*, 1985) and in ethanolic solutions to establish limits of detection and the linear response range of a video camera (Plant *et al.*, 1985).

Another use of solutions in fluorescence analyses is for the calibration of ratio measurements in situations in which the binding of a target ion alters the excitation or emission spectra of fluorescent probes. The underlying principle behind ratio measurements is that when a fluorescent probe undergoes a spectral shift as a function of its interaction with the target molecule, the degree of the shift is dependent upon the concentration of the target molecule which can theoretically be determined by dual wavelength ratioing; that is, by dividing the emitted intensities resulting from excitation at one wavelength by the emitted intensity resulting from excitation at a different wavelength. The advantage of the ratio method is that it normalizes the data for dye concentration, path length, and efficiency of the optical system.

An early use of ratio measurements was for the spectrophotometric measurement of pH in bacteria (Thomas *et al.*, 1976), lysosomes of macrophages (Ohkuma and Poole, 1978), and Ehrlich ascites tumor cells (Thomas *et al.*, 1979). This method was subsequently adapted for fluorescence microscopy by Heiple and Taylor (1980) and has since been used by a number of others (e.g., Tanasugarn *et al.*, 1984; Tycko *et al.*, 1983; Paradiso *et al.*, 1987, Bright *et al.*, 1987). In this work, two types of standards were used to develop calibration curves for the conversion of fluorescence ratios to pH. One involved solutions of known pH and the other the use of permeabilized cells. The principle behind this latter method is that in the presence of the ionophore nigericin and high extracellular levels of K^+, internal pH will equilibrate with external pH under circumstances in which most of the dye remains intracellular (Thomas *et al.*, 1979). Thus, measurements obtained by loading cells with an appropriate fluorochrome and exposing them to a graded series of buffers provide a calibration curve against which experimental data can be compared.

The other major use of the ratio technique has been for the measurement of intracellular free calcium ion concentrations ($[Ca^{2+}]_i$). There are now several probes available for the determination of $[Ca^{2+}]_i$, one of which is fura-2, which undergoes a blue shift in excitation spectrum and an increase in peak height as a function of $[Ca^{2+}]_i$ (see e.g., Tsien, 1980; Grynkiewicz *et al.*, 1985; Tsien *et al.*, 1985; Poenie *et al.*, 1985). The excitation peak shifts from 360 nm in zero Ca^{2+} to 339 nm in high Ca^{2+}. There is a wavelength, the isosbestic point (360 nm), which is insensitive to Ca^{2+} while at longer wavelengths fluorescence intensity decreases as a function of Ca^{2+}. Thus, by ratioing emitted intensity when fura-2 is excited at two different wavelengths, e.g., near 339 nm and at or beyond the isosbestic point, one can, in principle, determine the concentration of free intracellular Ca^{2+}. These measurements can be done in two ways. One is the calculation method of Tsien and co-workers (see e.g., Grynkiewicz *et al.*, 1985). The other is to convert ratios to free intracellular Ca^{2+} using a standard curve derived from measurements of standard solutions containing known concentration of free calcium ions (e.g., Ratan *et al.*, 1986).

B. Beads

Recent developments in the production of beads suggest that they can, at least theoretically, serve some of the same purposes as solutions. For example, beads are now available with fura-2 or indo-1 attached and in environments containing high or low levels of Ca^{2+}. Thus, the possibility exists that such beads may eventually be useful for calibrating ratio data for calcium concentrations.

Another recent innovation has been the availability of fluorescent beads for the calibration of absolute numbers of dye molecules bound to cells. Haaijman and Van Dalen (1974) pointed out that such beads could be mounted on slides in the same medium used for immunofluorescence so that the standard and the biological specimen could be observed in the same medium on the same slide and at the same pH. Thus, in analogy to Ploem's proposal (1970), the fluorescence of the cell could be expressed in terms of the number of fluorochrome molecules bound to the beads, provided that they were measured with the same optics.

More recently, Oonishi and Uyesaku (1985) synthesized a bead which could be dissolved at high pH so that measurements of the amount of bound dye could be determined spectrophotometrically. With such beads as standards, one could theoretically determine the number of FITC molecules, in their case on a labeled monoclonal antibody, that were bound to the cell. Fluorescently labeled beads are now commercially available for this purpose but, as Haaijman and Van Dalen (1974) have indicated, care is required in their use. For example, they pointed out that the emission spectra of FITC and TRITC bound to the Sephadex were shifted about 5 nm compared to the spectra of the same dyes bound to proteins and that the pH dependency of the spectra was also altered under such circumstances. Further, they reasoned that the quantum efficiency (QE) might also have been altered, suggesting that in the absence of data on the QE, the calibration could only be relative rather than absolute. Nevertheless, the potential of such beads is worth consideration by investigators.

IV. Concluding Statement

The intent of this chapter has been to describe the various kinds of standards that exist and how they may be used for system checks or for calibration of analytical data. Clearly, there are many available and, if the trend continues, others will probably appear in the future. The important thing is that their use can enhance one's confidence in the validity of measurements obtained. Further, it would seem reasonable to suggest that the publication of analytical data be accompanied by measurements of such standards where appropriate so that readers can evaluate the validity of the systems employed.

Acknowledgments

The author wishes to thank Dr. Abe Schwartz, of Flow Cytometry Standards Corporation, Charles Howe of Coulter Electronics, Dan Fannan of Polysciences, Inc., and Rick Galloway of Seragen Diagnostics, Inc. for test bead samples and/or some very informative discussions about bead standards. Thanks are also due to Dr. Richard Cross of the University of Kentucky for the sample of glutaraldehyde-fixed chick RBCs and to Dr Gary Bright for his critique of the manuscript. The work presented was supported, in part, by grant HD20655 from the National Institute of Child Health and Human Development.

References

Barrows, G. H., Sisken, J. E., Allegra, J. C., and Grasch, S. D. (1984). *J. Histochem. Cytochem.* **32,** 741–746.

Benson, D. M., Bryan, J., Plant, A. L., Gotto, A. M., Jr., and Smith, L. C. (1985). *J. Cell Biol.* **100,** 1309–1323.

Bright, G. R., Fisher, G. W., Rogowska, J., and Taylor, D. L. (1987). *J. Cell Biol.* **104,** 1019–1033.

Grynkiewicz, G., Poenie, M., and Tsien, R. Y. (1985). *J. Biol. Chem.* **260,** 3440–3450.

Haaijman, J. J., and Van Dalen, J. P. R. (1974). *J. Immunol. Methods* **5,** 359–374.

Heiple, J. M., and Taylor, D. L. (1980). *J. Cell Biol.* **86,** 885–890.

Herzenberg, L. A., and Hertzenberg, L. A. (1978). *In* "Handbook of Experimental Immunology" (D. M. Weir, ed.), 3rd Ed., pp. 22.1–22.21. Blackwell, Oxford.

Ohkuma, S., and Poole, B. (1978). *Proc. Natl. Acad. Sci. U.S.A.* **75,** 3327–3331.
Oonishi T., and Uyesaku, N. (1985). *J. Immunol. Methods* **84,** 143–154.
Paradiso, A. M., Tsien, R. Y., and Machen, T. E. (1987). *Nature (London)* **325,** 447–450.
Plant, A. L., Benson, D. M., and Smith, L. C. (1985). *J. Cell Biol.* **100,** 1295–1308.
Ploem, J. S. (1970). *In* "Standardization in Immunofluorescence" (E. J. Holborow, ed.), pp. 137–153. Blackwell, Oxford.
Ploem, J. S. (1975). *Ann. N.Y. Acad. Sci.* **254,** 4–20.
Poenie, M., Alderton, R., Tsien, R. Y., and Steinhardt, R. A. (1985). *Nature (London)* **315,** 147–149.
Ratan, R. R., Shelanski, M. L., and Maxfield, F. R. (1986). *Proc. Natl. Acad. Sci. U.S.A.* **83,** 5136–5140.
Ruch, F. (1973). *In* "Fluorescence Techniques in Cell Biology" (A. A. Thaer and M. Sernetz, eds.), pp. 51–55. Springer-Verlag. Berlin.
Sernetz, M., and Thaer, A. A. (1973). **In** "Fluorescence Techniques in Cell Biology" (A. A. Thaer and Sernetz, eds.), pp 41–49. Springer-Verlag, Berlin.
Smith, L. C., Jerecevic, Z., and Bryan, J. (1986). *Fed. Proc., Fed. Am. Soc. Exp. Biol.* **45,** 171.
Tanasugarn, L., McNeil, P. Reynolds, G. T., and Taylor, D. L. (1984). *J. Cell Biol.* **98,** 717–724.
Thomas, J. A., and Cole, R. E., and Langworthy, T. A. (1976). *Fed. Proc.; Fed. Am. Soc. Exp. Biol.* **35,** 1455.
Thomas, J. A., Buchsbaum, R. N., Zimniak, A., and Racker, E. (1979). *Biochemistry* **18,** 221–2118.
Tsien, R. Y. (1980). *Biochemistry* **19,** 2396–2404.
Tsien, R. Y., Rink, T. J., and Poenie, M. (1985). *Cell Calcium.* **6,** 145–157.
Tycko, B., Kieth, C. H., and Maxfield, F. R. (1983). *J. Cell Biol.* **97,** 1762–1776.

Chapter 5

Fluorescent Indicators of Ion Concentrations

ROGER Y. TSIEN

Department of Physiology-Anatomy
University of California
2549 Life Science Building
Berkeley, California 94720

I. Introduction

Ion concentration gradients across cell membranes are central to any understanding of biological energetics and signal transduction. The proton electrochemical gradient is generally accepted to be the key intermediary linking electron transport to ATP synthesis in bacteria, chloroplasts, and mitochondria. Cytoplasmic pH is a powerful regulator of enzyme activities (Busa, 1986), and is known to alkalinize in many types of cells undergoing mitogenic activation (Moolenaar, 1986). Gradients of Na^+ and K^+ concentration across the plasma membrane are responsible in most cases for the membrane potential, including action potentials of excitable cells. The Na^+ gradient also is the energy source for secondary active transport of nutrients into animal cells and many co- or countertransport systems for other ions. Cl^- gradients are vital to the function of epithelia, the control of cell volume, and to the actions of many inhibitory neurotransmitters. Ca^{2+} has the largest transmembrane gradient of any simple ion. Because it is normally kept so low inside cells, small changes in membrane permeability cause large percentage changes in cytosolic ionized Ca^{2+} concentrations, so that intracellular Ca^{2+} is able to act as a ubiquitous and dynamic messenger signaling cell activation triggered by plasma membrane potential or receptor occupancy (Berridge, 1987; Carafoli, 1987; Meldolesi and Pozzan, 1987).

Characterization of ion gradients obviously requires knowing both extracellular and intracellular ion activities or free concentrations. Determination of extracellular ions usually poses much less of a problem than intracellular, since the extracellular medium is much more accessible, and experiences smaller and slower concentration changes due to its larger volume. Extracellular concentrations of all the simple inorganic ions except K^+ are comparable to or higher than intracellular, so detection is easier extracellularly. Often extracellular media are under the complete control of the experimenter. Except where the local extracellular medium is barred from easy diffusional equilibrium with the bulk fluid, the problem of quantifying ion gradients usually reduces to measuring the intracellular ion activities.

Fluorescent probes are now available for all the ions mentioned above. These indicators are perhaps the most popular methods for measuring these ions for the following reasons:

(1) In most cells types tried, the probes can be incorporated into the intact functioning cells without any breaching of the plasma membrane. The plasma membrane has to be at least temporarily disrupted in other methods such as insertion of ion-selective microelectrodes, incorporation of photoproteins, or classical polysulfonate absorbance indicators. Such disruption leads to major concerns about the quality of resealing.

(2) Current fluorescent indicators measure the ion activities or free concentrations (Tsien, 1983), the values needed both to calculate thermodynamic gradients and to predict and explain binding equilibria. Many methods exist to measure total ion contents, but much or most of those totals consists of ions bound to sites irrelevant to the cellular process under study. Also, measurements of total ion contents usually require destroying the tissue, so that the time course of a physiological response cannot be followed continuously.

(3) Optical indicators typically work in the millisecond domain, much faster than the time resolution of other methods, for example, seconds for

ion-selective microelectrodes and minutes for nuclear magnetic resonance (NMR) of nonrepetitive changes.

(4) Fluorescent probes can be used at all levels of organization from whole organs to isolated tissue fragments to populations of disaggregated cells to single cells to subcellular domains of single cells. Combined with digital imaging (Tanasugarn *et al.*, 1984; Tsien and Poenie, 1986), fluorescent probes can give entire images of cytosolic ion levels with resolution down to microns. This resolution can be further enhanced, especially along the axis of view, by scanning confocal microscopy (White *et al.*, 1987; Brakenhoff *et al.*, Chapter 14, this volume) and digital image reconstruction algorithms (Agard, 1984; Agard *et al.*, Chapter 13, this volume). By contrast NMR needs lots of tissue, dense populations or intact organs, and generally is used without any spatial resolution. An ion-selective microelectrode inherently measures from one spot at a time in a single cell. Photoproteins and absorbance indicators were originally restricted to single large cells by the need for microinjection, though recent methods of mass lysis-resealing have permitted extensions to some cell populations (reviewed in Cobbold and Rink, 1987; and McNeil, 1988). However, spatially resolved images have been obtained only from single spread cells and are of relatively low definition.

(5) Fluorescence is the only ion measurement technique amenable to flow cytometry, which is the premier method to analyze statistics of a large population of cells as well as to fractionate them preparatively.

(6) Simple applications of fluorescent indicators only need a fluorometer, a versatile instrument routinely available in biological laboratories, whereas the other techniques require custom or highly expensive equipment at the outset. Once the experimenter becomes comfortable with a cuvet fluorometer, a seductive pathway of increasingly sophisticated instrumentation is available, leading to dual-wavelength ratio fluorometry (Heiple and Taylor, 1982; Tsien *et al.*, 1985), digital ratio imaging microscopy (Tanasugarn *et al.*, 1984; Williams *et al.*, 1985; Tsien and Poenie, 1986; Bright *et al.*, 1987; (see Bright *et al.*, Chapter 6 this volume), and scanning confocal microscopy (White *et al.*, 1987).

II. Proton Indicators

Current fluorescent pH-sensing indicators may be divided into two fundamental classes: (A) dyes that translocate across membranes in response to pH gradients across those membranes; (B) dyes that stay put in one or more aqueous compartments of the cell and signal the local pH by its direct effect on the excitation or emission spectrum.

A. Dyes Whose Accumulation Depends on pH

Dyes of class A are just weak acids or bases (usually the latter) which accumulate in alkaline or acidic compartments according to the same classic laws as govern bicarbonate or amine distributions (Roos and Boron, 1981). This accumulation occurs because the neutral form of the indicator is freely permeable through membranes, so its free concentration is equal in all compartments. However, the charged form (an anion for weak acids, a cation for weak bases) is not permeant as such, but is in continual local equilibrium (governed by the local pH) with the neutral species. Therefore, protonated base cations accumulate in acid compartments because the local pH demands a high ratio of [cation] to [neutral], the latter being the same everywhere. Acid anions correspondingly pile up in alkaline compartments. If the weak base cation is fluorescent, simple examination by fluorescence microscopy is sufficient qualitatively to reveal the acid compartments as bright spots. Quantitatively, however, the relative brightness of the acid compartments compared to the other parts of the cell or even the extracellular fluid would depend not only on their respective pHs and the pK_a of the indicator but also on the relative volumes of the compartments and whether either contained binding sites for the dye. Many of the most common dyes form cations that bind avidly to negatively charged proteins, membranes, and nucleic acids, enhancing or otherwise altering their fluorescence in the process. Another optical sensitivity mechanism arises from the propensity of many of these amines to quench or red-shift their fluorescence merely upon close-packing with fellow dye molecules. Such effects mean that, even without spatial resolution, the summated signal from all the compartments will be sensitive to changes in the pH gradients that redistribute dye from one compartment to another. Of course those signals become ever harder to interpret quantitatively, since they now are dependent not only on the factors listed above (pHs, pK_a, volumes, etc.) but on the absolute concentration of dye and the strength of dye–dye interactions. Nevertheless, they are highly useful in giving a continuous nondestructive rough estimate of the acidity of appropriate subcellular compartments. Acid compartments (such as endosomes, lysosomes, Golgi, epithelial luminal spaces, and vesicles) have received much more attention than basic organelles (chiefly mitochondria). The most commonly used weak bases are 9-aminoacridine and acridine orange (reviewed by Lee *et al.*, 1982; Simons *et al.*, 1982). An example of a weakly acidic dye that accumulates in mitochondria is fluorescein (Thomas *et al.,* 1979). Because the spectrum of fluorescein is itself pH-sensitive, fluorescein fits into both classes of pH probe.

B. Dyes Whose Spectra Depend on pH

Table I summarizes the properties of dyes whose fluorescence spectra are sensitive to pH and which have been useful in cell biology. Their structures are shown in Fig. 1. Their major application is to measure the pH of the cytosol, though a few are also used to examine the pH of endocytic vesicles (for example, Heiple and Taylor, 1982; McNeil *et al.,* 1983; Geisow, 1984). Since cytosolic pH is typically 6.5–7.5, the pK_a should be in the same range. This is most commonly achieved with a phenolic group attached to a chromophore which can delocalize the negative charge left by the proton ionization. Such delocalization stabilizes

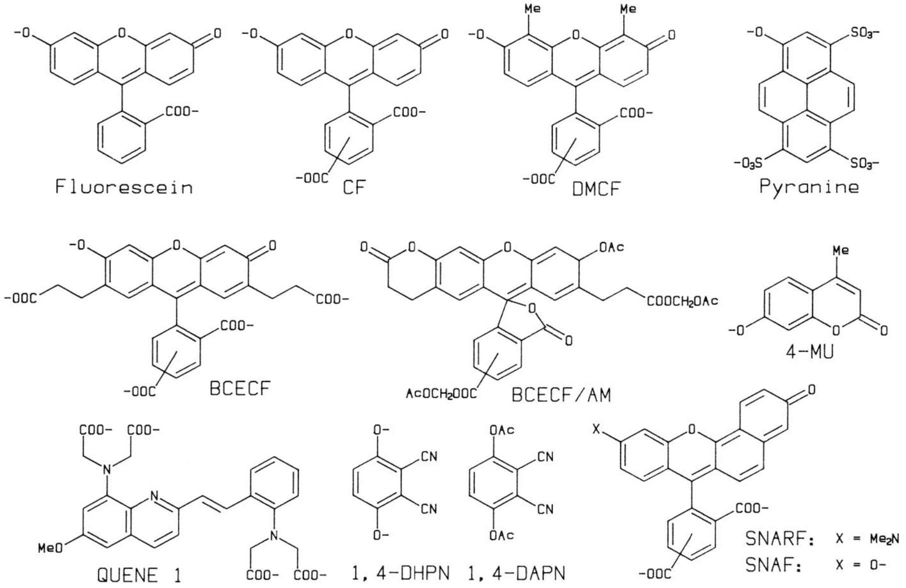

FIG. 1. Structures of biologically useful indicators whose fluorescence spectra are sensitive to pH, and membrane-permeant esters thereof. Each is drawn in deprotonated form as if ready to bind a proton. In CF, DMCF, BCECF, SNARF, and SNAF, the linkage of the carboxylate to the center of the lower benzene ring indicates that the carboxylate could be attached to either of the two adjacent positions (usually numbered 5 and 6), and that the usual preparation consists of a mixture of these two isomers. SNAF-2 actually has one extra chloro atom *ortho* and *para* to the phenol and ring oxygens, respectively.

TABLE I

PROPERTIES OF FLUORESCENT INDICATORS WHOSE SPECTRA DEPEND ON pH

Indicator[a]	λ_{exc}[b]	λ_{em}[b]	pK_a[c]	Excitation ratio[d]	Emission ratio[d]	Permeant ester	Leakage rate
Fluorescein	491	515	6.4–6.5	490/450	No	Yes	Fast
FD	490	515	6.4–6.5	489/452	No	No	Very slow
CF	492	518	6.4–6.5	518/464	No	Yes	Medium
DMCF	505	535	6.8–6.9	No	No	Yes	Medium
DMFD	512	538	6.75	No	No	No	Very slow
BCECF	502	528	6.97–6.99	490/439	[520/620]	Yes	Slow
Pyranine	405, 465	514	7.2	465/405	[465/514]	No	Slow
4-MU	330, 360	450	7.4–7.8	365/334	[450/550]	Yes	Very fast
Quene 1	390	530	7.3	No	No	Yes	Slow
DHPN	380, 405	453, 483	8.0	420/360	512/455 or 540/425	Yes	Fast
SNARF-1	546, 574	588, 637	7.5	590/500	636/568	Yes	?
SNAF-2	510, 546	539, 630	7.7	540/480	630/520	Yes	?

[a] FD, Fluorescein conjugated to dextran or other inert macromolecule; CF, 5(or 6)-carboxyfluorescein; DMCF, 4′,5′-dimethyl-5(or 6)-carboxyfluorescein; DMFD, 4′,5′-dimethylfluorescein conjugated to dextran; BCECF, 2′,7′-bis(carboxyethyl)-5(or 6)-carboxyfluorescein; Pyranine, 8-hydroxypyrene-1, 3, 6-sulfonate; 4-MU, 4-methylumbelliferone or 4-methyl-7-hydroxycoumarin; DHPN, 3, 6-dihydroxyphthalonitrile or 2, 3-dicyanohydroquinone; Quene 1, SNARF-1, SNAF-2, see Fig. 1 for structures.

[b] λ_{exc}, λ_{em}: Excitation and emission peak wavelengths in nanometers. Where two values are given, the first is for the protonated form, the second for the deprotonated species.

[c] pK_a: pH at which the protonated and deprotonated forms are equal in concentration.

[d] Excitation or emission ratio: Gives typical pairs of wavelengths whose excitation or emission ratio increases with increasing pH. Values in square brackets are pairs which are of doubtful usefulness due to insensitivity to pH or small amplitude at one wavelength.

the anion, shifts the absorbance (and sometimes the fluorescence emission) to longer wavelengths than the protonated, uncharged species, and lowers the pK_a to the physiological range.

1. Fluorescein and Its Derivatives

By far the most popular family of pH indicators has been the fluoresceins. Their brilliant fluorescence has long been exploited as a simple tag for antibodies or other proteins or as a marker for the integrity of membrane-enclosed compartments. Fluorescein fluorescence is also strongly pH dependent, increasing with deprotonation. The pK_a for the deprotonation of the parent compound is around 6.4 in physiological media, with some minor leeway (± 0.1). This pK_a does not interfere with use of fluorescein as a passive tag, since its fluorescence is already near maximal at physiological pHs near 7.0, but is unfavorable for its use as an indicator of cytosolic pH except under unusually acidotic conditions or in acidic organelles. Substituents on the xanthene chromophore itself markedly affect the pK_a. Electron withdrawing substituents such as halogens lower the pK_a, electron-donating groups such as alkyl groups raise the pK_a. Two methyls at the 4′ and 5′-positions, as in 4′,5′-dimethyl-5(or 6)-carboxyfluorescein (DMCF) or 4′,5′-dimethylfluorescein–dextran (DMFD) raise the pK_a to 6.8–6.9 (Simons *et al.*, 1982; Rothenberg *et al.*, 1983; Chaillet and Boron, 1985). Two —$CH_2CH_2COO^-$ groups as in 2′,7′-bis(carboxyethyl)-5(or 6)-carboxyfluorescein (BCECF) are even more effective, giving a nearly ideal pK_a of 6.97–6.99 (Rink *et al.*, 1982; Bright *et al.*, 1987). By contrast, substituents on the phthalein ring, such as the carboxyl groups in 5(or 6)-carboxyfluorescein (CF), have little effect because they are too remote and because the phthalein ring is twisted out of conjugation with the rigid tricyclic xanthene system.

The excitation maximum of fluorescein is in the blue-green, at 491 nm, and the emission peak is in the green, 515 nm. Much of the familiar yellowish hue of conventional photographs of fluorescein immunofluorescence is due to emission long-pass filters that emphasize the long-wavelength tail of the emission spectrum in order to reject the bluish excitation. In accordance with its name, the fully deprotonated anion of fluorescein is one of the most brightly fluorescent low-molecular-weight chromophores known, with an extinction coefficient of $9 \times 10^4\ M^{-1}\ cm^{-1}$ and a quantum efficiency of 0.9 (Demas and Crosby, 1971). The analogous extinction coefficients and quantum efficiencies for the alkyl-substituted fluorescein analogs are probably similar to the parent; for example, BCECF fully deprotonated has a quantum efficiency within 3% of that of fluorescein (R. Tsien, unpublished results). However, the excitation and emission maxima are each shifted about 10 nm to the red. All the fluoresceins respond to protonation by shifting their absorbance spectrum markedly to shorter wavelengths and decreasing their quantum efficiency. Fluorescein, carboxyfluorescein (CF), and BCECF retain some ability to fluoresce despite protonation, so that their excitation spectra, like their absorbance spectra, shift to shorter wavelengths as pH falls. Therefore each dye has a wavelength at which the excitation efficiency is independent of pH, i.e., at which all the spectra at different pHs cross. This crossover should perhaps be called the isoexcitation point, since it is the excitation spectrum analog of the isosbestic point, where all the absorbance spectra cross. The isoexcitation point is not identical with the isosbestic point, since the isoexcitation point depends not only on the absorbance spectra but also on the relative quantum efficiencies of the two dye species and on the wavelength chosen for monitoring the emission. The presence of an isoexcitation point is a valuable feature of these dyes, since the ratio of the excitation efficiency at a pH-sensitive wavelength to that at the isoexcitation point gives a measure of pH that cancels out the dye concentration and optical path length (Heiple and Taylor, 1982; Tanasugarn *et al.,* 1984; Alpern, 1985; Paradiso *et al.,* 1987). By contrast, acidification of DMCF over the range pH 6–8 simply scales the excitation spectra downward without a wavelength shift or isoexcitation point. Therefore acidification of DMCF is spectroscopically indistinguishable from loss of dye by bleaching or leakage from the cell or tissue, a significant disadvantage for microscopic observations (Simons *et al.,* 1982). The dye fluorescence can be monitored at a single wavelength when dye concentration and pathlength are constant, or its absorbance spectrum can be ratioed (Chaillet and Boron, 1985). In the latter case, careful correction needs to be made for the intrinsic scattering and absorbance of the tissue, which are more significant in comparison to the dye-related absorbance than autofluorescence is in comparison to BCECF fluorescence.

Upon acidification, the emission spectra of all the fluorescein dyes decrease in amplitude without a significant wavelength shift. A slight broadening of the peak can be observed with BCECF (R. Tsien, unpublished). This makes the ratio of 520 nm emission (near the peak) to that at 620 nm (in the long-wavelength tail) slightly pH sensitive. Though this ratio has been used in flow cytometry (Musgrove *et al.,* 1986), it is less pH sensitive and likely to be more noisy than the excitation ratio, so the latter is preferable whenever technically feasible.

The gentlest means of loading these dyes into cells is the use of membrane-permeant, hydrolyzable esters. Rotman and Papermaster (1966) originated the use of such esters by showing that uncharged colorless fluorescein diacetate is readily hydolyzed by cells. Intact cells retain enough of the fluorescein dianion to fluoresce brightly under a microscope, whereas damaged cells leak the dye so rapidly as to appear nonfluorescent. However, fluorescein does leak at a significant rate even from healthy cells, especially at higher temperatures and greater surface-to-volume ratios. Therefore fluorescein is generally useless in cells in a cuvette, where leaked extracellular dye cannot be spatially resolved from intracellular dye. An even more fundamental problem, pointed out by Thomas *et al.* (1979), is that fluorescein is a permeant enough weak acid to accumulate in alkaline intracellular organelles such as mitrochondria. Addition of another carboxyl group as in carboxyfluorescein substantially reduces leakage through membranes and migration into mitochondria. However, the carboxyl also hinders loading of the diacetate, which needs acid external pHs (~6.2) to partially repress ionization of the carboxyl. Carboxyfluorescein still leaks too fast for convenient use in small mammalian cells at 37° (Rink *et al.,* 1982; Moolenaar *et al.,* 1982). For such experiments, BCECF is preferable because it adds two more carboxyls and has a pK_a more closely matched to cytosolic pH. The extra carboxyls were added on the ends of 2-carbon spacer arms in the hope that they could form six-membered cyclic lactones with the phenolic hydroxyls in place of the customary acetate ester groups. Unfortunately, the lactone seemed insensitive to cytoplasmic esterases in lymphocytes, so the polar groups had to be protected as acetoxymethyl (AM) esters. In early preparations of the AM ester, the number and placement of the AM groups were not well defined, though enough were put on to make the ester soluble in weakly polar organic solvents and presumably membrane permeant. More recently, NMR and mass spectra have been used (R. Haugland, personal communication) to determine the structure of the main form of BCECF/AM at least as prepared by Molecular Probes Inc. This analysis indicates the surprisingly asymmetrical structure shown in Fig. 1, part lactone, part open-chain ester.

BCECF/AM loading seems to work on a wide variety of tissues ranging from cyanobacteria (Huflejt *et al.,* 1988) to mammalian cells. One exception is vesicles prepared from renal brush border membranes, where extracellular esterase activity seems strong enough to prevent intracellular accumulation (Verkman and Ives, 1986). The BCECF formed by hydrolysis gives a diffuse fluorescence consistent with equilibration throughout the nucleus and cytosol extending out the very edges of the cell. A few studies have checked for compartmentation into organelles such as mitochondria or acidic organelles but found no sign (Paradiso *et al.,* 1986; Bright *et al.,* 1987). BCECF is usually calibrated *in situ* by subjecting the cells to known internal pHs using nigericin, a K^+/H^+ ionophore (Thomas *et al.,* 1979).

When external K^+ equals intracellular, nigericin (1–10 μM) clamps intracellular pH to extracellular pH, which the experimenter sets to various levels while observing the dye fluorescence. In some tissues (e.g., lymphocytes, gastric glands, kidney), evidence has been found for interaction of the dye with cell constituents, in that the dye spectra from cells are shifted a few nanometers to longer wavelengths and the pK_a seems slightly higher (0.1–0.3 unit) than observed for dye in simple calibration buffers (Rink *et al.*, 1982; Paradiso *et al.*, 1986; Alpern and Chambers, 1987). However, a careful study in fibroblasts revealed neither a pK_a change nor any restricted mobility detectable by fluorescence recovery after photobleaching (Bright *et al.*, 1987). Excessive illumination gradually bleaches the dye and often seems to acidify the cells (Alpern, 1985; Paradiso and Tsien, unpublished results on gastric glands). This effect is probably due to photodynamic damage to the cells causing a real fall in pH_i since, when pH_i is clamped with ionophores, bleaching of dye causes no change in apparent pH measured by the excitation ratio (Bright *et al.*, 1987). Obviously it is always advisable to limit the excitation intensity to the minimum necessary for a satisfactory signal-to-noise ratio and then block it whenever a measurement is not actually in progress. With such precautions, there is generally no difficulty in maintaining observations for several hours, especially if leakage is reduced by working at temperatures somewhat lower than 37° and if the cells are attached to a substrate and superfused rather than stirred in a cuvette. Ultimately, the dye does leak out over a period of hours.

DMCF is loaded into cells via its diacetate ester at acidic external pH just as CF is. In cells, the absorbance of DMCF is shifted to the red by 5 nm, and its pK_a is higher by 0.3 units than dye *in vitro* (Chaillet and Boron, 1985), so the perturbation by cytoplasm appears quite similar to that encountered by BCECF. Leakage of DMCF might be expected to be faster than CF or BCECF since DMCF should be more hydophobic because of its two extra methyl groups with no extra carboxylates; however, actual comparative data are not available.

2. Fluorescein–Dextran Conjugates

In order to retain a fluorescein derivative inside cells for many hours or days, it seems necessary to conjugate the dye to a macromolecule such as ovalbumin (Heiple and Taylor, 1982) or dextran. Both fluorescein and 4′,5′-dimethylfluorescein have been linked to amino- or hydrazinodextran via isothiocyanate coupling (Rothenberg *et al.*, 1983). Of course, the labeled polymers cannot be loaded by hydrolysis of esters but must be introduced by direct microinjection or reversible permeabilization. Two main versions of the latter have been used, scrape loading and osmotic shock of pinocytotic compartments. Both techniques are described elsewhere in these volumes (McNeil, 1988; Swanson, 1988) and require the cells to be adherent and actively pinocytosing, respectively. It appears that both methods can leave significant quantities of indicator in acidic compartments such as endosomes or lysosomes (Bright *et al.,* 1987; Rothenberg *et al.,* 1983). In the osmotic shock method, these compartments might be pinosomes that evaded lysis. When the dextran has a molecular weight of 70,000, the conjugate seems excluded from the nucleus and the periphery of the cell (Bright *et al.,* 1987).

3. Pyranine (8-Hydroxypyrene-1,3,6-trisulfonate)

Pyranine (Fig. 1) is a pH indicator that has found some application in resealed vesicles (Kano and Fendler, 1978; Clement and Gould, 1981; Lee, 1985) and in fibroblasts amenable to reversible permeabilization (Giuliano and Gillies, 1987). Its fluorophore is the condensed hydrocarbon pyrene, made hydrophilic by the addition of three sulfonate substituents. As pH rises, the removal of the phenolic proton shifts the excitation peak from 405 to 465 nm. Emission is maximal at 514 nm and mainly changes amplitude without a major wavelength shift in response to changing pH. Emission at 465 nm is somewhat less pH sensitive than at the 514 nm peak, so that the ratio of 465 to 514 nm emission gives some measure of pH. However, this ratio increases only about 1.6-fold from complete protonation to deprotonation, and the 465 nm signal is very weak, so that emission ratioing is much less practical than 465/405 nm excitation ratioing, in which two strong signals give a ratio that increases 55-fold upon deprotonation. The pK_a is reported to be 7.22–8.04, depending on ionic strength (Kano and Fendler, 1978; Wolfbeis *et al.,* 1983; Giuliano and Gillies, 1987) *in vitro,* compared with 7.82 (Giuliano and Gillies, 1987) *in situ* in fibroblasts. These values are a bit higher than ideal for sensing pHs near 7.0, but the excellent responsiveness of the excitation ratio may compensate.

The greatest drawback of pyranine is that no permeant ester form is available, so that the dye must be introduced by reversible permeabilization, yet then leaks out surprisingly rapidly, with a half-life of about 1 hour in fibroblasts at 37° (Giuliano and Gillies, 1987). Pyranine thus offers neither the convenience of BCECF nor the intracellular longevity of a dextran–dye conjugate.

4. 4-Methylumbelliferone (7-Hydroxy-4-methylcoumarin)

4-Methylumbelliferone (4-MU, Fig. 1) is best known as a fluorescent molecule whose release from nonfluorescent esters or ethers is the basis for several popular enzyme assays. It also is a pH indicator, whose protonated and deprotonated forms have excitation peaks at 320 and 360 nm, respectively (Nakashima *et al.,* 1972). The emission peaks around 450 nm; increasing pH increases the emission intensity with little shift in wavelength. Gerson (1982) claimed that the ratio of the 450 to 550 nm emissions was a good measure of pH, but the ratio actually changes less than 1.5-fold over the entire pH range. More recent work (Graber *et al.,* 1986; Musgrove *et al.,* 1986) suggests that 4-MU emission ratios are not useful for pH measurement, partly because the 550 nm signal is weak and noisy. Graber *et al.* (1986) advocate 4-MU excitation ratioing (365/334 nm) instead; certainly this ratio is much more pH sensitive, increasing 6 to 10-fold from pH 6 to 8. One significant disadvantage is that 334 nm is poorly transmitted by standard microscope optics. The pK_a of 4-MU is usually stated to be 7.8 (Nakashima *et al.,* 1972; Wolfbeis *et al.,* 1985; Koller and Wolfbeis, 1985) though Graber *et al.* (1986) reported 7.40. Derivatives with electron-withdrawing substituents at the 3- or 4-position instead of the electron-donating 4-methyl are known to have lower pK_as and longer wavelengths of excitation and emission (Wolfbeis *et al.,* 1985; Koller and Wolfbeis, 1985). These derivatives would seem to be superior to 4-MU at least with respect to *in vitro* properties, but no biological applications have yet been reported.

4-MU is readily loaded by hydrolysis of its uncharged, permeable, relatively nonfluorescent acetate ester. However, it also leaks out extremely rapidly, since most of the molecules are uncharged at pH 7. About the only way it can be used on most tissues is by continual incubation in the ester (Graber *et al.,* 1986).

5. Quene 1

Quene 1 (Rogers *et al.,* 1983b) is a quinoline dye (Fig. 1) related to the Ca^{2+} indicator quin-2 but with a *trans*-double bond which disrupts the Ca^{2+}-binding site of quin-2. This double bond also allows conjugation of the quinoline ring nitrogen with the amino nitrogen on the right-hand benzene ring. This raises the pK_a of the former, since protonation now gives a delocalized cation, a vinylogous amidinium structure. Quene 1 is therefore different from all the other pH indicators discussed in this review, in that the protonation occurs on nitrogen rather than a phenolic oxygen. The excitation maximum is at 390 nm, with emission peaking at 530 nm. Increasing pH from 5 to 9 increases the amplitudes of both by more than 30-fold, with a pK_a of 7.30. Neither the excitation nor emission seems to undergo a wavelength shift useful for generating a ratio. One unusual characteristic of quene 1 is that it is somewhat divalent cation sensitive; thus 0.8 mM Mg^{2+}, a reasonable intracellular value, quenches the fluorescence to the same extent as 0.04 pH unit acidification and 0.1 mM Mn^{2+} is sufficient to quench the dye completely, a useful maneuver for checking dye leakage and cell autofluorescence.

Quene 1 was loaded into cells by hydrolysis of its AM ester. Leakage was variable but could be very low in good preparations (<0.1%/min). Responses to weak acids and bases could be detected in single thymocytes (Rogers *et al.,* 1983a), but in later work on mitogenic stimulation (Hesketh *et al.,* 1985), the same group of workers seem to have switched to BCECF. Because quene 1 seems not to be commercially available, its application is likely to remain limited.

6. 1,4-Dihydroxyphthalonitrile (2,3-Dicyanohydroquinone)

In all the pH indicators discussed above, the emission spectrum hardly changes wavelength in response to pH, so that the ratio of emissions at two wavelengths is relatively insensitive to pH. Some applications, particularly flow cytometry or scanning confocal microscopy with laser excitation, are much more convenient with a probe that shifts its emission spectrum. Then the ratio of the intensities measured simultaneously at two chosen emission wavelengths can signal the ion concentration. This preserves all the usual advantages of ratioing but eliminates the need to alternate two excitations or multiplex them by frequency modulation (Kurtz, 1987). Currently the only commerically available indicator for intracellular pH with a big emission shift is 1,4-dihydroxyphthalonitrile, 1,4-DHPN (Valet *et al.,* 1981: Kurtz and Balaban, 1985; Musgrove *et al.,* 1986). When excited at 375–407 nm, 1,4-DHPN shifts from prominently blue (~450 nm) to more greenish (~480 nm) emission as pH rises from 6 to 10, with a pK_a of 8.0 (Brown and Porter, 1977). 1,4-DHPN is readily loaded into cells by hydrolysis of its acetate diester, 1,4-diacetoxyphthalonitrile (1,4-DAPN, Fig. 1). The 1,4-DHPN leaks out quite readily, having only 1–2 negative charges, which moreover are partly delocalized. Fortunately, the ester precursor is relatively nonfluorescent, so that in many systems the cells may be observed during continuous incubation in the ester to establish a steady state between loading and leakage (Valet *et al.,* 1981). Kurtz and Balaban (1985) observed only diffuse fluorescence in cultured epithelial cells, but Valet *et al.* (1981) reported that some 10% of the dye in liver cells was associated with organelles.

7. Seminaphthorhodafluorescein and Seminaphthofluoresceins, SNA(R)F

Very recently, Molecular Probes Inc. has announced (R. Haugland, personal communication) a promising series of emission-shifting pH indicators with naphthofluorescein chromophores (Fig. 1). SNARF-1 excited at 514 nm emits at 588 nm (acid) versus 637 nm (base) with a pK_a of 7.5; SNAF-2 in acid is excited at 490–520 nm and emits at 540 nm, whereas in base the excitation and emission peaks are 586 and 630 nm, the pK_a being 7.65–7.7. Biological testing and modification to lower the pK_as are awaited with interest.

III. Sodium Indicators

Sodium-sensitive indicators are under development in at least two laboratories. Smith *et al.* (1986) have reported a probe whose ^{19}F NMR spectrum is sensitive to Na^+, and which may evolve into practical fluorescent indicators. Minta *et al.* (1987) have independently synthesized a variety of Na^+ indicators, the current favorite being "SBFI." It consists of a crown ether of the right size to form an equatorial belt around a Na^+ ion, with additional either oxygens capping both poles. Potassium rejection arises from the size of the crown ether cavity; expansion of the macrocyclic ring has been verified to convert Na^+ selectivity to K^+ selectivity. Divalent cations are rejected because there are no negative charges lining the cavity. The attached fluorophores are benzofurans rather similar to those in the Ca^{2+} indicator fura-2 (see below), so that the SBFI wavelengths and shift due to Na^+ binding are similar to fura-2 and its Ca^{2+} response. SBFI has two identical fluorophores mainly because the organic synthesis was eased by preserving the symmetry around the crown ether ring, though as a side benefit the extinction coefficient is doubled. In the presence of typical vertebrate intracellular K^+ levels, the effective dissociation constant for Na^+ is 17–18 m*M*, well suited to monitor $[Na^+]_i$ changes around the typical resting level of 10–20 m*M*. For example, SBFI has detected an increase of $[Na^+]_i$ by a few millimolar (A. T. Harootunian and R. Y. Tsien, unpublished results) in single fibroblasts stimulated with mitogens to activate

Na^+/H^+ exchange. As usual for polycarboxylate dyes, SBFI can be introduced into cells either by microinjection or by hydrolysis of its membrane-permeant AM ester. Calibration is most conveniently performed in intact cells with the pore-forming antibiotic gramicidin, which rapidly clamps $[Na^+]_i$ and $[K^+]_i$ equal to the extracellular levels of those ions.

IV. Chloride Indicators

Chloride ion fluxes are important in several types of inhibitory synapses and in pH regulation by Cl^-/HCO_3^- and related countertransport systems. Recently, Illsley and Verkman (1987) have shown that 6-methoxy-*N*-(3-sulfopropyl) quinolinium (SPQ, Fig. 2) can be used as a Cl^- indicator in vesicles and erythrocyte ghosts. SPQ differs in principle from all the other ion indicators, because in its ground state it does not associate with its target ion. Chloride interacts only with the excited state of SPQ, causing radiationless quenching of the dye fluorescence with no change in the

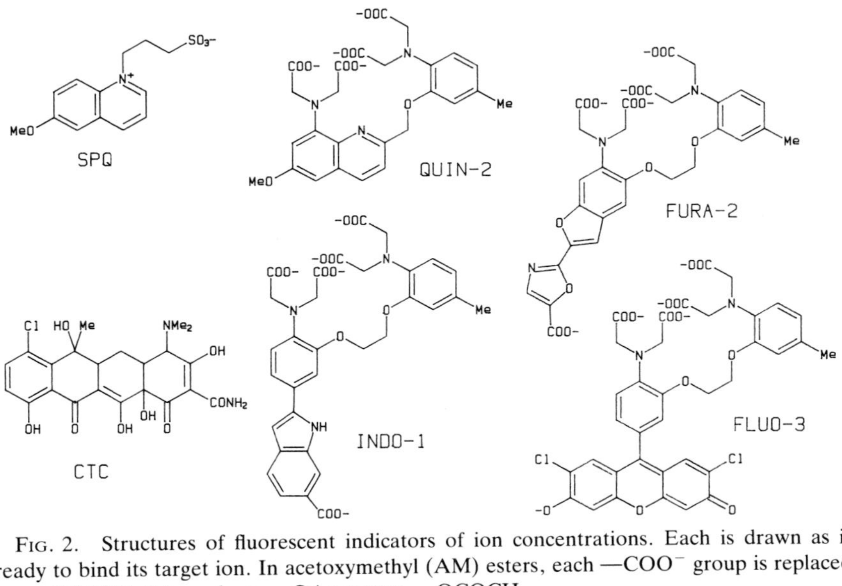

SPQ QUIN-2 FURA-2 CTC INDO-1 FLUO-3

Fig. 2. Structures of fluorescent indicators of ion concentrations. Each is drawn as if ready to bind its target ion. In acetoxymethyl (AM) esters, each —COO^- group is replaced by —$COOCH_2OAc$, where —OAc means —$OCOCH_3$.

absorbance spectrum. The mechanism for the quenching and the basis for halide selectivity are not really understood, though the effect has long been known in analogous heterocyclic cations such as diprotonated quinine. The Stern–Volmer equation for chloride-dependent quenching is mathematically equivalent to formation of a nonfluorescent Cl-complex with a dissociation constant of 8.5 mM, a value that should give good sensitivity to physiological alterations of $[Cl^-]_i$. Bromide, iodide, and thiocyanate are even better quenchers than chloride, but are usually absent from cells, whereas normally endogenous anions seem not to quench SPQ significantly. Since Cl^- simply quenches SPQ, there is no wavelength shift to enable ratioing of amplitudes at two wavelengths. Illsley and Verkman (1987) did not calibrate their biological records in terms of actual Cl^- concentrations, but one may speculate that measurement of the excited state lifetime could enable such calibration. The excited state lifetime τ is proportional to $(1 + [Cl^-]/8.5\ \text{m}M)^{-1}$ and is independent of dye concentration, path length, lamp intensity, and detector sensitivity, like wavelength ratios but unlike simple intensity at one wavelength.

SPQ is surprisingly permeable through membrances, considering that it is a zwitterion with a quaternary nitrogen cation and sulfonate anion, either of which alone is normally sufficient to prevent ready permeation. Perhaps the positive and negative charge nullify each other by ion-pairing. SPQ is loaded into cells and vesicles simply by soaking them in high concentrations of the dye; of course the dye also readily leaks out once the external excess is removed. This is a major current deficiency of SPQ, which may be fixable by additional carboxylate groups protected as AM esters.

V. Calcium Indicators

More work has been done with indicators for Ca^{2+} than for any other ion. This emphasis reflects the pivotal importance of Ca^{2+} in cellular signal transduction. Two basic classes of indicator exist for Ca^{2+} as they do for H^+. The first class is analogous to the amines that accumulate in acid compartments. The main example of this class is chlortetracycline, which accumulates in organelles containing high concentrations of Ca^{2+} and fluoresces preferentially when its Ca^{2+} complex binds to a hydrophobic site such as a membrane. The second class consists of molecules that reside in an aqueous compartment such as the cytosol and change their spectra when they bind Ca^{2+}. These molecules are hydrophilic and do not need to bind to membranes or follow Ca^{2+} across membranes.

A. Chlortetracycline

Chlortetracycline (CTC) (Fig. 2) is a well-known broad-spectrum antibiotic (Aureomycin) and is also used as a histological stain for bone tissue. Caswell (1972) showed that it gave interesting fluorescence signals (excitation at 380 nm, emission at 520 nm) associated with Ca^{2+} uptake into mitochondria. Many workers then applied CTC to a wide variety of organelles and intact cells (reviewed by Caswell, 1979), often with the misconception that they were measuring "membrane-bound Ca^{2+}." It now has become clear (Blinks *et al.*, 1982; Dixon *et al.*, 1984) that CTC does not measure Ca^{2+} actually bound to membranes, but that CTC accumulates and fluoresces in compartments in which a high concentration of free Ca^{2+} exists next to hydrophobic sites, particularly membranes. The membrane itself does not need any affinity for Ca^{2+}, only for the Ca^{2+}–CTC complex. Neutral uncomplexed CTC crosses membranes reasonably quickly (tens of seconds); it ionizes to an anion with a pK_a of 6.8 (Millman *et al.*, 1980) or 7.4 (Stephens *et al.*, 1954), then binds Ca^{2+} with a dissociation constant of 0.44 m*M* (Caswell and Hutchison, 1971a) to 2.6 m*M* (Millman *et al.*, 1980). Neither the anion nor the Ca^{2+} complex appears to cross membranes as such, but the latter binds to membranes and becomes more fluorescent as a result. The extent of the binding to membranes depends on the surface-to-volume ratio of the vesicle and on the properties of the lipid. Though the outer leaflet of the plasma membrane is also exposed to high Ca^{2+}, little of the CTC signal from intact cells arises from the plasma membrane because the internal organelles have a much greater surface-to-volume ratio than the plasma membrane. Also, CTC is normally removed from the extracellular medium during the observation period, so that CTC on the outer leaflet of the plasma membrane washes out (Blinks *et al.*, 1982), whereas the CTC inside the organelles bleeds out much more slowly. In carefully defined and reproducible suspensions of isolated vesicles, it is sometimes possible to calibrate CTC signals in terms of the free Ca^{2+} inside the vesicles (Dixon *et al.*, 1984). Such calibration is impossible in more complex systems such as intact cells, in which the fluorescence signals are at best qualitative hints as to the extent of Ca^{2+} sequestration into organelles. CTC itself actually binds Mg^{2+} more strongly than Ca^{2+} in aqueous media (Caswell and Hutchison, 1971a), though there is some difference in spectra between the two complexes and an increasing Ca^{2+} preference as organic solvents are added (Caswell and Hutchison, 1971b); instead it is organellar transport that is Ca^{2+} selective. At high (>100 μM) doses CTC can inhibit cell function; this may be because the CTC locks up Ca^{2+} in the stores (Caswell, 1979; Elferink and Deierkauf, 1984), or because photooxidation products or other contaminants in commercial

CTC samples are toxic (Caswell, 1979; Blinks *et al.*, 1982), or because CTC poisons mitochondria (Pershadsingh *et al.*, 1982).

B. Indicators of Cytosolic Free Ca^{2+} Concentrations ($[Ca^{2+}]_i$)

Earlier techniques for measuring cytosolic free Ca^{2+} (Blinks *et al.*, 1982; Tsien and Rink, 1983), such as the luminescent photoprotein aequorin, the absorbance dye arsenazo III, and Ca^{2+}–sensitive microelectrodes, all required microinjection or impalements, and were therefore applied mainly to giant cells. More recently, photoproteins have been loaded by various reversible permeabilization procedures (Cobbold and Rink, 1987), but the largest expansion in the range of cell types in which Ca^{2+} signals can be quantified has come from the development of new fluorescent indicators that can be loaded using hydrolyzable esters. The new dyes are not without their own difficulties and restrictions, but their generic structure is amenable to further optimization along reasonably rational chemical principles. Currently four fluorescent indicators are in use: quin-2, fura-2, indo-1, and fluo-3. Their structures (Fig. 2) share nearly identical binding sites, which are modeled (Tsien, 1980) on the well-known Ca^{2+}–selective chelator ethylene glycol bis(β-aminoethyl ether)*N*, *N*′-tetraacetic acid (EGTA). This octacoordinate binding site binds Mg^{2+} about five orders of magnitude more weakly than Ca^{2+} because Mg^{2+} is too small to contact more than about half the liganding groups simultaneously. Monovalent cations do not form detectable specific complexes, probably because their charge is inadequate to organize the binding pocket in the face of the electrostatic repulsion of the negative carboxylates. EGTA at pH 7 is normally occupied by two protons, but the incorporation of the aromatic rings in the fluorescent indicators lowers the pK_a of the amine nitrogens to 6.5 or below, thus eliminating almost all the proton interference for pH >6.8. Ca^{2+} binding diverts the nitrogen lone pair electrons away from the aromatic system, causing large spectral changes that mimic disconnection of the nitrogen substituents. Conversely, the more electron-donating or withdrawing the aromatic nucleus, the higher or lower the Ca^{2+} affinity (Tsien, 1980). This principle is exemplified in photochemically reactive chelators, in which photolysis increases or decreases the Ca^{2+} affinity by destroying or creating an electron-withdrawing ketone group *para* to the amine nitrogen (Tsien and Zucker, 1986; Gurney *et al.*, 1987; Adams *et al.*, 1988). Because a wide variety of substituents can be plugged in without changing the geometry of the binding site, the design of this family of tetracarboxylate Ca^{2+} indicators and chelators is quite versatile, rather like a household appliance that accepts a range of attachments for different jobs. Of course the actual organic syntheses are a little more difficult than just fitting pieces together.

1. Quin-2

Quin-2 (reviewed in Rink and Pozzan, 1985; Tsien and Pozzan, 1988) was the first practical fluorescent indicator of this family, with a simple 6-methoxyquinoline as its fluorophore. The small size of this group means that quin-2 is best excited at fairly short wavelengths (339 nm) and has only a modest extinction coefficient. The brightness of quin-2 fluorescence is not very great, so that relatively high intracellular concentrations, millimolar to tenths of millimolar, are needed to overcome cellular autofluorescence. These levels often buffer fast Ca^{2+} transients. Quin-2 binds Ca^{2+} with a dissociation constant of 60 nM or 115 nM, respectively, in the absence or presence of 1 mM Mg^{2+}, both measured under conditions intended to mimic mammalian cytoplasm, pH 7.05, 37°, ~140 mM ionic strength (Tsien *et al.*, 1982). The strong binding (low K_d) means that quin-2 is best at measuring submicromolar $[Ca^{2+}]_i$ and saturates near 1–2 μM. The effect of Mg^{2+} corresponds to a Mg^{2+} dissociation constant on the order of millimolar; quin-2 has poorer $Ca^{2+}:Mg^{2+}$ discrimination (only $10^4:1$) than its siblings due to its use of a quinoline nitrogen in place of one ether oxygen. At 339 nm excitation, Ca^{2+} binding increases the fluorescence intensity about sixfold whereas Mg^{2+} has no effect. At longer excitation wavelengths the fluorescence intensity drops off sharply and the proportional effect of Ca^{2+} decreases while the effect of Mg^{2+} increases. Therefore quin-2 does not show a useful Ca^{2+}-induced wavelength shift in either excitation or emission spectrum with which to generate a ratio signal (Tsien and Pozzan, 1988).

The short excitation wavelengths, modest fluorescence brightness, inadequacy of ratioing, and poor photostability of quin-2 make it unsuitable for single cell microscopy. Instead, quin-2 has been used mainly in suspensions of cells in a cuvette, though occasionally in cell monolayers attached to cover slips inserted diagonally in a cuvette. Its signal is calibrated by lysis of the cells and direct titration of the lysate to known levels of Ca^{2+}, or by using ionomycin first to raise $[Ca^{2+}]_i$ to saturating levels and then introducing Mn^{2+} into the cells to quench the dye and determine the autofluorescence level. Quin-2 does have advantages for some purposes over its more recent relatives. In particular, hydrolysis of quin-2/AM seems easier, reaches higher cytosolic concentrations of chelator, and is less often complicated by compartmentation into organellar compartments than is observed with AM esters of higher molecular weight and lesser water solubility. Loading with excess indicator (typically several millimolar intracellular concentration) in order to buffer cytosolic $[Ca^{2+}]_i$ is a powerful experimental tool, and quin-2 does it better than any of the other indicators. At a qualitative level, such buffering reveals which cellular responses are truly dependent on elevated $[Ca^{2+}]_i$. At a quantitative level, the dependence of $[Ca^{2+}]_i$ on amount of buffering can reveal the size of the net Ca^{2+} flux into the cytosolic compartment and the amount of the endogenous cellular Ca^{2+} buffering (Tsien and Rink, 1983).

2. Fura-2

Fura-2 (Fig. 2) is currently the most popular Ca^{2+} indicator for microscopy of individual cells. Compared to quin-2, the larger fluorophore of fura-2 gives it slightly longer wavelengths of excitation compatible with glass microscope optics, a much larger extinction coefficient, and a higher quantum efficiency, resulting in about 30-fold higher brightness per molecule. Ca^{2+} binding shifts the excitation spectrum about 30 nm to shorter wavelengths, so that the ratio of intensities obtained from 340/380 nm or 350/385 nm excitation pairs is a good measure of $[Ca^{2+}]_i$ unperturbed by variable cell thickness or dye content (Grynkiewicz *et al.*, 1985). The green emission from fura-2 peaks at 505–520 nm and does not shift usefully with Ca^{2+} binding. Fura-2 is also very much more resistant to photodestruction than quin-2. Though fura-2 can be degraded eventually (Becker and Fay, 1987), most investigators have found that by attenuating the excitation beam, blocking it whenever measurements are not actually in progress, and using efficient photodetectors, adequate signals can be obtained from single cells for tens of minutes to hours of observations.

Fura-2 binds Ca^{2+} slightly less strongly than quin-2 does. Dissociation constants of 135 n*M* (no Mg^{2+}, 20°, in 100 m*M* KCl), 224 n*M* (1 m*M* Mg^{2+}, 37°, 120 m*M* K^+, 20 m*M* Na^+) (Grynkiewicz *et al.*, 1985), and 774 n*M* (no Mg^{2+}, 18°, 225 m*M* K^+, 25 m*M* Na^+) (Poenie *et al.*, 1985) have been reported, implying that ionic strength not Mg^{2+} seems to be the most powerful influence on the apparent K_ds. The Mg^{2+} dissociation constants of 6–10 m*M* at 37° to 20° represent much better $Ca^{2+}:Mg^{2+}$ discrimination than that of quin-2. The absolute calibration of fura-2 inside cells is complicated by the fact that dye does seem to have somewhat different spectral characteristics in most cells than in calibration buffers. In cytoplasm the 380–385 nm excitation amplitude is increased 1.1- to 1.6-fold relative to that at 340–350 nm, shifting the ratios downward (Almers and Neher, 1985; Tsien *et al.*, 1985). This red-shift can be simulated by increasing the viscosity of the calibration medium with agents such as gelatin or sucrose; the amount of viscosity correction to be made can be estimated by fluorescence polarization measurements, or by the ratio of the intensity changes at the two excitation wavelengths when $[Ca^{2+}]_i$ changes but dye content does not (Poenie, 1989). The typical net effect of viscosity is to reduce the 350/385 nm ratio by about 15% (Poenie *et al.*, 1986). Ideally, one would calibrate the dye *in situ* by clamping the cell to known $[Ca^{2+}]_i$ values with an ionophore such as ionomycin (as in Williams *et al.*, 1985; or Chused *et al.*, 1987), but such ionophores do not mediate large fluxes of Ca^{2+} when the concentrations are only micromolar or less on both sides of the membrane, so such calibrations are more difficult than pH or Na^+ calibrations with nigericin or gramicidin, and are often impossible.

The kinetics of Ca^{2+} binding to fura-2 have been characterized *in vitro* by stopped-flow (Jackson *et al.*, 1987) and temperature jump (Kao and Tsien, 1988), giving an association rate constant k_a of $6 \times 10^8\ M^{-1}\ \text{sec}^{-1}$ and a dissociation rate k_d of 84–97 sec^{-1} at 20° in 0.1 M KCl. The exponential time constant τ for the response of an indicator of 1 : 1 stoichiometry is given by $\tau = (k_a[Ca^{2+}] + k_d)^{-1}$. Here too there is evidence for perturbation by cytoplasm, in that fura-2 seems to behave in skeletal muscle as if both rate constants were 4- to 8-fold lower than the *in vitro* values (Hollingworth and Baylor, 1987; Klein *et al.*, 1988).

Perhaps the worst problem with fura-2 is that is some tissues AM ester hydrolysis is incomplete (Highsmith *et al.*, 1986; Scanlon *et al.*, 1987), or the fluorescence becomes compartmentalized into organelles (Almers and Neher, 1985; Malgaroli *et al.*, 1987; Steinberg *et al.*, 1987) or extruded from the cell by anion transport mechanisms (DiVirgilio *et al.*, 1988). Many procedures have been empirically developed which can ameliorate these problems, but success on every tissue is not guaranteed. Loading is often much improved by mixing the fura-2/AM with amphiphilic dispersing agents before diluting into the incubation medium. Suitable agents include albumin, serum, and Pluronic F-127 (Poenie *et al.*, 1986; Barcenas-Ruiz and Wier, 1987). Endocytosis and compartmentalization can often be significantly slowed by a reduction in temperature, e.g., from 37° to 32° during observation (Poenie *et al.*, 1986) or to 15° just during loading (Malgaroli *et al.*, 1987). Anion extrusion can be inhibited by probenecid (DiVirgilio *et al.*, 1988) and sulfinpyrazone, which are well known clinically as blockers of uric acid transport. In guinea pig smooth muscle, fura-2 loading was improved by inhibition of serine proteases or extracellular cholinesterase (Maruyama *et al.*, 1987). Often introduction of the dye pentaanion by microinjection (e.g., Cannell *et al.*, 1987) or by perfusion with a patch pipet (Almers and Neher, 1985) or reversible permeabilization (e.g., Ratan *et al.*, 1986) gives better-behaved dye. Though these techniques are less convenient than AM ester hydrolysis, they are no worse with fura-2 than is always necessary with traditional Ca^{2+} indicators. There are some cells, e.g., sea urchin embryos (M. Poenie, J. Alderton, R. Steinhardt, and R. Y. Tsien, unpublished observations) and plant stamen hair cells (Hepler and Callaham, 1987), which gradually compartmentalize even injected dye; for such cells, molecular redesign or attachment of fura-2 to a macromolecule such as dextran (R. Haugland, personal communication) may be necessary.

Despite the above litany of cautions, fura-2 has provided much useful new information about the role and regulation of $[Ca^{2+}]_i$, especially when used with microspectrofluorometry (Tsien *et al.*, 1985; Poenie *et al.*, 1985) and digital image processing (Tsien and Poenie, 1986). $[Ca^{2+}]_i$ can be measured with good signal-to-noise ratio even in single cells as small as platelets with volumes of ~10 fl (Hallam *et al.*, 1986) and neuronal processes as thin as 1 μm (Thayer *et al.*, 1987). Signal rise times as fast as 7–10 milliseconds (measured from 10 to 90% of final amplitude) have been observed in optic nerve; dye bleaching and photodynamic damage were insignificant (Lev-Ram and Grinvald, 1987). In single pituitary cells, the $[Ca^{2+}]_i$ rises due to individual action potentials were clearly resolvable (Schlegel *et al.*, 1987).

Here is an arbitrary selection of recent biological findings with fura-2.

a. Mechanisms of $[Ca^{2+}]_i$ *Elevation.* Many workers have used fura-2 to demonstrate $[Ca^{2+}]_i$ rises associated with agonists that cause breakdown of inositol phospolipids, formation of inositol-1,4,5-trisphosphate, and release of Ca^{2+} stores (Berridge, 1987). The $[Ca^{2+}]_i$ reponses are often highly heterogeneous from cell to cell or even oscillatory (Tsien and Poenie, 1986; Wilson *et al.*, 1987; Ambler *et al.*, 1988). One particularly interesting example of receptor-mediated Ca^{2+} influx is in platelets, where stopped-flow kinetics (Sage and Rink, 1986) show that adenosine diphosphate, which causes little or no breakdown of inositol phospholipids, stimulates Ca^{2+} entry more rapidly than does thrombin, which does act through such lipid metabolism. Therefore adenosine diphosphate may directly open a receptor-operated Ca^{2+} channel. Evidence for such channels is even clearer in neurons that repond to transmitters such as glutamate (Kudo and Ogura, 1986; Connor *et al.*, 1987) or its analog *N*-methyl-D-aspartate (Murphy *et al.*, 1987). Even the inhibitory transmitter γ-aminobutyrate (GABA) can induce long-lasting moderate elevations of $[Ca^{2+}]_i$ (Connor *et al.*, 1987). Voltage-dependent channels and their pharmacology are also readily studied with fura-2 (e.g., Hirning *et al.*, 1988; Thayer *et al.*, 1987; Williams *et al.*, 1987; Wanke *et al.*, 1987; Lipscombe *et al.*, 1988a,b). Williams *et al.* (1987) saw that depolarization of smooth muscle cells caused self-limiting rises in $[Ca^{2+}]_i$, as if the latter exerted negative feedback on Ca^{2+} channels. Positive feedback in the form of Ca^{2+}-induced release of Ca^{2+} stores is suggested in other tissues such as heart (Barcenas-Ruiz and Wier, 1987) and sympathetic neurons (Lipscombe *et al.*, 1988a). For example, blockage of Ca^{2+} release from stores by ryanodine or prior depletion using caffeine greatly attenuates the $[Ca^{2+}]_i$ response to depolarization but not the plasma membrane Ca^{2+} currents. Depolarization and caffeine applied together trigger oscillations (Lipscombe *et al.* 1988a,b). However, Ca^{2+}-induced Ca^{2+} release (CICR) appears unlikely in skeletal muscle, since the more fura-2 is introduced to increase buffering and reduce the rise in $[Ca^{2+}]_i$, the more total Ca^{2+} is released from the sarcoplasmic reticulum during stimulation (Hollingworth and Baylor, 1987). Quite a different sort of CICR from the above has been revealed in parathyroid cells. Here the ability of fura-2 to give good signals with minimal buffering enabled Nemeth and Scarpa (1987) to demonstrate a remarkable release of Ca^{2+} stores in response to *extracellular* Ca^{2+} or other divalent cations.

b. Spatial Heterogeneity of $[Ca^{2+}]_i$. Fura-2 imaging is a powerful tool to study how $[Ca^{2+}]_i$ regulation varies from one part of a cell to another. Williams *et al.* (1985, 1987) found evidence that nuclear free $[Ca^{2+}]$ in smooth muscle cells might be different from cytosolic $[Ca^{2+}]$. Wier *et al.* (1987) showed rhythmic propagating waves of high $[Ca^{2+}]_i$ in a subpopulation of cardiac myocytes. Smith *et al.* (1987) observed that tetanization of the squid giant presynaptic terminal caused a wave of Ca^{2+} to spread from the side facing the synaptic cleft, as if the Ca^{2+} channels were co-localized with transmitter release. Connor (1986) and Cohan *et al.* (1987), working in cultured rat diencephalon and snail neurons, found that $[Ca^{2+}]_i$ in actively extending growth cones was moderately elevated compared to the levels in the soma or in stalled neurites. Excessive $[Ca^{2+}]_i$ resulting from action potentials or serotonin application also correlated with cessation of growth, suggesting that growth cone motility and extension require intermediate $[Ca^{2+}]_i$ values.

c. Difficult Cell Types. Fura-2 has also enabled Ca^{2+} measurement in preparations not amenable to previous methodologies. For example, Lev-Ram and Grinvald (1987) have shown that myelinated axons in rat optic nerve undergo fast-rising $[Ca^{2+}]_i$ transients during action potential conduction. This finding provides more direct evidence for voltage-sensitive Ca^{2+} channels in myelinated axons, and suggests that Ca^{2+} indicators may be useful for real-time imaging of CNS activity. Ratto *et al.* (1988) have loaded fura-2 into intact frog retina, which because of its light sensitivity might seem to be a most unsuitable preparation for experiments with a UV-excited, green-fluorescing dye. But by spreading the excitation and recording over the entire photoreceptor layer, and relying on rhodopsin to shield the deeper layers, rod $[Ca^{2+}]_i$ could be seen to drop from 220 to 140 nM within 1 to 2 seconds of turning on nonbleaching illumination. This is the first measurement of $[Ca^{2+}]_i$ in photoreceptors not poisoned by inhibitors of cyclic GMP phosphodiesterase, and provides further direct evidence against a rise in $[Ca^{2+}]_i$ as a step in vertebrate phototransduction.

d. Ca^{2+} and Secretion. Fura-2 measurements have been instrumental in recent modifications of the classical dogma that elevations in $[Ca^{2+}]_i$ are necessary and sufficient for regulated secretion. Schwartz (1987) has shown that depolarization can release GABA from fish retina horizontal cells even when a $[Ca^{2+}]_i$ rise is prevented. The proposed mechanism is voltage-dependent reversal of a Na^+/GABA cotransporter, not exocytosis of preformed vesicles. But even such exocytosis, at least in nonneuronal cells, may be much less Ca^{2+} dependent than previously thought. Neher (1987) observed that $[Ca^{2+}]_i$ spikes in perfused mast cells were neither necessary nor sufficient for exocytosis, which was precisely measured by capacitance changes. Instead, Ca^{2+} seemed at most to reinforce the effectiveness of more powerful stimulants such as GTP-γ-S. Poenie *et al.* (1987) monitored secretion of toxin granules from individual cytolytic T lymphocytes by observing the toxin's "postsynaptic" effect on closely apposed target cells. Though $[Ca^{2+}]_i$ transients did occur in the "presynaptic" T cell, they often peaked well before exocytosis and on the side of the cell remote from the target cell. These $[Ca^{2+}]_i$ images, together with experiments on phorbol esters and antibodies against protein kinase C (Poenie *et al.*, 1989), suggest that the kinase probably plays a much more important role than $[Ca^{2+}]_i$ in directing exocytosis in this system.

3. Indo-1

Indo-1 has a rigidified stilbene fluorophore like fura-2 but has the unique property that its emission and not just its excitation spectrum shifts to shorter wavelengths when the molecule binds Ca^{2+} (Grynkiewicz *et al.*, 1985; Tsien, 1986). Excitation can therefore be at a single wavelength, typically somewhere between 351 and 365 nm depending on whether an argon or krypton ion laser or a mercury lamp is used. The ratio of emissions peaking at 405 nm to that at 485 nm then measures the Ca^{2+} with the usual advantages of ratioing. The two wavelengths can, in principle, be separated by a dichroic mirror and measured simultaneously without any chopping. This methodology works well when the measurement is from one spatial location at any given instant, as in flow cytometry (Chused *et al.*, 1986, 1987; Rabinovitch *et al.*, 1986), non imaging photometry, or laser-scanning or specimen-scanning confocal microscopy, since photomultipliers can read the two emission bands without worries about spatial registration. The problem comes when using imaging devices, where it would be quite challenging to establish the necessary registration of corresponding pixels over the entire active regions of two separate low-light level television cameras. Although adaptive algorithms are available to interpolate one image to put its pixels in register with those of another image (Walter and Berns, 1986), the time required for the computation would negate the main advantage of emission ratioing over excitation ratioing, namely speed. The registration problem could be eased by alternating two filters in front of a single television camera, but again one might as well chop the excitation. Excitation chopping is easier because there is no need to maintain image quality, light transmission efficiency is noncritical, and the chopping apparatus can be located outside the microscopy, where its bulk and possible vibrations can be isolated.

Two other drawbacks of indo-1 compared to fura-2 should be mentioned. The blue and violet wavelengths of indo-1 emission overlap cellular autofluorescence from pyridine nucleotides (Aubin, 1979) more severely than the green of fura-2. Also indo-1 bleaches several-fold faster than fura-2 (S. R. Adams and R. Y. Tsien, unpublished observations). On the other hand, complaints about loading and compartmentation seem to crop up more frequently with fura-2 than with indo-1. Some of this difference may be that fura-2 has been tried by more people, on a much greater variety of cells, and examined more critically by microscopic imaging than indo-1, but some is probably real. For example, Bush and Jones (1988) have reported that plant cells (barley aleurones) can be loaded with indo-1 using acidic extracellular pH, whereas fura-2 concentrates in the vacuole. In endothelial cells, indo-1 is less prone to mitochondrial compartmentation than fura-2 is (Steinberg *et al.*, 1987). Also, Lee *et al.* (1987) have recorded indo-1 $[Ca^{2+}]_i$ signals from intact (not dissociated) beating mammalian heart, an organ whose pigmentation and pulsation are surely more severe than any other preparation.

4. Fluo-3

Fluo-3, the newest of our Ca^{2+} indicators (Minta *et al.*, 1987), has three main advantages over its predecessors: excitation at visible wavelengths rather than near-UV; a very large enhancement in fluorescence intensity, about 40 fold, upon binding Ca^{2+}; and a significantly weaker Ca^{2+} affinity, K_d ~450 nM, permitting measurement to 5–10 μM $[Ca^{2+}]_i$. These properties are particularly valuable when monitoring the effectiveness of photochemically reactive chelators, caged nucleotides (Gurney and Lester, 1987), or caged inositol phosphates (Walker *et al.*, 1987) at raising or lowering $[Ca^{2+}]_i$, since those compounds are photolyzed by the same near-UV irradiation (330–380 nm) that would be used to excite quin-2, fura-2, or indo-1. The visible excitation wavelengths of fluo-3, peaking at 503–506 nm, mean that there is no interference whatsoever between the actinic and monitoring wavelengths. Those wavelengths are close to the 488 nm output of argon lasers, so that fluo-3 is presently the only Ca^{2+} indicator usable with flow cytometers and laser confocal microscopes that lack UV capability. Another promising domain for fluo-3 will probably be investigations of the interactions of $[Na^+]_i$ and $[Ca^{2+}]_i$, since it should be usable simultaneously with UV-excited SBFI. However, binding of Ca^{2+} to fluo-3 causes negligible wavelength shifts in either excitation or emission spectra, so that fluo-3, like quin-2, is limited to intensity changes without wavelength pairs to ratio. Though changes in $[Ca^{2+}]_i$ are readily observed, absolute calibration requires treatment of the cells with ionophores, heavy metals, and/or detergent at the end of every experiment. For these reasons, fluo-3 is unlikely to displace ratio indicators like fura-2 and indo-1 from most single-cell imaging applications. In general, it is unlikely that there will ever be a single best Ca^{2+} indicator for all applications; every structure is a different compromise between many partially incompatible goals.

VI. Prospects and Conclusions

Due to limitations of space and expertise, this review cannot describe the thousands of biological experiments to which fluorescent indicators have been applied. They have become the dominant methodology for measuring free ion concentrations, thanks to their sensitivity, spatial and temporal resolution, nondestructive parallel readout, applicability to wide varieties of tissues, and versatile molecular specificity. But despite these strong points, currently available fluorescent probes have enough deficiencies so that future progress still requires better indicators. The design and production of such molecules exemplify yet another area in which improved molecular ingenuity and insight would strongly benefit cell biology.

Note Added in Proof. Smith *et al.* (1988) have described the synthesis of a promising new fluorescent Na^+ indicator. The Cl^- indicator SPQ has now been applied and calibrated in kidney cells [Krapf *et al.* (1988) and references therein]. Progress towards Mg^{2+} indicators is reported by Levy *et al.* (1988).

Acknowledgments

I thank F. DiVirgilio, A. Grinvald, R. Haugland, I. Kurtz, T. Pozzan, T. Rink, and A. Verkman for sending me preprints. The work in my laboratory has been supported by grants from NIH (GM31004 and EY04372), the Searle Scholars Program, and the Cancer Research Coordinating Committee of the University of California.

References

Adams, S. R., Kao, J. P. Y., Grynkiewicz, G., Minta, A., and Tsien, R. Y. (1988). *J. Am. Chem. Soc.* **110,** 3212–3220.

Agard, D. A. (1984). *Annu. Rev. Biophys. Bioeng.* **13,** 191–219.

Almers, W., and Neher, E. (1985). *FEBS Lett.* **192,** 13–18.

Alpern, R. J. (1985). *J. Gen. Physiol.* **86,** 613–636.

Alpern, R. J., and Chambers, M. (1987). *J. Gen. Physiol.* **89,** 581–598.

Ambler, S. K., Poenie, M., Tsien, R. Y., and Taylor, P. (1988). *J. Biol. Chem.* **236,** 1952–1959.

Aubin, J. E. (1979). *J. Histochem. Cytochem.* **27,** 36–43.

Barcenas-Ruiz, L., and Wier, G. W. (1987). *Circ. Res.* **61,** 148–154.

Becker, P. L., and Fay, F. S. (1987). *Am. J. Physiol.* **253,** C613–C618.

Berridge, M. J. (1987). *Annu. Rev. Biochem.* **56,** 156–193.

Blinks, J. R., Wier, W. G., Hess, P., and Prendergast, F. G. (1982). *Prog. Biophys. Mol. Biol.* **40,** 1–114.

Bright, G. R., Fisher, G. W., Rogowska, J., and Taylor, D. L. (1987). *J. Cell. Biol.* **104,** 1019–1033.

Brown, R. G., and Porter, G. (1977). *J. Chem. Soc. Faraday Trans. I* **73,** 1281–1285.

Busa, W. B. (1986). *Annu. Rev. Physiol.* **48,** 389–402.

Bush, D. S., and Jones, R. L. (1987). *Cell Calcium* **8,** 455–472.

Cannell, M. B., Berlin, J. R., and Lederer, W. J. (1987). *Science* **238,** 1419–1423.

Carafoli, E. (1987). *Annu. Rev. Biochem.* **56,** 395–433.

Caswell, A. H. (1972). *J. Membr. Biol.* **7,** 345–364.

Caswell, A. H. (1979). *Int. Rev. Cytol.* **56,** 145–181.

Caswell, A. H., and Hutchison, J. D. (1971a). *Biochem. Biophys. Res. Commun.* **42,** 43–49.

Caswell, A. H., and Hutchison, J. D. (1971b). *Biochem. Biophys. Res. Commun.* **43,** 625–630.

Chaillet, J. R., and Boron, W. F. (1985). *J. Gen. Physiol.* **86,** 765–794.

Chused, T. M., Wilson, H. A., Seligmann, B. E., and Tsien, R. Y. (1986). *In* "Applications of Fluorescence in the Biomedical Sciences" (D. L. Taylor, A. S. Waggoner, R. F. Murphy, F. Lanni, and R. R. Birge, eds.), pp. 531–544. Liss, New York.

Chused, T. M., Wilson, H. A., Greenblatt, D., Ishida, Y., Edison, L. J., Tsien, R. Y., and Finkelman, F. D. (1987). *Cytometry* **8,** 396–404.

Clement, N. R., and Gould, J. M. (1981). *Biochemistry* **20,** 1534–1538.

Cobbold, P. H., and Rink, T. J. (1987). *Biochem. J.* **248,** 313–328.

Cohan, C. S., Connor, J. A., and Kater, S. B. (1987). *J. Neurosci.* **7,** 3588–3599.

Connor, J. A. (1986). *Proc. Natl. Acad. Sci. U.S.A.* **83,** 6179–83.

Connor, J. A., Tseng, H. -Y., and Hockberger, P. E. (1987), *J. Neurosci.* **7,** 1384–1400.

Demas, J. N., and Crosby, G. A. (1971). *J. Phys. Chem.* **75,** 991–1024.

DiVirgilio, F., Steinberg, T. H., Swanson, J. A., and Silverstein, S. C. (1987). *Clin. Res.* **35,** 619A.

Dixon, D., Brandt, N., and Haynes, D. H. (1984). *J. Biol. Chem.* **259,** 13737–13741.
Elferink, J. G. R., and Deierkauf, M. (1984). *Biochem. Pharmacol.* **33,** 3667–3673.
Geisow, M. J. (1984). *Exp. Cell Res.* **150,** 29–35.
Gerson. D. F. (1982). *In* "Intracellular pH: Its Measurement, Regulation, and Utilization in Cell Functions" (R. Nuccitelli and D. W. Deamer, eds.), pp. 125–133. Liss, New York.
Giuliano, K.A., and Gillies, R. J. (1987). *Anal. Biochem.* **167,** 362–371.
Graber, M. L., and DiLillo, D. C., Friedman, B. L., and Pastoriza-Munoz, E. (1986). *Anal. Biochem.* **156,** 202–212.
Grynkiewicz, G., Poenie, M., and Tsien, R. Y. (1985). *J. Biol. Chem.* **260,** 3440–3450.
Gurney, A. M., and Lester, H. A. (1987). *Physiol. Rev.* **67,** 583–617.
Gurney, A. M., Tsien, R. Y., and Lester, H. A. (1987). *Proc. Natl. Acad. Sci. U.S.A.* **84,** 3496–3500.
Hallam, T. J., Poenie, M., and Tsien, R. Y. (1986). *J. Physiol.* (*London*) **377,** 123P.
Heiple, J. M., and Taylor, D. L. (1982). *In* "Intracellular pH: Its Measurement, Regulation, and Utilization in Cell Functions" (R. Nuccitelli and D. W. Deamer, eds.), pp 21–54. Liss, New York.
Hepler, P. K., and Callaham, D. A. (1987). *J. Cell Biol.* **105,** 2137–2143.
Hesketh, T. R., Moore, J. P., Morris, J. D. H., Taylor, M. V., Rogers, J., Smith, G. A., and Metcalfe, J. C. (1985). *Nature* (*London*) **313,** 481–484.
Highsmith, S., Bloebaum, P., and Snowdowne, K. W. (1986). *Biochem. Biophys. Res. Commun.* **138,** 1153–1162.
Hirning, L. D., Fox, A. P., McCleskey, E. W., Olivera, B. M., Thayer, S. A., Miller, R. J., and Tsien, R. W. (1988). *Science* **239,** 57–61.
Hollingworth, S., and Baylor, S. M. (1987). *Biophys. J.* **51,** 549a.
Huflejt, M. E., Negulescu, P. A., Machen, T. E., and Packer, L. (1988). *Biophys. J.* **53,** 616a.
Illsley, N. P., and Verkman, A. S. (1987). *Biochemistry* **28,** 1215–1219.
Jackson, A. P., Timmerman, M. P., Bagshaw, C. R., and Ashley, C. C. (1987). *FEBS Lett.* **216,** 35–39.
Kano, K., and Fendler, J. H. (1978). *Biochim. Biophys. Acta* **509,** 289–299.
Kao, J. P. Y., and Tsien, R. Y. (1988), *Biophys. J.,* **53,** 635–639.
Klein, M. G., Simon, B. J., Szucs, G., and Schneider, M. F. (1988). *Biophys. J.,* **53,** 955–962.
Koller, E, and Wolfbeis, O. S. (1985). *Monatsh. Chem.* **116,** 65–75.
Krapf, R., Berry, C. A., and Verkman, A. S. (1988). *Biophys. J.* **53,** 955–962.
Kudo, Y., and Ogura, A. (1986). *Br. J. Pharmacol.* **89,** 191–198.
Kurtz, I. (1987). *J. Clin. Invest.* **80,** 928–935.
Kurtz, I., and Balaban, R. S. (1985). *Biophys. J.* **48,** 499–508.
Lee, H. C. (1985). *J. Biol. Chem.* **260,** 10794–10799.
Lee, H. C., Forte, J. G., and Epel, D. (1982). *In* "Intracellular pH: Its Measurement, Regulation, and Utilization in Cell Functions" (R. Nuccitelli and D. W. Deamer, eds.), pp. 135–160. Liss, New York.
Lee, H. C., Smith, N., Mohabir, R., and Clusin, W. T. (1987). *Proc. Natl. Acad. Sci. U.S.A.* **84,** 7793–7797.
Lev-Ram, V., and Grinvald, A. (1987). *Biophys. J.* **52,** 571–576.
Levy, L. A., Murphy, E., Raju, B., and London, R. E. (1988). *Biochemistry* **27,** 4041–4048.
Lipscombe, D., Madison, D. V., Poenie, M., Reuter, H., Tsien, R. W., and Tsien, R. Y. (1988a). *Neuron,* **1,** 355–365.
Lipscombe, D., Madison, D. V., Poenie, M., Reuter, H., Tsien, R. Y., and Tsien, R. W. (1988b). *Proc. Natl. Acad. Sci. U.S.A.,* **85,** 2398–2402.

McNeil, P. (1988). *Methods Cell Biol.* **29,** in press.

McNeil, P., Tanasugarn, L., Meigs, J., and Taylor, D. L. (1983). *J. Cell Biol.* **97,** 692–702.

Malgaroli, A., Milani, D., Meldolesi, J., and Pozzan, T. (1987). *J. Cell Biol.* **105,** 2145–2155.

Maruyama, I., Oyamada, H., Hasegawa, T., Ohtsuka, K., and Momose, K. (1987). *FEBS Lett.* **220,** 89–92.

Meldolesi, J., and Pozzan, T. (1987). *Exp. Cell Res.* **171,** 271–283.

Millman, M. S., Caswell, A. H., and Haynes, D. H. (1980). *Membr. Biochem.* **3,** 291–315.

Minta, A., Harootunian, A. T., Kao, J. P. Y., and Tsien, R. Y. (1987). *J. Cell Biol.* **105,** 89a.

Moolenaar, W. H. (1986). *Annu. Rev. Physiol.* **48,** 363–376.

Moolenaar, W. H., deLaat, S. W., Mummery, C. L., and van der Saag, P. T. (1982). *In* "Ions, Cell Proliferation and Cancer" (A. L. Boynton, W. L. McKeehan, and J. F. Whitfield, eds.), pp. 151–162. Academic Press, New York.

Murphy, S. N., Thayer, S. A., and Miller, R. J. (1987). *J. Neurosci.* **7,** 4145–4158.

Musgrove, E. Rugg, C., and Hedley. D. (1986). *Cytometry* **7,** 347–355.

Nakashima, M., Sousa, J. A., and Clapp, R. C. (1972). *Nature (London) Phys. Sci.* **235,** 16–18.

Neher, E. (1988). *J. Physiol. (London)* **395,** 193–214.

Nemeth, E. F., and Scarpa, A. (1987). *J. Biol. Chem.* **262,** 5188–5196.

Paradiso, A. M., Negulescu, P. A., and Machen, T. E. (1986). *Am. J. Physiol.* **250,** G524–G534.

Paradiso, A. M., Tsien, R. Y., and Machen, T. E. (1987). *Nature (London)* **325,** 447–450.

Pershadsingh, H. A., Martin, A. P., Vorbeck, M. L., Long, J. W., Jr., and Stubbs, E. B., Jr. (1982). *J. Biol. Chem.* **257,** 12481–12484.

Poenie, M. (1989). In preparation.

Poenie, M., Alderton, J., Tsien, R. Y., and Steinhardt, R. A. (1985). *Nature (London)* **315,** 147–149.

Poenie, M., Alderton, J., Steinhardt, R., and Tsien, R. Y. (1986). *Science* **233,** 886–889.

Poenie, M., Tsien, R. Y., and Schmitt-Verhulst, A. -M. (1987). *EMBO J.* **6,** 2223–2232.

Poenie, M., Schmitt-Verhulst, A. M., and Tsien, R. Y. (1989). In preparation.

Rabinovitch, P. S., June, C. H., Grossman, A., and Ledbetter, J. A. (1986). *J. Immunol.* **137,** 952–961.

Ratan, R. R., Shelanski, M. L., and Maxfield, F. R. (1986). *Proc. Natl. Acad. Sci. U.S.A.* **83,** 5136–5140.

Ratto, G. M., Payne, R., Owen, W. G., and Tsien, R. Y. (1988). *J. Neurosci., in press.*

Rink, T. J., and Pozzan, T. (1985). *Cell Calcium* **6,** 133–144.

Rink, T. J., Tsien, R. Y., and Pozzan, T. (1982). *J. Cell Biol.* **95,** 189–196.

Rogers, J., Hesketh, T. R., Smith, G. A., Beaven, M. A., Metcalfe, J.C., Johnson, P., and Garland, P. B. (1983a). *FEBS Lett.* **161,** 21–27.

Rogers, J., Hesketh, T. R., Smith, G. A., and Metcalfe, J. C. (1983b). *J. Biol. Chem.* **258,** 5994–5997.

Roos, A., and Boron, W. F. (1981). *Physiol. Rev.* **61,** 296–434.

Rothenberg, P., Glaser, L., Schlesinger, P., and Cassel, D. (1983). *J. Biol. Chem.* **258,** 12644–12653.

Rotman, B., and Papermaster, B. (1966). *Proc. Natl. Acad. Sci. U.S.A.* **55,** 134–141.

Sage, S. O., and Rink, T. J. (1986). *Biochem. Biophys. Res. Commun.* **136,** 1124–1129.

Scanlon, M., Williams, D. A., and Fay, F. S. (1987). *J. Biol. Chem.* **262,** 6308–6312.

Schlegel, W., Winiger, B. P., Mollard, P., Vacher, P., Wuarin, F., Zahnd, G. R., Wollheim, C. B., and Dufy, B. (1987). *Nature (London)* **329,** 719–721.

Schwartz, E. A. (1987). *Science* **238,** 350–355.

Simons, E. R., Schwartz, D. B., and Norman, N. E. (1982). *In* "Intracellular pH: Its Measurement, Regulation, and Utilization in Cell Functions" (R. Nuccitelli and D. W. Deamer, eds.), pp. 463–482. Liss, New York.

Smith, G. A., Morris, P. G., Hesketh, T. R., and Metcalfe, J. C. (1986). *Biochim. Biophys. Acta* **889,** 72–83.

Smith, S. J., Osses, L. R., and Augustine, G. J. (1987). *Biophys. J.* **51,** 66a.

Smith, G. A., Hesketh, T. R., and Metcalfe, J. C. (1988). *Biochem. J.* **250,** 227–232.

Steinberg, S. F., Bilezikian, J. P., and Al-Awqati, Q. (1987). *Am. J. Physiol.* **253,** C744–C747.

Stephens, C. R., Conover, L. H., Pasternack, R., Hochstein, F. A., Moreland, W. T., Regna, P. P., Pilgrim, F. J., Brunings, L. J., and Woodward, R. B. (1954). *J. Am. Chem. Soc.* **76,** 3568–3575.

Swanson, J. (1988). *Methods Cell Biol.* **29,** in press.

Tanasugarn, L., McNeil, P., Reynolds, G., and Taylor, D. L. (1984). *J. Cell Biol.* **98,** 717–724.

Thayer, S. A., Hirning, L. D., and Miller, R. J. (1987). *Mol. Pharmacol.* **32,** 579–586.

Thomas, J. A., Buchsbaum, R. N., Zimniak, A., and Racker, E. (1979). *Biochemistry* **18,** 2210–2218.

Tsien, R. Y. (1980). *Biochemistry* **19,** 2396–2404.

Tsien, R. Y. (1983). *Annu. Rev. Biophys. Bioeng.* **12,** 91–116.

Tsien, R. Y. (1986). *In* "Optical Methods in Cell Physiology" (P. De Weer and B. M. Salzberg, eds.), pp. 327–345. Wiley (Interscience), New York.

Tsien, R. Y., and Poenie, M. (1986). *Trends Biochem. Sci.* **11,** 450–455.

Tsien, R. Y., and Pozzan, T. (1988). *In* "Methods in Enzymology" (R. Wu, ed.), Vol. 155. Academic Press, San Diego

Tsien, R. Y., and Rink, T. J. (1983). *Curr. Methods Cell. Neurobiol.* **3,** 249–312.

Tsien, R. Y., and Zucker, R. S. (1986). *Biophys. J.* **50,** 843–853.

Tsien, R. Y., Pozzan, T., and Rink, T. J. (1982). *J. Cell Biol.* **94,** 325–334.

Tsien, R. Y., Rink, T. J., and Poenie, M. (1985). *Cell Calcium* **6,** 145–157.

Valet, G., Raffael, A., Moroder, L., Wunsch, E., and Ruhenstroth-Bauer, G. (1981). *Naturwissenschaften* **68,** 265–266.

Verkman, A. S., and Ives, H. E. (1986). *Biochemistry* **25,** 2876–2882.

Walker, J. W., Somlyo, A. V., Goldman, Y. E., Somlyo, A. P., and Trentham, D. R. (1987). *Nature (London)* **327,** 249–252.

Walter, R. J., Jr., and Berns, M. W. (1986). *In* "Video Microscopy" (S. Inoue, ed.), pp. 327–392. Plenum, New York.

Wanke, E., Ferroni, A., Malgaroli, A., Ambrosini, A., Pozzan, T., and Meldolesi, J. (1987). *Proc. Natl. Acad. Sci. U.S.A.* **84,** 4313–4317.

White, J. G., Amos, W. B., and Fordham, M. (1987). *J. Cell Biol.* **105,** 41–48.

Wier, W. G., Cannell, M. B., Berlin, J. R., Marban, E., and Lederer, W. J. (1987). *Science* **235,** 325–328.

Williams, D. A., Fogarty, K. E., Tsien, R. Y., and Fay, F. S. (1985). *Nature (London)* **318,** 558–561.

Williams, D. A., Becker, P. L., and Fay, F. S. (1987). *Science* **235,** 1644–1648.

Wilson, H. A., Greenblatt, D., Poenie, M., Finkelman, F. D., and Tsien, R. Y. (1987). *J. Exp. Med.* **166,** 601–606.

Wolfbeis, O. S., Furlinger, E., Kroneis, H., and Marsoner, H. (1983). *Fresenius Z. Anal. Chem.* **314,** 119–124.

Wolfbeis, O. S., Koller, E., and Hochmuth, P. (1985). *Bull. Chem. Soc. Jpn.* **58,** 731–734.

Chapter 6

Fluorescence Ratio Imaging Microscopy

G. R. BRIGHT, G. W. FISHER, J. ROGOWSKA,
AND D. L. TAYLOR

Department of Biological Sciences
Center for Fluorescence Research in Biomedical Sciences
Carnegie Mellon University
Pittsburgh, Pennsylvania 15213

I. Introduction

Rapid advancements in low-light-level imaging technology have made it possible to apply fundamental principles of fluorescence spectroscopic analyses to living cells. The quantitation of an optical signal from living cells attached to a substrate is very difficult. Living cells move, change shape, and vary the local number density of membrane-bound organelles. Fluorescence ratio imaging provides a means of overcoming many of the problems associated with quantifying the complex temporal and spatial dynamics of molecules and organelles in living cells. The goal of this chapter is to describe issues of fluorescence ratio imaging microscopy including (1) basic requirements of the instrumentation, (2) validation of the basic tenets of ratioing as applied to microscopy, and (3) some methods and issues in the application of ratio imaging. A more general discussion has recently appeared (Bright *et al.,* 1987b).

II. Basic Principles of Fluorescence Ratio Imaging

Fluorescence ratio imaging microscopy is based on the principles of fluorescence microscopy and ratio spectroscopy and evolved from early ratio measurements made on cells in suspension (Ohkuma and Poole, 1978) and in single cells under the microscope (Heiple and Taylor, 1980, 1982a, b; Tanasugarn *et al.*, 1984; see Tsien and Poenie, 1986; Bright *et al.*, 1987b, for recent applications). A fundamental requirement for fluorescence ratio imaging is that the shape of the excitation or emission spectrum of the probe be sensitive to a specific parameter of interest, such as pH or pCa.

A. General Principles

Quantitation of the fluorescence emission of a probe is difficult due to factors involving instrumentation, sample geometry, and probe chemistry. Although the fluorescence intensity of a probe may be sensitive to the parameter of interest it is also affected by other factors. As stated in the standard fluorescence equation:

$$F = f(\theta)g(\lambda)\Phi_F I_0 \varepsilon bc \tag{1}$$

the fluorescence emission is related to a geometric factor $[f(\theta)]$, the quantum efficiency of the detector $[g(\lambda)]$, fluorescence quantum yield of the probe (Φ_F), excitation intensity (I_0), extinction coefficient of the probe (ε), optical pathlength (b), and fluorophore concentration (c). For this linear relation to hold, it is necessary that the absorbance of the specimen be low, and that there be no dye–dye interactions such as fluorescence resonance energy transfer or self-quenching. The living cell is a complex object where the optical pathlength varies from cell to cell, from one position to another within the cell, and also changes from one time to another. The probe concentration can also vary from cell to cell and over time due to labeling efficiency, photobleaching, dye leakage, and compartmentalization. In addition, the illumination intensity and detection efficiency can vary over the field of view of the microscope due to limitations of the instrumentation.

Ratio spectroscopy requires that the probe be differentially sensitive to the parameter of interest in at least two excitation or emission wavelengths. The excitation or emission at one wavelength may be either nonsensitive (i.e., isosbestic point), much less sensitive, or sensitive in the opposite direction compared to the excitation or emission of the other wavelength. It is the relationship of the changes in excitation or emission at each wavelength with respect to each other that defines the measured response. Because the excitation or emission originates from the same volume, the ratio relationship normalizes for optical pathlength, local probe concentration, and loss of signal due to photobleaching. Whole spectra are not required since measurements at two wavelengths can reflect changes in the shape of the spectrum. The basic premise of the ratio method is that the fluorescence represents a measure of the relative proportion of two fluorescent species. The population of species, however, is not always simple, and therefore, the exact species responsible for the signal are not necessarily obvious, particularly for intracellular dyes (see Section III,B,3).

B. Approaches

There are three basic approaches to ratio microscopy; ratio absorbance, ratio excitation, and ratio emission. Ratio absorption measurements have been limited so far to nonimaging approaches (MacDonald *et al.,* 1977; Thomas *et al.,* 1979; Chaillet *et al.,* 1986). Furthermore, absorption measurements lack sensitivity compared to fluorescence measurements because the signal is mixed with a high background which results in a poor signal-to-noise ratio (S/N).

Ratio excitation measurements are the easiest to implement and have been the standard approach. Only simple mechanical and optical modifications to the microscope excitation light path are necessary. Issues related to implementation of excitation ratios are discussed below. Ratio emission measurements have a great deal of potential, yet this approach is difficult to implement with a standard microscope since either rapid changes in the emission filters or two detectors are required. However, this latter approach has great potential and should be explored (see also Herman, Chapter 8, this volume).

C. Applications

To date, the applications of fluorescence ratio imaging have been to measure ion-concentrations such as pH (Tanasugarn, *et al.*, 1984; Tycko *et al.*, 1983; Bright *et al.*, 1987a; Paradiso *et al.*, 1987) and pCa (Keith *et al.*,

1985; Sawyer *et al.*, 1985; Williams *et al.*, 1985; Connor, 1986; Kruskal *et al.*, 1986; Poenie *et al.*, 1986, 1987; Ratan *et al.*, 1986; Wier *et al.*, 1987; see Tsien Chapter 5, this volume), and the mapping of local relative distributions of fluorescent analogs (Luby-Phelps and Taylor, 1988). In principle the approach is applicable to any parameter that can be measured as a function of the intensity at two or more wavelengths. Other possibilities include steady state polarization and resonance energy transfer (see Tanasugarn *et al.*, 1984)

III. Requirements for a Fluorescence Ratio Imaging Microscope System

The instrumentation required for fluorescence ratio imaging microscopy involves the integration of fluorescence spectroscopy, fluorescence microscopy, electronic imaging, and digital image processing.

A. Equipment

The ratio imaging microscope has usually been a modified standard fluorescence microscope. Many of the modifications are improvements in the specifications of standard components. The entire system should be mounted on a very stable platform such as an air-supported optical table to minimize vibrations since some components are attached to support arms mounted on the optical table. Mechanical devices such as excitation filter changers driven by stepper motors can be supported from the table, minimizing transmission of vibration to the optical system. The system includes a stable excitation light source, a shutter to control the duration of the excitation, a mechanism for changing excitation or emission wavelengths, an electromechanical stage, a low-light-level detector, an image processor, and a host computer (Fig. 1).

1. Microscope

Any standard research grade microscope can be used as the basis of a fluorescence ratio imaging system. The basic requirements for selecting a microscope and appropriate optics is beyond the scope of this discussion and can be found elsewhere (Inoue, 1986; Salmon and Taylor, Volume 29, this series). A fundamental need is a stand capable of supporting the weight, since several devices will be added to the basic microscope, Pe-

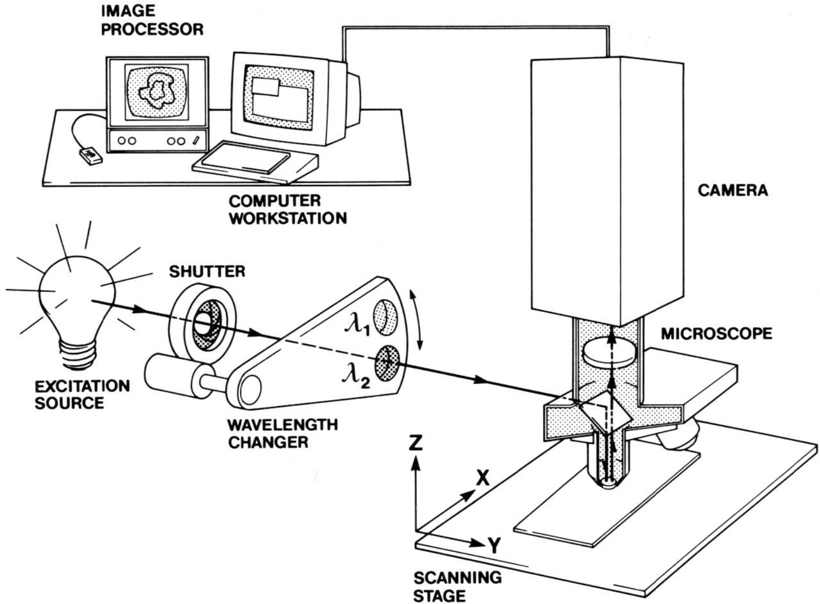

FIG. 1. The basic elements of a fluorescence ratio imaging microscope system. The elements include a light source, an excitation light control device (e.g., shutter), a wavelength tuning device (e.g., filter wheel), microscope, low-light-level camera, image processor, with computer coordinated control over all aspects of the system. From Bright *et al.* (1987b) with permission from *BioTechniques*.

ripheral devices should be supported from the table whenever possible to minimize stress on the microscope stand. Placing too much weight on the stand can dramatically affect the alignment of the optical path.

Since low-light level imaging is used, objectives with the highest numerical aperture (NA) at a given magnification are needed to maximize light collection efficiency. The collection efficiency of an objective increases with the square of the NA (Inoue, 1986; see Salmon and Taylor, Volume 29, this series). Since most current microscopes were designed for human vision, the use of probes with absorption or emission wavelengths in the ultraviolet or near-infrared may require special optics. The transmission efficiency at these wavelengths can vary from objective to objective and among different manufacturers, and these values should be checked. It is also necessary to consider the transmission efficiency of all optical components in the light path including lamp collector lenses, condenser, and intermediate windows.

2. Stage

The microscope stage serves two basic functions (1) support for the specimen and (2) movement of the specimen on *x, y,* and *z* axes. As a support for the specimen, the cellular environment must also be controlled to maintain the health of the cells since experiments can span many hours. Environmental control involves both temperature and gas (i.e., CO_2) when using bicarbonate buffers. The type of system used to control the environment ranges from small sensor feedback-controlled perfusion-based systems (Bright *et al.*, 1987a) to enclosure of the entire microscope in an environmentally controlled chamber (see Wang, Volume 29, this series).

The provision for computer-controlled movement of the stage allows the recording of specific coordinates of single cells or groups of cells. Once selected and coordinates recorded, the cells can be monitored repeatedly over time. Large numbers of cells can be recorded by low-light-level cameras at a fixed gain and high voltage by choosing cells that have the same general level of fluorescence intensity falling within the dynamic range of the detector. This is particularly convenient when using a bulk labeling method (see McNeil, Volume 29, this series) that results in a wide distribution of labeling efficiency. Motorized focus control provides convenience and the opportunity to automatically acquire images from several planes of focus. Finally, stage automation allows the researcher to back away from the microscope, thereby minimizing disturbances to the experiment.

3. Light Source

The two most important aspects of the excitation light source are spectral output and temporal as well as spatial stability of the emitted light. Tungsten–halogen filament lamps, mercury arc, and xenon arc lamps are the principle light sources used; however, lasers have recently been employed (Spring and Smith, 1987). Koehler illumination of a well-aligned light source is required to produce an even illumination of the microscope field no matter what type of illumination system is employed (see Salmon and Taylor, Volume 29, this series).

Filament lamps can provide a relatively cheap and simple excitation source. A tungsten–halogen lamp can often provide sufficient excitation energy in combination with the use of low-light-level cameras (Bright *et al.*, 1987a). Filament lamps, however, need very high quality power supplies to isolate them from line voltage fluctuations. The standard power supply provided by most manufacturers is usually inadequate for spectroscopy.

Tungsten–halogen lamps provide wavelengths from the visible to the infrared. However, the wavelength distribution (often called the color temperature, see Slayter, 1970) of filament lamps changes significantly with the operating current and as the lamp ages. Aging is due to the vaporization of the tungsten filament, which results in a change in resistance and thus the current.

Direct current high-pressure arc lamps provide a high-intensity illumination source. The available wavelengths range from the ultraviolet to the infrared. Xenon lamps provide a broader and more even distribution of wavelengths than mercury lamps. However, arc lamps may suffer from temporal instability and arc wander, part of which may be due to inadequate heat dissipation. Proper burn-in, possibly optical feedback circuits, and power conditioning may be needed (Oldham *et al.*, 1985). The higher intensities available from arc lamps are valuable for probes with low quantum yields and to reduce the time required to acquire an image by using excitation at a higher light level for a shorter amount of time.

The primary application of a laser is when very high excitation energy is needed for very short acquisition times. Lasers are the most expensive source both in purchase costs and maintenance and are also the most limiting in terms of available wavelengths. Two wavelengths are needed for an excitation ratio, thus the laser must be purchased with the option of operating in multiline mode; or two separate lasers can be used. One important advantage of ratio emission measurements is that only one excitation wavelength is required. The emitted light is incoherent since the fluorescent probe acts as a self-luminous object (see Salmon and Taylor, Volume 29, this series).

Regardless of the excitation source, a shutter is necessary to control the duration of the excitation pulse and is usually placed just in front of the light source (Fig. 1). This placement also prevents the constant exposure of interference filters to intense light and heat.

4. Wavelength Tuning

A fundamental need for ratio imaging is the ability to change either excitation or emission wavelengths quickly. The easiest and most common approach for changing wavelengths is to alternate bandpass filters into the excitation path with a filter wheel driven by a stepper motor (Fig. 1); 100 millisecond wavelength switching is possible using a filter wheel with low mass and a fast stepper motor.

It is important that the interference filters be of high quality and blocked for wavelengths outside the chosen bandpass. The appropriate bandpass will depend upon the probe (see Section III,B). It is critical that the filter be blocked since, for example, a xenon lamp emits light at wavelengths between 800 and 1000 nm at levels 10–20 times that emitted below 800 nm (see Inoue, 1986; Salmon and Taylor, Volume 29, this series). The use of a low-light-level detector that is sensitive to these longer wavelengths can result in a significant noise signal when using improperly blocked filters. A recent observation in this laboratory illustrates a dramatic manifestation of this effect. An attempt to image a rhodamine-labeled sample resulted in a confusing result. Direct observation of the sample gave a sharp red image, whereas electronic observation using an intensified video camera resulted in a usable image with a very poor S/N. Upon placing a 750-nm short pass filter into the excitation path, the rhodamine image was clearly visible with the camera. Subsequent analysis of the microscope filters indicated transmission of wavelengths greater than 800 nm. Thus, the rhodamine image was there all the time, just buried within the scattered light with wavelengths >800 nm (unpublished observations). Therefore, interference filters should be ordered blocked at all wavelengths outside the bandpass and within the spectral response of the detector.

A complex but more versatile approach to wavelength switching involves the use of a dual light source, dual monochrometer system with a rapid switching mirror (Tsien *et al.*, 1985). The monochrometers are set to the two wavelengths and the computer controls a switching mirror. This system provides high flexibility and allows very fast switching between the two wavelengths. This system also provides the ability to perform spectral scans and to simply vary the bandwidth of the excitation light. However, if more than two wavelengths are necessary, switching would be much slower, dictated by the speed of the monochrometer scanning motors. This type of system is now commercially available.

Recently, the ability of an acousto-optic modulator to act as an electronically adjustable diffraction grating and intensity regulator has been used to rapidly switch between laser lines on a microsecond time scale (Spring and Smith, 1987). The limitations on currently available modulators are the limiting aperture and the requirement of a highly collimated light source such as a laser. Recent development of very large aperture acousto-optic tunable filters (Gottlieb *et al.*, 1980) opens up the exciting possibility of using a broad-spectrum light source for microsecond frequency tuning and control of light intensity (our unpublished observations; Kurtz *et al.*, 1987).

Automated changing of emission wavelengths is a more difficult problem. Since the imaging optical path carries substantial information, changing anything in this path can lead to significant distortions. Changing the emission filters in an epifluorescence configuration involves not only changing the barrier filter but also the dichroic reflector. Current generation microscopes do not provide easy vibration-free mechanisms to change emission wavelengths. Moreover, an additional problem is the lack of sufficient degrees of freedom in the alignment of emission filters, resulting in the need for computationally expensive registration techniques. Improvements in microscope designs should be expected in the near future to meet the demands for automated electronic imaging.

5. Detector

The desire for spatial information dictates the need for a two-dimensional detection system. Intensified video cameras have been used extensively, since low-light-level excitation is generally used (Bright and Taylor, 1986). These cameras may contain from one to three stages of intensification. To use these devices as quantitative detectors, several limitations must be characterized and corrected. These include a limited intrascene dynamic range, shading, a nonlinear radiometric response, and geometric distortion (Bright and Taylor, 1986; see Spring and Lowy and Wampler and Kutz, Volume 29, this series; Jerecivic *et al.,* this volume). There are correction algorithms for some of these aberrations (see Jerĕcivić *et al.,* Chapter 2, this volume; Castleman, 1979; Baxes, 1984; Galbraith *et al.,* 1987). New detectors such as thermoelectrically cooled charge-coupled devices (CCD) show tremendous promise, with the advantage of a much larger dynamic range and a linear radiometric response which, in turn, simplifies some correction algorithms (see Aikens *et al.,* Volume 29, this series). For short integration times, a cooled CCD has about the same sensitivity as an ISIT camera operating at maximum gain in the green region of the spectrum, and is more sensitive at longer wavelengths (unpublished observations). The current limitation of CCDs is the rather long readout time (5 sec), although significant improvements are expected in the near future (see Aikens *et al.,* Volume 29, this series).

The frame transfer type CCD (see Collet, 1985) can be exploited for ratio imaging (see Aiken *et al.,* Volume 29, this series). A frame transfer CCD consists of a single chip with half of the surface covered by an opaque mask. The mask may be permanently attached to the chip or an optical mask can be used (unpublished observations). The device works by exposing the noncovered side, electronic transfer of the image to the masked side, and exposure of the noncovered side while the first image is read off the chip. This design allows the recording of two images in very rapid sucession. On-chip transfer of a 14-bit image takes less than 2 milliseconds. This type of CCD reduces the present negative aspects of the existing slow readout time.

6. Image Storage

Several types of devices exist for storing images, including standard digital computer disks, optical digital disks, digital video disks, optical video disks, and video tape recorders (see Inoue, 1986, for discussion). Quantitative image analysis requires a storage medium that does not compromise the accuracy of the image. Digital devices provide faithful reproduction of all aspects of the images since the pixels are stored as numeric data. Most modern computer workstations have high-performance disks available that are capable of very fast transfer rates, adequate for most applications. However, in some applications the temporal kinetics of the process demand faster recording. Video devices provide the ability to record images in real time, 30 images per second. Digital video disks allow real-time digital storage thus, providing the best of both worlds. Their disadvantage is the high cost of both purchase and maintenance. Optical video disks provide real-time storage at the cost of slightly reduced S/N. This is due to some processing of the signal. Manual control of the circuitry is essential. Recent price reductions in both the devices and the storage medium have made these a cost-effective alternative to video tape. Video tape recorders have limited application to quantitative imaging. They suffer from reduced spatial resolution, dynamic range, and S/N, as well as signal droppout. The video signal goes through a series of complex electronic processing steps (see Inoue, 1986). In addition, still-frame display is subject to image breakup and noise. Some recent recorders have improved resolution and manual control for some electronic processing, such as bias and equalization.

Digital disks are ideally for quantitative imaging applications. When real-time recording is necessary, optical video disks provide high capacity, video rate recording and playback, long term archival storage, and random access with a relatively small decrease in S/N. Use of these devices will require consideration of the reduced S/N on the analysis.

7. Image Processor

An image processor is required for acquisition, and contains specialized hardware that dramatically speeds the processing time over that of a conventional computer (see Baxes, 1984). Three basic components of this fast hardware are lookup tables (LUT), special arithmetic logic units (ALU), and convolvers. These devices are capable of processing an entire image at or near real time, 30 frames per second. An image processing system may contain many ALUs and LUTs, but usually only one or two convolvers. The ALU is a fast arithmetic and logic unit capable of performing simple operations, such as addition, subtraction, and Boolean operations, on input images. LUTs are fast random access memories which have as many addresses as there are possible pixel values. Each pixel value of the input image becomes an address in the LUT which holds the pixel value for the output image. LUTs are widely used for contrast manipulations, pseudocolor display, and threshold operations, etc. Convolvers perform neighborhood operations used for filtering and edge enhancement. A wide range of basic image processing commands is needed for image acquisition, real-time averaging; basic ALU operations, such as addition and subtraction; basic point operations, such as intensity remapping and thresholding; basic analysis operations, such as histogram and statistical calculations; and basic display operations, such as image scroll, zoom, and pseudocolor display. More sophisticated image processors may include an array processor capable of very fast floating point arithmetic operations that are needed for transform processing, such as application of the fast Fourier transform.

8. Computer Workstation

The central controller for the entire system is a host computer or workstation. This may be one of the more powerful personal computers or a sophisticated engineering workstation. This machine can control the microscope peripherals such as shutters, filter wheels, and scanning stage, and coordinates the acquisition, storage, and analysis of the images. It is also the platform for generation of final reports and graphics display. Several commercial systems are integrating the image processing system with a host computer into a light microscope workstation.

B. Fluorescent Probes

The fluorescent probe (see Tsien, Chapter 5, this volume (Waggoner, 1986) plays a central role in making meaningful fluorescence ratio measurements. The probe may be as simple as a single free dye, a single dye attached to a carrier molecule such as dextran, or more elaborate, such as a dextran molecule labeled with two separrate dyes, one sensitive and one insensitive to the parameter of interest. The appropriate choice of probe depends on the parameter of interest. Obviously, specificity for the parameter of interest is important. In some cases the list of available probes for a particular parameter is very short, as is the case for calcium-sensitive probes. Fura-2 (Grynkiewicz *et al.,* 1985) has been the only generally useful calcium-sensitive probe for ratio imaging applications, although others are under development. The situation is different for pH-sensitive probes. Several probes with variable pK_a and wavelength choices are now available (Table I), and the list of probes applicable for fluorescence ratio imaging is constantly expanding (see Tsien, Chapter 5, this volume). A wide choice of probes with varying spectral properties is highly desirable. The ultimate goal is to be able to monitor several parameters at the same time, allowing direct correlation of multiple parameters (DeBiasio *et al.*, 1987; see Waggoner *et al.*, Chapter 17, this volume). There is a general trend to develop probes with wavelengths further into the red and near infrared (see Waggoner *et al.*, Chapter 17, this volume). This effectively doubles the usable spectrum from a range of about 350–650 nm to a range from 350 to 1000 nm. Once a probe is chosen, there are issues of getting the probe into the right compartment, choosing appropriate wavelengths for ratioing, and calibration of the ratio with respect to the parameter.

1. Compartmentalization

There are two basic categories of probes (1) acetoxymethyl ester forms of dyes that can diffuse into cells and then are trapped inside when they become highly charged due to cleavage by cellular esterases (Tsien, 1981); and (2) probes that must be loaded into cells due to their highly charged nature or due to the presence of a carrier molecule (see Tsien, Chapter 5, this volume). A variety of methods are available for loading probes into cells, including microinjection, hypertonic lysis, and scrape loading (see McNeil, Volume 29, this series). The advantages of the acetoxymethyl ester probes include the simplicity of labeling large numbers of cells with relative uniformity. The disadvantages include the lack of complete specificity of labeling. The probe can be generated in any compartment that possess esterases, including mitochondria (Thomas, 1986). In addition, the acetoxymethyl ester form of the dye may exhibit some toxicity and a balance between concentration and incubation time is needed (Poenie *et al.*, 1986; Bright *et al.*, 1987a). The hydrolysis of the acetoxymethyl ester yields stoichiometric amounts of free dye, acetic acid, and methanol. In some cases the labeling process has been shown to cause cellular acidification (Spray *et al.*, 1984). BCECF [2′,7′-bis(2-carboxyethyl)-5(and 6)-carboxyfluorescein]-labeled fibroblasts exhibit an abrupt loss of dye about 3.5 hours post labeling (Bright *et al.*, 1987a). The rate of leakage of dye is related to the net charge on the dye. The higher the net charge, the slower the leakage. Furthermore, the incomplete hydrolysis of this class of probe by cellular esterases can potentially lead to a fluorescent species that is insensitive to the parameter, as, for example, has been shown for fura-2 (Highsmith *et al.*, 1986; Scanlon *et al.*, 1987). In

TABLE I

EXAMPLES OF RATIO pH PROBES

Probe	Label	pK_a	Wavelength[a]				Reference
			Exl*	Ex2	Em1	Em2	
4-Methylumbelliferone	4-MU	7.80	350		450	560	Musgrove *et al.* (1986)
		7.34	365	334	450		Graber *et al.* (1986)
5(and 6)-Carboxyfluorescein	CF	5.90	490	450	525		Graber *et al.* (1986)
		6.30	458	488	>570		Musgrove *et al.* (1986)
Carboxyfluorescein–dextran	CFD	6.33	489	452	520		Bright *et al.* (1987a)
			495	450	519		Ohkuma and Poole (1978)
			489	452	520		Tanasugarn *et al.* (1984)
1,4-Dihydroxyphthalonitrile	DHPN	8.00	385		455	512	Kurtz and Balaban (1985)
			350		425	540	Musgrove *et al.* (1986)
2′,7′-Bis(2-carboxyethyl)-5(and 6)-carboxyfluorescein	BCECF	6.70	490	450	525		Graber *et al.* (1986)
			488		520	620	Musgrove *et al.* (1986)
		6.99	489	452	520		Bright *et al.* (1987a)
BCECF–dextran	BCECFD	6.90	489	452	520		Bright *et al.* (unpublished data)
4′,5′-Dimethyl-5(and 6)-carboxyfluorescein	DMCF	5.80	490	450	525		Graber *et al.* (1986)
4′,5′-Dimethyl-5(and 6)-carboxyfluorescein–dextran	DMCFD	6.75	512		540		Rothenberg *et al.* (1983)

[a] Ex, Excitation, Em, emission.

fact, this calcium-insensitive species may have a distribution different from the other species of fura-2 (see Highsmith *et al.,* 1986). Fura-2 has a tendency to enter subcellular organelles such as secretory vesicles, the nucleus, and sarcoplasmic reticulum (unpublished observations; see Almers and Neher, 1985; Williams *et al.,* 1985; Highsmith *et al.,* 1986; Wier *et al.,* 1987). Also a specific esterase or class of esterases may be required for a particular probe. Quiescent BALB/c 3T3 cells do not label well with fura-2-AM (Scanlon *et al.,* 1987), whereas quiescent Swiss 3T3 cells label easier with BCECF-AM than nonquiescent cells (Bright *et al.,* 1987a). In some systems, extracellular esterases may hydrolyze the ester groups before the probe has chance to enter the cell (Maruyama *et al.,* 1987). Finally, the possibility of cell toxicity must be investigated by performing functional bioassays on the labeled cells.

A carrier-based probe has the advantage that the chemical and physical properties of the probe are generally dominated by the carrier molecule and not the dye (Heiple and Taylor, 1980). The most suitable carriers appear to be low-molecular-weight dextrans and ficolls since they do not bind to cytoplasmic compartments in contrast to "nonspecific" proteins (Luby-Phelps *et al.,* 1986; see Luby-Phelps, Volume 29, this series). Carrier-based probes stay in the original compartment unless actively transported to another by the cell. Since neutral dextrans and ficolls are available in various sizes, it is even possible to discriminate compartments based upon porosity. For example, dextrans > 40,000 Da are too large to pass through the nuclear pores (Tanasugarn *et al.,* 1984; Luby-Phelps *et al.,* 1986). Figure 10 illustrates the use of a 70,000-Da fluorescein-labeled dextran to avoid labeling the nucleus. The main difficulty is delivery of the probe into the desired compartment. Leakage does not occur and the probe concentration decreases only due to cell division. When using carrier-based probes, particularly some commercially available products, it is important to establish their purity. It has been shown that fluorescent impurities or by-products of their preparation can lead to misinterpretation (Preston *et al.,* 1987) and toxicity to the cells. Commercially available labeled dextrans must be extensively dialyzed and chromatographed to remove fluorescent and nonfluorescent impurities (Bright *et al.,* 1987a). Carrier-based probes are particularly useful for long-term experiments where retention of the probe is the limiting factor. However, there is often some autophagocytosis of both free and carrier-based probes. The advantage of carrier-based probes coupled with a broad range of available cell loading techniques makes these the probes of choice whenever possible. Figure 10 illustrates cells either labeled with fluorescein–dextran (a, b) or BCECF (c, d).

2. Selection of Appropriate Wavelengths

Appropriate selection of wavelengths for ratioing can be estimated visually from plots of the spectrum or calculated precisely from difference spectra (Kurtz and Balaban, 1985). In the case of BCECF there is an isosbestic point at about 440 nm. However, there is so little absorption that a wavelength of about 450 nm is necessary to provide a suitable signal-to-noise ratio for imaging (Bright *et al.*, 1987a). For maximum sensitivity the wavelengths with the largest difference magnitude should be selected. Figure 2 illustrates the effect of pH on the excitation spectra of BCECF. Autofluorescence can become significant due to excitation of flavins at wavelengths below about 500 nm, which decreases the S/N (Kohen *et al.*, 1975; Aubin, 1979). Potential photodestruction of flavins could also have detrimental effects on living cells, thus wavelengths below 500 nm should be used with caution and bioassays of labeled and unlabeled cells must be performed.

3. Calibration

The ability to use fluorescence ratio imaging as a quantitative method is limited by the quality of the calibration. Standard curves can be established in two ways, *in vitro* and *in situ.* An *in vitro* standard curve is generated using solutions of the probe in a series of buffers representing varying amounts of the ion or component of interest as the microscope sample. An *in situ* standard curve is generated using labeled cells. A basic requirement

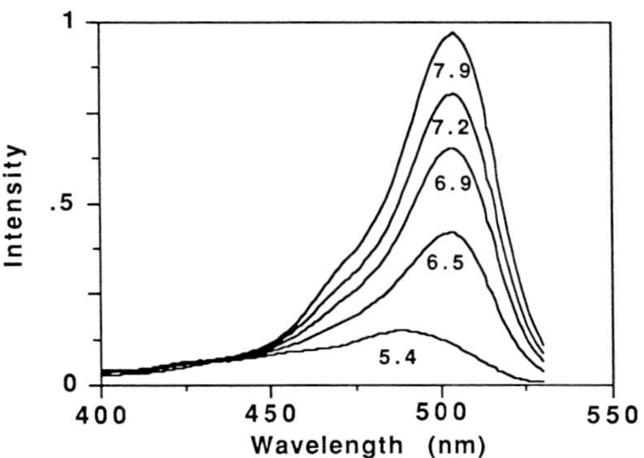

FIG. 2. Differential wavelength sensitivity of BCECF to pH. Excitation spectra of 10 μM BCECF in 130 mM KCl, 1 mM $MgCl_2$, 15 mM 2-(N-morpholino)ethanesulfonic acid (MES), 15 mM N-2-hydroxyethylpiperazine-N'-2-ethanesulfonic acid (HEPES) at the given pH.

is a means of selectively making the cell permeable to the ion, and then equilibrating the intracellular environment with the extracellular environment.

The assumptions implicit in the application of *in vitro* standards are (1) the intracellular probe possesses identical chemical and spectral properties to those exhibited in simple buffers, (2) the fluorescence intensity is proportional to the concentration of probe (i.e., no inner filter effect), and (3) that only two species exist.

These assumptions, however, may not be well founded. At least two pH-sensitive dyes, carboxyfluorescein (Thomas *et al.,* 1979) and BCECF (Rink *et al.,* 1982) show red-shifts of about 5 nm upon introduction into the cellular environment. *In virto* and *in situ* standard curves for both fluorescein-labeled dextran and fura-2 have also been shown to differ slightly (Tanasugarn *et al.,* 1984; Williams *et al.,* 1985). It is not clear whether the spectral shift of BCECF is due to environmental effects or to binding to cellular proteins. The cellular form of BCECF, however, has a translational diffusion coefficient smaller than that of a 5000-Da molecule based on measurements by fluorescence photobleaching recovery (FPR) measurements (Bright *et al.,* 1987a). An additional complication arises in the use of fura-2-AM. The hydrolysis of the acetoxymethyl ester of fura-2 has been shown to produce calcium-insensitive intermediates (Highsmith *et al.,* 1986; Scanlon *et al.,* 1987), resulting in multiple species with different spectral characteristics. In addition, the higher cytoplasmic viscosity has been claimed as altering the fluorescence of fura-2 (Peonie *et al.,* 1986).

Inner filter effects are potential problems when using probes that are concentrated in a given compartment, for example, fusing endosomes labeled with a dextran probe. As the compartment concentrates the probe, self-absorption and energy transfer could occur. In addition, the chemistry of the compartment could be altered by the concentration of the probe. The spectroscopic properties of the probes must be defined in each cellular compartment.

The *in vitro* calibration approach of Grynkiewiez *et al.,* (1985) has been used by most studies using the probe fura-2 for measuring calcium. The approach involves an *in vitro* determined standard curve and calibration based upon the equation:

$$[C_a{}^{2+}] = K_a\left(\frac{R - R_{min}}{R_{max} - R}\right)\left(\frac{F^{380}_{free}}{F^{380}_{sat}}\right) \qquad (2)$$

where R is the measured cellular ratio; R_{max} (340 nm/380 nm), the ratio at saturating calcium concentrations; R_{min}, the ratio at zero calcium concentration; F^{380}_{free}, the fluorescence intensity at zero calcium concentration with

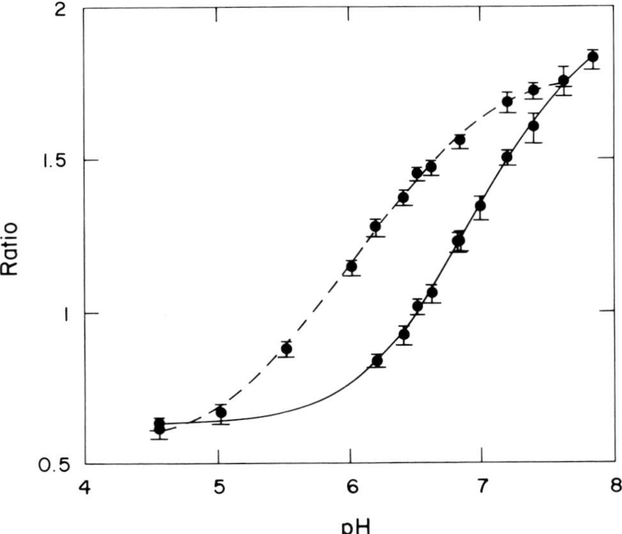

FIG. 3. *In situ* standard curves for fluorescein-labeled dextran and BCECF. Each point represents the mean and sample standard deviation of the ratio for 10 cells. The measured pK_a of fluorescein–dextran (----) is 6.33 and BCECF (——) is 6.99. Reproduced from *J. Cell Biol.* **104,** 1019–1033 (1987) with permission from the Rockefeller University Press.

excitation at 380 nm; F^{380}_{sat}, fluorescence intensity at saturating calcium concentration; and K_d, the *in vitro* dissociation constant. The last term of the equation evaluates to 1 if the wavelength of an isosbestic point (360 nm for fura-2) is used as the insensitive wavelength.

An alternative approach to calibration makes no assumptions about the K_d or limiting ratio values (Bright *et al.*, 1987a). The standard titration curve has a sigmoidal shape (Fig. 3). A four-parameter logistics equation provides a convenient fit to a sigmoidal curve (Rodbard and Hutt, (1974)). The equation:

$$y = \left(\frac{d - a}{1 + (x/c)^b}\right) + a \tag{3}$$

povides terms for defining the maximum asymptote (a), slope of the linear portion of the curve (b), center point of linear region (c), and minimum asymptote (d). Rearranging and translating into terms of, for example, pH gives:

$$x = C\left(\frac{d - a}{y - a} - 1\right)^{1/b} \tag{4a}$$

and

$$\mathrm{pH} = \mathrm{p}K_a\left(\frac{R_{max} - R}{R - R_{min}}\right)^{1/s} \tag{4b}$$

where s is the slope of the linear portion of the curve. This equation is fit to the standard curve data using standard nonlinear least squares methods (*see* Bevington, 1969). No assumptions are made other than the sigmoidal shape of the curve. All four parameters of the equation are determined by the best fit to the titration data. The measured $\mathrm{p}K_a$ can be used as a means of assessing the effects of introducing the probe into the cellular environment on the chemical properties.

The most suitable approach for pH calibration has been to perform an *in situ* calibration using the H^+/K^+ exchanger, nigericin, in the presence of high potassium ion concentration (Fig. 3; Thomas *et al.*, 1979; Tanasugarn *et al.*, 1984; Chaillet and Boron, 1985; Bright *et al.*, 1987a). Appropriateness of the nigericin approach has been confirmed directly using microelectrode measurements (Challet and Boron, 1985) and by comparison of cellular pH values made using a variety of fluorescent and nonfluorescent probes (*see* Bright *et al.*, 1987a).

Calcium measurements using fura-2, as mentioned above, have been primarily calibrated using *in vitro* standard curves. The need for *in situ* determination of R_{max} and R_{min} has recently been shown where the values can be lower in cells than in solution. (Almers and Neher, 1985; Scanlon *et al.*, 1987). Two studies have reported full *in situ* standard curves by either injecting calcium-EGTA buffers (Almers and Neher, 1985) or using external equilibration employing the ionophore ionomycin (Williams *et al.*, 1985). A report that the fura-2 signals must be corrected for viscosity effects has been made (Peonie *et al.*, 1986) and this point is discussed in this volume (Tsien, Chapter 5). Special calibration techniques may be necessary for some cells that respond to high levels of calcium, such as cardiac myocytes (Li *et al.*, 1987). One additional advantage of using indicators coupled to neutral carrier molecules is the potential for physically trapping the probe in the cytoplasm and then extracting the plasma membrane to expose the indicator to calibration buffer solutions.

IV. Validation of the Fluorescence Ratio Imaging Microscope

Fluorescence ratio imaging microscopy is relatively easy to perform. However, there are limitations in various components of the system. The integrated system should be tested to show that all the parts perform together as expected (*see* Jerečivić *et al.,* Chapter 2, this volume, Young, Chapter 1, this volume). We have characterized our system with respect to several of the fundamental issues involving operation of the microscope and image processor that were most likely to influence the performance of a ratio imaging system (Bright *et al.,* 1987a).

A. Microscopy

1. Excitation Level

The effect of varying intensity of the excitation light can be studied by inserting a series of neutral density filters into the excitation light path. Figure 4 illustrates that attenuation of the excitation light over the range of 0–90% resulted in no significant change in the mean ratio values for BCECF-stained cells. The lower and upper limits in excitation light levels were set by camera noise level and camera saturation, respectively. Typically, the excitation light levels used in our experiments were about 1.78×10^{-4} $\mu W/\mu m^2$. It is important to measure the intensity to the specimen plane and present this as part of the data (see Taylor and Salmon, Vol. 29, this series). Cellular autofluorescence and scattered light are generally not detectable at these excitation light levels in the cells that we have investigated.

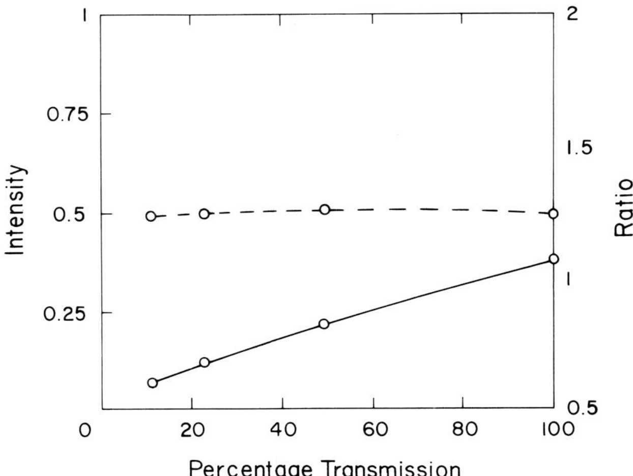

FIG. 4. Effects of varying excitation light levels on the ratio. The pH of a single cell labeled with BCECF was clamped to 6.98. The excitation light was attenuated with a series of neutral density filters. The solid line indicates changes in emission intensity using 489 nm excitation. The dashed line shows the mean ratios. Reproduced from *J. Cell Biol.* **104,** 1019–1033 (1987) with permission from the Rockefeller University Press.

2. Lamp Operation

Variation in the operating current of a tungsten–halogen lamp or the instability of an arc lamp can seriously affect the ratio value. Also, the distribution of wavelengths from a filament lamp changes with operating current. Figure 5 demonstrates the effect of lamp operating current on ratio values for a BCECF-stained cell. The steepness of the change illustrates the need for a highly regulated power supply with the lamp maintained at a constant current. The distribution of wavelength also changes with the age of the lamp and is made most evident by comparing standard curves recorded over a period of a few weeks. The long-term temporal stability of an excitation source will determine how often calibration curves need to be generated. The most accurate results will always result if the system is calibrated at the time of the experiment. Often full calibration can be performed weekly with one or two standard points used before or after each experiment (Bright *et al.*, 1987a).

3. Photobleaching

Photobleaching of a fluorophore results in a decrease in the effective concentration of the fluorophore and the release of photochemical by-products. Minimizing photobleaching is particularly important for long-

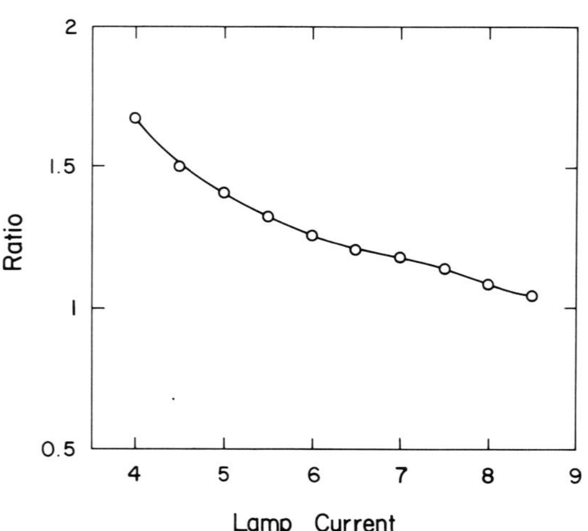

Fig. 5. Effects of the operating current of a tungsten lamp on the ratio. The mean ratio of a pH-clamped BCECF-stained cell was determined at each lamp current setting. Reproduced from *J. Cell Biol.* **104,** 1019–1033 (1987) with permission from the Rockefeller University Press.

term studies where multiple measurements will be made. The bleaching rate constant is proportional to the excitation intensity and thus can be kept to a minimum by reducing the total dose of excitation energy where dose equals intensity × time. This requirement can be satisfied by either short duration, high intensity illumination, or longer duration, lower intensity illumination. To date, most ratio images have been generated using image intensifiers with very low intensity illumination. These low light level signals have rather low S/N.

Photobleaching can affect the ratio calculations if the species of fluorophores present have different photobleaching decay constants. As stated earlier, a probe can be a single fluorophore or a complex of multiple fluorophores. Consider a probe consisting of a dextran molecule labeled with both fluorescein and rhodamine that could be used for pH measurements in a ratio emission protocol (Geisow, 1984; Murphy *et al.*, 1984). It is well known that fluorescein has a much higher bleaching rate than rhodamine. If photobleaching of the fluorescein occurs but not rhodamine, the ratio would not only be a function of the pH but also of the exposure history of the fluorescein. In the case of a single fluorophore pH probe, for example, the protonated species might have a different decay constant than the unprotonated species.

The effects of photobleaching on the ratio can be assessed by continuously exposing the cells at a given wavelength. Periodically the pairs of images can be recorded for calculation of the ratio. Figure 6 illustrates such an approach using BCECF-stained cells. Continuous exposure resulted in a dramatic decrease in the fluorescence intensity over time. However, the mean ratio remained relatively constant until the signal level decreased into the noise. This result demonstrates that even extensive photobleaching of this probe is normalized by the ratio measurement. However, it is important to note that single wavelength measurements of fluorescence are very sensitive to variable photobleaching decay constants (Benson *et al.*, 1985).

Photochemical breakdown of a fluorophore can result in free radicals and other reactive by-products (Picciolo and Kaplan, 1984). The reaction of these by-products with cellular components could be harmful to the cell. However, since the probes are used as tracers and therefore are present in very small quantities, the by-products should be present in minimal quantities. In addition, only a very small proportion of the tracer quantities of the fluorophore will be bleached. Good bioassays should be performed over an extended period to evaluate any potential for photodynamic damage and as a guide to minimizing the total dose of excitation light. Most cells will exhibit some photoinduced toxicity (phototoxicity) which can be eliminated or at least minimized by controlling the total dose.

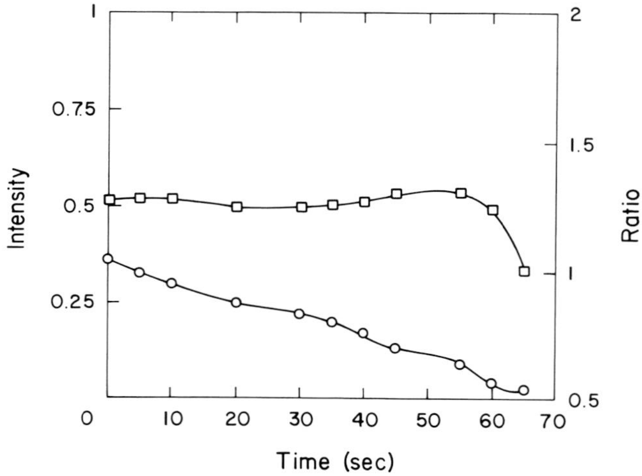

FIG. 6. Effects of photobleaching on the ratio. The pH of a single cell labeled with BCECF was clamped to 6.98. The cell was exposed to constant illumination at 489 nm. At periodic intervals images were recorded for the calculation of the ratio. Circles indicate the emission intensity at 489 nm excitation. Squares indicate the mean ratio. Reproduced from *J. Cell Biol.* **104,** 1019–1033 (1987) with permission from the Rockefeller University Press.

A second potential complication is the occurrence of photobleaching on the time scale of the measurement cycle. The excitation energy level, camera intensification, and dye concentration should be adjusted such that there is no significant photobleaching during the acquisition of each pair of images. If significant bleaching occurs, the ratio will be inaccurate. This can be tested by comparing the ratios calculated from the pair of images recorded in the order wavelength 1, wavelength 2 versus the order wavelength 2, wavelength 1. The sequence in which the two images are recorded should not be important for determining the ratio. If the sequence affects the ratio then the total excitation dose must be decreased.

4. FOCUS

The image represents a two-dimensional projection of a three-dimensional cell. Thus, the image includes both in-focus and out-of-focus components. The magnitude of the effect of out-of-focus components on the ratio will be related to the relationship between the thickness of the object and the depth of focus of the objective in an epifluorescence microscope. The effect of focus on the ratio measurement can be determined by calculating the ratio at various degrees of over and under focus.

For thin, substrate-attached, mammalian cells stained with BCECF the mean ratio was constant, within about 3% error, for 3 μm over or under focus (Bright *et al.,* 1987a). Thus, slight changes in focus did not significantly affect the measured values. Ratio images of pH-clamped cells were also uniform, indicating the lack of effect of larger amounts of out-of-focus fluorescence in the thicker nuclear region. For specimens much larger than the depth of focus of the objective, multiple images and deconvolution algorithms will be necessary to correct for the out-of-focus signals.

Focus also becomes a serious issue if the compartments are small with respect to the depth of field, for example, endosomes. Endosomes move quickly and in three dimensions, resulting in the need for very short image acquisition times to capture the ratio before they can go out of the plane of focus (Tycko *et al.,* 1983). In addition, out-of-focus signals from domains with a distinct value will affect the in-focus results.

B. Image Processing

1. Nonlinearity and Shading Correction

An ideal camera would produce signals that are linearly proportional to the intensity of the object. In addition, all pixels would have equivalent response functions. In practice, however, video cameras can have radiometric response curves with varying degrees of nonlinearity. Also, the camera response can vary from pixel to pixel depending on the unevenness of illumination of the field and uneveness of the pixel response (shading effect; Bright and Taylor, 1986; *see* Jerečivić *et al.,* Chapter 2, this volume; Young, Chapter 1, this volume).

Some degree of nonlinearity can be corrected by determining the radiometric response function, inverting that function, and applying the inverted function to an image (*see* Castleman, 1979; Green, 1983; Baxes, 1984). A fluorescent solution quantum counter provides an ideal flat field object (*see also* Sisken, Chapter 4, this volume). By dividing the pixels into groups having similar response functions, correction for shading can occur simultaneously.Establishing pixel response groups improves the shading–nonlinearity correction (Galbraith *et al.,* 1987). If a detector has a linear response, such as some vidicon-based cameras and particularly CCDs, the shading correction reduces to dividing the image to be corrected by the image of a fluorescent solution (Benson *et al.,* 1985) or the ratio image of a standard calcium solution (Poenie *et al.,* 1986). Issues of image correction and restoration are discussed elsewhere in this volume (*see* Jerečivić *et al.,* Chapter 2; Young, Chapter 1).

2. Geometric Distortion Correction

Cameras, particularly intensified versions, usually possess significant geometric distortion. This distortion manifests itself as pixels varying in size and shape over the area of the image. Intensified cameras are particularly prone to pin cushion distortion that varies with the level of intensification. Also, image rotation may occur as the intensifier high voltage is varied (Bright and Taylor, 1986). An advantage of solid-state detectors is their fixed geometric distortion due to the discrete structure of the chip (see Aikens *et al.,* Volume 29, this series). Correction of geometric distortion is not essential to the calculation of the ratio but is important for extracting quantitative spatial measurements, such as size and shape of compartments (see Jerečivić *et al.,* Chapter 2, this volume; Castleman, 1979; Green, 1983).

3. Image Registration

The calculation of the ratio of two images requires that the images be in register with each other. For ratio emission measurements this is a difficult requirement to fulfill as discussed earlier. Computationally expensive registration algorithms are required (see Jerečivić *et al.,* Chapter 2, this volume; Castleman, 1979; Green, 1983). For a filter wheel based ratio excitation microscope system there are few problems with misregistration. The excitation optical path is relatively insensitive to the alignment of the filters. Proper alignment of the excitation light and vibration-free filter changing are, however, very important.

4. Thresholding

An image contains both the object of interest and background. Removal of the background is beneficial for two reasons. (1) Since the ratio of noise to noise results in noise, background photocounts will result in an amplified noise signal. This noise can obscure the ratio image. Removal of the background via a threshold operation will make the information in the ratio image much more visible. (2) If calculations are performed only within the area of the cell, the number of calculations necessary can be dramatically reduced from that of a full 512 × 512 pixel image. The time to calculate the ratio will then be proportional to the area of the cell (however, see below).

A threshold operation involves making a mask that covers either the area to remain or the area to remove. Once defined, most image processors

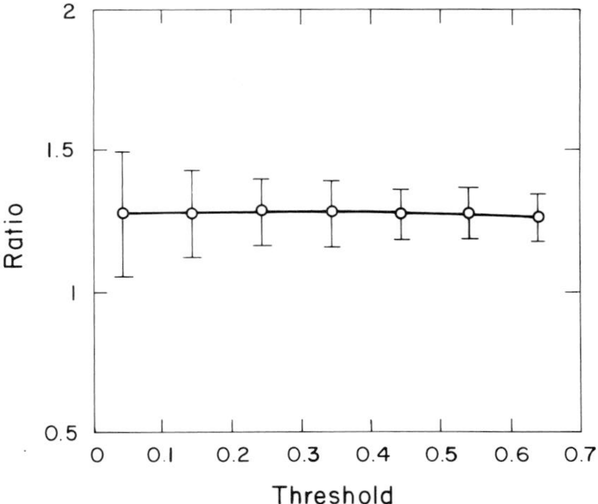

FIG. 7. Effects of threshold value selection on the ratio. The image (452 nm excitation) of a single pH-clamped BCECF-stained cell was divided into seven intensity levels. The mean and sample standard deviation of the ratio image was calculated for each level. The lower threshold value of each level is plotted on the x axis. Reproduced from *J. Cell Biol.* **104,** 1019–1033 (1987) with permission from the Rockefeller University Press.

contain commands for very fast application of the mask for separating the two parts. Defining the mask can be an interactive or a noninteractive process. It is extremely difficult to successfully apply noninteractive edge detection algorithms to low-light-level images of thin mammalian cells, since the S/N is usually very low at the edge (unpublished observations).

To date, the most useful approach has been to define the mask interactively, taking advantage of the many visual cues. Figure 7 illustrates the effectiveness of the interactive approach to determining a threshold value using a pH-clamped BCECF-stained cell as a model. A series of narrow masks was created by selecting lower and upper threshold values that divided the intensity range into seven levels. The lowest and highest threshold values generally represented the thin peripheral edges and the thick nuclear regions, respectively. The ratio at the edge of the cell should be the same as that at the center in pH-clamped cells. Figure 7 illustrates that the mean ratios were similar for all regions from the center to the edge of the cell. If the threshold value was too low and included regions with very low S/N, the ratio value at the edge would deviate significantly from the ratio at the center. The figure also illustrates the decrease in the S/N (i.e., increase in standard deviation) in the thin periphery of the cell as opposed to the center of the cell.

5. Numerical Methods

For a 512 × 512 pixel image, up to 262.144 operations may be necessary to calculate the ratio of two images. The time for a conventional computer to calculate a ratio can be several minutes. If large numbers of images need to be processed, the time can be prohibitive. Image processors contain special hardware to greatly accelerate calculation on images as discussed earlier. However, this special hardware can perform arithmetic operations with only limited numerical precision operations. The ALUs and LUTs in modern image processors typically perform only 8 or 12-bit fixed point operations.

The ratio of two numbers, A and B, can be calculated in three ways:

Division: A/B (5)

Log subtraction: $\exp(\ln A - \ln B)$ (6)

Inversion: $A(1/B)$ (7)

Image processors usually implement a divide command using either Eq. (6) or Eq. (7). Since logarithm and inverse operations can be implemented with LUTs, these operations can be performed at, or near, real

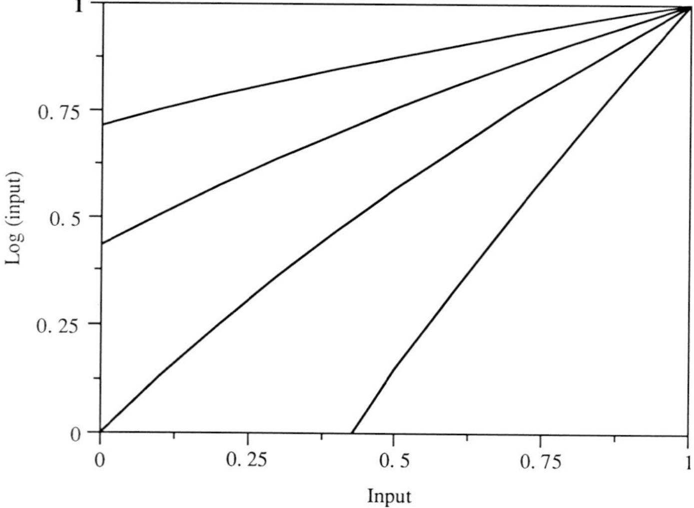

Fig. 8. An example of default logarithm LUTs with various scaling and offset factors for use in a 12-bit LUT in a particular image processor. These functions are very different from the expected functions. They work well for image enhancement purposes, but are inappropriate for quantitative analyses.

time. However, since LUTs have limited precision there can be a loss of information. Also, because of limited precision these operations are not implemented quite as expected. Figure 8 illustrates default logarithm LUTs available on one particular image processor. These functions include additional scaling and offset factors to compensate as much as possible for the limited precision of 12-bit LUTs. Appropriate selection of scaling and offset factors may allow the application of Eq. (6) or (7) to the accurate computation of the ratio (Rogowska *et al.*, 1988) but the most accurate calculation will come from the application of Eq. (5) using either a high-speed floating point coprocessor or an array processor (Bright *et al.*, 1987a; Rogowska *et al.*, 1988). The decreasing costs of computer memory will make it feasible to use 16-bit LUTS which will significantly decrease the dependency on floating point hardware in the future and allow many more real-time manipulations.

V. Measurements Using Fluorescence Ratio Imaging

Fluorescence ratio imaging microscopy involves five basic steps: (1) An image of the cell and background is recorded at each wavelength; (2) the background images are subtracted from the cell images; (3) image restoration algorithms for correction for shading, nonlinearity, etc, are applied to all images; (4) the ratio of the two resultant images is calculated; (5) the ratio images are calibrated in terms of the parameter under investigation. There are, however, specific issues that are important for the successful and accurate recording of temporal and spatial signals from living cells.

A. Temporal Analysis

The ability to faithfully record a signal is directly related to both the speed at which the signal is generated and the speed and accuracy of the measurement system. The sampling theorem states that a signal must be sampled minimally at twice the highest frequency present. If this is not done, aliasing will occur leading to a misrepresentation of the signal.

There are several temporal components to the ratio imaging method: time per frame, number of frames to average, time to switch wavelengths, time to store image onto a mass storage unit, and time to process the ratio. The ability to record a temporal process requires the appropriate selection of these parameters. The time per frame is typically fixed by the use of video cameras with a rate of 30 frames per second. The number of frames to average is a function of the S/N, input excitation energy, probe

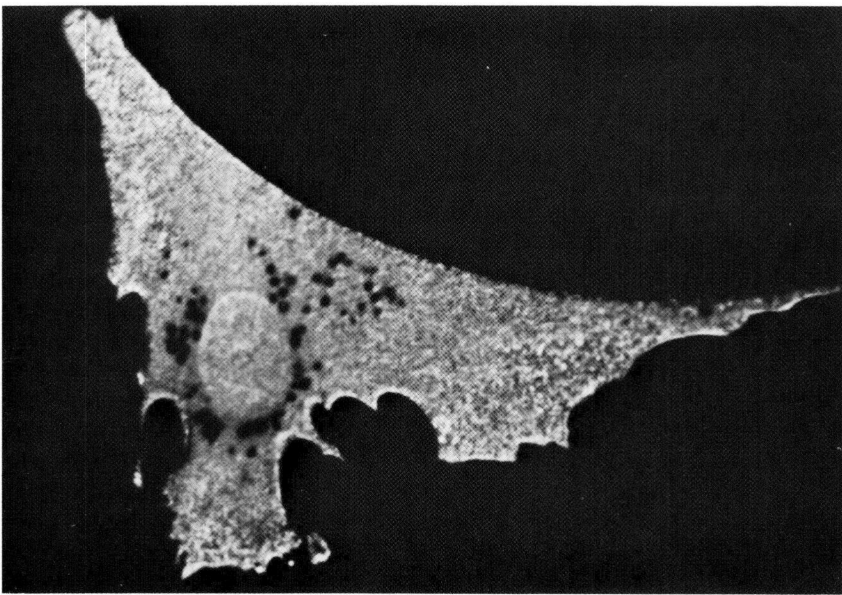

FIG. 9. Ratio image of fluorescein–dextran labeled cell illustrates presence of nuclear, vesicular, and cytoplasmic subcellular compartments. The nucleus is visible due to a slight translation between the image pairs. No difference in pH has been detected when comparing cytoplasm and nucleus. The acidic vesicles represent autophagosomes. Reproduced from *J. Cell Biol.* **104,** 1019–1033 (1987) with permission from the Rockefeller University Press.

photostability, dynamic range of the detector, and rate of movement of the cellular compartment. The time to switch wavelengths is a function of the hardware, method of implementation, and time scale of the cellular movement. The time to store images on a mass storage device is related to the particular computer hardware. This can range from many seconds or minutes on a personal computer to real time using a video disk.

Since current implementations of ratio imaging require the acquisition of images in rapid succession, any cellular movement during that acquisition can make analysis difficult or impossible. If the compartment is large and homogeneous, these slight movements may be tolerated if only the average value within the compartment is of interest. The edges can be ignored (see nucleus, Fig. 9). If the compartment is small, such as an endocytic vesicle, there are so few pixels representing the image that the only answer is to sample the data at a higher rate. For example, in Fig. 10, a cytoplasmic mean pH can be calculated, but the membrane ruffle moved during the acquisition sequence so no local measurement of pH in the ruffle is

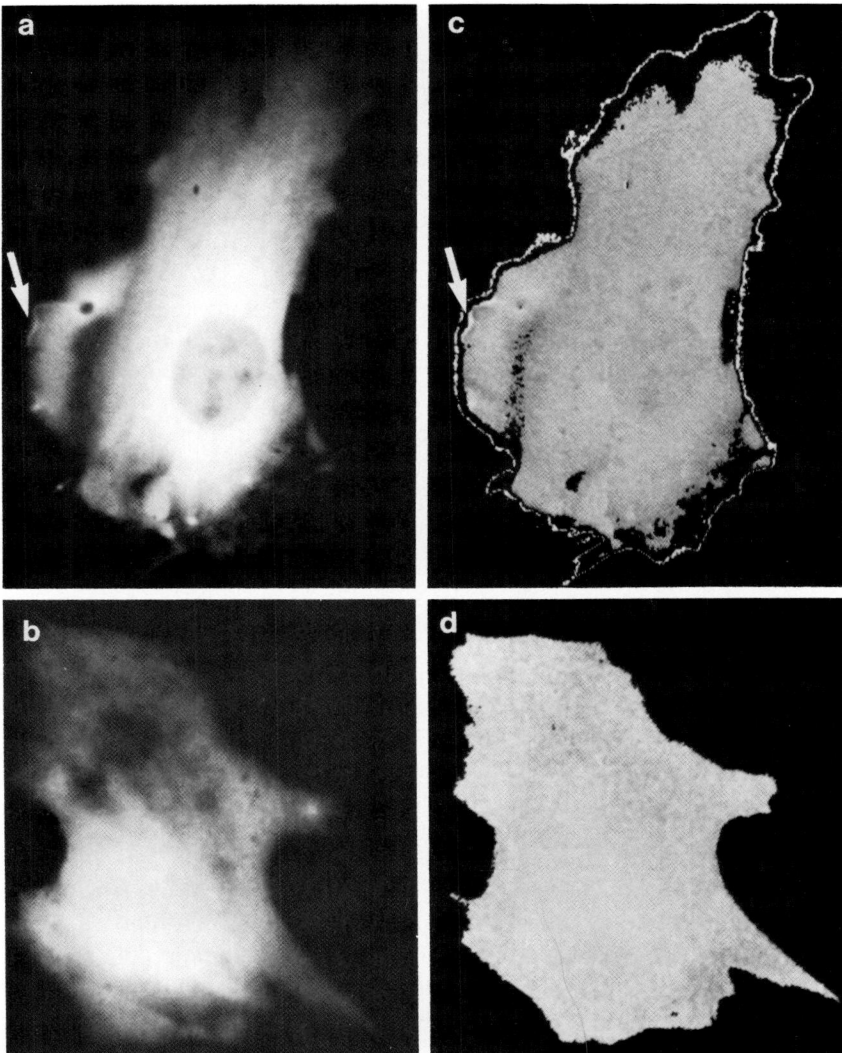

FIG. 10. Comparison of cells labeled with fluorescein–dextran and BCECF. (a, b) Fluorescein–dextran-labeled cell shows exclusion of the probe (70 kDa dextran) from the nucleus and limited penetration to the very peripheral edges of the cell. The outline in b defines the edge of the cell. The arrow in a illustrates an area of ruffling activity that results in the local misalignment of the image pairs. The high resultant ratios are, therefore, meaningless in this local region. (a) 489 nm excitation, (b) ratio. (c, d) BCECF-labeled cell shows penetration of the probe throughout the volume of the cell including the very peripheral edges. A large dynamic range is needed to image both the thick nuclear region and the thin peripheral edges of the cell. (c) 489 nm excitation, (d) ratio. Reproduced from *J. Cell Biol.* **104,** 1019–1033 (1987) with permission from the Rockefeller University Press.

possible. Areas exhibiting movement must be removed by proper threshold selection for calculation of the mean ratio. The ultimate test for finding the appropriate time scale of temporal kinetics is to use a very rapid wavelength switching mechanism and a photomultiplier. By limiting the measurement aperture to the small region of interest, the expected temporal kinetics can be determined using a photomultiplier. Appropriate configuration to the imaging system can then be applied.

The signal rates are usually limited by the time necessary to record an image with a sufficient *S/N*. Both the photobleaching rate constant and fluorescence intensity are linear functions of the excitation intensity. A higher excitation intensity means a higher signal but also a higher photobleaching rate. However, fewer numbers of frames are needed to attain a given *S/N*. Thus, by using a higher excitation intensity for a shorter period of time the same quality image can be recorded with no increase in photobleaching. A frame transfer, cooled CCD camera combined with a short-duration pulse of excitation should optimize the temporal resolution of ratio imaging.

B. Spatial Analysis

A basic goal of fluorescence ratio imaging is to determine a cellular map of the parameter of interest. This may be a variation in membrane-bound compartments such as endosomes (Tycko *et al.*, 1983; Tanasugarn *et al.*, 1984) or maybe a cytoplasmic gradient or uneven distribution (Paradisco *et al.*, 1987; Poenie *et al.*, 1987). The strength of fluorescence ratio imaging over other measurement techniques is its ability to provide quantitative spatial information. However, there are some trade-offs due to the use of intensified cameras. One of the more significant trade-offs is between the level of intensification and *S/N*. Although the area represented by a single pixel can be determined (Bright *et al.*, 1987a), the ultimate limit in spatial resolution is directly related to the spatial *S/N*. Excitation levels are generally kept low to avoid photobleaching, which requires cameras with high levels of intensification. As the level of intensification is increased, the spatial *S/N* decreases. This decrease can be offset to an extent by using image averaging. Image averaging improves the *S/N* proportional to the square root of the number of frames averaged (see Jerečivić *et al.*, Chapter 2, this volume; Young, Chapter 1, this volume). The improvement is possible because the noise is random and not correlated from frame to frame. It is also possible to improve the spatial *S/N* by spatial averaging. It is important to determine the spatial resolution under the same conditions that the experiments are to be performed. Figure 11 illustrates the relationship of spatial sample size and *S/N* using a 10 μM BCECF

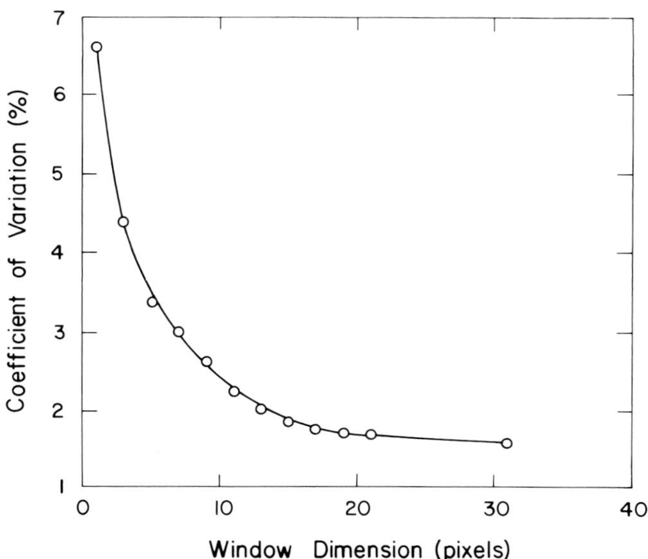

FIG. 11. Relationship between spatial sample size and S/N. A 10 μM solution of BCECF with a pH of 6.9 was used as an ideal specimen. A series of 60 regions was defined uniformly over the usable area of the field. Each area varied in size from 1 pixel (0.074 μm^2) to 31 × 31 pixel (71.6 μm^2) areas. The sample variance is used as a measure of the S/N. Reproduced from *J. Cell Biol.* **104,** 1019–1033 (1987) with permission from the Rockefeller University Press.

solution. This concentration results in a signal similar to that of BCECF-stained cells (Bright *et al.,* 1987a). Spatial averaging greatly improves the S/N at the expense of resolution. In this sample a pH could be reliably determined (<3% CV) if an area of 3.65 μm^2 (7 × 7 pixels) is averaged.

There are several methods for the detection of spatial variations. The simplest approach is to display the ratio image using pseudocolor mapping (see Tsien and Poenie, 1986). The human eye is more sensitive to color changes than to variations in grey scale. Thus, small differences in ratio may become visible as differences in color. The most common color scheme is the mapping of dark to bright as blue to red according to the visible spectrum.

If a cell contains areas of varying ratio, this will be reflected in the histogram of the pixel values. Comparing the histogram before and after clamping the cell to a given value will demonstrate any variations. The clamped cell will exhibit spatial variations due only to low S/N factors. Any deviation from this distribution in nonclamped cells indicates the possibility of a spatial variation in the ratio. Figure 12 shows histograms for

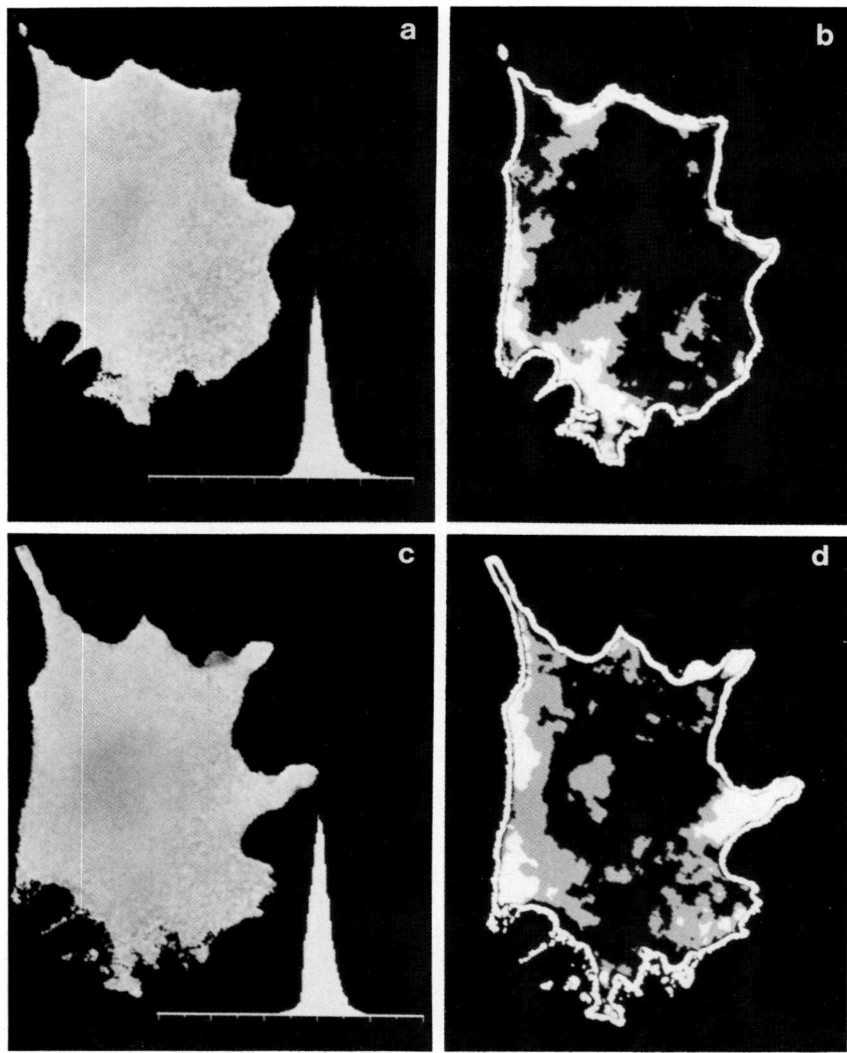

FIG. 12. Distribution of ratio values inside a single BCECF-stained cell before and after clamping pH. The histogram of ratio values in a pH-clamped cell (a) shows a narrow and approximately normal distribution, whereas the same cell before clamping (c) shows a slightly skewed distribution. The images are 7 × 7 spatial averages of the originals. Difference maps of these same images (b, d) show spatial variations that can be most easily explained by variations in S/N. Black (>5%), grey (5–10%), white (>10%). White edges are due to spatial filtering process. Reproduced from *J. Cell Biol.* **104,** 1019–1033 (1987) with permission from the Rockefeller University Press.

the same BCECF-stained cell after (a) and before (c) clamping the pH. The distribution is slightly skewed in the cell before clamping; however, no readily visible spatial variation above noise level could be found (Bright *et al.*, 1987a).

The difference map (previously referred to as an error map; Bright *et al.*, 1987a) is an alternate approach to the detection of variations which also provides spatial information. The difference map effectively compares each pixel to the mean ratio of the whole cell or of a subregion. The difference map represents the percent difference a pixel value is from a given value according to:

$$\text{Image dest}(i,j) = \frac{\text{Image src}(i,j) - \text{mean}}{\text{mean}} \times 100 \qquad (8)$$

The difference map will represent a combination of variations in S/N and any spatial variations in the ratio. Figure 12B and D shows magnitude difference maps (absolute value of difference) of the same pH-clamped (b) and nonclamped (d) cell discussed above. In this example, the spatial variations, particularly around the edges, are most easily explained by variations in S/N (Bright *et al.*, 1987a). The more effective display of the difference map is without the limitation of the absolute value and displaying it in color with positive differences distinguished from negative differences.

Organelles within the cytoplasm undergo both Brownian and directed translational movement. The organelle may be visible due to either specific labeling or to total exclusion of the probe. The movement of the organelle on the time scale of the acquisition sequence will affect the ratio calculations. Areas of local movement will give wrong values for the ratio. Local movement may be recognized by adding a third image to the acquisition sequence. The sequence becomes image recorded at the sensitive wavelength plus two successive images recorded at the insensitive wavelength. The duplicate images can be either subtracted or ratioed. Since these images are recorded on the same time scale, any variations are due to local movements, cell shape changes, and S/N. These variations can be compared to the variations in the ratio image. If the variations correlate, the ratio is likely to be a misrepresentation of the measurement parameter.

VI. Conclusions and Prospectus

Fluorescence imaging microscopy offers distinct advantages over other methods of measuring fluorescence signals. (1) Single cell spatial kinetic information is available using high magnification. Multiple cells can be monitored within the same experiment with the computer-controlled stage. (2) Low magnification allows single cell kinetics to be monitored with lower spatial information on larger numbers of cells in a single field. Multiple fields increase the total number of cells that can be analyzed in a single experiment. (3) The imaging approach allows debris, which may be labeled nonspecifically, to be removed either by selecting fields that do not have such artifacts or by removing them during the image analysis stage. (4) The imaging approach also allows the selection of particular types of cells for analysis (e.g., flat cells, round cells, dividing cells). Thus, a single cell analysis allows the collection and classification of individual cell types. (5) The ability to spectrally resolve multiple fluorescent probes allows the computer-controlled microscope to monitor multiple specific parameters within the same cell. The fluorescence imaging microscope provides the necessary instrumentation for correlating multiple physiological parameters within living cells (DeBiasio *et al.,* 1987; see Waggoner *et al.,* Chapter 17, this volume). (6) Spatially resolving signals provides much more information than mean values. For example, a particular signal may be localized in a small region of the cell and averaging over the whole cell would underestimate the signal.

With continued development and improvement in fluorescent probes for ion measurements (see Tsien, Chapter 5, this volume) as well as the development of probes for other important molecules, the potential applicability of ratio imaging is enormous. Fluorescence ratio imaging is a spectroscopic technique and is usable for any technique that can be expressed by the relationship of signals at two or more wavelengths. Other possible parameters include steady-state polarization, resonance energy transfer, and mapping the relative distributions of fluorescent analogs.

Acknowledgments

We thank Dr. Fred Lanni for his comments and suggestions on the manuscript. We also thank our many colleagues within the Center for Fluorescence Research for many helpful discussions during this work. This work was supported by grants from the NIH (AM-32461; 1-PO1-GM-34639); NSF (DMB-8414772); and Tobacco Council (1412a) to DLT. G.R.B. has been supported by a Leukemia Society Postdoctoral Fellowship and G.W.F. has been supported by a NIH Postdoctoral Fellowship (Al-07266).

References

Almers, W., and Neher, E. (1985). *FEBS Lett.* **192,** 13–18.

Aubin, J. (1979). *J. Histochem. Cytochem.* **27,** 36–43.

Baxes, G. A. (1984). "Digital Image Processing: A Practical Primer." Prentice Hall, New York.

Benson, D. M., Bryan, J., Plant, A. L., Gotto, A. M., Jr., and Smith, L. C. (1985). *J. Cell Biol.* **100,** 1209–1323.

Bevington, P. R. (1969). "Data Reduction and Error Analysis for the Physical Sciences," McGraw-Hill, New York.

Bright, G. R., and Taylor, D. L. (1986). *In* "Applications of Fluorescence in the Biomedical Sciences" (D. L. Taylor, A. S. Waggoner, R. F. Murphy, F. Lanni, and R. R. Birge, eds), pp. 257–288. Liss, New York.

Bright, G. R. Fisher, G. W., Rogowska, J., and Taylor, D. L. (1987a). *J. Cell Biol.* **104,** 1019–1033.

Bright, G. R., Rogowska, J., Fisher, G. W., and Taylor, D. L. (1987b). *BioTechniques* **5,** 556–563.

Castleman, K. R. (1979). "Digital Image Processing." Prentice Hall, New York.

Chaillet, J. R., and Boron, W. F. (1985). *J. Gen. Physiol.* **86,** 765–794.

Chaillet, J. R., Amsler, K., and Boron, W. F. (1986). *Proc. Natl. Acad. Sci. U.S.A.* **83,** 522–526.

Collet, M. (1985). *Photonics Spectra* **Sept.** 103–113.

Connor, J. A. (1986). *Proc. Natl. Acad. Sci. U.S.A.* **83,** 6179–6183.

DeBiasio, R., Bright, G. R., Ernst, L. A., Waggoner, A. S., and Taylor, D. L. (1987). *J. Cell Biol.* **105,** 1613–1622.

Fay, F., Fogarty, K., and Coggins, J. (1986). *Soc. Gen. Physiol.* **40,** 51–64.

Galbraith W., Ryan K. W., Gliksman N., Taylor D. L., and Waggoner A. S. (1987). *In* "Biomedical Imagery: Functional Mapping and Real Intelligence" (R. S. Ledley and D. L. McEachron, eds.). American Association for the Advancement of Science, Washington, D. C. (in press).

Geisow, M. J. (1984). *Exp. Cell. Res.* **150,** 29–35.

Gottlieb, M., Feichtner, J. D., and Conroy, J. (1980). *Soc. Photo-Opt. Instrum. Eng.* **232,** 33–41.

Graber, M. L., DiLillo, D. C., Friedman, B. L., and Pastoriza-Munoz, E. (1986). *Anal. Biochem.* **156,** 202–212.

Green, W. B. (1983). "Digital Image Processing: A Systems Approach." Van Nostrand Reinhold, Princeton, New Jersey.

Grynkiewicz, G., Poenie, M., and Tsien, R. Y. (1985). *J. Biol. Chem.* **260,** 3440–3450.

Heiple, J., and Taylor, D. L. (1980). *J. Cell Biol.* **86,** 885–890.

Heiple, J., and Taylor, D. L. (1982a). *J. Cell. Biol.* **94,** 143–149.

Heiple, J., and Taylor, D. L. (1982b). *In* "Intracellular pH: Its Measurements, Regulation, and Utilization in Cellular Functions" (R. Nuccitelli and D. W. Deamer, eds.), pp. 22–54. Liss, New York.

Highsmith, S., Bloebaum, P., and Snowdowne, K. W. (1986). *Biochem. Biophys. Res. Commun.* **138,** 1153–1162.

Inoue, S. (1986). "Video Microscopy." Plenum, New York.

Keith, C. H., Ratan, B., Maxfield, F. R., Bajer, A., and Shelanski, M. L. (1985). *Nature (London)* **316,** 848–850.

Kohen, E., Hirschberg, J. G., Kohen, C., Wouters, A., Pearson, H., Salmon, J. M., and Thorell, B. (1975). *Biochim. Biophys. Acta* **396,** 149–154.

Kruskal, B. A., Shak, S., and Maxfield, F. R. (1986). *Proc. Natl. Acad. Sci. U.S.A.* **83,** 2919–2923.

Kurtz, I., and Balaban, R. S. (1985). *Biophys. J.* **48,** 499–508.

Kurtz, I., Douelle, R., and Kutzka, P. (1987). *Biophys. J.* **51,** 287a.

Li, Q., Altschuld, R. A., and Stokes, B. T. (1987). *Biochem Biophys. Res. Commun.* **147,** 120–126.

Luby-Phelps, K., and Taylor, D. L. (1988). Submitted.

Luby-Phelps, K., Taylor, D. L., and Lanni, F. (1986). *J. Cell Biol.* **102,** 2015–2022.

MacDonald, V., Keiger, J., and Jobsis, F. (1977). *Arch. Biochem. Biophys.* **184,** 423–430.

Maruyama, I., Oyamada, H., Hasegawa, T., Ohtsuka, K., and Momose, K. (1987). *FEBS Lett.* **220,** 89–92.

Murphy, R. F., Powers, S., and Cantor, C. R. (1984). *J. Cell Biol.* **98,** 1757–1762.

Musgrove, E., Rugg, C., and Hedley, D. (1986). *Cytometry* **7,** 347–355.

Ohkuma, S., and Poole, B. (1978). *Proc. Natl. Acad. Sci. U.S.A.* **75,** 3327–3331.

Oldham, P. B., Patonay, G., and Warner, I. M. (1985). *Rev. Sci. Instrum.* **56.** 297–302.

Paradiso, A. M., Tsien, R. Y., and Machen, T. E. (1987). *Nature (London)* **325,** 447–450.

Picciolo, G. L., and Kaplan, D. S. (1984). *Adv. Appl. Microbiol.* **30,** 197–234.

Poenie, M., Alderton, J., Steinhardt, R., and Tsien, R. Y. (1986). *Science* **233,** 886–889.

Poenie, M., Tsien, R. Y., and Schmitt-Verhulst, A. M. (1987). *EMBO J.* **6,** 2223–2232.

Preston, R. A., Murphy, R. F., and Jones, E. W. (1987). *J. Cell Biol.* **105,** 1981–1987.

Ratan, R. R., Shelanski, M. L., and Maxfield, F. R. (1986). *Proc. Natl. Acad. Sci. U.S.A.* **83,** 5136–5140.

Rink, T. J., Tsien, R. Y., and Pozzan, T. (1982). *J. Cell Biol.* **95,** 189–196.

Rodbard, D., and Hutt, D. M. (1974). *Symp. RIA Related Procedures Med.* pp. 165–192.

Rogowska, J., Bright, G. R., Preston, Jr., K., and Taylor, D. L. (1988). Submitted.

Rothenberg, P., Glaser, L., Schlesinger, P., and Cassel, D. (1983). *J. Biol. Chem.* **258,** 12644–12653.

Sawyer, D. W., Sullivan, J. A., and Mandel, G. L. (1985). *Science* **230,** 663–666.

Scanlon, M., Williams, D. A., and Fay, F. S. (1987). *J. Biol. Chem.* **262,** 6308–6312.

Slayter, E. M. (1970). "Optical Methods in Biology." Wiley(-Interscience), New York.

Spray, D. C., Nerbonne, J., Campos De Carvalho, A., Harris, A. L., and Bennett, M. V. L. (1984). *J. Cell Biol.* **99,** 174–179.

Spring, K. R., and Smith, P. D. (1987). *J. Microsc.* **147,** 265–278.

Tanasugarn, L., McNeil, P., Reynolds, G. T., and Taylor, D. L. (1984). *J. Cell Biol.* **98,** 717–724.

Taylor, D. L., Amato, P. A., McNeil, P. L., Luby-Phelps, K., and Tanasugarn, L. (1986). *In* "Applications of Fluorescence in the Biomedical Sciences" (D. L. Taylor, A. S. Waggoner, R. F. Murphy, F. Lanni, and R. R. Birge, eds.), pp. 347–376. Liss, New York.

Thomas, J. A. (1986). *Soc. Gen. Physiol.* **40,** 311–325.

Thomas, J. A., Buchsbaum, R. M., Zimniak, A., and Racker, E. (1979). *Biochemistry* **18,** 2210–2218.

Tsien, T. Y. (1981). *Nature (London)* **290,** 527–528.

Tsien, R. Y., and Poenie, M. (1986). *Trends Biochem. Sci.* **11,** 450–455.

Tsien, R. Y., Pozzan, T., and Rink, T. J. (1982). *J. Cell Biol.* **94,** 325–334.

Tsien, R. Y., Rink, T. J., and Poenie, M. (1985). *Cell Calcium* **6,** 145–157.

Tycko, B., Keith, C. H., and Maxfield, F. R. (1983). *J. Cell Biol.* **97,** 1762–1776.

Waggoner, A. S. (1986). *In* "Applications of Fluorescence in the Biomedical Science" (D. L. Taylor, A. S. Waggoner, R. F. Murphy, F. Lanni, and R. R. Birge, eds.), pp. 3–28. Liss, New York.

Wier, W. G., Cannell, M. B., Berlin, J. R., Marban, E., and Lederer, W. J. (1987). *Science* **235,** 325–328.

Williams, D. A., Fogarty, K. E., Tsien, R. Y., and Fay, F. S. (1985). *Nature (London)* **318,** 558–561.

Chapter 7

Fluorescent Indicators of Membrane Potential: Microspectrofluorometry and Imaging

DAVID GROSS

Department of Biochemistry and Program in Molecular and Cellular Biology
University of Massachusetts
Amherst, Massachusetts 01003

LESLIE M. LOEW

Department of Physiology
University of Connecticut Health Center
Farmington, Connecticut 06032

I. Introduction

A. Background

The electric potential difference between the interior of a cell and the extracellular medium (the membrane potential) has long been known to play an important role in the mediation, regulation, and delivery of many transmembrane signals. For example, at the neuromuscular junction, the binding of acetylcholine released from the presynaptic axon terminal will bind and activate acetylcholine receptors located in the postsynaptic muscle cell, triggering a reduction in the magnitude of the postsynaptic cell membrance potential, which then mediates the further propagation of the electrical signal leading ultimatley to muscle contraction. Nerve and muscle cells most often come to mind when one considers membrane potential-mediated cellular events, but nonexcitable cells also demonstrate changes in membrane potential upon cell stimulation. Thus, it is desirable to be able to measure cell membrane potential in living cells in a noninvasive manner.

Fluorescent indicators of membrane potential have been developed over the last 15 years to achieve this end. The dyes have been broadly classified as either "fast" or "slow," depending on whether they are suitable for measurements of millisecond changes in excitable cells, or smaller cell physiological changes occuring in the minute time scale. In general, the fast dyes have been applied to cases where AC coupling and/or signal averaging can be employed to extract a small potential-dependent optical response from a large light background. The slow dyes are much more sensitive, but have been applicable mainly to experiments with large populations of cells in a fluorometer. These applications are the subjects of several excellent recent reviews (Salzberg, 1983; Grinvald, 1985; Waggoner, 1985; London *et al.*, 1986; Loew, 1988). In this chapter, we will outline the application of a "fast" dye in measurements of both the spatial and temporal distribution of membrane potential induced by an external electric field. We will also describe a new series of "slow" dyes which allow the measurement of membrane potential in individual cells through the microscope.

B. Membrane Electrical Properties

1. Nernstian Membranes

The term "membrane potential" is often used to elaborate bioelectric events at the level of the cell membrane, even though it provides only an incomplete picture of the distribution of electric fields which ultimately generate forces on the molecules in the vicinity of the membrane. These fields play a crucial role both in the control of bioelectric effects as well as in the generation of signals one can measure with fluorescent probes.

The nonequilibrium flux of ions across an ion-selective membrane leads to the generation of an electric field across the membrane such that the electric current due to the concentration gradient of ions across the membrane is balanced by an equal and opposite flow of electric current due to the electric field that is set up across the membrane. Assuming a uniform electric field E in the membrane, the electric potential difference V_m across the membrane of thickness d is $V_m = E_d$. At steady state, this membrane potential is constant and, for a single permeable ion species, is

$$V_m = (RT/ZF) \ln(C_o/C_i) \tag{1}$$

where R is the gas constant, T the absolute temperature, Z the valence of the permeable species, F the Faraday constant, and C is the concentration of the permeant ion either outside (subscript o) or inside (subscript i) the cell. For cations which are at higher concentration inside the cell than outside, V_m has a negative value. Most cell membranes at rest are predominantly permeable to potassium ions, thus the resting V_m is negative.

Real membranes are permeable to more than one ion species, leading to a more complicated relationship between V_m and ion concentrations than that described by the Nernst equation, Eq. (1). An extended treatment of the more complex situation leads to the Goldman–Hodgkin–Katz equation, which is of similar but expanded form compared to the Nernst equation (Aidley, 1971). The Goldman–Hodgkin–Katz equation is simplest to derive if one assumes that the electric field in the membrane is uniform. Although the field must be nonuniform at the molecular level, a uniform mean field at the macroscopic level provides a sensible model from which data can be modeled. The distribution of electric potential according to this model is shown in Fig. 1a. The electric potential profile is flat in the intracellular and extracellular aqueous media since no electric current flows through either phase. One should note that the potential difference which drops across the membrane is identical to that which drops between the intra- and extracellular aqueous phases.

2. Surface Charge Effects

It is well known that real biological membranes have associated with them fixed surface charges which usually are negative. The presence of surface charge alters the distribution of electric potential in the vicinity of the membrane since the negative surface charge must be balanced by opposing positive charge in the aqueous phases adjacent to the membrane in order that the whole membrane system be electrically neutral. This

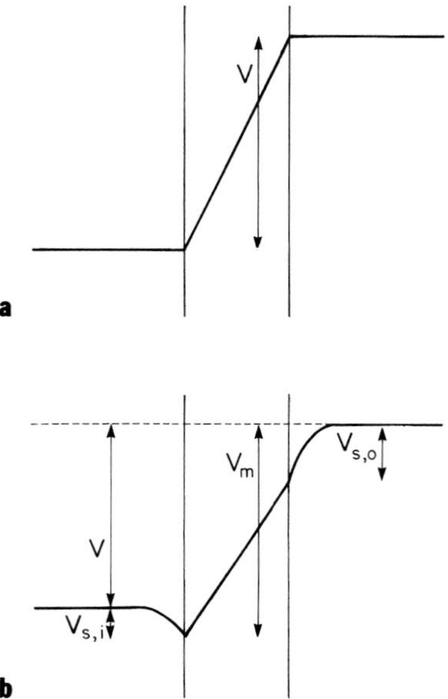

FIG. 1. (a) Simplified membrane potential profile for a Nernstian membrane with no surface charge. (b) Potential profile for a Nernstian or Goldman–Hodgkin–Katz membrane including surface charge effects.

opposition of two layers of charge is termed an electrical double layer, and is well characterized in many electrochemical systems (Bockris and Reddy, 1970). Although the theories which describe the electrical double layer are only approximate, they have been remarkably successful in explaining surface charge effects.

Gouy–Chapman double layer theory is often applied to biological membranes (McLaughlin, 1977). In this theory, the attraction of the negative surface charge of the membrane holds a cloud of cations near it while diffusion tends to disperse the cloud. The balance of these two forces produces a steady-state distribution of ions such that the concentration of cations is enhanced very near the membrane and drops to the concentration found in the bulk aqueous phase over a distance of a few angstroms. This charge distribution creates a surface potential V_s which is defined from the bulk aqueous phase to the membrane surface; thus, for a negatively charged membrane surface, V_s is negative.

The effect of this surface potential on the membrane potential profile is twofold. First, the electric potential drops from both of the membrane surfaces to the bulk of the aqueous phase will add to or subtract from the potential across the membrane proper due to ion gradients across the membrane, i.e.,

$$V = V_m + V_{s,o} - V_{s,i} \tag{2}$$

where V is the potential drop between the aqueous bulk phases, V_m the potential drop across the membrane itself, and $V_{s,i}$ ($V_{s,o}$) is the inner (outer) surface potential. Unless $V_{s,i}$ and $V_{s,o}$ are equal in sign and magnitude, the potential difference between the bulk aqueous phases will be different from that across the membrane itself.

The second effect of the surface potential upon the distribution of membrane potential in the vicinity of the membrane arises from the alteration of ion concentration at the surfaces of the membrane relative to that in the bulk aqueous medium. According to both the Nernst and the Goldman–Hodgkin–Katz equations, the potential difference generated across the membrane due to transmembrane ion gradients is related to the concentrations at the surface of the permselective membrane. It has been shown that, for simple cases, this effect leads to a relationship between the potentials in Eq. (2) such that V is equal to the Goldman–Hodgkin–Katz potential calculated from bulk ion concentrations and V_m is reduced or enhanced from that value according to Eq. (2) (Gross *et al.*, 1983). The distribution of electric potential in the vicinity of the membrane, based on Eq. (2), is shown in Fig. 1b. The differences between the potential drop across the plasma membrane itself, V_m, and that from bulk-to-bulk aqueous phases, V, are critical to the understanding of the response of fluorescent indicators of membrane potential.

In this chapter, we will discuss two different types of fluorescent indicators of membrane potential, membrane-bound probes which respond directly to the electric field within the membrane itself and charged membrane-permeable probes which redistribute passively across the membrane according to the Nernst equation. The latter probes, because they must redistribute to steady state upon a change in V_m, are relatively slow indicators, while the former are much faster since they respond directly at the molecular level to changes in the membrane electric field.

3. Applied Electric Field Effects

So far we have considered the effects of transmembrane ion concentration gradients and surface charge effects in the generation of membrane potential. A third mechanism by which the electric potential across the membrane can be altered involves the interaction of extracellular electric fields with the cell. If a spherical cell of radius a is placed in a uniform electric field E, a steady-state electric potential difference is generated across the membrane of value

$$\Delta V_m = -(3/2)Eaf \cos \theta \qquad (3)$$

where θ is the polar angle relative to the direction of the applied field and $f = 2\sigma_o\sigma_i/[a(\sigma_m/\delta)(2\sigma_o + \sigma_i) + (2\sigma_o + \sigma_m)(2\sigma_m + \sigma_i)]$ is a factor dependent on cell size, membrane thickness δ and the conductivities of the extracellular medium, the intracellular medium, and the membrane (σ_o, σ_i, and σ_m, respectively) (Gross *et al.,* 1986). For most healthy cells, $\sigma_o \approx \sigma_i \approx 10^4 a\sigma_m/\delta$, so the factor f reduces to unity (Jaffe and Nuccitelli, 1977; Poo, 1981).

This electric-field-induced membrane potential is different from the previously discussed contributions to membrane potential in two important aspects. First, this potential is a function of position around the cell, with maximal effects occuring at the sides of the cell where the membrane is perpendicular to the applied field direction and no effect at the poles. Second, this potential is directly proportional to the strength of the extracellular electric field, which can be manipulated easily with a simple power supply. These features allow one to directly calibrate membrane-bound probes in living cells as well as to demonstrate that a spatially varying membrane potential can be generated within a cell membrane.

Equation (3) is derived from Laplace's equation of electrostatics and the conditions which apply to the boundaries of the cell and in the medium far from the cell. When the cell membrane is poorly conductive with respect to the intra- and extracellular milieus, the magnitude of the induced electric field within the membrane proper is very much larger than that of the extracellular electric field. This effect results primarily from the thinness of the cell membrane; a potential drop across cellular dimensions when applied to the thin membrane is effectively amplified by the shift in dimension, i.e., by a factor a/δ. Thus, the membrane potential changes described by Eq. (3) have greatest effect directly within the membrane proper, and are very much attenuated in all other regions near the cell, including the region of the electric double layer.

Combining Eqs. (2) and (3) leads to an expression for the electric potential drop across the cell membrane from bulk extracellular aqueous phase to bulk intracellular aqueous phase:

$$V = V_m - (3/2)Eaf \cos \theta + V_{s,o} - V_{s,i} \qquad (4)$$

The responses of Nernstian and bilayer-bound fluorescent probes in cells exposed to an extracellular electric field are very different. The membrane-bound dye, by virtue of its direct response to the membrane electric field, will indicate the position-dependent electric field strength due to the Goldman–Hodgkin–Katz potential, the modifying effects of the two surface potentials, and that of the applied field as shown in Eq. (4). The Nernstian dye, however, will respond to the mean change in bulk-to-bulk membrane potential. Since the average of cos θ over the cell is zero, no change in the distribution of this type of probe will be found due to the direct action of the applied field. Of course, if the applied field induces an overall change in V_m as, for example, with a change in membrane ion permeabilities, the Nernstian dye will respond to it.

A straightforward extension of Eq. (3) leads to an expression for the spatial and temporal variation of V_m after a step change in E at $t = 0$ (Ehrenberg *et al.*, 1987a):

$$V_m = -(3/2)(1 - e^{-tf'/\tau})Eaf \cos \theta \tag{5}$$

where $\tau = aC_m(2\sigma_o + \sigma_i)/2\sigma_o\sigma_i$, C_m is the membrane capacitance per unit area, and $f' = [1 + \sigma_m(2\sigma_o + \sigma_i)a/2\sigma_o\sigma_i\delta]$. As was true for the factor f, the factor f' reduces to unity in most instances. For a cell of radius 10 μm with a membrane capacitance of 1 $\mu F/cm^2$ that is immersed in a medium of normal ionic strength for which the conductivity is 0.015 mho/cm, the characteristic membrane electrical time constant (τ) is 100 nsec.

C. Fluorescent Indicators

1. Nernstian Dyes

Most of the "slow" dyes operate via a potential-dependent redistribution mechanism and have been designed for measurements on large populations of cells in suspension. A hydrophobic ionic dye will have some equilibrium distribution between the external aqueous medium, the cell plasma membrane, the cytosol, and the membranes and aqueous compartments of intracellular organelles. The membrane potential plays a direct role in governing the distribution across and within the plasma membrane—the more negative the potential the greater the collection of positively charged dye. All these are coupled equilibria, so that the amount of dye which can associate with the organelles is also indirectly controlled. Furthermore, many dyes will form nonfluorescent aggregates at high concentrations. Thus, while it is easy to calibrate the dye response from a large population of cells at a given concentration, the heterogeneity of the fluorescence from an individual cell and its immediate surroundings makes it difficult to use these dyes to determine the membrane potential of a single cell through the fluorescence microscope. The "fast" dyes are less

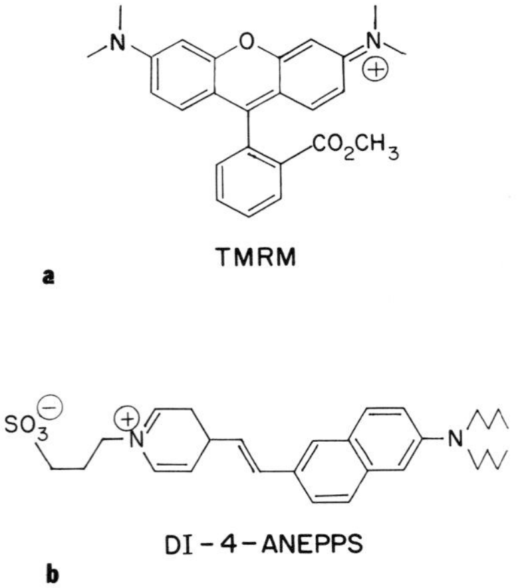

FIG. 2. (a) Molecular structure of a Nernstian dye, tetramethylrhodamine methyl ester (TMRM). (b) Molecular structure of a charge-shift dye, di-4-ANEPPS.

sensitive and are used to follow triggerable potential changes rather than measurements of absolute values of membrane voltage.

A method has recently been developed (Ehrenberg *et al.,* 1987b, 1988) which employs cationic dyes with the appropriate attributes to permit the microfluorometric measurement of membrane potential. Particular attributes include membrane permeability, low membrane binding, spectral properties which are insensitive to the environment, little tendency to aggregate, and, of course, strong fluorescence. These dyes should distribute into a cell according to Eq. (1), so that the relative fluorescence determined from a fixed volume inside the cytosol and in the surrounding medium should provide a direct measure of the membrane potential. The structure of one of these dyes, TMRM, is given in Fig. 2a.

2. Charge-Shift Probes

"Fast" dyes do not respond to potential changes by gross redistributions between differing chemical environments. Rather, they are usually membrane-staining dyes which sense the local potential drop or electric field. An example of one mechanism which may be employed by these probes is electrochromism (Platt, 1956). Here, the chromophore undergoes a large charge shift upon excitation; the energy difference between the ground and excited states is therefore sensitive to an external electric field oriented in the direction of the shifting charge. In principle, such dyes will respond instantaneously to potential changes and also permit the mapping of spatial variations in potential along the surface of a cell. They also may be designed via molecular orbital calculations (Loew *et al.,* 1978), and a number of such dyes have been synthesized and tested (Loew and Simpson, 1981; Hassner *et al.,* 1984; Fluhler *et al.,* 1985).

The best of the probes which has emerged from this effort is di-4-ANEPPS, whose structure is depicted in Fig. 2b. The fluorescence of this dye changes by 8–10% for a 100 mV change in membrane potential. This is less than the best sensitivity reported for a "fast" probe—21% for the dye RH421 on a neuroblastoma cell preparation (Grinvald *et al.,* 1981b, 1983; A. Grinvald *et al.,* 1983). However, di-4-ANEPPS has been remarkably consistent in a variety of preparations ranging from a number of cultured cells lines (Gross *et al.,* 1986; Ehrenberg *et al.,* 1987a) to red blood cells (Fluhler *et al.,* unpublished results) as well as the hemispherical bilayer model membrane (Fluhler *et al.,* 1985). It has also shown large and stable

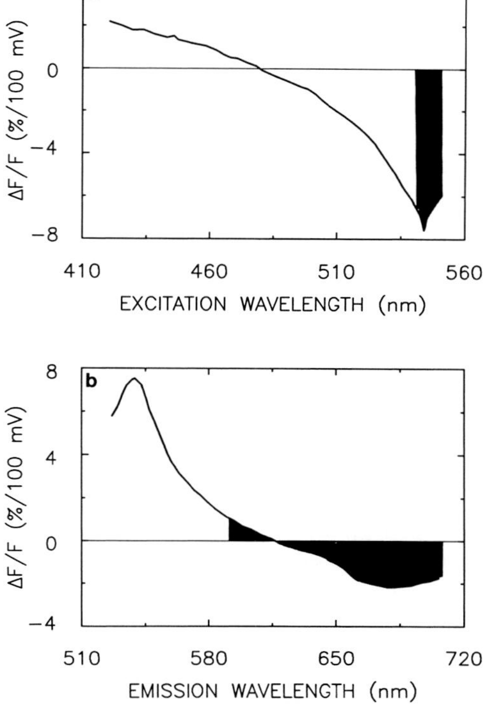

FIG. 3. Spectral response curves for di-4-ANEPPS fluorescence excitation (a) and emission (b) changes in response to a change in membrane potential.

fluorescence responses to voltage-clamp pulses on the squid axon preparation (L. B. Cohen, personal communication) and to action potential in heart muscle (G. Salama, personal communication), although the relative fluorescence changes could not be determined because of large background staining of nonexcitable tissue. The response spectra of di-4-ANEPPS from a voltage-clamped hemispherical bilayer (Fluhler *et al.*, 1985) are shown in Fig. 3. The shaded portions of the spectra indicate the best choice of excitation and emission filter wavelengths for the largest relative response.

II. Microspectrofluorometry with Nernstian Dyes

A. Methods

The three dyes that have been found to be most suitable for these measurements are TMRM (Fig. 2a), tetramethylrhodamine ethyl ester (TMRE), and rhodamine 6G. The latter is a commercially available laser dye while TMRM and TMRE were synthesized specifically for this purpose (Ehrenberg *et al.*, 1988). Other cationic dyes that were tried were either too slow to penetrate the cells or were irreversibly bound to the plasma and internal organelle membranes. Standard rhodamine optics (546 nm excitation, >590 nm emission) can be used in a fluorescence microscope equipped with a photomultiplier and measuring diaphragm to quantitate the light intensity.

Cells are incubated with the dye for 10 minutes at 37°C prior to the measurement. The specimen is mounted on a glass slide under a glass coverslip with medium containing the dye. A 100× NA 1.30 objective provides a shallow depth of focus and permits positioning of the measuring diaphragm over cellular regions which are free of mitochondria and other membranous organelles. Dye concentrations of 0.1 μM are sufficient for fluorescence which is bright enough so that a neutral density filter of OD 1 can be employed in front of a 50 W mercury excitation source. Thus, bleaching of the dye is insignificant during the time required to focus and take intensity readings.

Ideally, one should be able to measure the fluorescence of a Nernstian dye emanating from a fixed volume inside the cell and an equal volume outside the cell, take the ratio of these fluorescence intensities, and plug the ratio into Eq. (1) to obtain the potential across the plasma membrane. In reality, several additional factors must be considered which force a more complex analysis. First the background intensity, I_s, due to photomultiplier dark current, stray light, and any autofluorescence from the specimen must be subtracted (it is usually no more than 1% of any of the measured intensities in the presence of dye). A more serious correction involves the nonpotentiometric binding of the dye to the cell—mainly to its various membranous components. This may be determined by depolarizing the cell by subjecting it to a high potassium medium in the presence of valinomycin; the passive binding constant, K_b, is the ratio of the fluorescence intensities determined from the inside and outside of the cell under these conditions, less 1. Finally, the fluorescence from dye outside the cell can find its way through the microscope optics even when the plane of focus is set at the cell's center. This problem is minimized with a high NA objective but still can be appreciable. The magnitude of this depth of field error can be determined by bathing the cells in a medium containing an impermeable fluorescent marker such as FITC-dextran; the ratio, R_{DEX}, of fluorescence measured by focusing inside the cell and then off to its side should be zero if the depth of focus error was negligible, but R_{DEX} actually ranges from 0.5 to 0.9 for a variety of cells with diameters from 7 to 20 μm. The following equation (Ehrenberg *et al.*, 1988), then, relates the measured intensities of a Nernstian dye on the inside, I_i, and outside, I_o, of a cell to the ratio of concentration in Eq. (1):

$$\frac{C_i}{C_o} = \frac{(I_i - I_s) - (I_o - I_s)(R_{DEX})}{(I_o - I_s)(1 - R_{DEX})(1 + K_b)} \tag{6}$$

B. Results

The membrane potential of HeLa cells was determined with the three rhodamine dyes as well as amethyst violet and dimethyloxacarbocyanine [di-O-C_1(3)] (Ehrenberg, *et al.*, 1987b, 1988). A pair of fluorescence photomicrographs showing TMRM-stained HeLa cells in normal and depolarizing media, respectively, are shown in Fig. 4. For the measurements, the measuring diaphragm covered an approximately 1 μm square area in the region free of punctate fluorescence (which corresponded to the nucleus as visualized with phase contrast). All five dyes gave membrane potentials which were within their standard deviations of each other and within the range of values reported in the literature. The latter is not terribly impressive because the potentials reported for these cells from microelectrode impalements span an incredibly wide range. The former is noteworthy, however, because these dyes cover a wide range of K_b, from a low of 4.8 with TMRM to a high of 61.7 for rhodamine 6G. The importance of the depth of field correction is inversely related to the importance of the binding correction; therefore, the fact that all these probes yield similar results means that Eq. (6) deals correctly with both of these complicating factors.

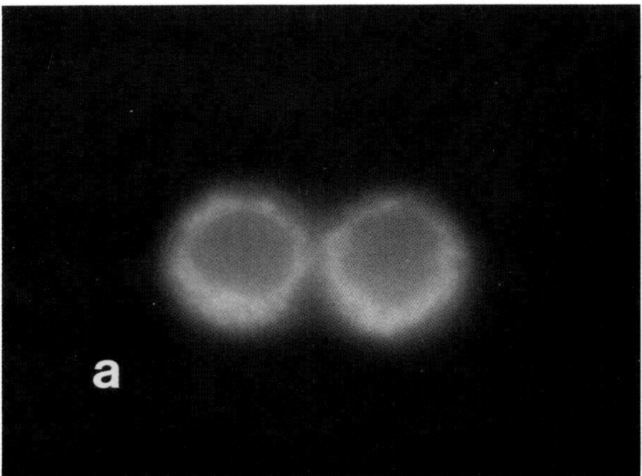

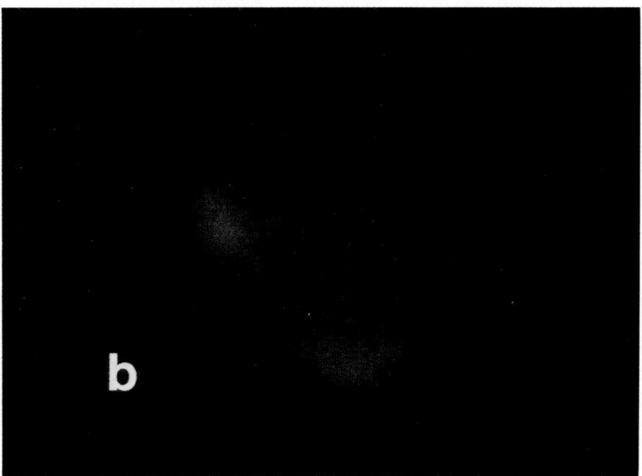

FIG. 4. Fluorescence photomicrographs of HeLa cells in the presence of TMRM. The cells in a are bathed in normal Earle's balanced salt solution. The two cells in b were depolarized by adding 0.5 μM valinomycin to the same medium in which the NaCl had been replaced with KCl. Both images were obtained with a Zeiss 100× NA 1.3 neofluar objective using a 2-second exposure on ASA 400 film.

Table I summarizes results obtained with several cell types using TMRE. All of these results are compatible with those available from published investigations employing other methods. The standard deviations are quite large but do not necessarily reflect poor precision in the measurements. It is likely that there is much more variation in physiological parameters such

TABLE I

MEMBRANE POTENTIAL DETERMINED WITH TMRE IN SEVERAL CELL LINES

Cell type	Potential (SD) (mV)
HeLa	−75.9 (17.1)
J774 macrophage	−60.2 (9.5)
Rabbit neutrophil	−74.1 (21.6)
Chick embryo muscle myotubes	−55.7 (9.4)
Neuroblastoma (clone Neuro-2a)	−34.4 (13.8)

as membrane potential than is commonly appreciated from measurements of large populations of cells. In the J774 cells, for example, it has been shown that the potential is highly dependent on the degree of adhesion of the cells to a substrate (Young *et al.*, 1983; Sung *et al.*, 1985), and this has been confirmed with the Nernstian dye method. Clearly, such factors as the position of the cell in the cell cycle, the degree of differentiation, cell viability, etc., may be other interesting factors that could affect the membrane potential of individual cells. In addition, the method should be ideally suited to kinetic measurements. Dyes with a low K_b, such as TMRM or TMRE, display readily reversible potential-dependent uptake. The probes should also be very useful for investigations employing fluorescence-activated cell sorting; additionally, they are currently being used in several studies of stimulus–response coupling with the aid of digital video microscopy. This imaging work is at too early a stage to be detailed here, but the combination of "fast" membrane-staining potentiometric probes with digital video imaging is the subject of the following section of this chapter.

III. Imaging with Charge-Shift Dyes

A. Methods

1. ILLUMINATION AND IMAGING SYSTEMS

As noted in Section I, the use of membrane-bound fluorescent indicators of membrane potential in combination with quantitative imaging allows measurement of the spatial distribution of membrane potential changes in individual cells. We performed the experiments described below with a video microscope and digital image analysis apparatus which has been described elsewhere (Gross and Webb, 1988). In brief, the fluorescence image of an epiilluminated specimen formed in a Zeiss Universal microscope was detected by a three-stage intensifier video camera manufactured by Venus Scientific. The camera output was digitized to 8-bit accuracy by a Grinnell Systems GMR 274 image processing system. Operation of the image processor and associated peripherals was under interactive computer control of a Digital Equipment PDP 11/73 minicomputer.

We chose to employ the fluorescence probe di-4-ANEPPS in this study for several reasons, many of which are illustrated in Figs. 2 and 3; (1) its response is linear with membrane potential change at the (maximal) level of about 9.5% fluorescence change per 100 mV membrane potential change, (2) its fluorescence spectral characteristics (Fig. 3) are well suited for microscopy, (3) the physical mechanism underlying the response of the dye appears to be electrochromic in nature (Loew *et al.*, 1979; Fluhler *et al.*, 1985), (4) the probe appears to incorporate reproducibly into only one leaflet of an artificial bilayer and it stains several types of cells with little apparent toxicity, and (5) it has no net charge at neutral pH. The utility of a linearly responding probe with well-characterized physical properties is clearly a desirable feature. The neutral charge of di-4-ANEPPS is important in these experiments in which cells are exposed to external electric fields since direct electrophoresis of the dye along the cell membrane is eliminated.

As shown in Fig. 3, the emission and excitation relative response spectra of di-4-ANEPPS are roughly inverted mirror images of each other, centered on their respective isosbestic points. In order to obtain a positive change in dye fluorescence with a positive change in membrane potential (if the dye is incorporated in the outer leaflet of the bilayer), both excitation and emission light should be collected on the short wavelength sides of the isosbestic points. Conversely, a negative-going response is obtained if the excitation and emission signals are collected from spectral regions of longer wavelength than the respective isosbestic points. Since we employed a mercury arc lamp as the excitation source, we chose to excite the dye with the strong green mercury line. This line, selected by an interference filter of center wavelength 546 nm and full width at half-maximum passband of 10 nm, excites the region in the excitation spectrum for which response is maximally negative ($\approx -7.7\%$ per 100 mV, stippled region in Fig. 3a). The corresponding maximum negative-going emission signal of $\approx -2.2\%$ is near 680 nm, but we chose to collect all the emitted light from 590 nm and above (stippled region in Fig. 3) in order to increase the amount of detected light and thus improve the signal-to-noise ratio in the image. All images collected for analysis were digital averages of 30 consecutive video frames (i.e., 1-second averages). Digital processing of single or multiple data frames on a pixel-by-pixel basis was accomplished in the image processor to allow rapid spatial analysis of the cell-associated fluorescence.

2. Cell Preparations

In this study, we employed four different cell types, A-431 human epidermoid carcinoma cells, rat basophilic leukemia (RBL) cells, protoplasts isolated from the root crown tissue of rye, *Secale cereale* L. cv Puma, and spores of the fungus *Uromyces appendiculatus* (Pers.) Ungar. The A-431 cells were cultured in Dulbecco's modified Eagle's medium (DMEM) supplemented with 10% calf serum in a 37°C, humidified 5% CO_2/95% air incubator. A-431 cells were plated onto # 1 glass coverslips 24–48 hours before experiments. RBL cells were grown in Eagle's minimum essential medium plus 20% fetal calf serum in a similar humidified atmosphere. These cells were trypsinized and kept in suspension for 1 hour before plating them onto coverslips for 10 minutes at 37°C immediately before an experiment. Rye protoplasts were isolated from whole root crown tissue in 0.53 OsM sorbitol solution. They were then pipetted onto a concanavalin A-coated coverslip to which some adhered after 5 minutes at 25°C. The fungus spores were aspirated onto a coverslip in 1 m*M* Tris buffer 30–60 minutes before an experiment.

A stock solution of 1 m*M* di-4-ANEPPS in ethanol was prepared and stored at −10°C. Dilution of this stock into suitable media (Medium 199 for A-431 cells, Tyrode's solution for the RBL cells, 0.53 OsM sorbitol for the rye protoplasts, and 1 m*M* Tris buffer for the fungus spores) to 1–10 μM final concentration was used to stain the cells. We found that saturating dye binding occurred within 10 minutes at 4°C, and thus we adopted these values for a standard staining protocol. Cells were then washed with the respective suitable medium and mounted for microscopic observation.

We found that these four cell types thus stained would retain the dye in their peripheral membranes for an hour or longer at room temperature. Many other cell types including normal rat kidney (NRK) cells, GM3348 and IMR90 human fibroblast cells, L5 myoblasts, and the germ tubes of germinating *U. appendiculatus* spores do not stain properly. In these cells we observe strong staining of internal membranes, indicating that the dye has breached the plasma membrane. In one experiment, we observed a fibroblast stained and held at 4°C. The cell retained peripheral membrane staining until we warmed the cell up to and through 18°C, at which point the dye rapidly entered the cell interior. Thus, di-4-ANEPPS may be more generally useful if the cells of interest can be observed at temperatures below 18°C. Since di-4-ANEPPS responds directly to the electric field vector (both magnitude and direction) in the membrane, it is imperative that the probe molecules be sequestered in only one leaflet of the membrane bilayer. With dye on both faces of the membrane, the opposing signals from the probes oriented antiparallel to each other would cancel.

We investigated the cell-staining properties of two charged styryl membrane potential dyes, di-10-ASPPS and di-4-ASPBS (Hassner *et al.*, 1984), and found that the time course of incorporation was too long for our purposes, although di-10-ASPPS appeared to stain only the plasma membrane of human fibroblasts. It has recently been shown that the rate of incorporation of di-10-ASPPS into membranes can be enhanced by two orders of magnitude by use of a surfactant (Lojewska and Loew, 1987). Thus, the use of membrane-bound potentiometric probes in a variety of cell types other than those reported here seems promising.

3. Applied Field Apparatus

In order to apply a uniform electric field to viable cells on a coverslip under a microscope, several criteria must be simultaneously satisfied. We designed an applied field microscope chamber for this purpose which has been described previously (Gross *et al.*, 1986). The chamber has two independent enclosed fluid channels, one for the medium which bathes the cells and one for cooling water. The former connects two access ports on either side of the plexiglass chamber to the cell side of the mounted coverslip. The coverslip is supported by a frame of rectangular sections of #1 glass coverslips glued to the chamber with a very thin film of Sylgard. The coverslip is held in place by gravity with the assistance of a thin coating of petroleum jelly or silicon grease on the coverslip spacers. It is important to thoroughly dry the chamber as well as the edges of the cell-containing coverslip in order to obtain a satisfactory seal with such an arrangement. The final spacing between cells and the bottom of the chamber is ~200 μm.

By passing current through two electrodes immersed in the two media wells at the ends of the chamber, a uniform electric field is produced at the central portion of the cell-containing coverslip since end and edge effects dissipate over distances of the order of the spacing between cells and chamber top. We found that chlorided silver wires proved suitable for the short-time electric field exposures reported here. Longer-term electric field exposures require agar bridges between the chamber wells and the electrolyte solutions in which the wire electrodes are immersed (T. Ryan and W. Webb, personal communication). Even with agar bridge connections, pH alterations in the cell medium are possible and must be carefully monitored.

To generate an electric-field-induced membrane potential change of 100 mV in a cell of 10 μm radius requires an electric field strength of 67 V/cm in the medium, according to Eq. 3. Since the electric conductivity of Medium 199 was measured to be 0.015 mho/cm, a current density of 1 amp/cm^2 will be present in the medium, leading to Joule heating of 67 W/cm^3. Thus 6.5 W of electrical heat must be dissipated from the chamber. If a temperature rise of 1°C is acceptable, a cooling water flow rate of 1.5 ml/sec is sufficient to remove this amount of heat. Convective cooling from the cover glass to the surrounding air is not sufficient to prevent excessive heating. In order to efficiently transfer the heat from the cell medium to the flowing cooling water, we used a sapphire plate between the two liquids since this material is optically transparent and has a much higher thermal conductivity than glass. In practice, we found that medium heating in chambers for which no cooling was provided produced severe focus drift (and concomitant large temperature rises) while the focus drift in the above chamber was much reduced.

The resistance of the chamber, when filled with Medium 199, is ~3.5 kΩ. Since 44 mA of current is needed to produce the 67 V/cm field strength, a power supply capable of at least 200 V at 45 mA output current is required. We used a vacuum tube high-voltage power supply in series with a 1–3 kΩ power resistor to provide reasonably uniform current to the chamber. The extra resistor helped to minimize impedance variations due to electrolytic processes at the electrodes. Two platinum wire electrodes immersed in the cell-bathing medium at the ends of the channel formed by the coverslip were connected to a voltmeter which indicated the actual potential drop across the coverslip.

B. Results

1. Individual Cells

The fluorescence image of an unusually large, ellipsoidal, di-4-ANEPPS-stained A-431 cell is shown in Fig. 5a. This cell is unusual in its shape, which is a nearly perfect ellipse, its large size (48 × 31 μm), and its isolation from the more common clusters of A-431 cells. In particular, the large size of this cell enhances the membrane potential changes induced in its membrane by an applied external electric field. The dye staining on this

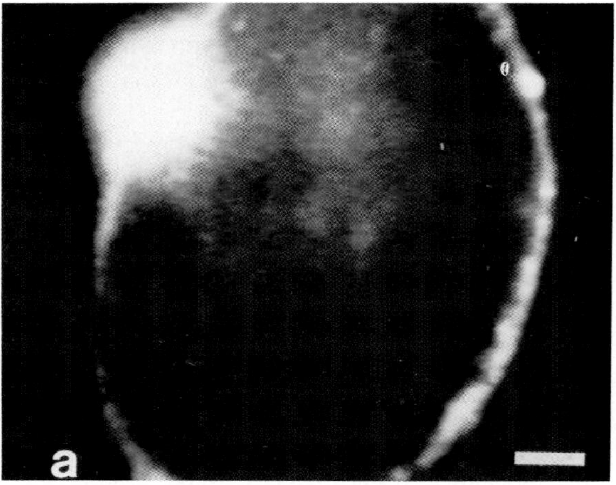

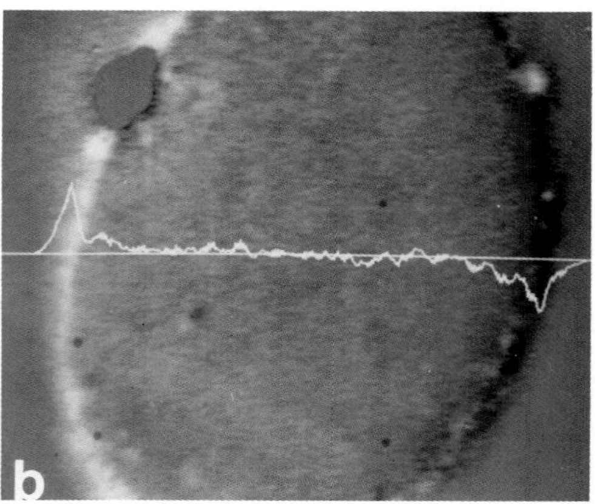

FIG. 5. (a) An A-431 cell stained with di-4-ANEPPS; bar = 10 μm. (b) The difference between cell fluorescence images collected at applied field strengths of +53.0 V/cm and −47.6 V/cm. Zero difference in fluorescence is a mid-gray level. The plot indicates the fluorescence difference across the line of pixels shown by the straight trace. (c) The relative fluorescence response of the cell responding to a −47.6 V/cm field compared to the zero-field response. Zero percent change is a mid-gray level, positive changes are brighter, negative changes are darker. (d) Data collected from this cell for several applied field strengths. Fluorescence responses were averaged over the left and right sides of the cell. Reproduced from the *Biophysical Journal* **50,** 339–348 (1986) by copyright permission of the Biophysical Society.

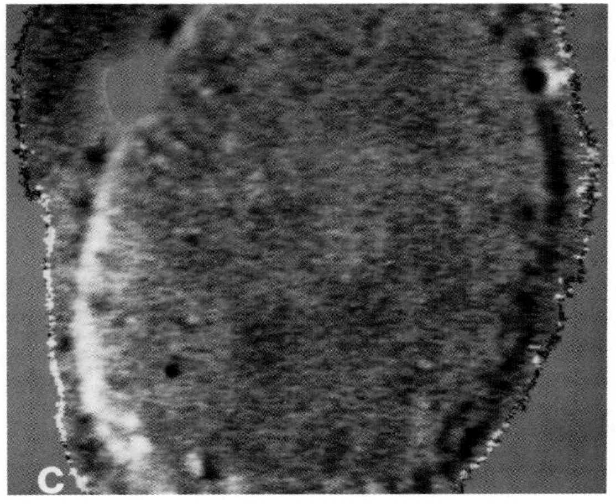

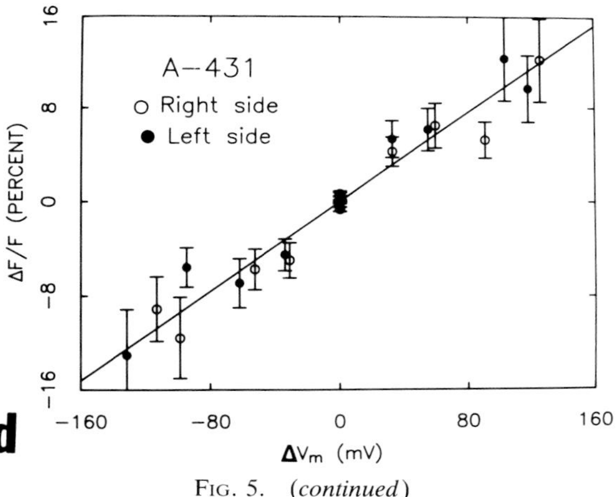

FIG. 5. (*continued*)

cell also illustrates another difficulty one can encounter. Bright spots of dye fluorescence can be seen at different points along the cell membrane. One large spot at the upper left of the cell, perhaps due to a microaggregate of dye, has saturated the camera tube. We noticed several of the smaller bright dots, but not the large bright spot, moving during the 15 minutes in which we observed this cell. The image in Fig. 5a is the fluorescence measured with the cell at rest, i.e. with no external field applied. The difference in cell fluorescence between an image collected with a uniform electric field of 53.0 V/cm applied from left to right and that with an applied field of 47.6 V/cm oriented right to left is shown in Fig. 5b. This image was produced simply by subtracting the second image from the first and thus represents changes in fluorescence not normalized to the local fluorescence intensity. A small misregistry between the images, due to a very slight focal drift, is seen at the lower right-hand corner of this image.

The relative fluorescence change between the cell image collected at zero applied field and that collected during application of a 47.6 V/cm field directed right to left is illustrated in Fig. 5c. In this image the difference between zero-field and applied field fluorescence is divided by the zero-field fluorescence on pixel-by-pixel basis, and the percent change is mapped to gray values such that 100% positive change is white, 100% negative change is black, and no change is a mid-gray level. Thus, this image more accurately reflects membrane potential changes since the fluorescence change has been normalized for variations in local levels of fluorescence due to different levels of dye incorporation or of amounts of plasma membrane which contribute to the images.

Both Fig. 5b and c illustrate the utility of imaging the fluorescence of di-4-ANEPPS to measure changes in membrane potential in individual cells. An applied field noticeably brightens the fluorescence intensity on the edge of the cell facing the anode and darkens it on the opposite face; this effect reverses upon reversing the field. This response is just as expected from the cell membrane potential changes predicted by Eq. (3) and the spectral characteristics of the dye shown in Fig. 3. In addition, it is clear from Fig. 5a, b, and c that considerable spatial information is contained in the images. We exploit this property to calibrate the response of the dye within this cell.

For several different images of the cell of Fig. 5 collected at different applied field strengths and directions, we measured the fluorescence response at 10° intervals around the cell periphery, avoiding the bright fluorescence inclusion at the upper left. We then plotted the measured fluorescence response, averaged over the left and right sides of the cell, versus the calculated average membrane potential. The theoretical response of an ellipsoidal cell is slightly different in detail from our result in Eq. (3) for a spherical cell, but retains the same qualitative behavior (Stratton, 1941; Fricke, 1924; Cole, 1968). The data are shown in Fig. 5d; the slope of the line indicates that the dye response in this cell is 9.5%/100 mV, very close to the response expected.

We have applied this technique to many other individual cells. In all cases in which the fluorescence signals are large enough to measure, the spatial variation of membrane potential changes induced by applied fields fits with that predicted by Eq. (3) or its appropriate extension to more

TABLE II

FLUORESCENCE RESPONSE OF DI-4-ANEPPS IN SINGLE CELLS RESPONDING TO APPLIED ELECTRIC FIELDS

Cell type	Response (%/100 mV)
A-431[a]	9.5
RBL[a]	12.0
Rye protoplast[a]	8.6
Fungus spore[a, b]	3.7
HeLa[c]	9.5

[a] Determined by the imaging method described in Section III.

[b] This spore was irregularly shaped and thus difficult to model. The spore was bathed in low-conductivity medium, which enhances the shunting effect of membrane conductivity in the factor f of Eq. (3). Thus, the value noted here may not accurately reflect the true response of the dye.

[c] Determined by a photometry method described by Ehrenberg *et al.* (1987a).

complicated cell geometry (Gross *et al.*, 1986). A summary of the measured di-4-ANEPPS fluorescence response in these cells is given in Table II. Although the dye response is often near the 9%/100 mV value found in artificial membranes, this value should be used with caution. Preferably, the response value should be determined for each cell type and condition.

2. CELL CLUMPS

The response of clumps of A-431 cells stained with di-4-ANEPPS to an applied electric field was examined to explore the applicability of this technique to cell systems more complicated than isolated individual cells. One example is a clump of four A-431 cells shown in Fig. 6a. The region of contact between cells is more than twice as bright as the peripheral membranes, which leads to difficulty in capturing the full dynamic range of the fluorescence on film as is the case for Fig. 6a. Although the reason underlying this differential staining is not understood, it may be due to a larger total membrane area resulting from unresolved membrane folds, summated contributions of fluorescence along a vertical membrane segment, or heavier staining in the cell contact area.

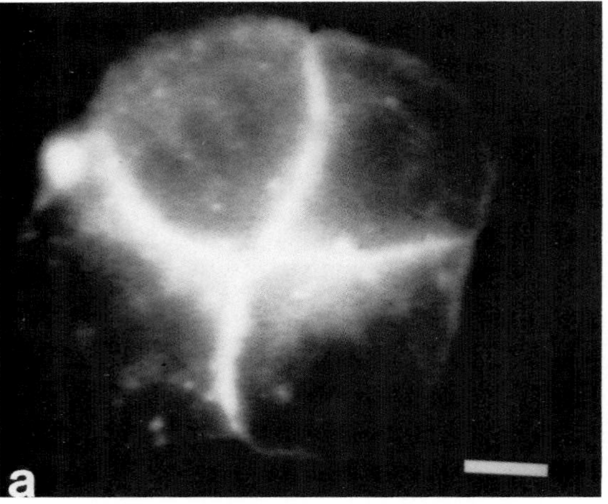

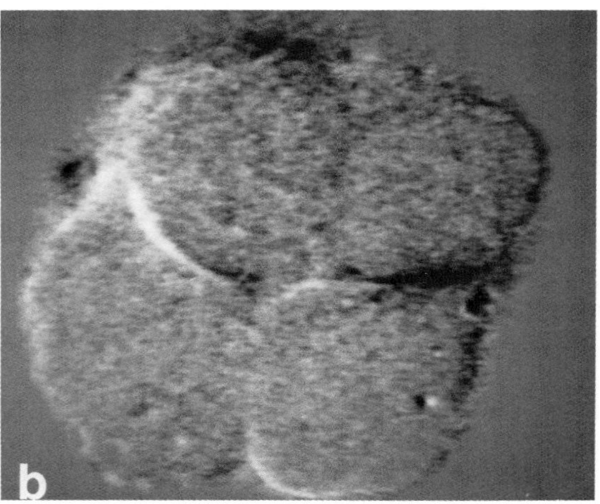

FIG. 6. (a) A clump of four A-431 cells stained with di-4-ANEPPS; bar = 10 μm. (b) The difference fluorescence image as in Fig. 5b for applied fields of +24 V/cm and 0 V/cm. Reproduced from the *Biophysical Journal* **50,** 339–348 (1986) by copyright permission of the Biophysical Society.

The response of this cell clump to an applied field of 24 V/cm directed left to right is shown in Fig. 6b, which is a difference image similar to that in Fig. 5b except it displays the difference between field-on and field-off images. The apparent response of these cells is not entirely intuitive. Although the anode- and cathode-facing edges of the clump (left and right side, respectively) appear to hyperpolarize and depolarize as expected, regions at the top and bottom of the clump seem to respond strongly as well. If these cells were of rectangular cross section, maximal response would be expected at the electrode-facing membranes, with the response at the other two membranes decreasing linearly from the edges to the center of the cell. At vertical junctions between equal-size cells, the net fluorescence change should be zero since changes in apposing membranes would be equal in magnitude but opposite in sign. Membrane junctions running left to right should show the summation of linear gradients of fluorescence change in the apposed cells. Although these general features are apparent in Fig. 6b, it is also clear that the strongest fluorescence changes occur at cell junctions oriented parallel to the applied field direction.

Data from other large clumps of A-431 cells (Gross *et al.*, 1986) indicate that some groups of cells within multicellular clumps are collectively hyperpolarized while others are collectively depolarized. At the peripheral membranes at the clump edges, the magnitudes of the maximal fluorescence changes induced by applied fields appear independent of clump size and are of similar magnitude as those in isolated cells. Thus, cell–cell electrical connections are not strong enough to form an effective electrical syncytium which would increase the induced change in V_m at terminal membranes at cluster edges, an effect suggested by Cooper (1984). A-431 cells are not known to form such intercellular connections, thus this result is expected.

The nonintuitive distribution of changes in V_m of cells within clumps induced by applied uniform electric fields likely reflects the tortuous path taken by the extracellular electric current through the cell clump. This likelihood, coupled with the limited two-dimensional map obtained by microscopy, could explain how some cells in a clump could appear entirely depolarized in response to an applied field even though the mean change in V_m in such a case is necessarily zero. For example, if the clump geometry is such that the applied field is redirected from the bottom to the top of a particular cell and if the fluorescence image focal plane is at the top of the clump, the signal detected would necessarily indicate depolarization at the top of the cell.

IV. Discussion

The use of fluorescence indicators of membrane potential provides a unique alternative to invasive micropipet impalement or patch-clamp techniques. The optical techniques can be applied to very small cells, and they involve no complicated micromanipulation. Different types of fluorescent indicators are available, each possessing properties which can be suited for a wide variety of measurements. We have described some specific applications of both redistribution and membrane-bound potentiometric probes. As with any technique, one must characterize the properties of each probe in the system of study in order to understand the signals generated by that probe. Nernstian redistribution dyes are simple to use and they provide a large signal, but binding of these probes to membranes complicates their use. Membrane-bound dyes provide only a weak signal, but their ability to respond rapidly to changes in the electric field within the membrane as well as the intrinsic capability to provide spatial information makes this class of dye promising for future uses.

We have described methods whereby both temporal and spatial changes in membrane potential in individual cells can be measured by employing a membrane-bound potentiometric probe. The data-handling capabilities of presently available equipment forces one to trade spatial resolution for temporal resolution and vice versa. Although we have described one of the extremes (high spatial resolution with low time resolution via digitial video imaging), other alternative solutions are possible. Photodiode array detectors (Grinvald *et al.*, 1981a; Salzberg *et al.*, 1977) have been shown to provide modest spatial information with fast time response while charge-coupled device (CCD) imagers can be configured to generate variable amounts of spatial resolution which inversely correlates with the speed of signal acquisition. Ehrenberg *et al.* (1987a) have shown that the temporal response of membrane potential changes in hemispherical bilayers in response to step changes in applied electric field strength follows Eq. (5) as predicted. This high temporal resolution is obtained by point photometry, i.e. at the lowest spatial resolution.

We have described an applied-field methodology which allows one to calibrate the response of a membrane-bound potentiometric probe within a particular cell membrane or artificial bilayer. This methodology is useful in that it allows one to dissect cell-to-cell variations in measured electrical properties from variations in probe response. One point which has not been addressed is that of calibration of the dye signal on an absolute scale of membrane potential. The major difficulty with such calibration arises from the inability to resolve the potential-dependent fluorescence of the probe from fluorescence dependence due to variable dye concentration in the membrane, amount of (unresolvable) membrane area present, and spatial variation in the fluorescence excitation and detection systems. Since the charge-shift dyes respond to the electric field strength directed along their long axis by shifts of both the excitation and emission spectra, these probes should be amenable to the self-referencing ratiometric methods presently in use for other fluorescent probes (see Chapter 5 by Tsien, and Chapter 6 by Bright *et al.,* this volume). Specifically, dividing the fluorescence image or signal excited with a detection wavelength by that excited at a reference wavelength should provide a unique ratio value as a function of membrane potential which is independent of dye concentration, membrane area, or optical artifacts. We are currently exploring the utility of this approach to the measurement of spatial variations of absolute membrane potential in single cells.

As noted by Ehrenberg *et al.* (1987a), the rapid risetime of V_m in small cells coupled with the extremely fast response of membrane-bound potentiometric dyes should allow one to explore the electrical behavior of cells, perhaps even at the level of the single ion channel, with a temporal resolution much greater than that presently available by the patch-clamp method. Thus it appears that the future uses of potentiometric probes in cell biology is limited only by the imagination of the investigator, and is bright indeed.

Acknowledgments

We are pleased to acknowledge the help of Mei-de Wei, Val Montana, and Anna Maria Vites. Watt W. Webb provided valuable discussions and support. This work was supported by the U.S. Public Health Service through NIH grants GM35063 and AI22106, the NSF through grant PCM-83-03404, and the ONR through grant N00014-84-K-0390.

References

Aidley, D. J. (1971). "The Physiology of Excitable Cells." Cambridge Univ. Press, London and New York.

Bockris, J. O'M., and Reddy, A. K. N. (1970). "Modern Electrochemistry." Plenum, New York.

Cole, K. S. (1968). "Membranes, Ions and Impluses." Univ. of California Press, Berkeley.

Cooper, M. S. (1984). *J. Theor. Biol.* **111,** 123–130.

Ehrenberg, B., Farkas, D. L., Fluhler, E. N., Lojewska, Z., and Loew, L. M. (1987a). *Biophys. J.* **51,** 833–837.

Ehrenberg, B., Wei, M.-D., and Loew, L. M. (1987b). *In* "Membrane Proteins" (S. C. Goheen, ed.), pp. 279–294. Bio-Rad Laboratories, Richmond, California.

Ehrenberg, B., Montana, V., Wei, M.-D., Wuskell, J. P., and Loew, L. M. (1988). *Biophys. J.* **53,** 785–794.

Fluhler, E., Burnham, V. G., and Loew, L. M. (1985). *Biochemistry* **24,** 5749–5755.

Fricke, H. (1924). *Phy. Rev.* **24,** 575–587.

Grinvald, A. (1985). *Annu. Rev. Neurosci.* **8,** 263–305.

Grinvald, A., Cohen, L. B., Lesher, S., and Boyle, M. D. (1981a). *J. Neurophysiol.* **45,** 829–840.

Grinvald, A., Ross, W. N., and Farber, I. (1981b). *Proc. Natl. Acad. Sci. U.S.A.* **78,** 3245–3249.

Grinvald, A., Fine, A., Farber, I. C., and Hildesheim, R. (1983). *Biophys. J.* **42,** 195–198.

Gross, D., and Webb, W. W. (1988). *In* "Spectroscopic Membrane Probes" (L. M. Loew, ed.), Vol. 2. CRC Press, Boca Raton, Florida (in press).

Gross, D., Williams, W. S., and Connor, J. A. (1983). *Cell. Mol. Neurobiol.* **3,** 89–111.

Gross, D., Loew, L. M., and Webb, W. W. (1986). *Biophys. J.* **50,** 339–348.

Hassner, A., Birnbaum, D., and Loew, L. M. (1984). *J. Org. Chem.* **49,** 2546–2551.

Jaffe, L. F., and Nuccitelli, R. (1977). *Annu. Rev. Biophys. Bioeng.* **6,** 445–476.

Loew, L. M. (1988). *In* "Spectroscopic Membrane Probes" (L. M. Loew, ed.), Vol. 2 CRC Press, Boca Raton, Florida.

Loew, L. M., and Simpson, L. (1981). *Biophys. J.* **34,** 353–365.

Loew, L. M., Bonneville, G. W., and Surow, J. (1978). *Biochemistry* **17,** 4065–4071.

Loew, L. M., Scully, S., Simpson, L., and Waggoner, A. S. (1979). *Nature* (*London*) **281,** 479–499.

Lojewska, Z., and Loew, L. M. (1987). *Biochim. Biophys. Acta* **281,** 479–499.

London, J. A., Zecevic, D., Loew, L. M., Ohrbach, H. S., and Cohen, L. B. (1986). *In* "Fluorescence in the Biological Sciences" (D. L. Taylor, A. S. Waggoner, F. Lanni, R. F. Murphy, and R., Birge, eds.,), pp. 423–447. Liss, New York.

McLaughlin, S. L. (1977). *Curr. Top. Membr. Transp.* **9,** 71–144.

Platt, J. R. (1956). *J. Chem. Phys.* **25,** 80–105.

Poo, M.-M. (1981). *Annu. Rev. Biophys. Bioeng.* **10,** 245–276.

Salzberg, B. M. (1983). *Curr. Methods Cell. Neurobiol.* **3** 139–187.

Salzberg, B. M., Grinvald, A., Cohen, L. B., Davila, H. V., and Ross, W. N. (1977). *J. Neurophysiol.* **40,** 1281–1291.

Stratton, J. A. (1941). "Electromagnetic Theory" McGraw-Hill, New York.

Sung, S. S. J., Young, J. D.-E, Origlio, A. M., Heiple, J. M., Kaback, H. R., and Silverstein, S. C. (1985). *J. Biol. Chem.* **260,** 13442–13449.

Waggoner, A. S. (1985). *In* "The Enzymes of Biological Membranes" (A. N. Martonosi, ed.), pp. 313–331. Plenum, New York.

Young, J. D.-E, Unkeless, J. C., Kaback, H. R., and Cohn, Z. A. (1983). *Proc. Natl. Acad. Sci. U.S.A.* **80,** 1636–1640.

Chapter 8

Resonance Energy Transfer Microscopy

BRIAN HERMAN

Laboratories for Cell Biology
Department of Cell Biology and Anatomy
University of North Carolina School of Medicine
Chapel Hill, North Carolina 27599

I. Introduction

Scientists have long sought technical approaches which would allow them to study the real-time behavior of cellular functions at the subcellular level. Fluorescence, which is the emission of photons that occur as the electrons in a chromophore decay from an excited (higher energy) state back to the ground (lower energy) state, has provided a tool by which scientists can examine various cellular properties. In this chapter, we will explore how fluorescence can be used as a spectroscopic ruler to measure distances between cellular components and gain information about interactions of these components on the molecular level. This method, known as resonance energy transfer (RET) is founded on the fact that a fluorophore (donor) in an excited state may transfer its excitation energy to a neighboring chromophore (acceptor) nonradiatively through dipole–dipole interactions. This process requires some spectral overlap between the emission spectrum of the donor and the absorption spectrum of the acceptor and, for a given donor–acceptor pair, the efficiency of the transfer process is dependent on their relative orientation and on the distance between them.

II. Fluorescence Theory

A. Principles of Fluorescence

The quantum yield of a fluorescent molecule is the ratio of the number of photons emitted to the number absorbed and is determined by the competing rates of the different deactivation processes:

$$Q = \frac{K_f}{K_f + K_g + K_X} \tag{1}$$

where Q is the quantum yield, K_f the rate constant for fluorescence emission, K_q the rate constant for radiationless energy loss and K_x the rate for intersystem crossing. If $K_q + K_x \ll K_f$, $Q = 1$.

The lifetime of the excited state is defined as the average time a fluorescent molecule spends in the excited state prior to return to the ground state:

$$\tau_0 = 1/K_f + K_q + K_x \tag{2}$$

In the absence of nonradiative decay processes or intersystem crossing, the

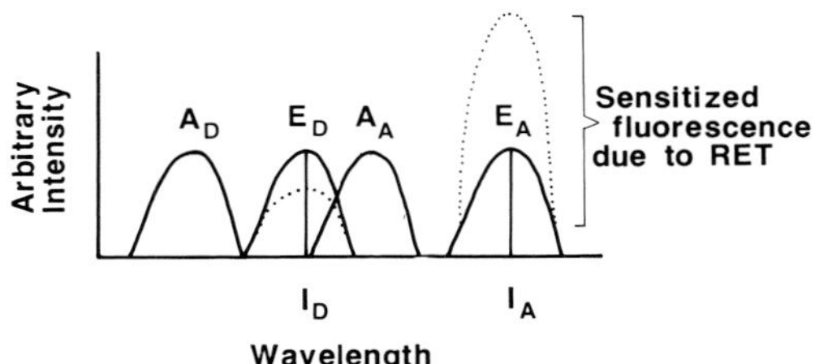

FIG. 1. Sensitized fluorescence. Resonance energy transfer (RET) results in a decrease (quenching) of the donor emission (dotted line of E_D) when measured at I_D, the maximum of the donor emission, and a concomitant increase (sensitization) of the acceptor emission (dotted line of E_A) when measured at I_A, the maximum of the acceptor emission. The ratio of I_A/I_D can be taken as a measure of RET.

intrinsic lifetime (τ_0) equals

$$\tau_0 = 1/K_f \tag{3}$$

$$Q = \tau/\tau_0 \tag{4}$$

The decay process is thought to obey first order kinetics and therefore the intensity of fluorescence emission would be

$$I = I_0 e^{-t/\tau} \tag{5}$$

where I_0 is the intensity at $t = 0$ and τ in the fluorescent lifetime. Measurements of RET using lifetime analysis will be discussed below.

For resonance energy transfer to occur, the donor must be fluorescent and of sufficiently long lifetime. The transfer does not involve the actual reabsorption of light by the acceptor but requires the distance between the chromophores to be relatively close (usually not exceeding 100 Å). The process of RET results in an increase in acceptor emission, so called sensitized fluorescence (Fig. 1). Energy transfer varies most importantly as the inverse of the sixth power of the distance separating the chromophores. The dependence of the energy transfer efficiency on donor–acceptor separation provides the basis for the utility of this phenomenon in the study of cell component interactions.

The choice of a donor–acceptor pair should be governed by several parameters:

1. The maximum absorption of the donor should coincide with the maximum output of the light source.
2. The excitation spectrum of the acceptor should maximally overlap the emission spectrum of the donor.

3. There is little direct excitation of the acceptor at the excitation maximum of the donor.

4. The emission of both the donor and acceptor occur on a wavelength in which the detector has reasonably high quantum efficiency.

5. The donor has small overlap between its own emission and absorption spectra, thus minimizing donor–donor self transfer.

6. The donor emission results from several overlapping transitions and thus exhibits low polarization. This minimizes uncertainties associated with the K^2 orientation factor (see below).

B. Overlap Integral

The overlap integral $[J(\lambda)]$, which expresses the degree of spectral overlap between the donor emission and the acceptor absorption (Fig. 2) is given by

$$J(\lambda) = \int_0^\infty \varepsilon(\lambda)\, f(\lambda)\lambda^4\, d\lambda \tag{6}$$

where $\varepsilon(\lambda)$ is the decadic molar extinction coefficient of the acceptor at wavelength λ, and $f(\lambda)$ is the normalized fluorescence spectrum of the donor defined as

$$f(\lambda)\, d\lambda = \frac{F(\lambda)\, d\lambda}{\int_0^\infty F(\lambda)\, d\lambda} \tag{7}$$

C. Orientation Factor (K^2)

The orientation factor (K^2) which describes the orientation between the emission dipole of the donor (θ_D) and the absorption dipole of the acceptor

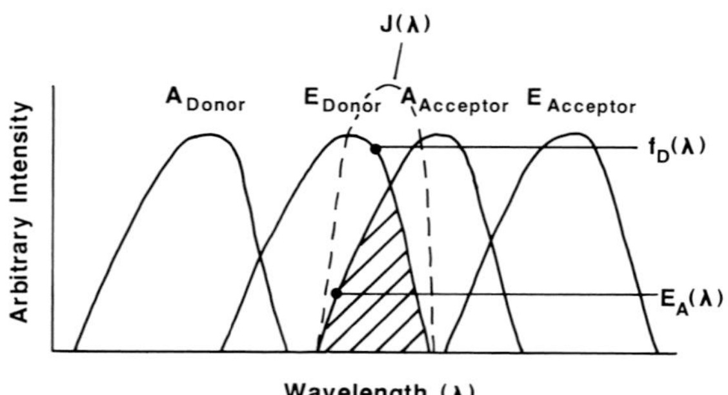

FIG. 2. Overlap integral $[J(\lambda)]$. The area of overlap between the donor emission spectrum $[f_D(\lambda)]$ and the absorption spectrum $[E_A(\lambda)]$ of the acceptor defines the overlap integral.

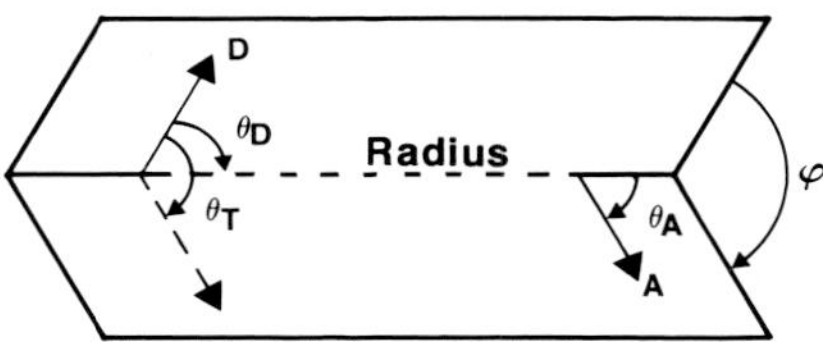

FIG. 3. Orientation factor. D, donor; A. acceptor; ϕ_D, angle between donor and radius connecting donor to acceptor; ϕ_A, angle between acceptor and radius connecting acceptor to donor; ϕ_T, angle between emission dipole of donor and absorption dipole of acceptor.

(θ_A) (Fig. 3) is given by

$$K^2 = (\cos \theta_T - 3 \cos \theta_D \cos \theta_A) \tag{8}$$

where θ_T is the angle between the emission dipole of the donor and the absorption dipole of the acceptor, and θ_D and θ_A are the angles between these dipoles and the vector joining the donor and acceptor. This factor can range from 0 to 4. Since the sixth root is taken to calculate the distance between donor and acceptor, variation of the K^2 from 1 to 4 results in an error of at most 26% in r. However, if the dipoles are oriented perpendicular to one another, K^2 equals zero and serious errors could result in r. By measurements of the fluorescence antisotropy of the donor and the acceptor, one can set limits on K^2 and thereby minimize uncertainties in the calculated distance. If the polarization of the donor or acceptor is low (less than 0.3), the errors in the distance (r) due to errors in K^2 are likely to be less than 10%. K^2 is generally assumed to be 2/3, which is the value for donors and acceptors which randomize by rotational diffusion prior to energy transfer. For fluorophores bound to macromolecules, the segmental rotations of the donor and acceptor should randomize the orientation.

D. Critical Distance (R_0)

A given donor–acceptor pair will only be useful over a certain range of donor–acceptor separations. The range of intermolecular distances for which a given donor–acceptor pair is useful is related to their R_0 value, the critical Forster distance (Forster, 1959);

$$R_0 = [8.75 \times 10^{-25} \eta^{-4} q_0 K^2 \, \mathrm{J}(\lambda)]^{1/6} \tag{9}$$

where η is the refractive index of the medium in the range of overlap, q_0 is the quantum yield of the donor in the absence of acceptor, K^2 is the orientation factor, and $J(\lambda)$ the overlap integral. R_0 is the donor–acceptor distance for which the efficiency of RET is equal to the donor-deexcitation rate in the absence of acceptor. When the separation of the donor–acceptor reaches R_0, there exist equal probabilities for transfer and for intramolecular deactivation of the excited state of the donor by radiative and nonradiative processes. If the donor–acceptor separation is much greater than R_0, little if any transfer will take place.

E. Rate of Energy Transfer (K_T)

The RET rate from an excited donor (D) to a suitable acceptor (A) in the very weak coupling case under prevailing dipole–dipole interaction (Forster, 1959) is given by

$$K_T = 8.7 \times 10^{-28} \frac{q_D K^2}{\eta^4 \tau_D R^6} J(\lambda) \tag{10}$$

where q_D and τ_D are the donor emission quantum yield and lifetime, respectively, in the absence of acceptor R, is the distance between down and acceptor, and the other variables assume the same meaning as those defined previously.

Since

$$R_0^6 = 8.8 \times 10^{-28} \eta^{-4} q_0 K^2 J(\lambda) \tag{11}$$

$$K_T = \frac{1}{\tau_D} \frac{R_0^6}{R} \tag{12}$$

Thus if one can measure τ_D and R_0, one can calculate R and K_T. At $R_0 = R$, $K_T = K_f + K_q + K_x$.

F. Efficiency of Energy Transfer (E_T)

The efficiency of energy transfer (E_T), which is the proportion of photons absorbed by the donor that are transferred to the acceptor, is given by

$$E_T = R_0^6/R^6 + R_0^6 \tag{13}$$

or

$$R = R_0(1/E_T - 1)^6 \tag{14}$$

In terms of lifetimes:

$$E_T = (\tau_{DA}/\tau_D) \tag{15}$$

where τ_{DA} is the lifetime of the donor in the presence of an acceptor. In terms of relative fluorescent yield in the presence (F_{DA}) or absence (F_D) of acceptor:

$$E_T = 1 - (F_{DA}/F_D) \tag{16}$$

Thus, if we could measure E_T, one could calculate the distance between the donor and acceptor. E_T can be calculated in three ways:

1. Measurement of quantum yield of the donor in the presence and absence of the acceptor:

$$E_T = 1 - q/q_0 \tag{17}$$

2. Measurements of donor lifetime in the presence and absence of acceptor:

$$E_T = 1 - \tau/\tau_0 \tag{18}$$

The advantage of lifetime measurements is that with present technology they are easier and more accurately measured. However, there is a problem with these two approaches in that, if the acceptor molecule introduces quenching in addition to RET, erroneous values of E_T may be obtained. It is also important that measurements of τ_0 and q_0 be performed in the same environment as those of τ or q.

3. Sensitized fluorescence: The problem of spurious quenching may be resolved when the acceptor is fluorescent by looking at sensitized fluorescence. This, however, requires a good separation between the donor and acceptor (Fig. 1). In the case in which the acceptor does not absorb in the excitation range of the donor,

$$E_T = \frac{A_A(\lambda_A)\, I(\lambda_D)}{A_D(\lambda_D)\, I(\lambda_A)} \tag{19}$$

where $I(\lambda_D)$ is the emission of the acceptor when excited at the donor absorption maxima and $I(\lambda_A)$ is the emission of the acceptor when excited at the acceptor absorption maximum. $A_A(\lambda_A)/A_D(\lambda_D)$ is a normalization term necessary because the donor and acceptor may not absorb the same amount of energy or more donor may be excited than acceptors. When direct excitation of the acceptor does occur during excitation of the donor:

$$E_T = \frac{A_A(\lambda_D)}{A_D(\lambda_D)}\left[\frac{I(\lambda_D)}{I_A(\lambda_D)} - 1\right] \tag{20}$$

G. Polarization

There is a technique to study transfer between like molecules utilizing fluorescence polarization measurements. Fluorophores preferentially absorb photons whose electric vectors are aligned parallel to the transition moments of the fluorophore. Using polarized light, one selectively excites those fluorophore molecules whose absorption transition dipoles are parallel to the electric vector of the excitation. The transition moments for absorption and emission have fixed orientations within each fluorophore, and the relative angle between those moments determines the maximum measured anisotropy. Rotational motion (that is further displacement of the emission dipole from its starting position) which occurs during the lifetime of the excited state will further lower the observed emission anisotropy.

When the concentration of a fluorescent compound in solution is increased, eventually a point is reached where a decrease in the quantum yield becomes noticeable. At a lower concentration (sometimes much lower), an effect of the interaction between solute molecules may be observed, which is a drop in the polarization of the fluorescence. This phenomenon is referred to as concentration depolarization. The reason for this depolarization, which takes place even in a solvent of infinite viscosity, is that the absorbed quantum of light is transferred from one oscillator to another, with a certain angle between them, before it is reemitted as radiation. One or more steps may be involved. The transfer may occur in two different ways: radiative or nonradiative (resonance transfer).

The relationship between polarization and concentration (Weber, 1954) is given by

$$1/p = 1/p_0 + Ac\tau_0 \tag{21}$$

where p is the polarization at concentration c of the fluorescent species, p_0 is the polarization at infinite dilution, τ_0 is the lifetime of the excited state (which does not change untill a concentration is reached where concentration quenching takes place), and A is a constant. A full derivation which applies to the whole concentration range, expressed in terms of the average number of transfers between like oscillators (Weber, 1954), yields

$$\frac{1}{p} \pm \frac{1}{3} = \left(\frac{1}{p_0} \pm \frac{1}{3}\right)\left(1 + \frac{4N\pi c R_0^6}{15(2a)^3} \times 10^{-3}\right) \tag{22}$$

where the plus sign apply to excitation with natural (unpolarized) light and the minus signs to excitation with completely polarized light (excitation and emission at right angles), c is the concentration in moles/liter, R_0 the critical distance, a the effective molecular radius, and p and p_0 are defined as in Eq. (2). From the slope s of the straight line obtained when $1/P^{\pm}\ 1/3$ is plotted against c, we get

$$R_0 = (2a)^{1/2}\left[\frac{15s \times 10^3}{4N\pi\left(1/p_0 \pm \dfrac{1}{3}\right)}\right]^{1/6} \tag{23}$$

The linear dependence upon concentration may be understood as follows: The excited donor molecule D is capable of transferring excitation energy to another molecule A, which may or may nor be similar to D. A is located within a spherical shell, the inner radius of which equals the sum of the radii of D and A (in the case where D and A are identical molecules, the radius equals $2a$ and the outer radius is dependent upon R_0, the critical distance). The probability of a molecule A being within this spherical shell surrounding D is then directly proportional to the concentration of A. Clearly, it is possible to define a critical concentration in analogy with the critical distance R_0 such that

$$C_0 \alpha \frac{1}{(R_0)^3} \tag{24}$$

For fluorescent dyes having relatively high quantum yields of fluorescence and relatively extensive overlap between their absorption and emission spectra, C_0 is on the order of 10^{-3}.

III. Techniques of Measurement in Microscopy

A. Lifetime Analysis (τ)

The fluorescence lifetime of a substance usually represents the average amount of time the molecule remains in the excited state prior to its return to the ground state. There are two widely used methods for measuring fluorescent lifetimes; these are the pulse method and the harmonic or phase modulation methods. In the pulse method, the sample is excited with a brief pulse of light and the time-dependent decay of the fluorescent intensity is measured. In the harmonic method, the sample is excited with sinusoidally modulated incident light and the phase shift and demodulation of the emission relative to the incident light are used to calculate the lifetime.

In the pulse method, measurements are performed with a pulsed microspectrofluorometer (i.e., ORTEC 9200 nanosecond spectrometer) coupled to a microscope [fluorescence relaxation microscopy (FRM)—Herman and Fernandez, 1978] employing the time-correlated single-photon technique. The system (Fig. 4) operates as follows: when the exciting lamp flashes, a photomultiplier tube in direct view of the lamp generates a START signal for the time to amplitude converter (TAC). The first single photon (fluorescence) subsequently detected by a second photomultiplier tube located at the detector end of the system generates a

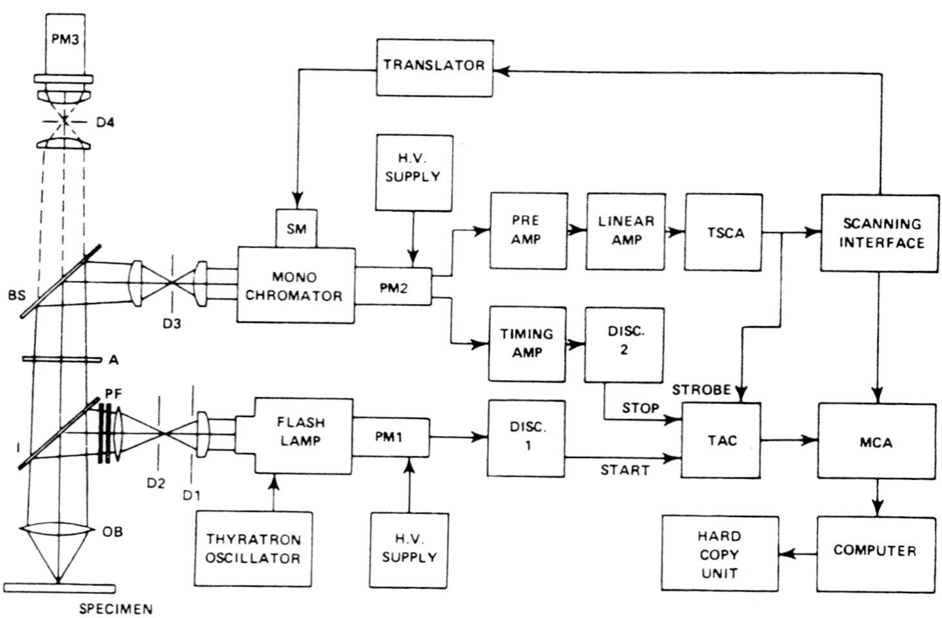

FIG. 4. Schematic of fluorescence relaxation microscopy (FRM) system. The exciting light is focussed through the appropriate excitation filter onto an appropriate dichroic mirror. Light is reflected downward and focussed on the sample through the microscope objective (either a planchromat 1.25NA × 100 or a neofluor 1.30NA × 100). Specimen fluorescence is focussed onto a grating monochromator and then onto a RCA 8850 photomultiplier tube.

STOP signal. The TAC in turn provides a pulse of amplitude proportional to the time elapsed between the excitation and the detection of the first single fluorescent photon. The pulse is then stored in the multichannel analyzer (MCA). After repeating this cycle a large number of times ($\sim 10^7$), a decay of intensity distribution function is accumulated in the MCA [a photon-counting histogram of accumulated counts (ordinate) versus time (abcissa)] which is related to the decay of intensity as a function of time. Multiple photon events are eliminated by energy discrimination of the pulses. This is accomplished by strobing the TAC with a pulse derived from a linear channel consisting of a preamplifier, amplifier, and a single-channel analyzer (SCA) connected to the twelfth dynode of the photomultiplier.

Several procedures have been developed to calculate fluorescent lifetimes using the pulse method. These methods use deconvolution to extract the lifetime values from the decay curves and include Laplace transform methods, a methods of moments, and a method of least square (O'Connor *et al.*, 1979; Grinvald and Steinberg, 1974).

Deconvolution of total intensity arising from a population of chromophore conjugated proteins may be complex due to heterogeneity in the population of dye-binding sites. The analysis of these data may, therefore, be of a multicomponent nature; when n components are present, n decay constants and $n - 1$ ratios of amplitudes of the decays must be found. We have obtained our time decay data utilizing pulse techniques (Isenberg and Dyson, 1969) in which the emitted fluorescence in response to discrete flashes is measured. If the decay function is of short duration, significant distortions of the experimental data occur because of the finite duration of the exciting pulse and the finite resolving time of the detector photomultiplier tube and associated electronics.

In the phase modulation method (Spencer and Weber, 1969), the sample is excited with light whose intensity is modulated sinusoidally. The emission is a forced response to the excitation, and therefore the emission is modulated at the same circular frequency ($\omega = 2\pi \times$ frequency in Hz) as the excitation. Because of the finite lifetime of the excited state, the modulated emission is delayed in phase by an angle ϕ relative to the excitation. Furthermore, the emission is less modulated (demodulated) relative to the excitation, i.e., the relative amplitude of the variable portion of the emission is smaller for the emission than for the excitation. The phase angle (ϕ) and demodulation factor (m) are both measured and used to calculate the phase (τ_p) and modulation (τ_m) lifetimes using

$$\tan \phi = \tau_p, \qquad \tau_p = \omega^{-1} \tan \phi \tag{25}$$

$$m = [1 + \omega^2\tau_m^2]^{1/2}, \qquad \tau_m = \omega^{-1} [(1/m^2) - 1]^{1/2} \tag{26}$$

The phase angle increases and the demodulation factor decreases as the lifetime increases. This procedure is very sensitive to errors when measuring fluorophores of very short or long lifetimes, or fluorophores with multiexponential decays.

RET depletes the population of excited donors and thus shortens the lifetime. Additional information on the distribution of donors and acceptors can be obtained from analysis of the fluorescence spectra taken at intervals during the lifetime of the excited donor population [so called time-resolved spectra (TRS)]. For donors and acceptors separated by statistically variable distances the probability of RET between a donor and the population of acceptors that surround it will also vary. Consequently, immediately after excitation, the probability of RET will be very high between close donors and acceptors and will decline rapidly as the remaining donors are separated from acceptors by greater distances.

Nanosecond (TRS) are fluorescence decay emission spectra obtained at discrete times during the fluorescence decay. The complete data set obtainable, $f(\lambda, t)$, is a surface representing the intensity at different wavelengths and times during the fluorescence decay. TRS can be useful to probe the dynamics of RET. In practice, TRS are obtained by collecting fluorescence decay curves at all wavelengths of interest. These decay curves, once deconvoluted, represent the fluorescence intensity at each wavelength as a function of time.

B. Spectral Measurements

Since RET results in an increase in acceptor emission (sensitized fluorescence) and a decrease in donor emission fluorescence, the ratio of the intensity of the acceptor-to-donor fluorescence emission (I_A/I_D) can be taken as a convenient experimental measure of RET. The value of this ratio depends on the average distance between donor–accepor pairs, and thus on the state of aggregation of donor–acceptor pairs. As the average distance betwen pairs increases, this ratio approaches a limiting value which corresponds to the absence of RET. This limiting value of I_A/I_D depends on the shape of the donor emission spectrum and on the amount of direct excitation of the acceptor at the exciting wavelength of the donor. Likewise, an upper limit for $I_A I_D$ would be expected to exist corresponding to maximal packing density of donor–acceptor fluorophores. It is assumed that the value of K^2 does not change significantly (see above), and thus changes in I_A/I_D are considered to reflect changes in the average degree of proximity of donor and acceptor. Measurements of I_A/I_D can be made through the microscope with either a photomultiplier or video detector (see below). To use a photomultiplier tube one only needs a mechanism to count the number of photons impinging on the photomultiplier/ unit time at both I_A and I_D. In our studies of cell surface receptor topography (Herman and Fernandez, 1982) the MCA was operated in a multiscaling mode, and it receives output pulses from the single-channel analyzer (TSCA). Dwell time and channel advance were controlled externally by a logic interface which was coupled to a stepping motor which drives a monochromator. Thus, information stored in the MCA in this configuration represented intensity versus wavelength. The sensitivity of this instrument is considerably greater than that of a conventional macrofluorometer. Because of the high numerical aperture of the microscope objective (which also acts as a condenser in a vertical illumination system), a microfluorometer is capable of utilizing a much larger fraction of the fluorescence emitted by the sample; and likewise, much higher fluxes of exciting light per unit volume of sample can be achieved.

C. Video Detection

The use of video detectors to monitor RET falls along two lines. The first approach consists of placing a spectrograph at the exit port of the microscope and in front of the video detector. The spectrograph disperses the emitted fluorescence into its component wavelengths (Wampler, 1986) and, via the proper optics, the fluorescence spectrum of the emitted fluorescence is focussed across the faceplate of the video camera (Fig. 5).

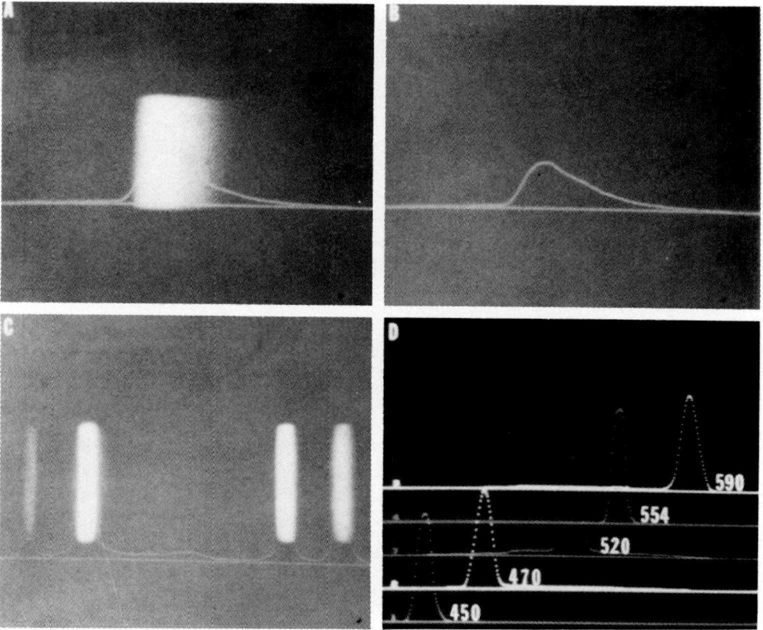

Fig. 5. Video spectrograph. RET measurements can be made using a video spectrograph. (A) Image seen by ISIT camera using FITC-labeled concanavalin A (Con A) cross-linked and smeared on a slide. (B) Intensity versus wavelength of FITC-labeled Con A visualized in A. (C) Image and scan (intensity versus wavelength) of the spectral output of a Zeiss mercury lamp. (D) Spectral plot (intensity versus wavelength) of the output of a Zeis XBO lamp as seen through 450, 470, 5200, 554, and 590 nm (10 nm FWHM) interference filters.

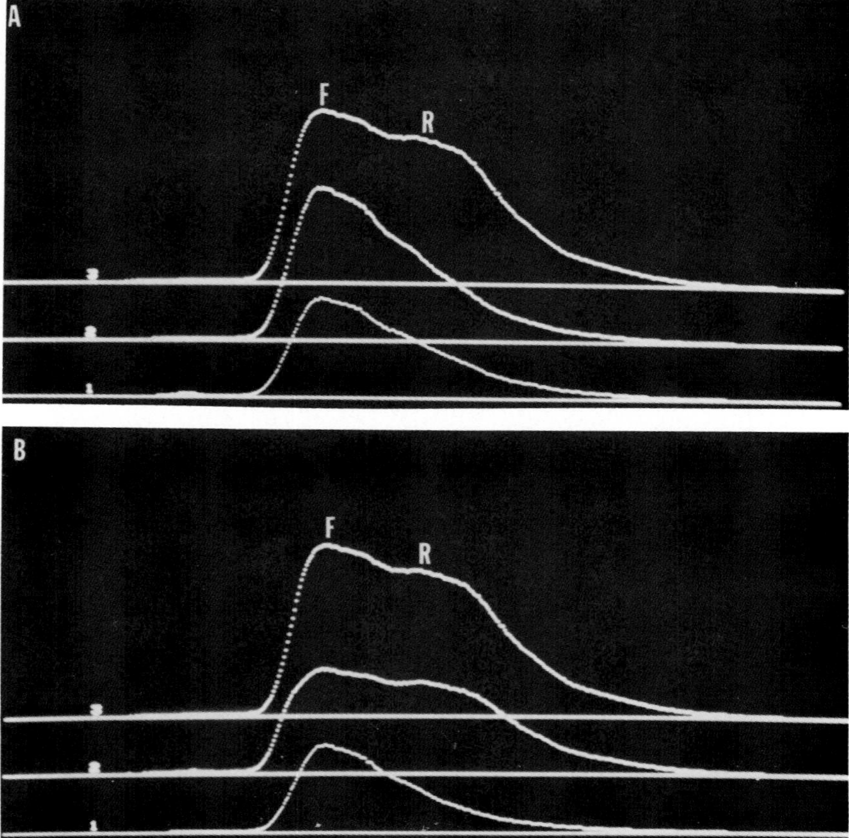

FIG. 6. RET between fluorescein- and rhodamine-labeled Con A. Fluorescein- and rhodamine-labeled Con A were mixed together (at constant Con A concentrations) but with varying donor to acceptor (D/A) ratios. These mixtures (D/A of 1 : 1, 1 : 5, and 1 : 10) were then cross-linked and precipitated with anti-Con A antibody, smeared on a microscope slide, andd illuminated at the excitation wavelength of fluorescein. (A) Illustration that, as the D/A ratio decreases [line 1 D/A = 1 : 1 (1), line 3 D/A = 1 : 10 (0.1)], the amount of RET observed (as detected by an increase in the acceptor fluorescence) goes up. (B) Demonstration that, as more unlabeled Con A is added to a mixture of fluorescein and rhodamine (keeping the total Con A concentration constant), the amount of RET decreases (compare line 3 with line 1). The letters indicate the location of the emission peaks for (F) luorescein and (R) hodamine. (Reprinted with permission of Eaton Publishing Co.)

Thus a two-dimensional fluorescence emission spectrum is recorded by the video detector. The digitization of the video image then enables the extraction of quantitative information from the image; resolution will be determined by the digitized pixel size relative to the wavelength range dispersed across the faceplate of the video detector. An example of the use of a spectrograph to measure RET between cell surface lectin receptors is shown in Fig. 6.

The other approach for using a video detector to measure RET allows the investigator to gain a two-dimensional image of RET between cellular components of interest. This is a relatively new application of video detectors in fluorescence studies and uses the approach of monitoring sensitized emission from the acceptor. RET is quantitated as the ratio of digitized acceptor fluorescence image (I_A) to digitized donor fluorescence image (I_D), using direct excitation of the donor. The ratio (I_A/I_D) is measured at the peak of the acceptor and donor emission intensities, respectively. Thus changes in the intensity of the I_A image divided by the I_D image (I_A/I_D) are considered to reflect changes in the average degree of proximity of donors and acceptors. This approach has been used to monitor interactions between cellular membrane phospholipids (see below) and provides a two-dimensional map of RET in cellular components.

IV. Experimental Parameters Measured in the Microscope

A. R_0 and $J(\lambda)$

R_0 and $J(\lambda)$ are calculated according to Eqs. (II) and (6), respectively; $[J(\lambda)]$ is calculated by integrating the area under appropriate absorption and normalized emission spectra.

B. Donor–Acceptor Ratio.

In the ideal situation, one would like to choose the ratio of acceptor to donor such that, at the density of fluorescent molecules in given cellular component(s), an approximation of donor and acceptors would be observed as an increase in RET. When working with fluorescently labeled protein the dye-to-protein ratio of both the donor and acceptor conjugated proteins must be considered. Differences in the dye-to-protein ratio between donor and acceptor would affect the statistical number of donors and acceptors near one another and thus the efficiency of RET (see below). To determine the proper donor–acceptor ratio, one experimentally varies the donor-acceptor ratio over a wide range (i.e., 10 : 1 D/A to 1 : 10 D/A) under conditions where maximal transfer is expected (Fernandez and Berlin, 1976). For proteins, a mixture of the donor- and acceptor-labeled proteins are cross-linked (either with antibodies or glutaraldehyde), followed by acetone precipitation, and the cross-linked precipitate is smeared on a slide under a coverslip and examined in the microscope. The donor–acceptor ratio which gives the largest change in I_A/I_D is then used experimentally as the donor–acceptor ratio.

C. Limiting Values of I_A/I_D

The procedure cited in Section IV,B can be employed to empirically determine the magnitude of the limiting values of I_A/I_D: Solutions of fluorophore compound conjugates are mixed such that the ratio of donor to acceptor is varied while the total concentration is kept constant. Once the proper donor–acceptor ratio is identified, solutions of the fluorophore compound conjugates are serially diluted with native unlabeled compound in such a way as to vary the relative proportions of labeled and unlabeled species while maintaining the total compound concentration and the donor–acceptor ratio constant. These mixtures of the compound are then precipitated and smeared on glass slides for observation in the microscope. I_A/I_D values are then determined for the various samples. The results of this procedure show that as the average donor–acceptor distance is increased by the presence of greater amounts of unlabeled compound, I_A/I_D decreases and approaches a constant value. This empirical calibration curve thus provides a semiquantitative basis for interpreting RET data. As an example, in order to interpret changes in the proximity of lectin receptors during myogenesis (Herman and Fernandez, 1982), solutions of concanavalin A (Con A) containing donor–acceptor derivatives in the same ratio as used for cellular studies (1 : 6.8) were serially diluted with native unlabeled Con A in such a way as to vary the relative proportions of labeled and unlabeled species while maintaining the total Con A concentration and the donor–acceptor ratio constant. These mixtures of Con A were then precipitated with anti-Con A antibody and the precipitates smeared on glass slides for observation in the microscope. I_A/I_D values were then determined for the various samples. The results of this procedure are presented in Fig. 7, which shows that the maximum value of I_A/I_D obtained is 1.75, and that, indeed, as the average donor–acceptor distance is increased by the presence of greater amounts of unlabeled Con A, I_A/I_A decreases and approaches a constant value of 0.750. This limiting value of I_A/I_D is reflective of the shape of the donor fluorescence spectrum and any direct excitation of the acceptor.

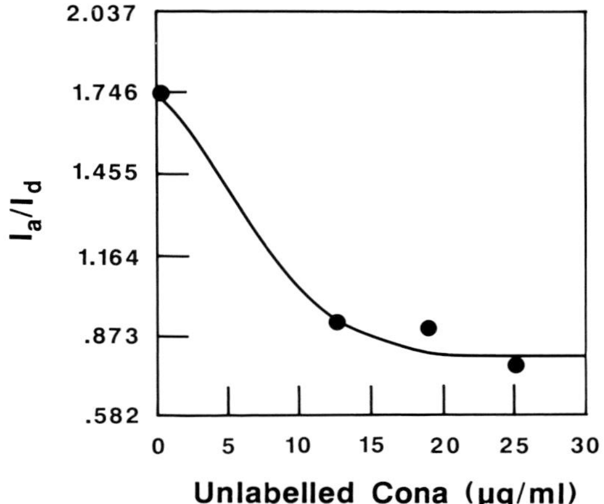

FIG. 7. Ratio of acceptor (FITC) to donor (pyrene) fluorescence, I_A/I_D, in aggregated Con A mixtures as a function of concentration of unlabeled Con A. The peak intensities at 510 and 460 nm were taken as a measure of acceptor and donor fluorescence, respectively. The separately labeled Con A was mixed to give the desired acceptor-to-donor ratio of 1:6.8. Aliquots of this mixture were then diluted with appropriate concentrations of unlabeled Con A, keeping the total concentration of Con A constant. The mixtures were then cross-linked with anti-Con A antibody for 30 minutes at 37°C. The cross-linked precipitate was pelleted, smeared on a glass slide, and examined in the FRM.

V. Biological Applications

A. Receptor Clustering and Topography.

RET has been used to examine alterations in cell surface receptor topography that accompany many biological processes. The use of the RET approach allows the investigator to ask questions such as: (1) What types of alterations in surface topography occur during differentiation, cell cycle, membrane fusion, and cell division? (2) Does the *local* topography of receptors at regions of cell contact/cell fusion, leading edge of cells, or cleavage furrow differ from elsewhere on the cell? (3) Do changes in membrane lipid composition occur during various cellular processes, e.g., apoptosis, cell–cell fusion, receptor-mediated endocytosis? (4) Do changes in lateral mobility of surface receptors accompany differentiation or other biological processes? If so, what are the mechanisms which regulate the lateral mobility? Does the cytoskeleton play a role in the modulation of lateral mobility of cell surface receptors?

Various groups have used RET to study the proximity of lectin receptors on cell surfaces under a variety of conditions. Initial studies examined the proximity of Con A receptors on Friend virus-transformed erythroleukemia cells using flow cytometry and measurements of acceptor emission anisotropies (Chan *et al.*, 1979). These investigators concluded that Con A receptors existed in micropatches on the cell surface. Subsequently, other investigators, using RET, confirmed these findings by showing that receptors for lentil lectins also exist as microclusters on cell surfaces (Schreiber *et al.*, 1981), and that the cytoskeleton regulated the formation and dispersion of lectin microclusters. (Herman and Fernandez, 1982).

The first studies to make use of microscopic RET measurements come from the work of Fernandez and Berlin (1976), who quantitated the distribution of lectin receptors on normal or SV40-transformed fibroblasts. These studies led to the conclusion that lectin–receptor complexes are more clustered on transformed cells than on normal cells. Additional findings showed that a nonrandom distribution of lectin–receptor complexes existed on normal (nontransformed) cells, and that the lectin–receptor complexes are capable of movement within the plane of the membrane over fairly large distances.

Building upon these early studies, we used the technique of RET to study the dynamics and topographical distribution of surface glycoproteins during myoblast fusion (Herman and Fernandez, 1982). All measurements were performed through a microscope on single cells. Substrate-attached cultured cells were labeled in a solution of Con A which contains a mixture of two populations of Con A separately conjugated to donor and acceptor fluorophores. Pyrene- and fluorescein-labeled Con A conjugates were employed as donor and acceptor, respectively.

In order to interpret the value of I_A/I_D obtained from single cell studies, we employed the procedure outlined in Sections IV,B and C to empirically determine the magnitude of the limiting values of I_A/I_D. Since energy transfer also results in a decrease in the fluorescence lifetime of the donor fluorophore, we carried out donor lifetime determinations from single cells as an alternative method to quantitate the extent of RET. Lastly, we examined the emission anisotropy of the fluorescein-labeled Con A as a measure of RET. Two findings emerged from these studies.

1. Surface Con A receptors undergo a transient marked reorganization during the period of myoblast fusion. This was documented as a large decrease in: (a) I_A/I_D, (b) emission anisotropy from donor–acceptor labeled cells, (c) the I_A/I_D ratio in the nanosecond time range (as demonstrated by time-resolved emission spectra), and (d) an increase in the donor lifetime when I_A/I_D decreased, all ocurring at the time of myoblast fusion.

Changes in I_A/I_D were taken as an indication of alterations in the average proximity of the Con A receptors as a function differentiation. The interpretation in terms of an "average proximity", however, is somewhat ambiguous. Different types of topographical distribution could lead to the same values of I_A/I_D. For example, one could assume that cell receptors always exist in clusters (i.e., no isolated receptors). In this case, changes in RET would reflect different degrees of packing of the clusters. Alternatively, one could envision receptors as existing in either of two extreme states: free or aggregated. In this case, changes in RET would reflect changes in the relative population of these two states. A third possibility would be to assume that receptors are uniformly distributed: in this case, changes in RET may simply represent changes in the uniform surface density of receptors. Our RET data (I_A/I_D ratios) by themselves do not permit distinction between these various possibilities.

2. Alterations in membrane fluidity and components of the cytoskeleton are important in the regulation of lectin receptor topography during myoblast fusion. Previous studies from our laboratory (Herman and Fernandez, 1978) had demonstrated that myoblast fusion is accompanied by a transient increase in membrane fluidity at the point of myoblast fusion. We examined the kinetics of change in I_A/I_D ratios as a function of time during the process of myoblast fusion (Fig. 8), and found that, during the period when the membrane fluidity was at its highest, the rate of change in I_A/I_D (receptor aggregation) occurs with the fastest time course. These results provide a correlation between the rotational diffusion mobility (membrane fluidity) of a lipid probe [anilonaphthalenesulfonic acid (ANS)] and the lateral translational mobility of a surface glycoprotein, and give credence to the notion that the endogenous developmentally regulated change in membrane fluidity associated with myoblast fusion may serve to modulate the mobility and display of cell surface components. Additionally, pharmacologic agents which disrupt cytoplasmic microtubules were found to inhibit both alterations in I_A/I_D and myoblast fusion.

A problem that arises in using any of these approaches for measurements of RET in studies in intact cells is the geometry of the cell surface (i.e., the presence of microvilli or other specialized membrane areas), nonrandom distribution of receptors, and the position of the fluorophore relative to the molecule that it labeled, which itself has dimensions commensurate with R_0.

B. Endocytosis and Intracellular Membrane Traffic

This is a relatively new application of RET techniques and has been pioneered by Pagano and colleagues (Ulster and Pagano, 1986). In

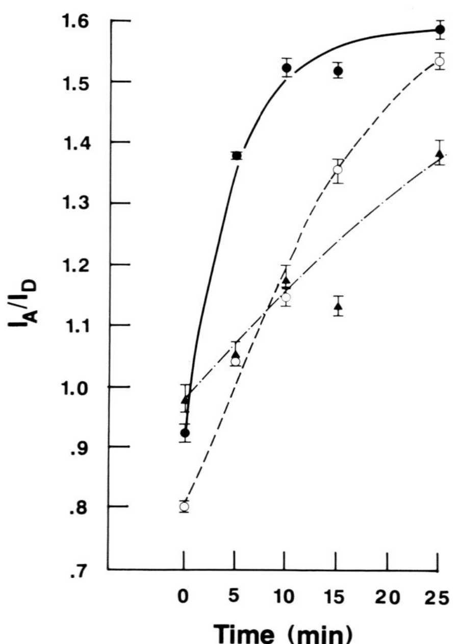

FIG. 8. RET (I_A/I_D) during aggregation of Con A on the surface of myogenic cells at 24 (open circles), 48 (closed circles), and 72 hours (closed triangles) *in vitro*. Cells were labeled with pyrene–Con A : FITC-Con A at a donor–acceptor ratio of 1 : 6.8 for 1 hour at 4°C. The final concentration of Con A was 20 μg/ml. At the end of the 1 hour labeling incubation, the cells were placed at 37°C for varying times and subsequently fixed with 2% paraformaldehyde for 20 minutes at 37°C. (Reprinted with permission of The American Chemical Society.)

addition to their work, other investigators have used RET to monitor the kinetics of fusion between biological membranes (Struck *et al.*, 1981) and calcium-promoted fusion of isolated chromaffin granules (Morris and Bradley, 1984).

The uniqueness of the work by Pagano and colleagues in that they have used RET as a visual microscopic tool to monitor directly the movement and interactions of membrane-bond fluorescent probes in model membranes and in living cells. This later study has been conducted on living cells through a microscope. Microscopic assessment was first used to examine RET between *N*-4-nitrobenzo-2-oxal-1,3-diazole (NBD) or fluorescein as a donor and sulforhodamine as an acceptor in liposomes where the geometry of the transfer process is known in great detail. They modified their microscope epiilumination system such that they had three filter excitation–emission combinations:

1. Excitation–maximum absorbance of donor
   Emission–maximum emission of donor (Donor)
2. Excitation–maximum absorbance of acceptor
   Emission–maximum emission of acceptor (Acceptor)
3. Excitation–maximum absorbance of donor
   Emission–maximum emission of acceptor (Transfer)

Thus, by visually inspecting the specimen for quenching of donor emission (1) or an increase is sensitized acceptor fluorescence (3), these investigators could detect transfer between the different lipid molecules. RET microscopy was then used in cultured fibroblasts to identify intracellular organelles labeled with *N*-sulforhodamine-decylamine (*N*-Shr-decylamine as the acceptor) using transfer between *N*-Shr-decylamine and NBD-labeled lipid (donor). The endoplasmic reticulum and mitochondria were observed in the transfer channel, suggesting colocalization of these two probes. Subsequent studies using lectin-conjugated donor and sulforhodamine-conjugated phosphatidycholine (acceptor) demonstrated intracellular sorting of these two probes as documented by RET. Thus, these studies show that RET microscopy can colocalize different membrane-bound components at high resolution; coupling of this approach with low-light digitized video microscopy will allow temporal and spatial changes in the distribution of fluorescent membrane molecules and their interactions to be visualized in real time in cells.

VI. Problems

A. Quenching

Trivial reabsorption can usually be avoided by keeping the optical densities low enough. Self-transfer can also occur, especially under conditions where the excitation and emission spectram of the donor show a large degree of overlap (see below); this would lead to lower efficiency of RET, depending on the local concentration of donors to themselves and relative to the number of acceptors.

It is usually found that the fluorescence intensity of a given molecule increases linearly with increasing concentration (at relatively low concentrations). At higher concentrations, the fluorescence intensity may reach a limiting value and even decrease with further increases in concentration (quenching). Quenching processes fall into two general types: (1) collisional quenching and (2) quenching by formation of a complex which has zero or small quantum yield (Pesce *et al.*, 1971). In addition, there are various intermediate states between (1) and (2). Quenchers of the first type effectively shorten the lifetime of the excited state and, since more collisions are possible the longer the excited state exists, molecules with long excited lifetimes are more sensitive to this type of quenching. Compounds forming dark complexes (2) do not affect the lifetime, but simply reduce the concentration of potentially fluorescent molecules. Last, because the low-frequency tail of the absorption spectrum of a chromophore often overlaps the high-frequency end of its fluorescence spectrum, fluorescence from an excited molecule can be reabsorbed by a ground-state molecule of the same solute. The probability of such an event increases with increasing solute concentration. "Self-absorption" distorts the shape of the fluorescence spectrum since only the higher frequencies in the spectrum are reabsorbed. Self-absorption ultimately reduces the fluorescence intensity. Thus, RET can be differentiated from reabsorption, complex formation, and collisional quenching by examining whether the recorded change in fluorescence shows any dependence on sample volume or viscosity, and whether the donor lifetime, donor fluorescence spectrum, or absorption spectrum are altered at all. True RET shows no dependence on sample volume or viscosity, no change in the donor fluorescence or absorption spectrum, but is accompanied by a decrease in the donor lifetime.

B. Concentration Depolarization

The term "concentration depolarization" suggests a phenomenon occurring between solute molecules in a solution of sufficient concentration, but basically the situation would be similar for any system in which two or more molecules of a fluorescent compound are brought close enough together for transfer to take place. Thus, fluorescence depolarization through energy transfer could be used much the same as other forms of RET for measurements between different compounds. Conversely, should a situation arise where the number of fluorophores/unit volume is exceedingly high (i.e., a large local concentration of acceptors), then erroneous values of RET would be obtained due to the phenomenon of concentration depolarization.

C. Photobleaching

Photobleaching is the irreversible destruction of the ability of a fluorophore to emit fluorescence following excitation. Almost all fluorescent molecules are sensitive to photobleaching; thus photobleaching can alter the donor–acceptor ratio (by changing the absolute numbers of fluorescing donors or acceptors) and hence the efficiency of transfer. With the advent of single-photon-sensitive detectors [photomultiplier tubes or charge-coupled device (CCD) cameras], intensity levels of exciting light should be able to be kept low enough to prevent photobleaching. Checks should be performed on the stability of donor and acceptor fluorescence emission (in the absence of RET) under identical conditions to that of the experiment (number of exposures to illumination, length of each exposure, intensity of excitation illumination, etc.).

D. Signal/Noise (S/N).

Photomultiplier (PMT) tubes and image intensifiers intensify the impinging fluorescence emission by electronic amplification; therefore, there is a potential for a noise component to be present in the recorded signal. If the noise is a relatively large component of the recorded intensity, then erroneous values will be obtained. The use of cooled PMT tubes as well as single-photon-counting techniques greatly reduces the noise component during data acquisition. One would like the noise to be 0.1–1% of the total intensity. Averaging of individual video frames greatly enhances the S/N in video images, due to the fact that the signal is relatively constant in strength and location within the image while the noise is random. Thus averaging tends to strengthen the signal component of the image while averaging out the noise. Of course, averaging frames (each one-thirtieth of a second in duration) might cost the investigator temporal resolution. The application of intensified CCDs in this respect (where noise is greatly reduced) should help circumvent this problem.

E. Detector Response

As already mentioned, there are various types of detectors which one can use to measure RET through the microscope. These fall into two classes, PMT tubes and video detectors. Already addressed is the concern about dark current noise found with most PMT tubes and ways to diminish this problem. Another concern regarding PMT tubes (and video detectors) is the spectral response of the photocathode. Not all tubes are alike, and there is great variation in spectral response of detectors. Corrections in detection efficiencies as a function of wavelength need to be applied in calculation of the true efficiency of RET. Additional problems of video detectors include shading, lag, geometric response, spatial signal to noise, radiometric response, dynamic response, and resolution. The reader is referred to the work of Bright and colleagues (Chapter 6, this volume) for a discussion of these problems.

F. Statistical Problems

1. Dilution of Donor and Acceptor by Unlabeled Molecules

As the number of unlabeled molecules increases in a population of donor-and acceptor-labeled molecules, the average proximity between donor and acceptor will decrease, resulting in a decrease in RET. We have already discussed (see Section IV,B and C) how one can use this approach to gain a lower limiting value of energy transfer, the amount of acceptor fluorescence observed in the absence of RET. There are certain instances where it is possible (i.e., exchange of labeled and unlabeled cytoskeletal proteins within a cell), where the relative populations of labeled donor–acceptor molecules to unlabeled molecules would change while the actual interrelationship between the molecules remains unchanged. Ideally, then, one would like to perform experiments where incorporation of unlabeled molecules would be prohibited (fixed cells) but, in most fluorescent studies employing microinjected proteins, only 10% of the endogenous protein pool is injected, and thus most molecules will not be labeled.

2. Multiplicity of Donor Environments

A major problem in the analysis of RET from a fluorescent donor to a statistical distribution of acceptors is the presence of a multiplicity of donor environments. Donors in these different environments experience different rates of quenching, with the results that the fluoroscence decay function measured for an ensemble of donors is nonexponential, even though the decay for the individual donors within the ensemble is exponential. In principle, information about the distribution of donor–acceptor distances can be obtained from analysis of the fluorescence decay of the donor (Grinvald *et al.*, 1978). Additionally, two-exponential decay of the donor can further complicate this type of analysis. The decays become significantly nonexponential, especially at higher acceptor concentrations. The origin of this nonexponential decay is the time-dependent distribution of acceptors around the excited donors. At short times following excitation, there exist more donors with nearby acceptors, and these donors decay more rapidly because of the nearby acceptors. At later times, the donors with nearby acceptors have decayed, and the emission results from donors without nearby acceptors. The decay time of these donors is longer, due to a slower rate of energy transfer. The distance between the donor and acceptor can vary both as a result of a range of distances and by diffusion. While the diffusion of donors and/ or acceptors can enhance the rate of energy transfer and complicate the interpretation of the data, the situation becomes simplified when we consider the case of the rapid diffusion limit. To use the Forster equation, we must assume that the donors and acceptors do not change their average displacement during the lifetime of the donor.

VII. Conclusions

RET is a powerful tool with which the biologist can study interactions between cellular constituents at molecular resolution. The coupling of a fluorescence microscope with the RET approach enhances this ability, as it allows observation of the spatial and temporal changes in RET at the subcellular level. With the advent of better imaging detectors, coupled with better fluorescent probes, the technique of microscopic RET microscopy promises to provide a more detailed look at the inner workings of the cell.

References

Chan, S. S., Arndt-Jovin, D., and Jovin T. (1979). *J. Histochem. Cytochem.* **27,** 56–64.

Fernandez, S. M., and Berlin, R. D. (1976). *Nature* (*London*) **264,** 411–415.

Foster, T. (1959). *Discuss. Faraday Soc.* **27,** 7–29.

Grinvald, A., and Steinberg, I. (1974). *Anal. Biochem.* **59,** 583–598.

Grinvald, A., Haas, E., and Steinberg, I. (1978). *Proc. Natl. Acad. Sci. U.S.A.* **69,** 2273–2277.

Herman, B., and Fernandez, S. M. (1978). *J. Cell. Physiol.* **94,** 253–263.

Herman, B., and Fernandez, S. M. (1982). *Biochemistry* **21,** 3275–3283.

Isenberg, I., Dyson, R., and Hanson, R. (1973). *Biophys. J.* **13,** 1090–1115.

Morris, S. J., and Bradley, D. (1984.). *Biochemistry* **23,** 4642–4650.

O'Connor, D. V. O., Ware, W. R., and Andre, J. C. (1979). *J. Phys. Chem.* **83,** 1333–1343.

Pesce, A. J., Rosén, C.-G., and Pasby, T. L. (1971). *In* "Fluorescence Spectroscopy, An Introduction for Biology and Medicine," pp. 50. Dekker, New York.

Schreiber, A. B., Holbeke, J., Vary, B., and Strosberg, A. D. (1981). *Exp. Cell Res.* **132,** 273–280.

Spencer, R. D., and Weber, G. (1969). *Ann. N. Y. Acad. Sci.* **158,** 361–376.

Struck, D. K., Hoekstra, D., and Pagano, R. E. (1981). *Biochemistry* **20,** 4093–4099.

Uster, P. S., and Pagano, R. E. (1986). *J. Cell Biol.* **103,** 1221–1234.

Wampler, J. E. (1986). *In* "Applications of Fluorescence in the Biomedical Sciences" (D. C. Taylor, A. S. Waggoner, R. F. Murphy, F. Lanni, and R. R. Birge, eds.), pp. 301–319. Liss, New York.

Weber, G. (1954). *Trans. Faraday Soc.* **50,** 552–

Chapter 9

Total Internal Reflection Fluorescence Microscopy

DANIEL AXELROD

Biophysics Research Division and Department of Physics
University of Michigan
2200 Bonisteel Boulevard
Ann Arbor, Michigan 48109

I. Introduction

Total internal reflection fluorescence (TIRF) provides a means to selectively excite just those fluorophores in an aqueous environment very near a glass (or plastic) interface. As applied to biological cell cultures, TIRF allows selective visualization of cell/substrate contact regions, even in samples in which fluorescence elsewhere would otherwise obscure the fluorescent pattern in contact regions. TIRF can be used qualitatively to observe the position, extent, composition, and motion of these contact regions, or quantitatively to measure concentrations of fluorophores as a function of distance from the substrate, or to measure binding/unbinding equilibria and kinetic rates at a biological surface. Figure 1 shows a photographic comparison of a single fluorescent labeled cell in culture as viewed by either TIRF of standard epiillumination.

TIRF is conceptually simple. An excitation light beam traveling in a solid (e.g., a glass coverslip or tissue culture plastic) is incident at a high

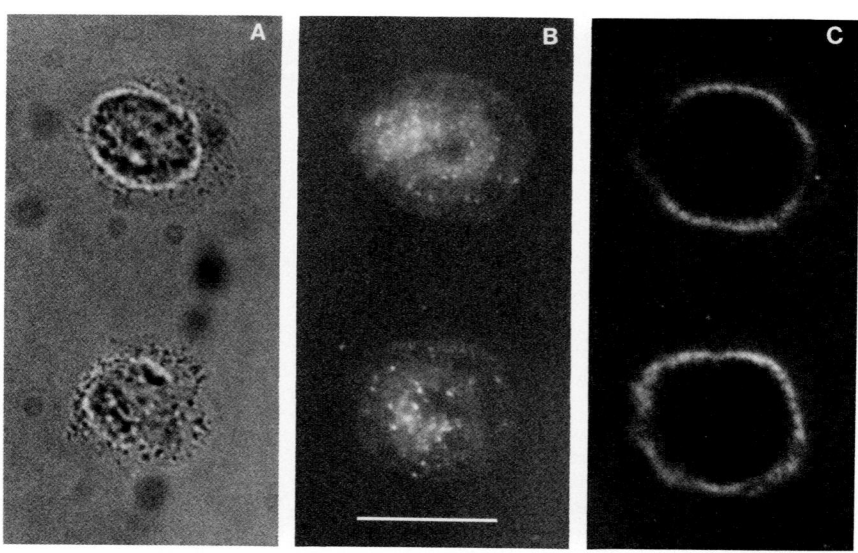

FIG. 1. Three types of illumination of rhodamine–epidermal growth factor (EGF) labeling of A431 (human carcinomoid) cells. (a) Bright field, photographed at low illumination aperture; (b) epiillumination at $\lambda = 514.5$ nm; (c) TIRF at $\lambda = 514.5$ nm. Note that TIRF only illuminates the periphery where the cell is in close contact with the bare glass coverslip substrate. A 50×, 1.00 NA water immersion objective was used. Scale bar = 20 μm.

angle θ upon the solid/liquid surface to which the cells adhere. That angle θ, measured from the normal, must be large enough for the beam to totally internally reflect rather than refract through the interface, a condition that occurs above some "critical angle." TIR generates a very thin (generally less than 200 nm) electromagnetic field in the liquid with the same frequency as the incident light and exponentially decaying in intensity with distance from the surface. The field is called the "evanescent wave" and is capable of exciting fluorophores in the liquid near the surface while avoiding excitation of a possibly much larger number of fluorophores farther out in the liquid.

TIRF is easy to set up on a conventional upright or inverted microscope with a laser light source. (A conventional xenon or mercury arc of filament light source would work in principle, but in practice it is difficult to direct a collimated conventional beam at the proper angle while still maintaining sufficient intensity.) TIRF is completely compatible with standard epifluorescence, bright-field, dark-field, or phase-contrast illumination, so that these methods of illumination can be switched back and forth readily. This chapter is intended as a guide to setting up TIRF with a laser/microscope combination. Section II is a brief summary of the theory of TIRF. Section III describes optical configurations appropriate for microscopy. Section IV discusses some novel applications of TIRF to biology and biochemistry. It is intended as a brief survey of the kinds of measurements that can be made, not as a complete review nor an assessment of the biological results.

A closely related subject to TIR evanescent wave excitation is how the emission light (rather than the excitation light) behaves near an interface. The interface causes the emission to be rather anisotropic, so exactly how much fluorescence is actually observed from a fluorophore depends in a complicated manner on its orientation, its distance from the surface, and on the numerical aperture of the objective. That subject is discussed by Axelrod and Hellen, Chapter 15, this volume.

II. Theory

A. Single Interface

The critical angle θ_c for TIR is given by:

$$\theta_c = \sin^{-1}(n_1/n_3) \tag{1}$$

where n_1 and n_3 are the refractive indices of the liquid and the solid, respectively, with the assumption that $n_1 < n_3$. For incidence angle θ less than θ_c, much of the light propagates through the interface with a refraction angle (also measured from the normal) given by Snell's Law. (Some of the incident light internally reflects back into the solid.) For $\theta > \theta_c$, all of the light reflects back into the solid. However, even with TIR, some of the incident energy penetrates through the interface and propagates parallel to the surface in the plane of incidence. (For a finite-width beam, this propagation in the nearby liquid can be pictured as the beam's partial emergence from the solid into the liquid, travel for some finite distance along the surface, and then reentrance into the solid. The distance of propagation along the surface is measureable for a finite-width beam and is called the Goos-Hänchen shift). The field in the liquid, the evanescent wave, is capable of exciting fluorescent molecules that might be present near the surface.

For an infinitely wide beam (i.e., a beam width many times the wavelength of the light, which is a very good approximation for our purposes), the intensity of the evanescent wave (measured in units of energy/area/second) exponentially decays with perpendicular distance z from the interface:

$$I(z) = I(0)\, e^{-z/d} \tag{2}$$

where

$$d = \frac{\lambda_0}{4\pi}[n_3^2 \sin^2 \theta - n_1^2]^{-1/2} \tag{3}$$

where λ_0 is the wavelength of the incident light in vacuum. Depth d is independent of the polarization of the incident light and decreases with increasing θ. Except at $\theta = \theta_c$ (where $d \to \infty$), d is in the order of λ_0 or smaller. A physical picture of refraction at an interface shows TIR to be part of a continuum, rather than a sudden new phenomenon appearing at $\theta = \theta_c$. For small θ, the light waves in the liquid are sinusoidal, with a certain characteristic period noted as one moves normally away from the surface. As θ approaches θ_c, that period becomes longer as the refracted rays propagate increasingly parallel to the surface. At $\theta = \theta_c$ exactly, that period is infinite, as the wave fronts of the refracted light are normal to the surface. This situation corresponds to $d = \infty$. As θ increase beyond θ_c, the period becomes mathematically imaginary; physically, this corresponds to the exponential decay of Eq. (3).

The factor $I(0)$ in Eq. (2) is a function of θ and the polarization of the incident light. $I(0)$ is different for each polarization component of the evanescent wave. To write the $I(0)$ intensities that are observed polarized along three orthogonal directions, we first set up a coordinate system such that the plane of incidence is the x–z plane, with x parallel to the surface and z perpendicular to it. The y-direction is then normal to the plane of incidence.

There are two independent incident polarizations possible: p and s, for the electric field vectors parallel or perpendicular, respectively, to the plane of incidence defined by the paths of the incident and reflected rays. For s-polarized incident light, the evanescent wave is also s-polarized, which in our coordinate system, is entirely along the y-direction. Its intensity is

$$I_y(0) = \mathcal{I}_s \frac{4\cos^2\theta}{1 - n^2} \tag{4}$$

For p-polarized incident light, the evanescent wave is entirely p-polarized. But the situation is a little more complex, because the p-polarization can contain both an x- and z-component. The x- and z-directed polarized intensities for p-polarized evanescent waves are

$$I_x(0) = \mathcal{I}_p \frac{4\cos^2\theta(\sin^2\theta - n^2)}{n^4\cos^2\theta + \sin^2\theta - n^2} \tag{5}$$

$$I_z(0) = \mathcal{I}_p \frac{4\cos^2\theta\sin^2\theta}{n^4\cos^2\theta + \sin^2\theta - n^2} \tag{6}$$

In Eqs. (4–6), $\mathcal{I}_{p,s}$ are the polarized intensities of the incident electric field in the glass, and n is the ratio n_1/n_3, assumed less than unity. For s-polarized incident light, the total evanescent intensity $I_s(0) = I_y(0)$. For p-polarized incident light, the total evanescent intensity $I_p(0) = I_x(0) + I_z(0)$. The I_y intensity is linearly polarized, but the I_p intensity is elliptically polarized because the x- and z-components of the electric fields are 90° out of phase with each other.

Intensities $I_{p,s}(0)$ are plotted versus θ in Fig. 2, assuming the incident intensities $\mathcal{I}_{s,p}$ are set equal to unity. The plots can be extended without breaks to the subcritical angle range (based on calculations with Fresnel coefficients), again illustrating the continuity of the transition to TIR. The evanescent intensity approaches zero as $\theta \rightarrow 90°$. On the other hand, for supercritical angles within 10° of θ_c, the evanescent intensity is as great or greater than the incident light intensity.

B. Intermediate Layer

In actual experiments, the interface may not be a simple one between two media but rather more stratified with at least one intermediate layer. One example is the case of a biological membrane or lipid bilayer

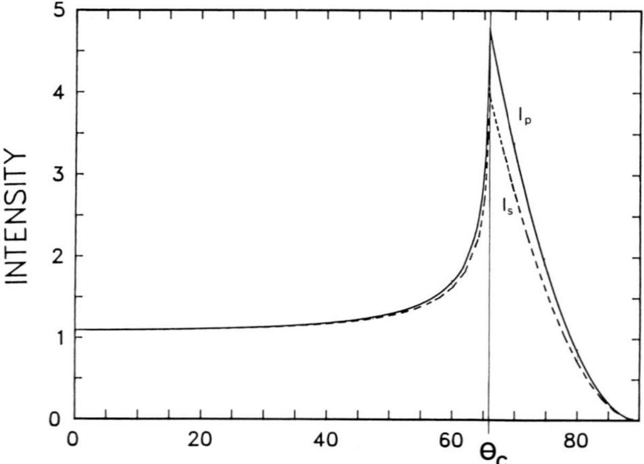

FIG. 2. Intensities $I_p(0)$ (solid line) and $I_s(0)$ (dashed line) at a fused silica/water interface ($z = 0$) versus incidence angle θ. Note that I_p is always the sum $(I_x + I_z)$. Refractive indices $n_3 = 1.46$ and $n_1 = 1.33$ are assumed, corresponding to $\theta_c = 65.7°$.

interposed between glass and aqueous media. Another example is a thin metal film coating, which has useful features to be discussed later.

The rather complicated expressions for the evanescent intensities at stratified interfaces can be found in Axelrod *et al.* (1989). Certain qualitative features can be noted for a three-layer system in which incident light travels from medium 3 (refractive index n_3) through the intermediate layer (n_2) toward medium 1 (n_1):

1. Insertion of an intermediate layer never thwarts TIR, regardless of the intermediate layer's refractive index n_2. The only question is whether TIR takes place at the $n_3 : n_2$ interface or the $n_2 : n_1$ interface. Since the intermediate layer is likely to be very thin (no deeper than several tens of nanometers) in many applications, precisely which interface supports TIR is not important for qualitative studies.
2. Regardless of n_2 and the thickness of the intermediate layer, the evanescent wave's characteristic depth in medium 1 will be given by Eqs. (2) and (3). However, the overall distance of penetration of the field measured from the surface of medium 3 is affected by the intermediate layer.
3. Irregularities in the intermediate layer can cause scattering of incident light, which then propagates in all directions in medium 1. This subject has been treated theoretically by Chew *et al.* (1979). Experimentally, scattering appears not to be a problem on samples even as in-

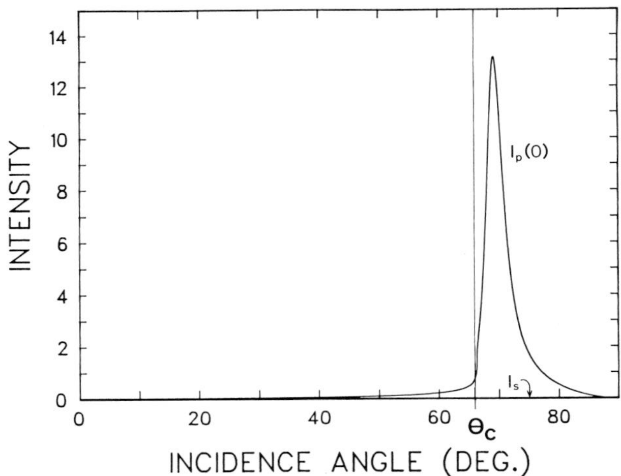

FIG. 3. Intensity $I_p(0)$ at a fused silica/20 nm thick aluminum film/water interface versus incidence angle θ. $I_s(0)$ is too small to appear on this graph. Refractive indices $n_3 = 1.46$ and $n_1 = 1.33$ are assumed, corresponding to $\theta_c = 65.7°$. The aluminum film has a complex refractive index given be $n_2 = 0.73 + 5.75i$.

homogeneous as biological cells. Direct viewing of incident light scattered by a cell surface lying between the glass substrate and an aqueous medium confirms that scattering is many orders of magnitude dimmer than the incident or evanescent intensity, and will thereby excite a correspondingly dim contribution to the fluorescence.

4. $I(0)$ is affected by n_2. If the intermediate layer is a thin (~20 nm) film of metal, the effect is dramatic (Fig. 3). The s-polarized evanescent intensity $I_y(0)$ becomes negligibly small. But the p-polarized intensity is actually enhanced for a very narrow band of incidence angles, becoming an order of magnitude brighter than the incident light at the peak. This resonance-like effect is due to excitation of a surface plasmon mode at the metal/water interface. The peak is at the "surface plasmon angle," which, for an aluminum film at a glass/water interface, is greater than the critical angle for TIR. This enhancement feature is rather remarkable since a 20 nm thick metal film is almost opaque to the eye.

Several practical consequences arise from the metal-film effects. First, metal films can be used to quench fluorescence within 10 nm of the surface (see Axelrod and Hellen, Chapter 15, this volume). Figure 3 shows that TIR can still be used to selectively excite fluorophores in the 10 to 200 nm distance range from metal-film-coated glass. Second, a light beam incident on a 20 nm aluminum film from the glass side at a glass/aluminum film/water interface evidently does not have to be collimated to produce TIR. Those rays that are incident at the surface plasmon angle will create a strong evanescent wave; those rays that are too low or high in incidence angle will create a negligible field in the water. This phenomenon may ease the practical requirement for a collimated incident beam in TIRF. Third, the metal film leads to a highly polarized evanescent wave (provided $\mathcal{I}_p \neq 0$), regardless of the purity of the incident polarization.

III. Optical Configuration for a Microscope

A. Inverted Microscope

Figure 4 shows a schematic drawing of the sample chamber region in a possible TIRF for an inverted microscope. In general, a fixed-stage microscope (in which the objective moves during focusing) is more convenient than a moving-stage microscope, so that the alignment of the beam with respect to the sample is sure to remain fixed during focusing. However, moving-stage microscopes are more common and either type will suffice. The laser beam first enters a focusing lens positioned obliquely above the microscope stage. The purpose of that lens is to concentrate the illumination in much the same manner as does the objective in epiillumination and also to narrow the beam width for easier alignment. The lens' focal length is not critical (but approximately 50 mm will suffice). The

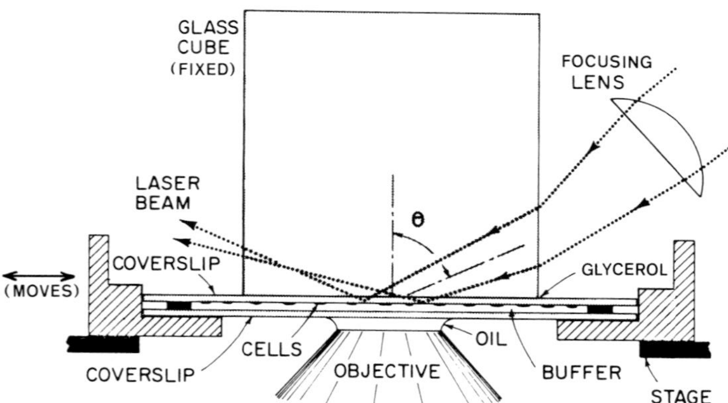

FIG. 4. TIRF optical configuration for an inverted microscope, detail of the sample chamber region.

focusing lens should be mounted on its own *x-y-z* translator with one of the axes along the direction of the beam. The translator itself may be fixed either to the stage or to the table, preferably the former for a moving-stage microscope.

The beam then enters a glass prism. The only purpose of the prism is to ensure that the beam will ultimately strike the TIR surface below the prism with an incidence angle $\theta > \theta_c$. Neither the size nor the shape of the prism is critical; a cube is depicted here, but it could be rectangular, or 45°-45°-90° triangular, or equilateral triangular in cross-section. The latter two are standard commercial items. (A hemicylinder or hemisphere is also useable; the latter will be discussed in the subsection on intersecting beams.) A prism with a flat top, such as a cube or a truncated triangle, allows placement of a tungsten lamp and condenser above the prism, thereby permitting conventional illumination techniques such bright field, dark, field, and phase contrast.

The prism should be mounted on a single-axis translator for some limited motion in the vertical direction to allow the prism to be smoothly lowered over the sample. That translator *must* be mounted onto the stage (rather than onto the table) if a moving-stage microscope is used, so that the act of focusing does not separate the prism from the sample below it.

The sample, perhaps a monolayer of living cells, adheres, facing down, to a glass substrate plate, either a coverslip or microscope slide. On the upper side of this substrate is placed a droplet of pure glycerol, and then the prism is lowered onto it, thereby spreading the droplet into a thin layer. The glycerol serves two purposes: to ensure optical contact between the prism and the sample substrate, and to mechanically lubricate the region between the two glass surfaces. Note that, as the sample below is laterally scanned by the standard stage translators, the prism remains laterally fixed. The incident laser beam should propagate obliquely through the bottom of the prism, through the glycerol, and through the substrate, totally reflecting at the bottom surface of the substrate directly over the optical axis of the microscope. The substrate thickness is not critical, except that thick substrates (e.g., those approaching 1 mm) may constrain the beam from meeting the bottom surface over the microscope's optical axis.

Glycerol should be used sparingly, because an excess will bead up at the bottom edges of the prism and interfere with the incident beam as it enters the prism. The clearance between the glycerol bead and the incident beam increases with larger prisms. On the other hand, a large prism may inhibit lateral motion of the sample below. In practice, a prism with a bottom face area of about 1 cm^2 seems suitable.

The sample chamber contains the buffer needed to maintain the sample (e.g., the cells' viability). The buffer is sandwiched between the substrate surface and a glass coverslip below by a Neoprene or Teflon ring spacer. Teflon rings are commercially available with thicknesses down to less than 50 μm. The depth of solution must be quite thin if high-power, high-aperture, short-working-distance objectives are to be used. The downward pressure of the prism upon the substrate plate may be adequate to seal the sample chamber without additional clamping. Both the substrate plate and the coverslip below should be the same size and placed in a holder that fits both of them snugly and can be clamped by the microscope stage translator.

Alignment of the system mainly involves centering the TIR region directly over the objective while the sample is in focus. First, assemble the chamber and prism combination, and bring the sample into focus using bright-field illumination. Then, with the TIR focusing lens removed, direct the "raw" laser beam so that it totally reflects approximately over the objective: this can be seen by looking directly (i.e., not through the microscope eyepieces) for scattered incident light in the prism, glycerol layer, and TIR substrate plate. Now insert the focusing lens and adjust its position so that the focused beam enters the side surface of the prism approximately 1 mm above the bottom edge. Direct viewing again should now reveal three colinear dots of scattered light at the bottom of the prism: the two outside ones where the beam traverses the glycerol layer, and the middle one where the beam undergoes TIR on the sample. Still under direct viewing, adjust the focusing lens so this middle dot appears to be over the objective. Now, view the sample through the microscope eyepieces. TIR fluorescence excited by the middle dot should be visible as an elongated cylindrical area. Fine-tune the position of the area with the focusing lens translators, and adjust its size as desired with the focusing lens' longitudinal position adjustment.

The shape of the TIRF area will depend on the orientation of the side surface of the prism where the incident beam entered it. If it is vertical, as in Fig. 4, the area will be fairly elongated. In fact, if the incident light is made highly convergent (determined by the ratio of the beam width at the focusing lens to its focal length), then the focused TIRF area will be a thin elongated line. This elongation is due to the aberration introduced by the large refraction at the first surface of the prism. On the other hand, if the side surface of the prism is angled so that the beam enters it more normally, then the TIRF area will resemble a less-elongated conic section expected from slicing a cylindrical beam at an angle.

In the optical configuration described here, the illuminated region will remain stationary in the center of the field of view as the sample is scanned laterally. However, with a moving microscope stage only, the illuminated region may move laterally as one focuses up and down, depending on whether the focusing lens and the beam director preceding it (described below) were mounted on the table or the stage.

The beam path "upstream" from the microscope consists of three elements in addition to those normally found in a laser/microscope system: a beam expander, a beam splitter, and a beam director (see Fig. 5).

The beam expander (commercially available) is used to control the width of the laser beam at the focusing lens, and hence the angle of convergence onto the sample and the width of the focused spot. The wider the beam at the focusing lens, the thinner the width of the TIRF-illuminated area. With the optical configuration being considered, a half-width (measured from the center to the e^{-2} intensity) as small as about 2.5 μm can be attained.

The beam splitter is a mirror that diverts the horizontal beam into the vertical direction up past the level of the microscope stage for TIRF. The mirror should be mounted on a slide so that simple removal enables the beam to enter the microscope's field diaphragm port for quick and

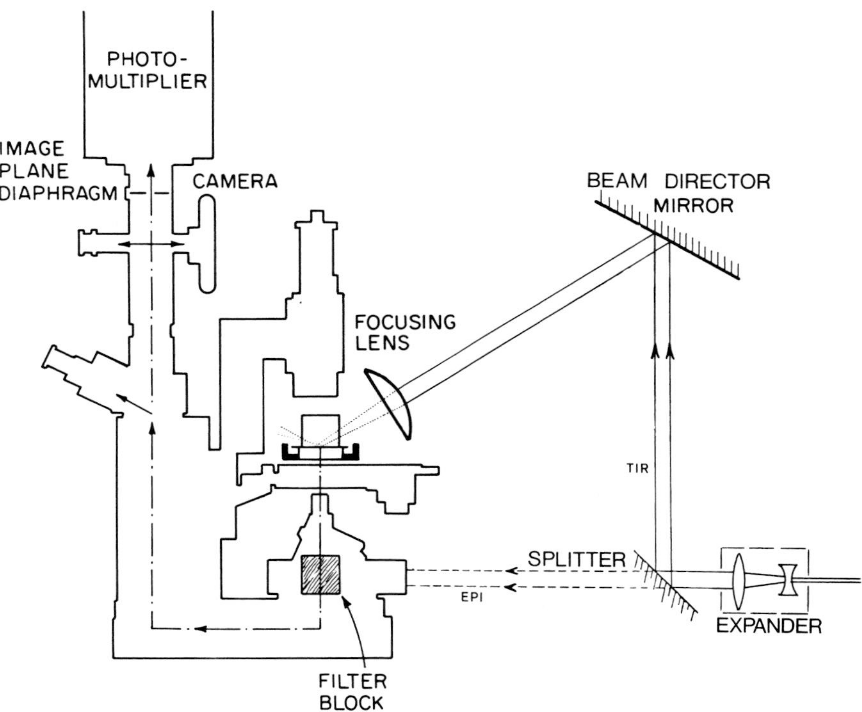

FIG. 5. TIRF optical configuration for an inverted microscope, incident and emission light paths.

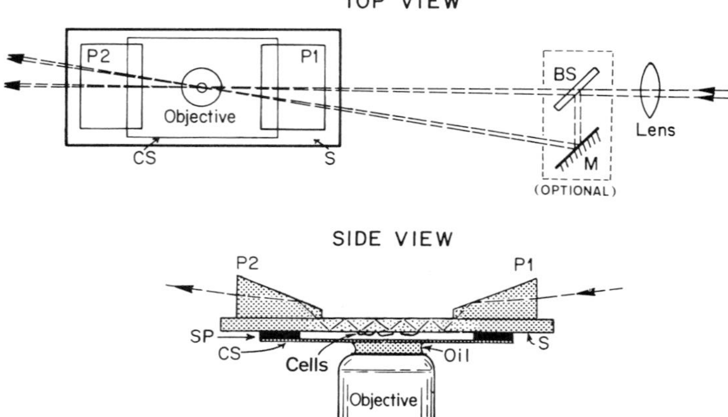

FIG. 6. TIRF microscopy with prisms fixed to substrate plate, adapted from Weis *et al.* (1982). An optional second beam to create interference fringes is shown. Abbreviations are BS, beam splitter (optional); CS, coverslip; M, mirror (optional); P1, entrance prism; P2, exit prism; S, glass slide; SP, spacer.

reversible conversion to epiillumination. By using a partially silvered mirror or even a plain glass slide as the beam splitter, both epiillumination and TIR can be viewed simultaneously.

The beam director is a mirror that angles the beam from the vertical direction obliquely down toward the focusing lens. The height of the mirror determines the incidence angle θ at the TIRF surface. In addition to that linear height adjustment, the mirror should be mounted on a biaxial rotational mount to direct the "raw" beam properly as described above. If the mirror mount is attached to the stage of a moving-stage microscope, rather than to the table, then the position of the TIRF-illuminated region can be rendered insensitive to changes in focusing.

Other TIRF configurations for inverted microscopes have been employed. Figure 6 shows an alternative system (Weis *et al.*, 1982). Instead of a prism fixed with respect to the beam as above, the prism is fixed with respect to the sample. The substrate glass slide propagates the incident beam toward the microscope's optical axis via multiple internal reflections. The illuminated TIRF area will move with translation of the sample in this system.

B. Upright Microscope

A very simple setup for an upright microscope requiring few external mounts and mirrors is shown in Fig. 7. The beam-expanded laser beam

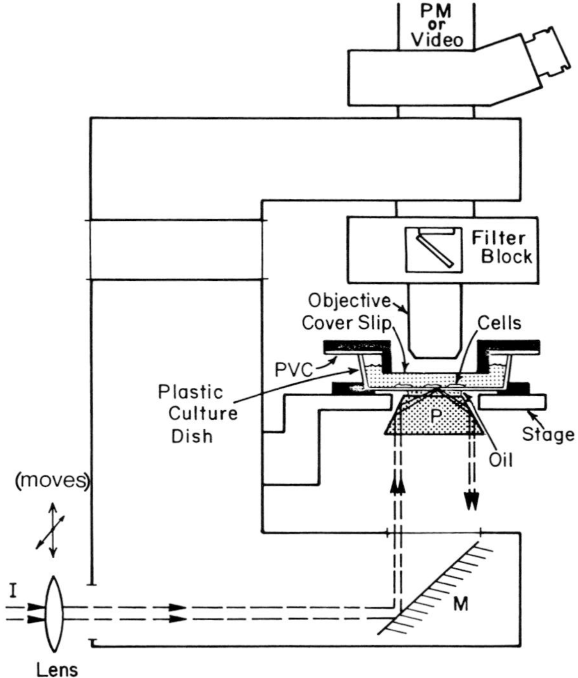

FIG. 7. TIRF adapted to an upright microscope, shown here with a special sample chamber for long-term viewing of cells in a plastic tissue culture dish. The prism is a truncated equilateral triangle. The region of the sample chamber is shown enlarged relative to the rest of the microscope for pictorial clarity. Abbreviations are as follows: I, incident light; M, mirror or prism already installed in microscope base; P, flint glass prism; PM, photomultiplier; PVC, polyvinylidene chloride ("Saran wrap" used to seal in a 10% CO_2 atmosphere over culture medium for long-term viewing.

passes through a long-focal-length (~250 mm) converging lens and enters the base of the microscope stand through the port normally reserved for the transmitted illumination light source. The converging lens is mounted on an *x-y* translator so that the position of the beam can be finely adjusted laterally in both directions.

The beam reflects up at the internal mirror or prism already installed in the microscope base. It then enters a specially designed prism made of flint glass ($n_3 = 1.62$) mounted on the holder that would normally carry the microscope condenser. The prism has a rhombic cross-section and is manufactured by truncating and polishing the top of a commercial equilateral triangle prism. The "raw" beam position (i.e., without the focusing lens) is adjusted to enter the bottom of this prism off-center, so that it totally internally reflects at the sloping side and heads toward the upper surface at an incidence angle of 60°. With the focusing lens in place, the beam follows the same course but is much thinner and more intense.

The sample substrate plate is simply placed on the microscope stage. Then the prism, with a spot of glycerol on its top, is brought up into optical contact with the lower surface of the substrate. The laser beam position is then adjusted by the focusing lens translator to totally internally reflect at the upper surface of the substrate and follow a symmetrical course back down toward the microscope base on the opposite side of the prism. This setup is suitable for any sort of substrate (microscope slides, coverslips, etc.), but is particularly well-adapted for viewing cells growing on the bottom of plastic culture dishes. The cells may thereby be viewed in the same dish in which they incubated without any remounting step.

The fluorescence may be viewed from above through a microscope coverslip which is held off from the cell monolayer surface by a spacer. Alternatively, a water immersion objective can be placed directly in the buffer solution bathing the cells. This upright microscope arrangement is extremely convenient, since samples may be switched as easily as they would with any standard illumination system. The position of the TIRF illumination area is also stationary and unaffected by focusing, even on a moving-stage microscope. The disadvantage is that the incidence angle is not adjustable without switching to another prism with a different slope angle.

With a 60° incidence angle, the refractive index of the prism must be at least 1.53 in order for TIR to take place at the substrate/water interface (regardless of the substrate material). This is why flint glass, rather than ordinary optical glass or fused silica, is the prism material in the example here.

C. Prismless TIRF

It is possible to utilize a form of epiillumination to obtain TIRF without any prism at all, but the scheme requires a very high numerical aperture (NA) objective (generally at least 1.4), as depicted in Fig. 8. The laser beam, collimated to a narrow width by an appropriate focusing lens placed before the field diaphragm plane of the epiilluminator, is positioned to propagate along the very edge of the objective's aperture. It emerges into the immersion oil ($n_3 = 1.52$) at a maximum angle θ given by:

$$\mathrm{NA} = n_3 \sin \theta \tag{7}$$

For total internal reflection to take place at the sample surface, θ must be greater than the critical angle θ_c given by

$$n_1 = n_{\mathrm{water}} = 1.33 = n_3 \sin \theta_c \tag{8}$$

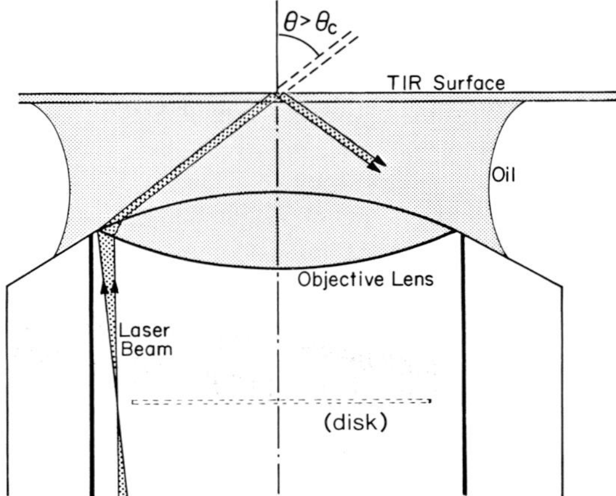

FIG. 8. Prismless TIR illumination through the periphery of a high aperture objective. An optional opaque disk that would block subcritical angle excitation light is depicted. The vertical dimension is somewhat stretched here for pictorial clarity.

From Eqs. (7) and (8), it is evident that NA must be greater than 1.33, preferably by a substantial margin.

In practice, this system works but the TIR illumination occurs over only part of the field of view. In principle, a more even illumination might be attainable by expanding the source to cover the entire objective aperture but blocking those rays that would emerge with $\theta < \theta_c$ with a concentric opaque disk inserted into the objective at its back focal plane.

D. Intersecting Beams

Because of the geometrical contradiction between producing a highly convergent incident beam while ensuring that no rays propagate toward the TIR surface at less than the critical angle, it is difficult to produce a very small, submicron spot of TIR illumination as can be done with standard epiillumination. However, for some types of TIRF experiments, in particular those combining TIR with fluorescence photobleaching recovery for measuring surface diffusion rates, it may be desirable to produce a variegated intensity pattern with a very small characteristic distance in the evanescent wave.

A convenient variegated pattern compatible with TIRF with a very small characteristic distance is an interference fringe pattern created by two coherent TIR laser beams intersecting at the same region on the surface. The internode spacing d in the pattern is given by:

$$d = \lambda_0/(2n_3 \sin\theta \sin\phi/2) \tag{9}$$

where ϕ is the angle between the planes of incidence of the intersecting beams. In principle, d can be made as small as $\lambda_o/2$ for oppositely directed beams.

One experimental setup for intersecting TIR beams is based on the prism/chamber arrangement shown in Fig. 6. Another arrangement, which allows for a full range of incidence angles θ, intersecting beam angles ϕ, and a TIRF illumination area which is fixed in position as the sample is moved, is shown in Fig. 9. This latter arrangement is described in some detail as follows.

The laser beam first must be split into two equal-intensity, coherent beams. With a standard argon laser not equipped with an etalon, the coherence length is only on the order of a centimeter, which means that the difference of pathlength of the two beams to the intersection point of the samples must be less than a few millimeters. To accomplish this, the single laser beam first passes through a long-focal-length converging lens. Then it reflects upward at the front and back surfaces of a clear microscope slide (1 mm thickness) oriented at 45°; these two coherent reflections generate

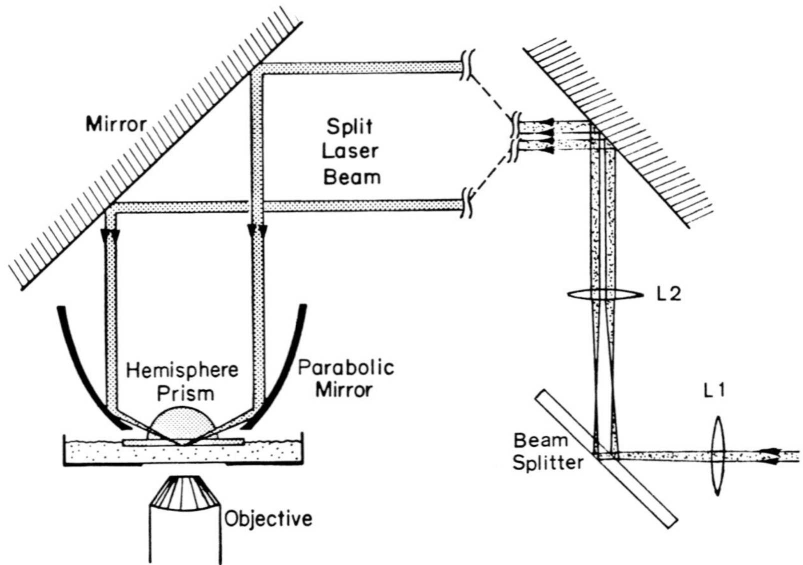

FIG. 9. Intersecting TIR beams split from the same laser to produce interference fringes for viewing in an inverted microscope. The TIR surface is positioned vertically to be at the focus of the parabola. The positions and focal lengths of lenses L1 and L2 determined the widths and separation of the two beams. The beams need not reflect at opposite sides of the paraboloid as depicted, but rather at any azimuthal angle, thereby allowing adjustment of the intersection angle ϕ.

the split beam. Another converging lens placed after the focal point of the first lens converts the two focal spots into two nonoverlapping coherent collimated beams. The exact focal lengths of the two lenses determine the width and separation of the two final beams.

The two beams are directed vertically down over the microscope stage by two 45° mirrors whose rotationally adjustable mounts are (preferably) fixed to the microscope stage (for a moving-stage microscope). The beams then strike two points of a parabolic mirror (of focal length ~2 cm) at the same height but different azimuthal angles. (The exact positions are adjusted by the 45° mirror mounts.) The parabolicity ensures that any such beam will focus and intersect at the same point regardless of where it first strikes the parabolic mirror. To permit TIRF, the two beams enter a glass hemispherical prism, analogous in function to the cube used for single-beam TIRF. The glycerol contact and the sample chamber construction are also similar to those previously described. The whole system—45° mirrors, parabolic mirror, and hemispherical prism—is a single unit and mounted on the microscope's vertically adjustable condenser holder (which moves up and down as a unit during focusing on a moving-stage microscope).

This arrangement successfully produces very distinct and closely spaced interference fringes, suitable for use on even small biological cells (see Fig. 10). However, because of the number of optical elements involved, the adjustment and alignment procedure is somewhat arduous, and the

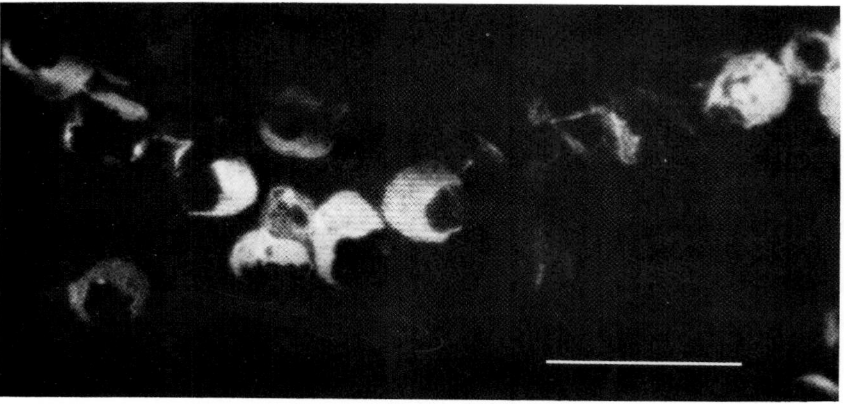

FIG. 10. Fringe pattern created by the intersecting beam configuration of Fig. 9. The sample consists of flattened human erythrocyte ghosts on an aluminum film/poly-L-lysine substrate as described in the text; the TIR-excited fluorescence is from fluorescein-epidermal growth factor (F-EGF) that is nonspecifically adsorbed to the ghost membrane in equilibrium with bulk-dissolved F-EGF. The distinctive membrane tear creating a "bite" and "crescent" appearance has been confirmed by other labels to expose the cytoplasmic surface in the "bite" and the outer surface in the "crescent." Scale bar = 20 μm; λ_0 = 488 nm.

whole structure should be constructed solidly to avoid tiny vibrations which can easily blur the fringes.

E. Emission Beam Path

For local quantitative measurements of TIRF intensity, delimitation of the area from which emission light is gathered may be desired. Because the TIR-illuminated area is not likely to be very small, a small surface area must be defined by an adjustable diaphragm placed at an image plane in the microscope's emission light path.

F. Chemical Derivatization of the TIR Surface

A glass surface can be chemically derivatized to yield special physi- or chemiabsorptive properties. Covalent attachment of certain specific chemicals are particularly useful in cell biology and biophysics, including poly-L-lysine for enhanced adherence of cells; hydrocarbon chains for hydrophobicizing the surface in preparation for lipid monolayer adsorption; and antibodies, antigens, or lectins for producing specific reactivities.

Derivatization generally involves pretreatment by an organosilane (see the catalog of Petrarch Systems). The protocol for poly-L-lysine attachment to planar glass slides is similar to that described for treatment of spherical glass beads (Jacobson *et al.,* 1978). The protocol for preparing lipid monolayers on hydrophobic glass is given by Von Tscharner and McConnell (1981).

G. TIR on Flattened Biological Membranes

To examine reversible binding/unbinding of fluorescently labeled molecules on biological membranes using TIRF, the membranes must be flattened against the TIR surface. This has been successfully accomplished with erythrocyte ghosts (Axelrod *et al.,* 1986). After the glass substrate is covalently coated with poly-L-lysine, erythrocytes are allowed to adhere, followed by hypoosmotic shock. Rather than floating away or crumpling up on the surface, the membrane ghosts flatten into circular disks on the glass with a characteristic tear that exposes the outer surface and the cytoplasmic surface to the solution in their own distinct regions (Fig. 10). This technique or a modification of it may also work for other cell types.

H. Metal-Film Coating

If the TIR surface is precoated with a thin (~20 nm) layer of aluminium certain remarkable alterations in the emission pattern occur (see Axelrod and Hellen, Chapter 15, this volume) while still permitting effective TIR illumination at certain incidence angles. The ability of a metal film to almost completely quench fluorescence of those fluorophores within about 10 nm allows suppression of the signal from labeled protein nonspecifically adsorbed to the substrate, while permitting fluorescence from the slightly more distant labeled protein adsorbed to an adherent membrane. Aluminum coating can be accomplished in a standard vacuum evaporator; the amount of deposition can be made reproducible by completely evaporating a premeasured constant amount of aluminum.

After deposition, the upper surface of the aluminum film spontaneously oxidizes in air very rapidly. This aluminum oxide layer appear to have some similar chemical properties to the silicon dioxide of a glass surface; it can be derivatized by organosilanes in much the same manner.

I. General Experimental Suggestions

Regardless of the optical configuration chosen, the following suggestions may be helpful.

1. The prism used to couple the light into the system and the (usually disposable) slide or coverslip in which TIR takes place need not be matched exactly in refractive index.
2. The prism and slide may be optically coupled with glycerol, cyclohexanol, or microscope immersion oil, among other liquids. Immersion oil has a higher refractive index (thereby avoiding possible TIR at the prism/coupling liquid interface at low incidence angles) but it tends to be more autofluorescent (even the "extremely low" fluorescence types).
3. The prism and slide can both be made of ordinary optical glass for many applications, unless shorter penetration depths arising from higher refractive indicies are desired. Optical glass does not transmit light below about 310 nm and also has a dim autoluminescence with a long (several hundred microsecond) decay time, which can be a problem in some photobleaching experiments. The autoluminescence of high quality fused silica (often called "quartz") is much lower. Tissue culture dish plastic (particularly convenient as a substrate in the upright microscope setup) is also suitable, but tends to have a significant autofluorescence compared to ordinary glass. More exotic high n_3 materials such as sapphire, titanium dioxide, and strontium titanate can yield exponential decay depths (d) as low as $\lambda_0/20$.
4. The TIR surface need not be specially polished: the smoothness of a standard commercial microscope slide is adequate.
5. Illumination of surface-adsorbed proteins can lead to apparent photochemically induced cross-linking. This effect is observed as a slow,

continual, illumination-dependent increase in the observed fluorescence. It can be inhibited by deoxygenation (aided by the use of an O_2-consuming enzyme/substrate system such as protocatachuic deoxygenase/protocatachuic acid) or by 0.05 *M* cysteamine.

6. Virtually any laser with a total visible output in the 0.5 W or greater range should be adequate. The most popular laser for cell biological work with a microscope appears to be a 3 W continuous-wave argon laser.

7. The only absolute requirement for the microscope objective is that it have a long enough working distance to see the TIR surface across a thin water gap defined by the spacer thickness. Since the spacer can be made thinner than any standard working distance while still appearing infinitely deep relative to the evanescent wave, this is not normally a restriction. However, the aberration corrections of some high-magnification objectives are crucially dependent upon the sample being *immediately* on the far side of a coverslip rather than across an additional thin water layer. The result can be an image which, although sharply focused, may be cast with a diffuse haze which lowers contrast. No general rule for which objectives will work best can be provided, but water-immersion objectives generally work well and success has been achieved with certain objectives of every commercially available magnification.

8. For incidence angles near θ_c, it may be difficult to determine by direct viewing whether or not TIR has been achieved. However, TIRF observed through the microscope eyepieces is quite distinct from fluorescence excited by a subcritical angle incident beam propagating at a skimming angle through the aqueous medium. TIRF originates from what appears to be a single plane, since the evanescent wave (unlike a propagating beam) is thinner than the microscope's depth of focus. Epiillumination, on the other hand, usually excites fluorophores in a range of focal planes. At subcritical angles near θ_c, the cells appear to cast a shadow so that the whole field of view appears streaked.

IV. Applications of TIRF

A. Qualitative Observation of Labeled Cells

The most straightforward application of TIRF is to observe the location and (with time-lapse video) the motion of cell/substrate contacts. Cells may be labeled by a membrane lipid fluorescent analog such as 3, 3′-dioctadecylindocarbocyanine (diI), or specific ligands such as rhodamine α-bungarotoxin (targeted to acetylcholine receptors), or fluorescent antibodies against cell surface and cytoplasmic components. (See Axelrod,

1981; Axelrod *et al.*, 1984, 1988; and Bloch *et al.*, 1989, for photographic examples.) TIRF seems particularly advantageous for long-term viewing of cells, since the evanescent wave minimizes exposure of the cells' organelles to excitation light.

In all cases, some probe becomes internalized on living cells. Under epiillumination, this tends to obscure the fluorescent pattern of the membrane or submembrane labeling. TIRF "optically sections" the sample, allowing observation of a distinct surface pattern even in the presence of a large amount of internalized label. The optical sectioning is particularly useful in viewing submembrane cytoplasmic filaments on thick cells. Epiillumination excites fluorescence from out-of-focus planes and leads to a diffuse fluorescence that obscures detail. Although TIRF cannot view deeply into the cell, it can display the submembrane filament structure with high contrast.

B. Negative TIRF for Viewing Cell/Substrate Contacts

A variation of TIRF to observe cellular morphology, introduced by Gingell *et al.* (1985), produces essentially a negative of the standard fluorescence view of labeled cells. The solution surrounding the cells is doped with a nonadsorbing and nonpermeable fluorescent volume marker, fluorescein-labeled dextran. Focal contacts then appear as dark areas and other areas appear brighter, depending on the depth of solution illuminated by the evanescent wave in the cell/substrate gap. A quantitative theory for converting fluorescence intensities into cell/substrate contact distances has been developed (Gingell *et al.*, 1987).

C. Adsorption Equilibria of Proteins

TIRF has been used to study equilibrium adsorption of blood proteins to artificial surfaces (e.g., Lowe *et al.* 1986), mainly to learn about the surface properties of various biomaterials that have medical applications (e.g., as nonthrombogenic medical prostheses). Much of this work has been done in nonmicroscopic configurations, but the approach can usually be easily adapted to microscopy for proteins extrinsically labeled with visible or near-UV fluorophores such as fluorescein, rhodamine NBD (7-chloro-4-nitrobenzo-2-oxa-1,3-diazole), or dansyl groups.

It is possible that such extrinsic groups might affect the physical adsorption process under investigation. To avoid this possible disruption, some nonmicroscopic studies have monitored the intrinsic fluorescence of tryptophan residues on unlabeled proteins, excited at $\lambda_0 = 280$ nm (Hlady *et al.*, 1985). In standard epiillumination microscopy, working with intrinsic protein fluorescence requires expensive fused silica objectives to transmit the excitation light. TIR intrinsic fluorescence microscopy would require only regular objectives (which should transmit the emission at $\lambda = 330$ nm) along with fused silica excitation focusing lenses and prism, and a very bright (usually pulsed) laser source coupled with a dye laser and frequency doubler to achieve $\lambda_0 = 280$ nm.

Calibration of a TIRF intensity to derive an absolute concentration of adsorbate is a nontrivial problem, mainly because fluorescence quantum efficiencies are apt to change upon adsorption to a surface. One route around this problem is to measure the depletion of bulk solute (in epiillumination mode or in a standard spectrofluorimeter) when it is allowed to adsorb onto a known surface area (e.g., Burghardt and Axelrod, 1981). Another route involves the use of a radioactively labeled derivative of the protein adsorbate (Hlady *et al.*, 1986).

D. TIRF Combined with Fluorescence Photobleaching Recovery (FPR)

Consider a labeled molecule in equilibrium between a surface-bound state and a free solute state:

$$\text{free solute} + \text{vacant surface site} \underset{k_2}{\overset{k_1}{\rightleftarrows}} \text{surface-bound solute}$$

If the evanescent wave intensity is briefly flashed brightly, then some of the fluorophores associated with the surface will be photobleached. Subsequent exchange with unbleached dissolved fluorophores in equilibrium with the surface will lead to a recovery of fluorescence, excited by a continuous but much attenuated evanescent wave. The time course of this recovery is a measure of the desorption kinetic rate k_2 (Thompson *et al.*, 1981).

If the evanescent wave intensity is variegated over a short characteristic distance on the surface, then surface diffusion coefficients can also be measured (at least if surface diffusion is fast enough to carry the molecule through the characteristic distance before desorption). Two studies have utilized TIR/FPR in this manner. The adsorption/desorption kinetics and surface diffusion of rhodamine bovine serum albumin at a glass surface have been examined by a TIR illumination area focused into a thin line (Burghardt and Axelrod, 1981). Similar parameters have been examined for fluorescein-epidermal growth factor reversibly adsorbed to erythrocyte membrane ghosts with a TIR illumination area of interference fringes created by intersecting beams (Fig. 10).

E. TIRF Combined with Fluorescence Correlation Spectroscopy (FCS)

The volume defined by the depth of the evanescent wave in the area defined by the image plane diaphragm of a microscope can be extremely small, down to about 0.01 μm^3. Within this volume, the entrance or exit of a single fluorophore can cause a significant change in the fluorescence intensity. In fact, these TIRF fluctuations are clearly visible to the "naked eye" through the microscope. By autocorrelating (on-line) the random noise arising from such statistical fluctuations (a technique called fluorescence correlation spectroscopy or FCS; see Elson and Qian, Chapter 11, this volume), one can obtain information about three parameters: the mean time of surface binding ($=1/k_2$), the surface diffusion coefficient, and the absolute mean number of fluorescent molecule bound per surface area (even without one knowing anything about quantum efficiencies or light collection efficiencies).

Two investigations have combined TIR with FCS. The first (Hirschfeld *et al.*, 1977) adapted TIR/FCS to measure the absolute concentration of virions in solution. The other (Thompson and Axelrod, 1983) measured the adsorption/desorption kinetics of immunoglobulin on a protein-coated surface on the millisecond time scale.

F. Concentration of Molecules versus Distance from Surfaces

The concentration of a solute or adsorbate may be a nontrivial function of the distance to the surface, a function which contains information about the thermodynamics of the surface interaction. To explore the fluorophore concentration $C(z)$ as a function of distance z from the surface, one can record the observed fluorescent intensity F as the characteristic depth d of the evanescent wave is varied. Mathematically, the relationship is

$$F(d) = \int_0 I(0)\, e^{-z/d}\, C(z)\, g(z)\, dz \tag{10}$$

where the exponential arises from the evanescent intensity. Term $g(z)$ is, in general, a rather complicated function arising from the nonisotropic and z-dependent emission pattern from fluorophores near an interface (see Axelrod and Hellen, Chapter 15, this volume). However, if we approximate $g(z)$ as a constant, then $C(z)$ can be computed simply as an inverse Laplace transform of an experimentally measured $F(d)$.

To vary d, the angle of incidence θ can be varied. Experimentally, this is not trivial, because d is a very strong function of θ within only a few degrees greater than θ_c, and therefore θ must be measured to fractions of a degree. In addition, the presence of a solute (or the cytoplasm of a biological cell) alters the refractive index n_1 from its pure water value, and this must also be known accurately.

Another method of obtaining $C(z)$ involves varying the observation angle of emission in a nonmicroscope setup (see Axelrod and Hellen, Chapter 15, this volume).

Deduction of the absolute distance from the surface to a labeled cell membrane at a cell/substrate contact region can be based on the variation of $F(d)$ with θ (Lanni *et al.*, 1985). This effort is challenging because corrections have to be made for θ-dependent reflection and transmission through four stratified layers (glass, culture medium, membrane, and cytoplasm), all with different refractive indices.

G. Orientation and Rotation of Molecules near Surfaces

The polarization properties of the evanescent wave can be used to excite selected orientations of fluorophores, e.g., fluorescent-labeled phosphatidylethanolamine embedded in lecithin monolayers on hydrophobic glass (Thompson *et al.*, 1984). When interpreted according to an approximate theory, the total fluorescence gathered by a high-aperture objective for different evanescent polarizations gives a measure of the probe's orientational order.

Some first experimental steps in combining TIRF with time-resolved polarized anisotropy decay to measure molecular rotation rates have been tested on sapphire prisms coated with fluorescence-doped polystyrene films (Masuhara *et al.*, 1986).

H. Fluorescence Energy Transfer and TIRF

TIRF can be combined with fluorescence energy transfer to measure distances between fluorophores on surface in the presence of a large background of bulk-dissolved fluorophores. Watts *et al.* (1986) explored whether helper T cells could force two nonidentical antigens in a target membrane into closer proximity with each other. These two antigens, one labeled with a fluorescence energy transfer donor and the other with an acceptor, were incorporated into a planer lipid bilayer on a TIR hydrophilic glass surface. Significant amounts of one of them remained in solution, so microscopic TIRF was needed to limit excitation to the region near the glass and overlaying lipid bilayer. TIRF also served to reduce the autofluorescence normally observed from the T cells that were allowed to settle on the lipid bilayer. It was found that fluorescence energy transfer occurred only in those microscope lipid bilayer regions where the T cell surface came into close apposition with bilayer. The conclusion was that the T cell surface forces the two membrane antigens to which it binds to within a distance of 4 nm of each other.

V. Conclusions

TIRF is an experimentally simple technique for selective excitation of fluorophores on or near a surface. It requires a laser source, but otherwise can be set up on a standard upright or inverted microscope. It is compatible and rapidly switchable with bright field, dark field, phase contrast, and epiillumination and accommodates a wide variety of common microscope objectives without alteration. Applications in cell biology and surface chemistry include (1) localization off cell–substrate contact regions in cell culture. (2) High-contrast visualization of submembrane cytoskeletal structure on thick cells. (3) Measurement of the kinetic rates and surface diffusion of reversibly bound biomolecules at flattened biological membrane surfaces. (4) Measurement of concentration and orientational distributions of fluorescent molecules as a function of distance from the surface. (5) Measurement of intermolecular distances between fluorescent surface-bound molecules in the presence of a large excess of fluorophore or background fluorescence in the bulk. (6) Reduction of the contribution of cell internal autofluorescence relative to cell surface fluorescence excited at cell–substrate contacts.

Acknowledgments

The author is grateful to Robert M. Fulbright and Edward H. Hellen for the photography work in two figures. This work was supported by USPHS NIH grant NS 14565.

References

Axelrod, D. (1981). *J. Cell Biol.* **89,** 141–145.

Axelrod, D., Burghardt, T. P., and Thompson, N. L. (1984). *Annu. Rev. Biophys. Bioeng.* **13,** 247–268.

Axelrod, D., Hellen, E. H., and Fulbright, R. M. (1989). *In* "Fluorescence Spectroscopy: *the Biomedical Sciences*" (D. L. Taylor, A. S. Waggoner, F. Lanni, R. F. Murphy, and R. Birge, eds.), pp. 461–467. Liss, New York.

Axelrod, D., Hellen, E. H., and Fulbright, R. M. (1988). *In* "Fluorescence Spectroscopy: Principles and Applications, Vol. 2: Biochemical Applications" (J. Lakowicz, ed.). Plenum, New York, in press.

Bloch, R. J., Velez, M., Krikorian, J., and Velez, M. (1989). Submitted.
Burghardt, T. P., and Axelrod, D. (1981). *Biophys. J.* **33,** 455–468.
Chew, H., Wang, D., and Kerker, M. (1979). *Appl. Opt.* **18,** 2679–2687.
Gingell, D., Todd, I., and Bailey, J. (1985). *J. Cell. Biol.* **100,** 1334–1338.
Gingell, D., Heavens, O. S., and Mellor, J. S. (1987). *J. Cell Sci.* **87,** 677–693.
Hirschfeld, T., Block, M. J., and Mueller, W. (1977). *J. Histochem. Cytochem.* **25,** 719–723.
Hlady, V., Van Wagenen, R. A., and Andrade, J. D. (1985). *In* "Surface and Interfacial Properties of Biomedical Polymers, Vol. 2: Protein Adsorption" (J. D. Andrade, ed.), pp. 81–119. Plenum, New York.
Hlady, V., Reinecke, D. R., and Andrade, J. D. (1986). *J. Colloid Interface Sci.* **111,** 555–569.
Jacobson, B. S., Cronin, J., and Branton, D. (1978). *Biochim. Biophys. Acta* **506,** 81–96.
Lanni, F., Waggoner, A. S., and Taylor, D. L. (1985). *J. Cell Biol.* **100,** 1091–1102.
Lowe, R., Hlady, V., Andrade, J. D., and Van Wagenen, R. A. (1986). *Biomaterials* **7,** 41–44.
Masuhara, H., Tazuke, S., Tamai, N., and Yamazaki, I. (1986). *J. Phys. Chem.* **90,** 5830–5835.
Thompson, N. L., and Axelrod, D. (1983). *Biophys. J.* **43,** 103–114.
Thompson, N. L., Burghardt, T. P., and Axelrod, D. (1981). *Biophys. J.* **33,** 435–454.
Thompson, N. L., McConnell, H. M., and Burghardt, T.P. (1984). *Biophys. J.* **46,** 739–747.
Von Tscharner, V., and McConnell, H. M. (1981). *Biophys. J.* **36,** 421–427.
Watts, T. H., Gaub, H. E., and McConnell, H. M. (1981). *Nature (London)* **320,** 176–179.
Weis, R. M., Balakrishnan, K., Smith, B. A., and McConnell, H. M. (1982). *J. Biol. Chem.* **257,** 6440–6445.

Chapter 10

Designing, Building, and Using a Fluorescence Recovery after Photobleaching Instrument

DAVID E. WOLF

Worcester Foundation for Experimental Biology
222 Maple Avenue
Shrewsbury, Massachusetts 01545

I. Introduction

In this chapter, it is my intention to describe how to design, construct, and properly use a fluorescence recovery after photobleaching (FPR) instrument. One cannot overemphasize the phrase "properly use." As is the case with any sophisticated biophysical technique, in FPR there are parameters which must be properly set and experimental criteria which must be strictly adhered to, if one wishes to avoid obtaining erroneous measurements. In this chapter, I will emphasize the design of spot FPR instruments and the development of intensified video imaging FPR instruments. Of course, the design possibilities of such instruments are essentially infinite, and as such, I will principally describe the design of instruments which I have either developed myself or been involved in the development of. I will further emphasize instruments which are interfaced to personal computers. Such interfacing brings, at reasonable cost to the individual laboratory, the ability to perform both nonlinear least squared data fitting and image processing. Of course, such interfacing requires compromise. However, the rapid evolution of the personal computer in the past 10 years promises continued growth of power and flexibility in the future.

FPR is a technique for measuring the lateral diffusibility of macromolecules in membranes and aqueous phases. Early FPR measurements were performed in the laboratory of Reiner Peters (Peters *et al.,* 1974a,b). The essentials of current FPR technology was developed in the early 1970s in the laboratories of Watt Webb and Elliot Elson (Axelrod *et al.,* 1976; Schlessinger *et al.,* 1976, 1977), Michael Edidin (Edidin *et al.,* 1976), and Kenneth Jacobson (Jacobson *et al.,* 1976a,b). In FPR the molecule whose diffusion is to be measured must be fluorescently labeled directly in a noncross-linking manner. A laser beam is focussed using a modified fluorescence microscope to a small spot on the sample. The light from this spot is monitored and found to be essentially constant with time. The incident light level is then momentarily raised approximately 10,000-fold so as to irreversibily photobleach a significant fraction of the fluorescence within the spot. Thus, when one returns to the monitoring intensity, the fluorescence intensity is found to be significantly reduced. If there is no freedom of motion of molecules in and out of the spot, the fluorescence intensity will remain at this level ad infinitum. This is the condition of no diffusibility. If, however, there is complete freedom to diffuse in and out of the spot, the fluorescence intensity will recover to the prebleach level. The diffusion coefficient is obtained by fitting the recovery data to diffusion theory (Axelrod *et al.,* 1976). An intermediate condition can also occur where only a fraction of the molecules is free to diffuse. In that case, only partial recovery is observed. FPR thus provides two measures of diffusibility; the fraction or percent of the molecules free to diffuse, $\%R$), and the diffusion coefficient of that fraction D.

That many membrane proteins are not completely free to diffuse, and that their diffusion rates are too slow to be controlled by lipid fluidity were early and important discoveries of FPR measurements (for review see Peters, 1981; Edidin, 1981). These discoveries have led to refinement of the fluid mosaic model (Singer and Nicolson, 1972), and point to the power of techniques which, like FPR, can be applied to measurement of a wide range of macromolecules. Particularly with the increased spatial information obtainable by imaging FPR, the latter can be used for measuring motions other than diffusion, such as membrane flows (Axelrod *et al.*, 1976) and cytoskeletal treadmilling (Wang, 1985). In addition, FPR is not limited to studying diffusion in membranes but has also been used to study motion in the cytoplasm (Wojcieszyn *et al.*, 1981; Lubyphelps *et al.*, 1987; Wang, 1985).

II. Designing and Building a Spot FPR Instrument

A. Before Thoughts for Fitting the Instrument to the Experiment

Before considering the specifics of designing an FPR instrument, it is important to consider the specific purpose for which you intend to use this instrument. Several aspects of application govern design. Here we will consider four issues of application: concentration of molecule to be measured, diffusion rates to be measured, sample geometry, and fluorescent probe to be used.

1. Concentration

As a before thought, macromolecular concentration addresses two significant issues: first, whether measurement is feasible at all, and second, the required sensitivity and nature of the detection system.

If one wants to obtain reasonable statistics from an FPR measurement, it is generally necessary to measure the signal from at least 100 molecules. In most biological systems the intrinsic variability precludes doing signal averaging. One can, in some instances, help the situation by increasing the beam radius. However, for a constant power level, intensity falls off with the square of the beam radius, just compensating for the increase in particle number with radius squared. Thus, the overall light intensity at best stays the same. Another consequence of the square law is that, in order to hold the amount of bleach constant with increasing radius, one must either increase the bleach time or power. While one cannot apply an absolute rule to this issue, a good rule of thumb is that the macromolecular concentration should be at least 100 labeled molecules/μm^2 in order to obtain reasonable FPR data.

As noted above, a further issue which molecular concentration raises is the necessary sensitivity of the detection system. If one wants to work with probes at low concentration or of low quantum efficiency, it will be necessary to use a sensitive photomultiplier tube as well as single photon counting. On the other hand, if one's principal interest is in measuring high-efficiency, relatively high-concentration probes, less-expensive, less-sensitive photomultipliers can be used. In such cases, it may also be possible to use analog detection systems which measure the current produced by these tubes in response to light. This issue will be discussed in greater detail below in Section II,H.

2. System Response Time—Diffusion Rates to Be Measured

Fitting system response time to an experiment is an extremely critical issue. At the design point, one must consider both the fastest and the slowest processes which one wishes to measure. Let us begin by defining a quantity called the diffusional correlation time, t_c. We define t_c as

$$t_c = w^2/4D, \tag{1}$$

where w is the exp(−2) beam radius. In Table I we have calculated t_c for a 1 μm laser beam for diffusion coefficients ranging from 10^{-8} to 10^{-12} cm^2/second. We next need to define system response time. In the language of analog measurement systems, the system response time is defined by the resistance-capacitance (RC) time constants of the detection circuitry. A less confusing definition is afforded by digital counting systems. In a digital system, the electronics has a minimum counting interval during which it counts photons. If one wants to accurately count for a period on the order of t_c, then one needs to break it up into a minimum of about 10 intervals. Thus, the longest the counting interval should be is 0.1 t_c. This maximum allowable response time is also given in Table I for each of the range of diffusion coefficients. Table I also shows the kinds of macromolecules and systems which typically exhibit a given diffusion coefficient. If one wishes to measure lipid and protein diffusion rates in most biological membranes, then a reasonable minimum system response time or counting interval would be about 5 milliseconds. If, on occasion, one needed to push the system so as to be able to measure diffusion rates near 10^{-7} cm^2/second, then one could increase the beam radius to about 3 μm [see Eq. (1)].

One must also consider the maximum time over which one wishes to be able to take data. This is particularly an issue in digital systems, where

TABLE I

DIFFUSIONAL CORRELATION AND MINIMUM SYSTEM RESPONSE TIMES FOR A 1 μm BEAM

D(cm^2/sec)	$t_c{}^a$(sec)	Minimum response time^b (sec)	Molecules
10^{-8}	0.250	0.025	Membrane lipids and unrestricted membrane proteins. Proteins diffusing in cytoplasm
10^{-9}	2.5	0.250	Slow lipids and fastest proteins
10^{-10}	25	2.5	Membrane proteins
10^{-11}	250	25	Membrane proteins
10^{-12}	2500	250	Slowest membrane proteins

[a] Calculated from $t_c = w^2/4D$.

[b] Calculated as $0.1t_c$ limits of t_c, and therefore minimum response time can be increased by increasing beam radius w.

incoming data is stored in some kind of data buffer. In that case there is an maximum number of data points which one can store. A useful approach is to define two counting intervals, a short and a long one. The first 50 or 100 data points are taken at the short counting interval, so as to accurately define the important initial phase of the recovery. The remaining data are then taken at the longer time interval. This enables one to measure the end point for the curve and to determine if there is a second and slower component to the recovery. For most applications, being able to take data for up to about 10 minutes is sufficient. One should bear in mind that accurate measurement for that length of time requires that both the laser beam and the sample remain positionally stable to a fraction of a micron for the 10 minutes. Depending upon the particular laboratory, this may require vibrational isolation of the system. In many cases, this isolation can be accomplishd either by using a sand table or by mounting the system on a metal plate which floats on motorcycle inner tubes. In extreme cases, or when accurate determination of slow diffusion coefficients is important, it may be necessary to use a pneumatic isolation table.

3. SAMPLE GEOMETRY

The specifics of sample geometry and experimental protocol are also important considerations in instrument design. If one needs to have access to the sample for microinjection or electrophysiology, the system is best built around an inverted microscope. On the other hand, if one will need to use water-immersion objectives, an upright microscope will be preferable.

Sample thickness may require that one use long-working-distance, low-magnification, low-numerical-aperature objectives. This in turn will limit both the beam size and available light fluxes.

4. Fluorescent Probes

The probes to be used also pose important design considerations. A very critical issue which probes determine is the laser to be used. Most FPR applications use continuous wave visible light tunable ion lasers, usually argon or krypton. Table II lists the wavelengths available with the major forms of ion lasers. It should be noted that it is often the case that, in order to obtain the full range of wavelengths from a given laser, one may have to change optics. From a practical standpoint this is something that is best avoided.

FPR can also be done using near-ultraviolet light. However, this presents a number of problems. Not the least of which are the hazards presented by possible exposure of eyes to ultraviolet radiation, and of lungs and electronics to ozone. This latter problem must be prevented by installation of a customized ventilation system. Operation of lasers in the ultraviolet also tends to shorten length of life of plasma tubes by producing color centers in the Brewster windows. Standard microscope optics absorb strongly in the ultraviolet and also tend to fluoresce when exposed to these wavelengths. Thus, ultraviolet operation may require special ultraviolet

TABLE II

Wavelengths[a] Available from Common Continuous Visible Light Lasers

Argon	Krypton	Helium neon	Helium cadmium
3511 w	3507 w	6328 s	3250 w
3638 w	3564 w		4416 w to s
4579 s	4762 s		
4658 s	4825 w		
4727 s	5208 s		
4765 m	5309 s		
4880 m	5682 s		
4965 s to m	6471 m		
5107 s	6764 s		
5145 m	7525 s		
5287 s to m	7931 w		
	7993 w		

[a] Wavelengths given in angstroms. w, Weak; s, strong; m, major.

optics, which can be quite expensive. It should also be remembered that most cells absorb in the ultraviolet and also autofluoresce when excited at ultraviolet wavelengths.

For specialized applications, where only one wavelength will be used, it may be possible to use other types of visible light lasers, such as helium cadmium or even the workhorse of laserdom, the helium neon laser. Another option, albeit an expensive one, where maximum wavelength flexibility is desired, is to pump a dye laser with an ion laser. For most applications a good choice of laser is a 2 W argon or krypton. Between these, one must decide between the greater choice of wavelengths available from krypton and the intrinsically longer lifetime of argon. It should be noted, that, since argon lasers have an extremely powerful line at 514.5, they can be used to excite most green-sensitive probes such as rhodamine and indocarbocyanines. However, if you plan to do double-labeling experiments with fluorescein and rhodamine, this 514.5 nm line will not enable you to excite rhodamine exclusively. For such applications the 568.2 nm line of krypton would be preferable.

The question of what makes a probe usable for FPR needs also to be addressed. There are several important criteria. First of all, the probe should show irreversible photobleaching. That is, when bleached the fluorescence should not recover spontaneously. Second, the probe should not be so light sensitive that it is bleached by the monitoring light. This is an important criterion in determining experimental parameters, since failure to do so can lead to erroneous diffusion results. Third, both the native probe and the combination of probe plus light should not be lethal to cell, and should not produce products which alter diffusion results. This issue of photodamage has been extensively addressed with the standard probes used (Axelrod, 1977; Jacobson *et al.,* 1978; Sheetz and Koppel, 1979; Wolf *et al.,* 1980a; Koppel and Sheetz, 1981). Wolf *et al.* (1980a) have developed a general approach which can be used to test probes. It is extremely important to test for photodamage. A final criterion, one which alas is "more practiced in the breach than in the observance" is that the bleaching process needs to be first order in probe concentration. If it is not at least approximately so, then one has no business applying the diffusion theory for FPR developed by Axelrod *et al.* (1976).

B. An Overview of the Instrument

Before considering the individual components of the FPR instrument, let us consider the instrument as a whole. Figure 1 shows such an overview. Let us follow the light path through the system.

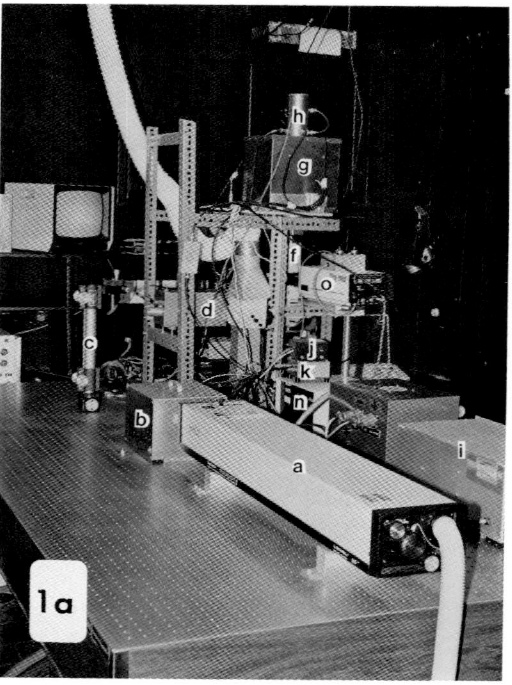

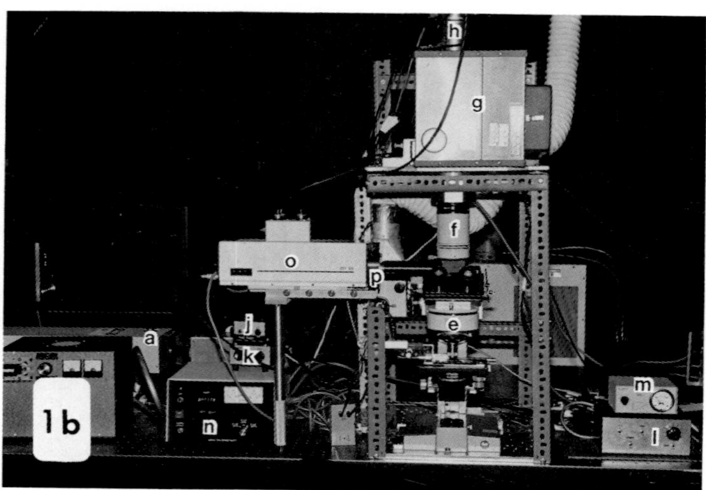

FIG. 1. Overview of FPR instrument. (A) Side view and (B) front view; (a) ion laser, (b) beam-splitter attenuator, (c) beam steerer, (d) focussing optics, (e) fluorescence microscope, (f) depth-of-field-limiting aperture and shutters, (g) photomultiplier housing, (h) amplifier/discriminator, (i) high-voltage supply for photomultiplier, (j) window heater supply, (k) bleach detector supply, (l) shutter and amplifier/discriminator power supply, (m) tungsten lamp supply, (n) xenon lamp supply, (o) SIT camera, (p) SIT camera shutter.

Coherent light is produced by the ion laser, in this case a Lexel 95-2 argon laser, and then enters the beam-splitter attenuator. The output of the beam-splitter attenuator is either the unattenuated laser beam (during the bleach) or a 10,000-fold attenuated beam (during monitoring). Significantly, at both intensities the output light path is identical. A photodiode detector within the beam-splitter attenuator serves to monitor the bleach. The light then strikes the beam steerer, which directs the beam toward the microscope along the microscope's optical axis. The beam then enters a box that contains optics which cause the beam to ultimately be focussed at the object plane. In the absence of these optics, the parallel laser light would be brought to a focus at the microscope objective's focal rather than object plane. The beam then enters the microscope, where it is reflected off the dichroic mirror into the objective. The objective then focusses the beam onto the object. The fluorescence light emitted by the sample begins the reverse route. However, when it strikes the dichroic mirror it is transmitted rather than reflected. The light then has several possible paths that it can take. One of these is into the eyepiece for observation. During measurement, however, the light is brought to a real focus by the objective at the microscope's secondary image plane. A depth-of-field-limiting aperture at this image plane assures that one is measuring fluorescence from a thin plane. The light then passes through a shutter which protects the detector from light when one is not doing an FPR measurement. In addition, a second shutter blocks the eyepieces during the experiment to prevent stray light from entering along that path. After this aperture the light continues to the photocathode of the photomultiplier. The light is expanded so that when it strikes the photocathode the image of the laser spot nearly fills the cathode surface. This image need not be in focus. This filling of the surface serves to maximize signal to noise. Signal to noise is also aided by magnetic lenses. The photomultiplier is contained in a housing. In the case of the instrument shown, this housing cools the tube by surrounding it with dry ice. Cooling the tube reduces thermal noise known as dark current. Some form of cooling is necessary with very sensitive photomultipliers, particularly in the red. Another popular form of cooling is Peltier effect coolers. Directly above the housing is a metal cannister which contains the amplifier/discriminator circuitry, as well as an electronic shutter which prevents the photomultiplier from operating during the bleach. The photocurrent from the photomultiplier's anode is a series of low variable amplitude and width pulses. These pulses are amplified by the amplifier. The discriminator then rejects pulses below a certain preset amplitude and converts the remaining pulses to a fixed amplitude and duration. The signal is thus digitized. It is necessary that the amplifier/discriminator be as close as possible to the photomultiplier so as to reduce radio frequency noise. The pulses from the amplifier/discriminator are then fed into a scalar or counting board customized for the International Business Machines personal computer (IBM PC). In our system this board was made for us by Searls and Cheevers. This board takes data, digitizes it, as will be described later, and stores it in the random access memory (RAM) of the computer. The computer then analyzes and stores the data curves.

Several other components of the system should be identified. A high-voltage power supply powers the photomultiplier. The bleach shutter inside the beam-splitter attenuator is controlled and powered on signal from the computer. The window of the photomultiplier is kept from fogging over by a window defroster which has its own power supply. There is also an external detection circuit for the bleach detector. An additional custom circuit box controls the shutters and powers the amplifiers. The catastrophe of exposing the photomultiplier to the incandescent lamp of the microscope is avoided by automatically switching this lamp off whenever the photomultiplier shutters are open. Notice that our system has a xenon lamp which inputs the back of the microscope at right angles to the laser beam. Choice of illumination is controlled by raising and lowering a right-angle mirror. Notice also that the entire system rests on a pneumatic vibration isolation table from NRC and that a rigid metal framework surrounds the microscope and supports the various elements of the system.

C. The Laser

We have already discussed some of the factors involved in choosing a laser for the FPR instrument. A couple of other points should be made about the laser.

Many of the usable laser systems will need to be water cooled. You will then have to decide between using tap water for cooling and using a closed cooling system. If you decide to use a closed system, as we have done, consider the amount of cooling required. For the kinds of power required you can very easily need 10,000 W of cooling or even more. With so much heat, it may be necessary to have the heat exchanger outside the building. Whatever type of cooling system you use, you should consider redundant fail-safing of the system against a water failure. The laser is a very expensive component of the system and deserves this extra level of protection.

In addition to being monochromatic, as well as spatially and temporally coherent, laser light has a very defined mode structure. At a practical level, this refers to the intensity profile of the laser beam in a plane perpendicular to the beam. This profile takes on several very defined modes. Some of the modes are circularly symmetric others are not. The fundamental mode, referred to as the TEM_{00} mode, is Gaussian in intensity profile. Another common mode is a "doughnut" profile with a node in the center. The point is that the diffusion theory for spot photobleaching developed by Axelrod *et al.* (1976) assumes a Gaussian TEM_{00}. If you do not have this, you cannot apply that theory. In a sense then, mode structure from the laser is the most critical aspect of the illumination. Care must be taken to check that the laser is in the right mode, and, if not, to retune the laser until it is. Often mode structure is the first thing to deteriorate in the plasma tube. Sometimes the inevitable replacement may be delayed by tuning the laser below maximum power to obtain the correct mode structure.

D. The Beam-Splitter Attenuator

Figure 2A shows a photographic view of the beam-splitter attenuator, and Fig. 2B is a schematic diagram showing the light path. This design, which uses two $\frac{1}{2}$-in. thick, 1-in. diameter BK7 glass parallel optical flats, was originally developed by Koppel (1979). A similar approach using four mirrors has been developed by Barisas and Leuther (1977). In early FPR experiments, beam attenuation was accomplished by moving a neutral density filter in and out of the beam. This has the serious disadvantage that the filter can cause the monitoring beam to deviate from the path of the bleaching beam. In the beam-splitter design, the first parallel flat splits the beam in two. One ray is essentially nonattenuated while the other, by virtue of the fact that it has undergone two 10% reflections, is 100-fold attenuated. The nonattenuated beam is shuttered during monitoring. The two beams are then rejoined by the second parallel flat. The monitoring beam undergoes two additional 10% reflections before exiting. Note that the approximately 10% reflectance is a property of glass-air interfaces when the angle of incidence is near 45° and when the polarization vector is parallel to the flat (i.e., in this case perpendicular to the table). Laser light is intrinsically polarized, but it is useful to determine its orientation. Notice that a photodiode detects the reflection of the nonattenuated beam off the second parallel flat. This photodiode is used to monitor the actual bleach pulse. In designing a beam-splitter attenuator it is important to trace the ray path through the flats to be sure that you do not exceed the boundaries of the glass. The displacement, d, of the two beams following the first flat depends on both the thickness of the flat, t, and the angle of rotation from perpendicularity to the beam, a. The displacement tends to place extreme

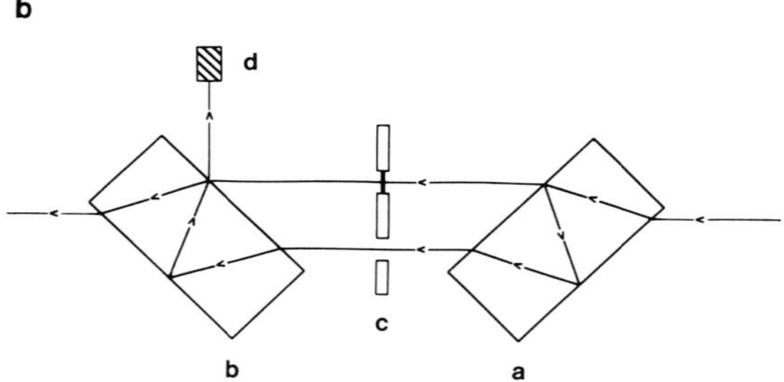

FIG. 2. Closeup of beam-splitter attenuator. (A) Photograph and (B) schematic of light path; (a, b) parallel optical flats, (c) shutter, (d) photodiode bleach detector. Laser beam enters from the right.

tolerances on designing the shutter, since it often requires drilling a hole in a nonshuttered region of the shutter. Using Snell's law one calculates this displacement to be,

$$d = 2t\left\{\frac{n_1 \sin a \cos a}{n_2[1 - (n_1^2/n_2^2)\sin^2 a]^{1/2}}\right\} \tag{2}$$

where n_1 is the index of refration of air and n_2 that of the glass.

E. The Beam-Steerer

The beam steerer consists of two mirrors with positional control which directs the laser beam into the microscope along the optical axis. This

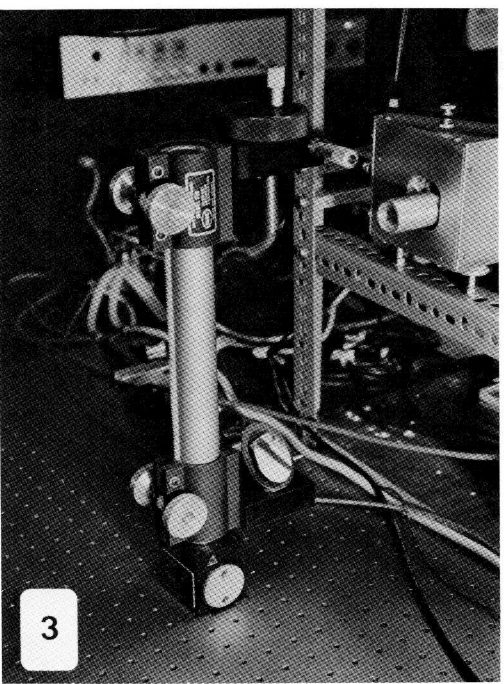

FIG. 3. Closeup of beam steerer. Notice magnetic base clamp and focussing optics box (beam expander).

component is shown in Fig. 3. It is important that the beam run along the optical axis rather than crossing it at one point on the object plane. These two conditions can be distinguished by moving the microscope stage up and down through the focus while observing the image of the spot on a piece of fluorescent paper through the microscope. When the beam travels along the optical axis, the image will appear round and will expand symmetrically above and below the focal plane. If you are not on axis, the spot will expand as an ellipse skewed to one side above and to the other side below the plane of focus.

F. The Microscope

As discussed above, the decision of which microscope to use is often guided by the kinds of samples you intend to use and the kinds of experiments you intend to perform. Several microscope-related issues need to be discussed in greater detail.

1. Microscope Objectives

The spot size will be a function of both magnification and numerical aperture. Since one is often intensity limited, in fluorescence one often needs to work with the highest power, highest numerical aperture possible. Objectives designed for fluorescence are preferable because they have less autofluorescence. Unfortunately, not all microscope manufacturers sell phase and other special illumination optics that are designed for fluorescence use as well. In my experience, there is not a problem in using phase objectives for FPR. It may, however, be necessary to check objectives for autofluorescence using the xenon or mercury lamp. It is also useful to test objectives before purchasing them to determine the quality of the laser spot they produce. From time to time one runs into an objective which creates a poor-quality laser spot. We have also found that the optical axis of some objectives deviates so much from the others that it becomes a nuisance to realign the system each time an objective is changed.

2. Focussing the Laser on the Object Plane

As discussed above, it is necessary to alter the microscope optics to bring the parallel laser beam to a focus at the object instead of the focal plane. Unfortunately, it is difficult to obtain exact optical diagrams of microscopes. So, to a large extent, accomplishing this becomes a hit-or-miss matter. This is especially true in situations where you want to be able also to use the xenon or mercury lamp, and therefore cannot remove any of the optics from the vertical illuminator.

The object is, then, to find a lens or series of lenses which will create a sharp laser spot on the object plane. The microscope contains two secondary image planes. The first of these we have previously discussed as being where the objective focusses the specimen to a real image. This is usually about 170 mm above the objective in an upright microscope. A second secondary image plane is located also about 170 mm beyond the objective, only along the imaginary path of a ray reflected back by the dichroic mirror. Usually, this plane is identifiable by the presence of a field diaphragm at this plane. This is the iris which can close off the fluorescence illumination. If the laser beam is focussed onto this image plane, it will also be in focus on the object plane. Realize that any focussing lens of any focal length can accomplish this. A long-focal-length lens could be placed near the laser. A short-focal-length lens could be placed near or in the microscope. However, if the purpose is to obtain the smallest spot possible, which is usually the case, then the shorter the focal length the better. This is essentially an issue of numerical aperture. However, you will find that really short-focal-length lens tend to defeat the purpose by introducing abberations. In my experience, focal lengths of around 130 mm seem to work best. In my current instrument I found that, in the absence of adding any additional lenses, the laser was nearly in focus, but was too large. To make the spot smaller at the object plane, I created a beam expander by placing two lenses separated by approximately their focal length from one another. This beam expander is shown in Fig. 4. It is mounted at the back of the microscope where the xenon lamp is usually placed. The two lens have a total focal length of 150 mm. Screws independently adjust the lenses along the two axes perpendicular to the beam. The lens nearest the laser is mounted on a fine thread which enables one to adjust the separation of the two lenses and thus to bring the beam to a focus at the object plane. As indicated, following the above guidelines,

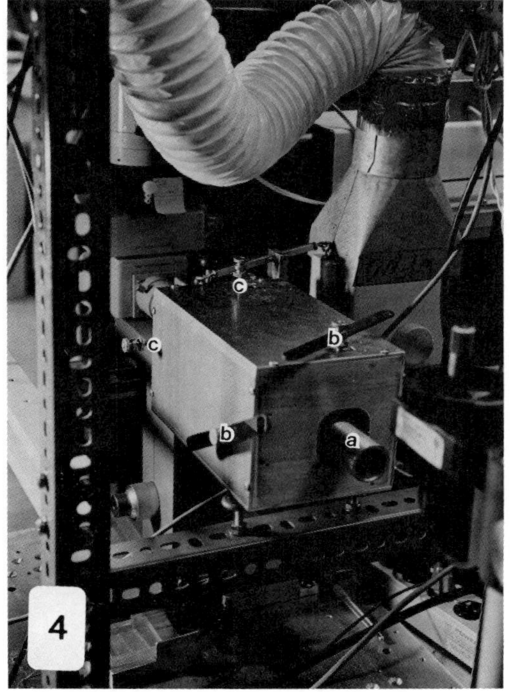

FIG. 4. Closeup of beam expander. (a) Screw allows motion of rear lens along the optical axis, (b) screws allow motion of rear lens horizontal and vertical to optical axis, (c) screws allow motion of front lens horizontal and vertical to optical axis.

one has to determine the best lenses empirically. The best approach is to purchase a cheap series of simple lenses with which to experiment. Once a suitable arrangement is determined, one should then purchase spherically corrected coated optics.

3. Beam Splitting to Allow Normal Fluorescence Illumination

The positioning of these lenses before the normal components of the epiilluminator allows one to still use the mercury or xenon lamp with the microscope. As discussed above, we have added a mirror to select between the two sources of illumination. This is shown in closeup in Fig. 5. Notice that a seesaw shutters the lamp when the mirror is raised to allow the laser beam to enter. This avoids stray lamp light from entering the system dur-

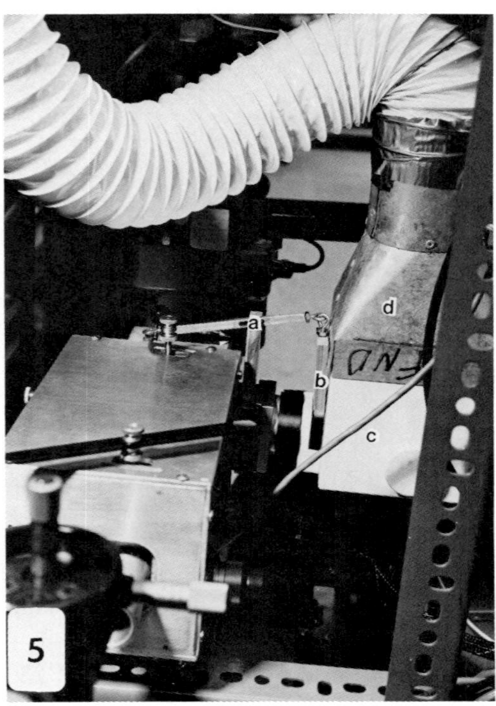

Fig. 5. Closeup of mirror used to select between xenon and laser illumination. (a) Seesaw which moves mirror up and down and simultaneously controls lamp shutter, (b), (c) xenon lamp, (d) ozone vent on xenon lamp.

ing measurement. Removing the mirror for laser illumination avoids the problem of having to return the mirror to a critically precise position, which would be necessary if the beam had to strike the mirror to reach the microscope. The lamp illumination is much less demanding.

4. Choice of Dichroic Mirrors

The dichroic mirrors and accompanying filters to be used in an FPR experiment are also dictated by the probe to be used. Fundamentally, these wavelength selectors consist of three components. First, the light passes through a low pass filter which transmits light below a certain wavelength. A wide band pass filter can also be used. Second, the light strikes the dichroic mirror, which reflects light below a certain wavelength and transmits light above that wavelength. Third, the transmitted fluorescence light then strikes a long pass or a band pass filter which selects the wavelengths to be measured. The filter arrangement to be used needs to be fitted to the excitation and emission characteristics of the probe. Greatest flexibility is offered by systems which allow each of the filters to be changed independently. Currently, many microscope manufacturers supply the combinations as fixed units. For some probes, custom arrangements may be necessary. In any event, you should carefully consider the options offered to you by the manufacturer as well as your specific application. For instance, some manufacturers offer two options for rhodamine illumination, one which stringently eliminates any fluorescein fluorescence, and one which is less stringent but which offers brighter illumination. If you are not going to do double labelling, the less stringent arrangement may be advantageous.

5. Depth-of-Field-Limiting Aperture

As I have already indicated, one of the most important components of the FPR system is the depth-of-field-limiting aperture. This aperture makes sure that you are monitoring fluorescence from only one plane. One needs to be able to close this aperture down to nearly the size of the image of the laser spot in the secondary image plane, that is, to the beam diameter times the objective magnification times the power of any extra lenses which are in the path. In addition, you need to be able to position the spot in this aperture. In the system discussed here, this positioning is accomplished by means of the positioning screws on the beam expander. In most systems this will require a mirror and eyepiece arrangement above the image plane. On my microscope I have used the Leitz photometry system, which back projects the bright image of the aperture into the microscope eyepieces. This avoids the problem of having to have an additional eyepiece and of having to attempt to get the laser beam into a smaller transparent area of an opaque field.

Once this aperture is in place, to determine if it is sufficiently closed down, one focuses on a labeled round cell and monitors the fluorescence intensity as one focuses from above down through the cell. One should see two sharp peaks of fluorescence corresponding to the two labeled membranes, and background level fluorescence in-between. How well you can confine yourself to a given plane is also limited by the numerical aperture of the objective. Under the best of circumstances one can hope for several tenths of a micron. The point of all of this is to eliminate signal from the cell's interior and from the opposite membrane, which is illuminated by a much expanded beam when the laser is focussed on the other membrane. This latter condition can, of course, result in erroneously slow diffusion rates and, possibly, low percent recoveries.

G. Photomultiplier Shutter

It is useful to protect the photomultiplier of the FPR instrument from light when not in use. This is accomplished by adding a shutter, either mechanical or electrical, to the detection system. Depending on the system geometry, this shutter can be placed either before or after the depth-of-field-limiting aperture. The most detrimental situation from which to protect the photomultiplier is illumination by the incandescent source of the microscope. In the instrument shown in Fig. 1, this is accomplished by controlling the light source with the same switch that controls the shutter. The light can then only be on when the shutter is closed. In addition, one needs to consider the possibility of stray light entering the system during measurement through the eyepieces. In my system, a maximum of 90% of the light can be directed to the phototube with the remainder going to the eyepieces. To avoid light from entering the eyepieces during measurement, a shutter, provided as part of the Leitz photometry system, covers the eyepieces. Depending on the system, if such a shutter cannot be incorporated, it may be necessary to cover the eyepieces when making a measurement.

H. Photomultiplier

There are a number of excellent discussions on choosing, setting up, and using photomultipliers, not least of which are the catalogs of major

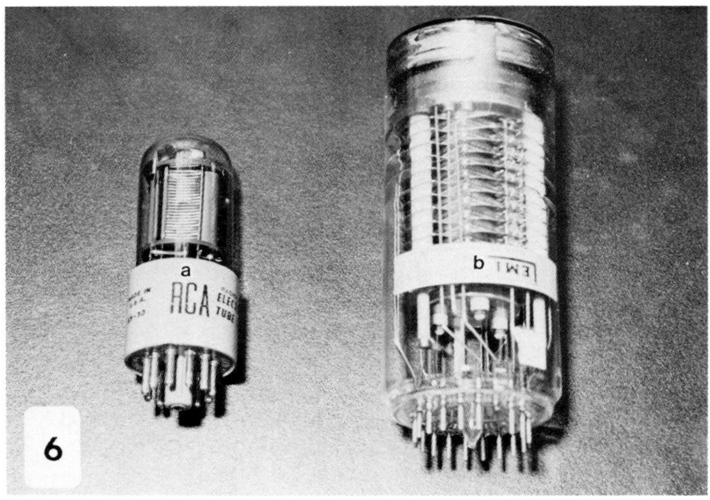

FIG. 6. Closeup of photomultiplier tubes. (a) Side window (window is up) and (b) front window tubes.

photomultiplier manufacturers. A comprehensive discussion is also given by Engstrom's "Photomultiplier Handbook," which is an RCA publication. This book also contains a fascinating history of these devices. I will thus refer you to these other sources for details, and make only a few general points here.

A number of factors go into tube selection. The first of these is necessarily spectral response. For the usual blue to yellow requirements S20 or S20 extended red make excellent photocathode choices. Galluim arsenide is also a good choice, one with amazing flatness of spectral response, if this is important to you. You will have to choose between side window and end window geometry (see Fig. 6) and you will have to choose the sensitivity you want. Increased sensitivity tends to weigh against increased cost and ease of accidental destruction. Increasing sensitivity also tends to increase dark current, especially when one has a red-sensitive tube.

If you decide on a sensitive tube, you will have to cool it to reduce dark current and then a decision about the method of cooling must be made. The usual choices are no cooling, dry ice cooling, and Peltier cooling. Peltier effect coolers must in turn be cooled by air, water, or other coolants, depending on the temperatures to be reached. When pumping a coolant through your system, you should consider the possibility of creating detrimental vibrations. Cooling systems may also require a window heater to prevent icing. In any event, the housing window must be designed to thermally isolate the tube. You must finally decide if you will use the tube as an analog device or if it will be used for single photon counting.

The system shown in Fig. 1 was designed for single photon counting. It employs a 9658RA tube from EMI GENCOM Inc., which was specially selected (a service provided by the manufacturers) for extended red sensitivity and low dark current. The tube is cooled by a dry ice-cooled housing purchased from Products for Research, Inc. This housing is traditionally designed to operate horizontally, and we have thus added a ice tray which facilitates vertical operation (Fig. 7). The amplifier/discriminator circuits, along with the tube base, were also purchased and installed by Products for Research. In addition, the system contains an electronic "shutter," actually a gating circuit, which is attached to the tube base. This circuit is available from Thorn EMI GENCOM, Inc. and shuts off the photomultiplier during the bleach by driving the first two dynodes negative relative to the photocathode. An equivalent approach is to use a programmable power supply and to drop the high voltage to the photomultiplier. This circuit required modification to adjust it for the time scales

Fig. 7. Closeup of ice tray arrangement to facilitate use of housing in a vertical arrangement. (a) Spring-loaded plunger, (b) shadeable wall, (c) ice tray. Note beveled edge (d) to facilitate insertion against ice, and tray control (e), which wraps around the photomultiplier.

of FPR measurement. These modifications were again performed by Products for Research. Realize that this approach to gating assumes that, given the light levels involved, the danger to the tube results from multiplying the signal rather than from exposing the photocathode to the bleaching light. Whether this is the case must be determined, on a case-by-case basis, in consultation with the manufacturers. Your specific tube requirements may require physically shuttering the system. The fact that all of these issues must be dealt with on a case-by-case basis is a disadvantage, but the manufacturers involved are used to this, and, in my experience, are enormously helpful in designing these detection systems with you.

Once you have gone to all of the trouble of designing your system and getting the best tube you can afford, remember that the system will perform only as well as you set it up. *Follow the directions of the manufacturers in determining the ideal operating voltage and temperature for your tube.* Photomultipliers are magnificiently linear devices. However, they are only linear if you have taken the trouble to set them up correctly. Similarly, you can only achieve optimal signal to noise by properly setting up.

I. Output Devices

The simplest method for outputing FPR data is to use an analog signal fed to a chart recorder. The critical issue becomes whether the pen response time, which is quite different from the maximum chart speed, is much less than the diffusion times to be measured. Basically, recorder response time is the time for half or full pen deflection. Realize that, if one wants to measure diffusion times of, for example, 300 milliseconds, you will want this response time to be at least 30 milliseconds. Such response times are available from many of the recorders used for electrophysiology, and of course, from storage scopes. However, most run-of-the-mill recorders do not approach such response times. A way around this problem is to input the data to a storage buffer, such as a transient recorder, and then to output it at a slower rate to the chart recorder.

Of course, once your data are in a storage buffer you are halfway to computerized data acquisition with its advantage of computer analysis. Computers allow you not only to analyze data more rapidly, but also to fit all, rather than just three points, of the data (see below) to any given model for recovery. In recent years, a number of companies have specialized in providing interface devices for the IBM PC and other laboratory computers. Unlike in years past, these interfaces are addressed through high-level languages such as BASIC, FORTRAN, PASCAL, and C, rather than machine language. These devices also allow for computer-driven control of the experiment. The flexibility and ease of these systems have also been increased by the availability of large amounts of low-cost RAM memory for the PCs. To illustrate this, consider that, in 1979, along with Michael Edidin at Johns Hopkins, I interfaced a Technico SS19 Computer to an FPR instrument. This computer had 32K of memory. Because of memory restrictions we designed a scalar which counted photons for a fixed interval and then shuttled the total into computer RAM. This could be done a maximum of 1024 times. Before the data file was stored on disk for subsequent analysis, it was compressed by leaving the first 50 points alone, but averaging the remainder. The current system which we employ was developed by Searles and Cheevers around 1985. Here the 256K memory of an IBM PC is available and the data are stored every 25 milliseconds directly into RAM. The counting interval of 25 milliseconds may be increased by factors of two when appropriate. The files are stored as two fields. The first contains the first 6000 set of 25-millisecond data points and the second is a compressed field of up to 6000 points. The degree of compression by a power of two up to five is determined by the experimenter.

J. Pulse Sequence in an FPR Experiment

I will now describe the pulse sequence in the electronics of the FPR machine currently in use in my laboratory. Modifications of this sequence will be required for instruments of different design. However, the considerations in determining this sequence remain the same.

At the beginning of an FPR experiment one must determine and specify several parameters. One must first of all decide upon the counting interval, CI, which is the duration of a single data point. That is, CI is the time over which the photons for a single data point are counted. One must also determine the nominal bleach time, NBT. The NBT is the duration of the pulse which we will send to the bleach shutter to open. NBT is to be distinguished from the actual duration which the shutter will be open. We will refer to the latter as the actual bleach time, ABT. As we have already discussed, it is essential that CI, NBT, and ABT all be much less than the characteristic diffusion time, t_c. One must also specify to the system the number of data points to be taken before bleaching, N_{pre}, in order to determine the prebleach level, and the maximum number of points to be taken postbleach, N_{post}. In the system under discussion, N_{post} is specified as a power of 2 times 6000. The system need not reach this level, but can be prematurely stopped by a keyboard interrupt.

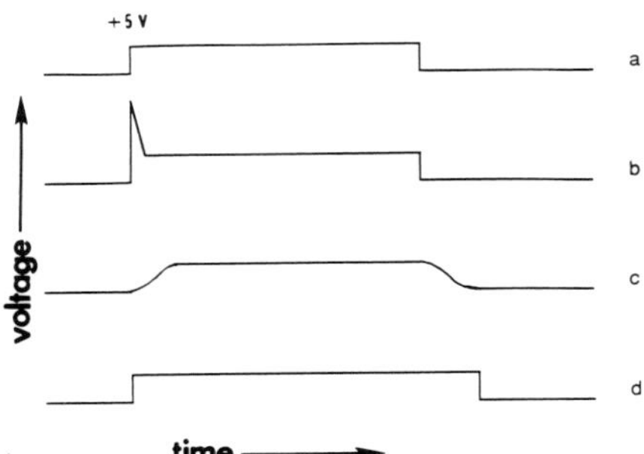

FIG. 8. Schematic diagram of pulse sequence in FPR experiment. (a) Command to bleach controller, (b) bleach controller output to bleach shutter, (c) actual bleach profile, (d) command to gate photomultiplier.

After specifying these parameters the sequence is initiated by a keyboard command. The system acquires N_{pre} data points of duration CI. It then sends a positive 5 V command to bleach (Fig. 8a). In our system we are using Uniblitz shutters from Vincent Associates, which require an approximately positive 70 V to open, with positive 5 V to hold. Thus the 5 V square pulse is fed into a Uniblitz shutter controller which converts it to the necessary output (Fig. 8b). The bleach signal is then sent to the shutter. Of course, the shutter is delayed in both opening and closing, and, because the beam is of finite width, the bleach time profile has a finite rise and fall time. The actual opening and closing of the shutter is monitored by the photodiode in the beam-splitter attenuator (Fig. 8c). This signal is fed back to the computer where it serves two purposes. The first is to determine the duration and time of the bleach. The second is to protect the photomultiplier from the bleach. Realize that the photomultiplier is potentially vulnerable from the instant when you send the command to bleach until the end of the actual bleach. Thus, it must be shuttered for that entire period. The computer therefore creates an additional output of positive 5 V which it sends to the photomultiplier gating circuit to shut off multiplication during that period (Fig. 8d). One can thus see the advantage of electronic rather that electrical mechanical shuttering. If one had a physical shutter protecting the photomultiplier, one would again have to worry about delays in opening and closing.

During the bleach, the computer stops counting photons and instead counts the bleaching interval. The first counting interval postbleach is initiated immediately at the end of the actual bleach. Thus, the all-important early part of the recovery curve is determined as accurately as possible. Note, that as a result, the counting intervals postbleach are shifted in time relative to those prebleach. This is not a problem, since the taking of prebleach data was just to determine a DC level. The computer stores the actual bleach duration, ABT, and since the bleach ends as the first CI begins the computer can easily extrapolate back to time zero, which is simply ABT/2. Thus the *i*th counting interval is centered at time ABT/2 + *i**CI.

K. Algorithms for Data Fitting

Assume that the computer now contains a file with a series of data points of intensity, y_i, at times t_i. The issue is how to fit them to diffusion theories for recovery. It is useful to return to the three-point method of hand-fitting FPR curves developed by Axelrod *et al.* (1976). This is illustrated in Fig. 9. One determines the prebleach intensity $I\ (t < 0)$, the extrapolated intensity at the time of bleach, $I(0)$, the asymptotic intensity at infinite time, $I(\infty)$, and the time for half recovery, $t_{1/2}$. $t_{1/2}$ is the time to recover to $I(0) + (I(\infty) - I(0)]/2$. The percent recovery is then given by

$$\%R = [I(\infty) - I(0)]/[I(t < 0) - I(0) \tag{3}$$

The diffusion time is then calculated from the relationship:

$$t_c = t_{1/2}/\gamma \tag{4}$$

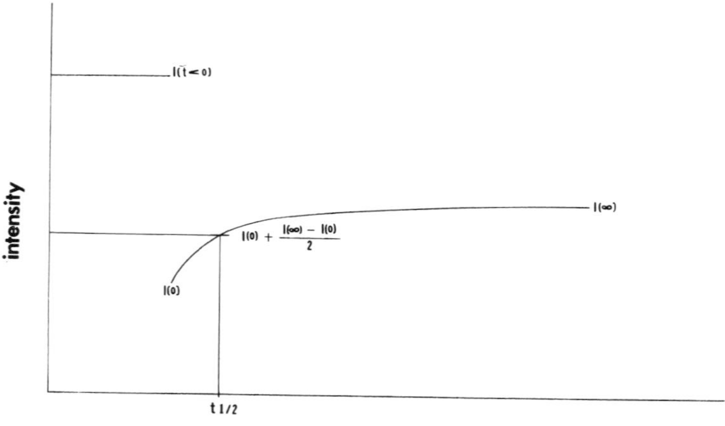

FIG. 9. Three-point method of fitting FPR data. $I(t < 0)$, prebleach intensity; $I(0)$, after bleach intensity extrapolated to midpoint of bleach; $I(\infty)$, asymptotic intensity; $I(t_{1/2}) = I(0) + [I(\infty) - I(0)]/2$.

γ is defined by Eq. (4). It is a function of the unbleached fraction, namely,

$$F(0) = I(0)/I(t < \text{o}) \tag{5}$$

Theoretical tables of γ as a function of $F(0)$ can be numerically generated by determining the ratio of $t_{1/2}/t_c$ from theoretical recovery curves for Gaussian spot recovery (Axelrod *et al.*, 1976; Wolf and Edidin, 1981), the equation for which is

$$F(t) = \sum_{N=0}^{\infty} \frac{(-K)^N}{N!} \frac{1}{1 + N(1 + 2t/t_c)} \tag{6}$$

K is a constant which defines the bleach (Axelrod, 1976). If one solves Eq. (6) for time zero, one obtains

$$F(0) = (1 - e^{-K})/K. \tag{7}$$

The point of all this is that one generates a table of γ as a function of $F(0)$ using Eqs (6) and (7). One then uses a three-point hand-fitted curve to determine $t_{1/2}$ and $F(0)$, divides $t_{1/2}$ by the γ appropriate to $F(0)$ to obtain t_c, and then determines D from $D = w^2/4t_c$. This approach is simple and straightforward. It suffers from the fact that one is fitting the curve to only three points, ignoring essentially the rest of the curve.

To get around this problem, one fits the entire curve using nonlinear least squares analysis. An excellent discussion of this technique, including usable Fortran programs is to be found in Bevington (1969). While approximate methods which linearize FPR recovery curves have been proposed (Van Zoelen *et al.*, 1983; Yguerabide *et al.*, 1982), the availability of nonlinear least squares analysis on PCs make these approximations unnecessary.

The principle of nonlinear least squares is that, if one has a set of experimental data points (y_i, t_i) which one wishes to fit to an arbitrary function $F_{A_1 \ldots A_m}$ defined by m parameters A_j, then one needs to determine the values of the A_js which will minimize the chi-squared deviation of the function from the experimental data points. Chi-squared deviation is defined by

$$X^2 = \sum_{i=1}^{N} w_i[Y_i - F_{A_1 \cdots A_m}(t_i)]^2/N \tag{8}$$

The w_is are the weighting factors for the points, and will be discussed later. The minimization is achieved by differentiating chi-squared with respect to the A_js and setting the derivatives equal to zero. That is, one obtains m

simultaneous equations of the form:

$$\frac{\partial X^2}{\partial A_{\mathrm{j}}} = \sum_{i=1}^{N} \frac{2W_{\mathrm{i}}}{N} [Y_{\mathrm{i}} - F_{A_1 \cdots A_m}(t_{\mathrm{i}})] \qquad F_{A_1 \cdots A_m}(t_{\mathrm{i}}) = 0 \tag{9}$$

For straight-line and polynomial functions, closed-form solutions of Eq. (9) are readily obtained. For more involved functions, numerical approaches to minimizing chi-squared must be employed (Bevington, 1969).

It is thus possible (although slow) to fit FPR data directly to Eq. (5) using the programs from Bevington (1969). A somewhat empirical approach developed by Barisas and Leuther (1977) speeds calculation up considerably. Let us expand Eq. (6) to its first three terms. One obtains

$$F(t) = 1 - \frac{K}{2 + 2t/t_{\mathrm{c}}} + \frac{K^2}{3 + 4t/t_{\mathrm{c}}} \tag{10}$$

Equation (10) does not, of course, do a very good job of fitting the experimental data. It does, however, suggest a function which will fit the data well, namely,

$$F(t) = A_1 - \frac{A_2}{1 + A_3 t} + \frac{A_4}{(1 + A_3 t)^2} \tag{11}$$

In Eq. (11) we have allowed A_1 to be nonunity to allow for fractional recovery.

As it turns out, Eq. (11) will usually fit the data as well as Eq. (6), as judged by the chi-squared obtained for the fit. The problem then becomes one of relating the A_{j}s back to $\%R$ and t_{C}. Since $F(0) = A_1 - A_2 + A_4$ and $F(\infty) = A_1$, we immediately see that

$$\%R = (A_2 - A_4)/(1 - A_1 + A_2 - A_4) \tag{12}$$

The issue of determining t_{C} is a bit more complicated. It is convenient to define a variable,

$$z = 1/(1 + A_3 t) \tag{13}$$

Then Eq. (11) becomes

$$F(t) = A_1 - A_2 z + A_4 z^2 \tag{14}$$

Thus, if at time $t_{1/2}$ we have $z_{1/2}$, then

$$F(z_{1/2}) = A_1 - A_2 z_{1/2} + A_4 z_{1/2}^2 \tag{15}$$

Alternatively since $F(z_{1/2}) = F(0) + [F(\infty) - F(0)/2$, we have that

$$F(z_{1/2}) = A_1 + (A_4/2) - (A_2/2) \tag{16}$$

Equating Eqs (15) and (16), solving the quadratic, and solving in turn for $z_{1/2}$ and then $t_{1/2}$ one obtains

$$t_{1/2} = 2\,A_4/\{A_3A_2 \pm A_3[A_2^2 - 2A_4(A_2 - A_4)]^{1/2}\} - 1/A_3 \qquad (17)$$

Having thus determined $t_{1/2}$, one then determines t_C by referring to a table of γs. Finally, one calculate D from t_c. All of this is done by the computer, so that it is necessary to have a table of γs within the program. Note that, if one assumes $A_4 < A_2$, then the correct sign to choose in Eq. (17) is negative since this gives the correct value $t_{1/2} = 1/A_3$ in the limit of $A_4 = 0$. It should also be noted that typically a three-point data fit is used to provide the initial guesses for the chi-squared minimization programs.

The issue of weighting factors needs also to be considered. Probably the most rational approach is to weight the points inversely as the square of their uncertainties (Bevington, 1969). If one assumes that the uncertainty of measurement is stochastic, then if one has N counts in an interval the uncertainty is $(N - 1)^{1/2}$ and since N is typically large, this can be approximated as $N^{1/2}$. Thus, it is reasonable to weight the points inversely with their intensity. This has the effect of weighting the earlier points more heavily than the later ones. It also proves advantageous as discussed earlier to fit the first 50 or so points, and then to fit every tenth point or the average of every 10 points. This improves speed without compromizing accuracy, since the purpose of including later points is to obtain the end of recovery.

The change of variables in Eq. (14) also suggests a useful trick which has been employed by Searles and Cheevers to greatly speed up analysis. What they do is assume a value of A_3 and then use a closed form least squares polynomial fit to Eq. (14) to obtain A_1, A_2, and A_4. They then systematically vary A_3 and repeat the polynomial fit until chi-squared is minimized. As I say, this approach is very fast and gives the same answer as fitting all the A_js by nonlinear least squares. The purist will, however, object to this approach, since it violates the mathematical requirement that the A_js be truly independent.

L. Measuring the Beam Radius—Absolute versus Relative Diffusion Coefficients

Before data can be expressed as an absolute diffusion coefficient it will be necessary to measure the exp(−2) radius of the laser beam. A variety of methods exists for performing these measurements. These have been extensively described and compared by Schneider and Webb (1981). One of the most direct approaches is to attach a motorized differential micrometer to drive the microscope stage in one direction and to measure the intensity profile as a submicron fluorescent bead is driven through the laser beam. The profile obtained should be Gaussian and the exp(−2) radius can be determined directly.

It should be noted that a 20% uncertainty in radius corresponds to a 40% uncertainty in diffusion coefficient, because of the rules of error propagation. It is difficult to produce an absolutely consistent laser beam from day to day. As a result, the determination of accurate absolute diffusion coefficients can be difficult. It is therefore important to always measure a standard control situation against which experimental situations may be compared. In that way relative results, at least, will always be correct.

III. Artifacts of FPR

In the above discussion, I have considered several potential sources of artifactual error in FPR measurements. At this point, I would like to summarize these and to consider some additional sources of error.

1. The fluorophore used exhibit approximately first order bleaching kinetics. If it does not, then one can not accurately apply the theory of Axelrod *et al.* (1976). See chapter 11 by Elson and Qian for further discussion (this volume).
2. The fluorophore used should not recovery spontaneously in the absence of diffusion. Such recovery would obviously lead to erroneous results.
3. The fluorophore and light conditions should not result in photodamage. Several aspects of photodamage should be considered: alteration of diffusion values due to localized heating or cross-linking, morphological damage to the cell, and damage to cell function. One should be aware of potentially absorbent molecular species within a given cell type. Each cell and experiment represents a separate case. The fact that a given fluorophore is "safe" to use when measuring surface diffusion does not mean that it is "safe" to use when measuring cytoplasmic diffusion.
4. The fluorophore and light conditions used should be such that they do not result in bleaching under monitoring conditions. This will lead to erroneous diffusion coefficients and percent recovereis.
5. The laser beam must be either Gaussian or homogeneous in profile if one intends to use the theory of Axelrod *et al.* (1976). It should be noted that production of a homogeneous laser spot is less than trivial.
6. The depth-of-field-limiting aperture must be set so that recovery within only a thin plane is being measured. In this way, recovery due to

material in the cytoplasm or the opposite membrane is excluded. An obvious exception is with extremely thin cells one is sure that there is no internal fluorescence (see Elson and Qeian, Chapter 11, this volume).

7. The counting interval must be much less than the characteristic recovery time.

8. The bleach time must be much less than the characteristic recovery time.

9. The photomultiplier's high voltage must be set so that the photomultiplier is operating in a linear range.

10. The background signal in an EPR measurement must be accurately determined and taken into account. In general, it is best to avoid situations where the background also exhibits recoveries, which must therefore be subtracted from the experimental curves. However, even when background is a DC level it must be accurately determined. Often one observes a different DC level pre- and postbleach. If these differences are not correctly determined, erroneous values of both $\%R$ and $F(0)$ will result. The incorrect values of $F(0)$ in turn will result in incorrect values of D.

11. Carefully consider the geometry of your experiment. Is it appropriate to treat it as planar two-dimensional diffusion? Would a one-dimensional or other solution be more appropriate? For example, when studying membranes, is the surface really planar? (See Aizenbud and Gershon, 1982; Wolf *et al.*, 1982.)

12. The beam area must be much smaller than the area of the surface being measured. If it is not, then the reservoir of material for recovery will not be sufficient to apply the theory of Axelrod *et al.* (1976). One can take finite area into account. If one does not, incorrect values of both D and $\%R$ will result.

13. Are you certain that the cell does not move during a measurement? If it does, erroneous values of both D and $\%R$ will result.

14. When using lipid probes, you need to be certain that they are properly intercalated into the bilayer.

15. When using lipid probes, it is important to know that the probes are not altering the properties of the membrane. As a rule of thumb, it is important to keep probe concentrations to $\ll 1\%$ of the membrane lipid, to test for morphological and physiological alteration due to insertion of the probe, and to compare the behavior of several probes.

16. When using ligands as probes of proteins, antigens, or receptors it is important to use these under noncross-linking conditions (Wolf *et al.*, 1977; 1980b; Dragsten *et al.*, 1979; Wolf and Edidin, 1981). When using antibodies for instance, it is usually necessary to prepare either directly labeled F_{ab} fragment or to use a labeled F_{ab} fragment against an F_{ab} fragment (see Edidin, Volume 29, this series).

17. When using ligands, it is essential to make sure that one is measuring membrane diffusion rather than "hopping" of ligand from receptor to receptor along the surface. The most direct approach is to fix cells prior to labeling and to demonstrate that no recovery is observed on these cells. Another approach is to demonstrate that surface label and diffusion are unaltered in the presence of excess multivalent ligand in solution.

IV. Beyond Spot FPR—Obtaining Spatial Information

In recent years, a number of alternative geometries to spot FPR have been developed. Some examples of these are pattern photobleaching (Smith and McConnell, 1978), scanning spot and normal mode analysis (Koppel, 1979), and total internal reflection EPR (Thompson *et al.,* 1981; Burghardt and Axelrod, 1981). Each of these is especially well suited for particular situations. I refer the reader to the original references for a further discussion of these modifications. It is worth noting that some of these approaches enable one to get around the difficulty inherent in spot FPR, of uncertainty in absolute diffusion coefficients.

From my perspective, the major limitations of spot FPR stem from an inherent loss of spatial information, because one is integrating over the intensity within the laser spot. As a result, it becomes difficult to determine in a single measurement whether diffusion is anisotropic, whether directional motions or multicomponent diffusional motions are present, and whether diffusion over long distances is the same as diffusion over short distances.

An extremely versatile approach to obtaining this additional spatial information is to use intensified video imaging FPR (Kapitza *et al.,* 1985). In recent years, digital imaging boards of sufficient depth, resolution, and speed have become available for PC computers. It is therefore, now becoming possible to develop such intensified video imaging FPR instruments at reasonable cost. Of course, this has the added benefit of providing image processing capabilities at the same time. Indeed, many of the same issues discussed elsewhere in this volume apply to the development of an intensified video imaging FPR instrument. I will here connect these issues as they relate to FPR. For more complete discussion of video imaging issues, I refer the reader to these other chapters.

The choice of camera for video imaging FPR is the same as in other fluorescence imaging applications. SIT, ISIT, and cooled intensified CCD cameras are probably the best choices. It is also possible to use an intensifier screen, such as a multichannel plate, and a nonintensified camera. Where quantitative information is to be obtained, one will want to consider the issue of response linearity. In addition, it will be necessary to be able to override the automatic controls of the camera to set manually black level, high voltage, and gain. Since, in many cameras, the frequency filtering is tied to the gain control, it is also useful to remove this feature and to replace it with a passive low pass filter.

When one operates the camera in manual mode, the camera is in mortal danger of being blasted by intense light sources. Since it is difficult to apply the voltage gating approaches discussed above for photomultipliers to video cameras, an electric mechanical shutter must be installed. As discussed above, this presents timing problems for the FPR experiment. In particular, since large shutters tend to open and close more slowly than smaller ones, it becomes necessary to employ delay and detector circuits which ensure that the bleach does not begin until the camera shutter closes, and that the bleach is over before the shutter opens. In our system the bleach is monitored by the photodiode in the beam-splitter attenuator, and the opening and closing of the shutter are monitored by an infrared diode switch which we have incorporated into the shutter.

When qualitative information only is to be obtained, then a simple video imaging FPR system can be constructed with just a camera and a VCR. Such a system was employed by Wolf and Voglmayr (1984) to show that lipid diffuses between the morphologically distinct regions of the ram sperm surface. Another example of such an experiment is shown in Fig. 10, which is taken from Wang (1985). This experiment demonstrates potential treadmilling of actin filaments in the lamellipodium of fibroblasts. Figure 10a shows prebleached labeled actin filaments. The leading edge of

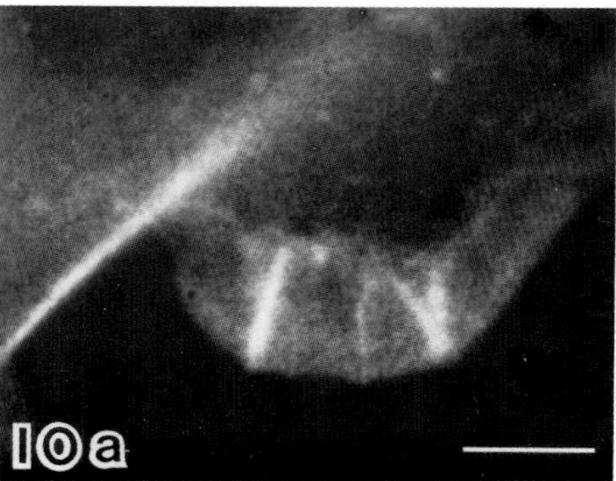

FIG. 10. Video imaging FPR experiment showing possible treadmilling of labeled actin filaments in the lamelli podium of a fibroblast. (a) Prebleached labeled filaments, (b) 24 seconds after leading edge of a filament was bleached, (c) 159 seconds postbleach, (d) 241 seconds postbleach. We observed that the bleached sport migrates along the filament away from the leading edge. From Wang (1985); bar = 5 μm.

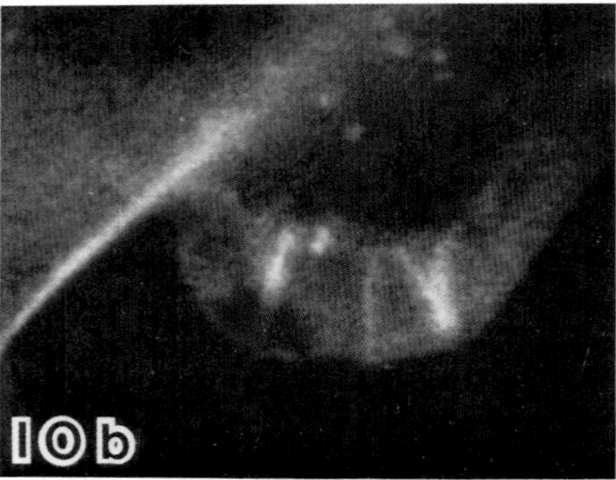

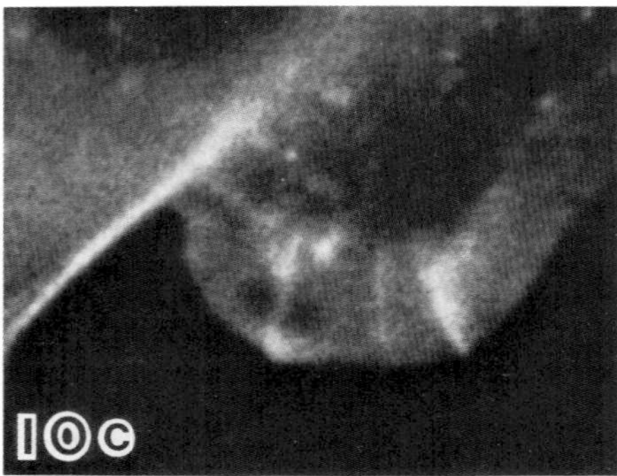

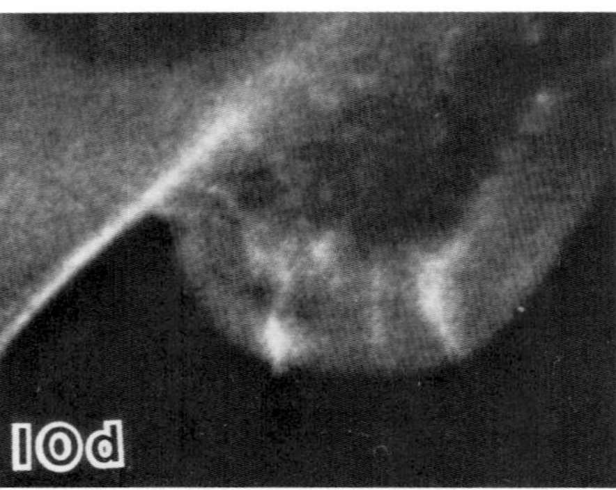

FIG. 10. *(continued)*

a filament was then bleached. Figure 10b–d shows the bleached filament 24, 159, and 241 seconds postbleach. We observe that the bleached spot migrates along the filament away from the leading edge of the cell.

Quantitating video imaging FPR experiments is, of course, a much more complex issue. There are several critical factors to consider in choosing a digital imaging board.

Always begin by considering how good you can expect software and hardware support to be. How probable is it that the company you are buying from will be around in 2 years? Also, since you must develop customized software, how can you do this with a given board? The simplest approach is always to address the board from a higher level language. Ideally, standard imaging routines should be callable from these languages. Can this be done with a given board? Some manufacturers offer "macro" languages with their boards. While these can be quite useful, they are not always suitable for the kind of speed and software interfacing which will be required.

What level of depth do you need? Eight-bit depth is probably minimal with 16-bit preferable. What is the resolution of your board? While images can be magnified before reaching the camera, a pixel array size of 512 by 512 is preferable to a smaller array size.

How many image buffers does your board have? For the kinds of manipulations required two buffers are minimal. A much more flexible system can be developed with, say, four buffers. Coprocessors which allow for rapid hardware performance of standard imaging operations like addition, subtraction, multiplication, convolutions, etc. are extremely helpful, and add speed to processing.

Speed is often the limiting factor with PC-based imaging systems. A good test of speed is to determine whether your board can do real time frame averaging. That is, can it average video frames without loosing any frames? It also becomes important to determine how such averaging routines handle image depth.

It can be useful to be able to digitize recorded video images. If this is your intent, how will you accomplish this? Many boards claim to be able to accept VCR signals directly without the addition of a time base corrector. Test this, preferably with your own VCR and computer, before purchasing a given board.

Many of the same considerations discussed above for spot FPR must be considered in quantitative video imaging FPR. You must determine that your camera is operating in a linear range. If this cannot be done, you will have to digitally correct your images for linearity. Note also, that you must make sure, in a given experiment, that none of the important elements of the image are at saturation.

A number of the corrections discussed elsewhere in this volume must also be applied to the image. The image must be corrected for spatial distortion, typically for image pin cushioning. The image must be corrected for nonhomogeneous illumination and detection. This can be accomplished by dividing each image by the image of a homogeneous fluorescent film. Recognize that such image division should be done as real rather than integer division if one wants to preserve the dynamic range of the image. In the same vein, you may need to correct for lamp intensity fluctuations between frames during measurement. The resolution of the video system is limited by both the optical resolution and the video resolution. In principle, if one wants the resolution to be limited by the optics, rather than the camera, then one makes sure that a distance on the video detector surface corresponding to the optical resolution fills many pixels. Recognize that the ultimate resolution of the microscope depends on the objective. Eyepiece magnification is empty magnification. Thus, assuming you are using eyepieces or other lenses to produce a real image on the video detector, this distance is the composite eyepiece magnification times the resolution of the objective. (For a high-quality, high-power objective this is typically a few tenths of a micron.) The above discussion assumes that a change in light from black to white over one pixel is detectable by the camera. This is usually not the case (see Chapter 1, this volume, by T. Young on image transfer functions). Functionally, the issue becomes one of determining the limit of resolution of your system, and not attempting to get spatial information below this limit. Empty magnification enables one to set the system so that optical resolution issues dominate over video resolution issues.

With PC-based systems one is, as already noted, limited by speed and space. These limitations extend also to data storage and to available RAM for analysis. Often the situation can be greatly helped by judicious choice of the region of the image to be considered. It is usually not necessary to measure and process the entire video image. Indeed, in many experiments, one may only want to consider data along a line or along two orthogonal lines. Each imaging board has its own pecularities which must be explored. Grabbing a rectangular segment of an image can actually be faster than grabbing data along two orthogonal lines or data from a set of grid vertices.

V. Conclusions

In this chapter, I have attempted to emphasize what one must do, not only to make FPR measurements, but to make them correctly. In preparing this chapter I was struck by the large number of papers which have employed the photobleaching technique over the past 10+ years. This, I suppose, points to enormous versatility of FPR as a technique for measuring both surface and cytoplasmic diffusion. This versatility in turn stems from the general applicability of the technique to many probes and macromolecules. In a sense, one is only limited by one's ability to specifically label the molecule that one wants to study in a way which does not alter the cell. The continued use of FPR demonstrates that it continues to expand our fundamental understanding of the molecular structure and dynamics of cells.

Acknowledgments

The author would like to thank first of all the pioneers of FPR with whom he has had the pleasure, both personal and intellectual, of interacting over the years: Drs. Watt W. Webb, Michael Edidin, Elliot Elson, Kenneth Jacobson, Reiner Peters, Dan Axelrod, Dennis Koppel, and Yossie abba David Schlessinger. I would also like to thank the members of my own laboratory who have contributed to this chapter: Drs. Valerie Maynard and Richard Cardullo, Ms. Margaret Moynihan McManus, Ms. Christine McKinnon, Mr. Carlos Flores, and Mr. Robert Mungovan. I am also enormously grateful to the technical staff of the Worcester Foundation who have made instrument design such a pleasure: Mr. Walter Zydlewski, Mr. David Bernklow, Mr. Richard Cassidy, and Mr. Richard Barthelmas. This work was supported in part by NIH Grants HD 17377 and HD 23294 and by private grants from the A. W. Mellon Foundation and Educational Foundation of America to the Worchester Foundation.

References

Aizenbud, B. M., and Gershon, N. D. (1982). *Biophys. J.* **38,** 287–293.

Axelrod, D. (1977). *Biophys. J.* **18,** 129–131.

Axelrod, D., Koppel, D. E., Schlessinger, J., Elson, E., and Webb, W. W. (1976). *Biophys. J.* **16,** 1055–1069.

Barisas, B. G., and Leuther, M. D. (1977). *Biophys. Chem.* **10,** 221–229.

Bevington, P. R. (1969). "Data Reduction and Error Analysis for the Physical Sciences." McGraw-Hill, New York.

Burghardt, T. P., and Axelrod, D. (1981). *Biophys J.* **33,** 455–408.

Dragsten, P. R., Henkartt, P., Blumenthal, R., Weinstein, J., and Schlessinger, J. (1979). *Proc. Natl. Acad. Sci. U.S.A.* **76,** 5163–5167.

Edidin, M. (1981). "Comprehensive Biochemistry," Vol. 1, pp. 37–82. Elsevier, Amsterdam.

Edidin, M., Zagyansky, Y., and Lardner, T. J. (1976). *Science* **191,** 466–468.

Engstrom, R. W. (1980). "The Photomultiplier Handbook," p. 180. RCA Corp., Lansing, Pennsylvania.

Jacobson, K., Derzko, Z., Wu, E.-S, Hou, Y., and Poste, G. (1976a). *J. Supramol. Struct.* **5,** 565–576.

Jacobson, K., Wu, E.-S., and Poste, G. (1976b). *Biochim. Biophys. Acta* **423,** 215–222.

Jacobson, K., Hou, Y., and Wojcieszyn, J. (1978). *Exp. Cell Res.* **116,** 179–189.

Kapitza, H. G., McGregor, G. N., and Jacobson, K. A. (1985). *Proc. Natl. Acad. Sci. U.S.A.* **82,** 4122–4126.
Koppel, D. E. (1979). *Biophys. J.* **28,** 281–292.
Koppel, D. E., and Sheetz, M. P. (1981). *Nature* (*London*) **293,** 159–160.
Lubyphelps, K., Castle, P. E., Taylor, D. L., and Lanni, F. (1987). *Proc. Natl. Acad. Sci. U.S.A.* **84,** 4910–4913
Peters, R. (1981). *Cell Biol. Int. Rep.* **5,** 733–760.
Peters, R., Peter, J., and Tews, K. H. (1974a). *Pflugers Arch.* **347,** R36.
Peters, R., Peters, J., Tews, K. H., and Bahr, W. (1974b). *Biochim. Biophys. Acta* **367,** 282–294.
Schlessinger, J., Koppel, D. E., Axelrod, D., Jacobson, K., Webb, W. W., and Elson, E. L. (1976). *Proc. Natl. Acad. Sci. U.S.A.* **73,** 2409–2413.
Schlessinger, J., Axelrod, D., Koppel. D. E., Webb, W. W., and Elson, E. L. (1977). *Science* **195,** 307–309.
Schneider, M. B., and Webb, W. W. (1981). *Appl. Opt.* **20,** 1382–1388.
Sheetz, M. P., and Koppel, D. E. (1979). *Proc. Natl. Acad. Sci. U.S.A. 76,* 3314–3317.
Singer, S. J., and Nicolson, G. L. (1972). *Science* **175,** 720–731.
Smith, B. A., and McConnell, H. M. (1978). *Proc. Natl. Acad. Sci. U.S.A.* **75,** 2759–2763.
Thompson, N. L., Burghardt, T. P., and Axelrod, D. (1981). *Biophys. J.* **33,** 435–454.
Van Zoelen, E. J., Tertoolen, L. G. J., and DeLaat, S. W. (1983). *Biophys. J.* **42,** 103–108.
Wang, Y.-L. (1985). *J. Cell Biol.* **101,** 597–602.
Wojcieszyn, J. W., Schlegel, R. A., Wu, E.-S., and Jacobson, K. A. (1981). *Proc. Natl. Acad. Sci. U.S.A.* **78,** 4407–4410.
Wolf, D. E., and Edidin, M. E. (1981). *Tech. Cell. Physiol.* **105,** 1–14.
Wolf, D. E., and Voglmayr, J. K. (1984). *J. Cell Biol.* **98,** 1678–1684.
Wolf, D. E., Schlessinger, J., Elson, E. L., Webb, W. W., Blumenthal, R., and Henkart, P. (1977). *Biochemistry* **16,** 3476–3483.
Wolf, D. E., Edidin, M., and Dragsten, P. R. (1980a). *Proc. Natl. Acad. Sci. U.S.A.* **77,** 2043–2045.
Wolf, D. E., Henkart, P., and Webb. W. W. (1980b). *Biochemistry* **19,** 3893–3904.
Wolf, D. E., Handyside, A. H., and Edidin, M. (1982). *Biophys. J.* **38,** 295–297.
Yguerabide, J., Schmidt, J. A. and Yguerabide, E. E. (1982). *Biophys. J.* **40,** 69–76.

Chapter 11

Interpretation of Fluorescence Correlation Spectroscopy and Photobleaching Recovery in Terms of Molecular Interactions

ELLIOT. L. ELSON AND HONG QIAN

Department of Biological Chemistry
Washington University School of Medicine
St. Louis, Missouri 63110

I. Introduction

Fluorescence photobleaching recovery (FPR) and fluorescence correlation spectroscopy (FCS) are two closely related methods for measuring rates of diffusion, flow, and chemical reactions (Elson, 1985). Although the two methods share similar theoretical bases and experimental methods, they differ in that FPR measures the dissipation of a macroscopic concentration gradient generated by pulse photolysis, whereas FCS measures the average time course of spontaneous microscopic concentration fluctuations that are constantly occurring even though the system is in equilibrium.

Both FPR and FCS are based on measurements of the fluorescence emitted from a small open observation subregion of the sample. The measured intensity varies with the number of fluorescent molecules and so increases or decreases as these molecules enter or leave the subregion. The rate at which molecules enter or depart depends both on their rate of motion (diffusion coefficient or flow or drift velocity) and on the size of the observation region. Hence, knowing the size of the observation region, one can determine diffusion coefficients and drift or flow velocities from measurements of the rate of change of fluorescence. The intensity could also change as a result of intermolecular interactions of conformational changes that affect the absorption coefficient or fluorescent quantum yield of the observed molecules. Then, measured fluorescence changes could also directly characterize the kinetics of these reactions. Information about chemical kinetics, however, can come not only directly from reaction-dependent changes in fluorescence (e.g., Magde *et al.,* 1974; Icenogle and Elson, 1983a,b) but also and more commonly, from the effects of reactions on the diffusion or flow rates of the reactants. Hence, in principle, FCS and FPR can be used to characterize both transport properties (diffusion and flow or drift) and chemical reaction kinetics.

The open observation subregion of the sample is defined by a narrow beam of light that excites the observed fluorescence. The measurement is typically carried out using a microscope (see Chapter 10, by Wolf, this volume). Since beam radii can be smaller than 1 μm, both FPR and FCS can be applied to small systems such as cells. The methods have the additional advantages of being quantitatively precise, rapid, applicable to individual cells and to different regions of the same cell, and relatively nonperturbing. Photochemical artifacts and cellular physiological perturbations are possible, however, and their possible effects should be controlled.

FPR (Axelrod *et al.,* 1976; Koppel, *et al.,* 1976; Edidin *et al.,* 1976; Jacobson *et al.,* 1976) [also known as "fluorescence recovery (or redistribution) after photobleaching" and "fluorescence microphotolysis" (Peters, 1981)] has been extensively used during the last 10 years to measure the rates of diffusion of fluorescent molecules inside and on the surface of cells and in model systems (e.g., Petersen, 1986; McCloskey and Poo, 1984). FCS (Magde *et al.,* 1972; Elson and Magde, 1974; Magde *et al.,* 1974) is more difficult to apply to cells but can provide important complementary information in some circumstances. In this chapter, we discuss the conceptual bases, theory, capabilities, and interpretation of these two approaches, emphasizing their application to measure molecular interactions.

Knowing the rate of diffusion of specific molecules in cells is important for cell biologists for several reasons. Although it is not clear that diffusion limits their rates (Axelrod, 1983; McCloskey and Poo, 1986), some cellular processes may require the diffusional encounter of reactants as, for example, in synaptogenesis (e.g., Poo, 1982; Axelrod, 1983) in receptor-mediated endocytosis (e.g., Keizer *et al.*, 1985; Schlessinger, 1983) or in oxidative phosphorylation (Hackenbrock *et al.*, 1986). Rates of diffusion can be influenced both by specific interactions between the diffusant and other fixed and mobile molecules and by nonspecific or steric interactions with structural components of the cell (Jacobson and Wojcieszyn, 1984; Luby-Phelps *et al.*, 1986). Hence measurements of diffusion can help to characterize both specific interactions and more general structural properties. Finally, in addition to the effects of intermolecular interactions, a molecule's diffusion rate also depends on its shape. Hence diffusion measurements could in principle be used to characterize major conformational changes.

II. Characteristic Responses

The number of fluorescent molecules in the open observation subregion of the system changes more rapdily the faster they diffuse, drift, or flow into or out of the subregion and the faster they are created or destroyed by chemical reactions. Hence the rates of the transport processes and reactions can be determined from the rates of change of the numbers of molecules in the region. In FCS and FPR experiments the numbers of molecules in the observation region are measured by their fluorescence. The observation region is defined by intensity profile $I(\mathbf{r})$ which specifies the excitation intensity at position $\mathbf{r}$ in the sample. The total measured fluorescence intensity, $F(t)$, is proportional to the concentration of the fluorescent molecules $c(\mathbf{r}, t)$ at position $\mathbf{r}$ and time t multiplied by $I(\mathbf{r})$ summed over the sample space: $F(t) = Q \int I(\mathbf{r})\, c(\mathbf{r}, t)\, d\mathbf{r}$ where Q accounts for the absorption coefficient and fluorescence quantum yield of the fluorophore and for photon losses in the measurement optical system. Hence, for both FCS and FPR the central experimental task is to measure $F(t)$ over time. In an FPR experiment, the photobleaching pulse generates a macroscopic concentration gradient $\Delta c(\mathbf{r}, t) = \bar{c} - c(\mathbf{r}, t)$ which determines a macroscopic difference in fluorescence, $\Delta F(t) = \bar{F} - F(t)$, where $\bar{c}$ and $\bar{F}$ are the equilibrium values of concentration and measured fluorescence. Diffusion coefficients, drift and flow velocities, and chemical reaction rates can be determined directly from measurements of $\Delta F(t)$. In contrast, FCS measures microscopic changes in fluorescence, $\delta F(t)$, which result from spontaneous microscopic concentration fluctuations. Many fluorescence fluctuations must be measured and analyzed statistically to determine transport coefficients and chemical rate constants accurately. This is not only because each $\delta F(t)$ is too small to be measured with high precision but also because the time course of an individual fluctuation is only stochastically determined by the desired transport coefficients and rate constants. Hence even if each $\delta F(t)$ were measured with high accuracy, it would still be necessary to analyze statistically many fluctuations to obtain accurate values of transport coefficients and rate constants. The statistical analysis of microscopic fluorescence fluctuations is usually carried out by calculating the fluorescence fluctuation autocorrelation function, $G(\tau) = \langle \delta F(0)\, \delta F(\tau) \rangle$, where $\langle \cdots \rangle$ denotes an average over many fluctuations (cf. Elson and Webb, 1975).

The characteristic behavior of $\Delta F(t)$ or of $G(\tau)$ for diffusion, drift or flow, and chemical reaction seen in FPR and FCS measurements depends on the form of $I(\mathbf{r})$. The simplest experimental arrangement uses a cylindrical Gaussian excitation profile for FCS experiments and for both photobleaching and monitoring the fluorescence recovery in FPR measurements. While monitoring fluorescence recovery, the beam is attenuated 10^3- to 10^4-fold to avoid unwanted photobleaching. The excitation profile can be represented as $I(\mathrm{r}) = BI_0 \exp(-2r^2/w^2)$ where $B = 1$ for the monitoring beam and $B = 10^3$–10^4 for the photobleaching pulse and w is the radius at which the beam intensity has fallen to $\exp(-2)$ of its maximum central value, I_0. In FPR experiments the response also depends on the extent of photobleaching, measured by $\kappa = \lambda B I_0 T$, where λ and T are the rate constant for photolysis (assumed to be a first order irreversible process) and the duration of the photobleaching pulse, respectively. Then for two-dimensional (planar) systems containing only a single fully mobile fluorescent component:

$$G(\tau) = G(0)[1 + \tau/\tau_D]^{-1} \tag{1a}$$

and

$$F(t) = QP\bar{c} \sum_{n=0} [(-\kappa)^n/n!][1 + n(1 + 2t/\tau_D)]^{-1} \tag{1b}$$

for diffusion where $\tau_D = w^2/4D$, $G(0) = (PQ)^2\bar{c}/(\pi w^2)$, and $P = \int I(r)\, d^2r = \pi w^2 I_0/2$, the laser power.
Further,

$$G(\tau) = G(0) \exp[-(\tau/\tau_V)^2] \tag{2a}$$

and

$$F(t) = QP\bar{c} \sum_{n=0} [(-\kappa)^n/(n + 1)!] \exp\{[-2n/(n + 1)](t/\tau_V)^2\} \tag{2b}$$

for uniform drift or flow where $\tau_V = w/V$.

Also

$$G(\tau) = G(0) \exp[-(\tau/\tau_V)^2/(1 + \tau/\tau_D)]/[1 + \tau/\tau_D] \tag{3a}$$

and

$$F(t) = QP\bar{c} \sum_{n=0} (-\kappa)^n \exp\{-2n(t/\tau_V)^2/[1 + n(1 + 2t/\tau_D)]\} \times \{n![1 + n(1 + 2t/\tau_D)]\}^{-1} \tag{3b}$$

for simultaneous diffusion and flow or drift. It is evident that for $\kappa \ll 1$, $G(\tau)$ and $\Delta F(t)$ will have the same form. In particular for $\kappa \ll$ $\Delta F(t)/(QP\bar{c}\kappa/2 = G(t)/G(0)$. Examples of these characteristic patterns of behavior for small κ are shown in Fig. 1. In this and subsequent figures, except Fig. 2 and 6, we shall plot the characteristic time behavior

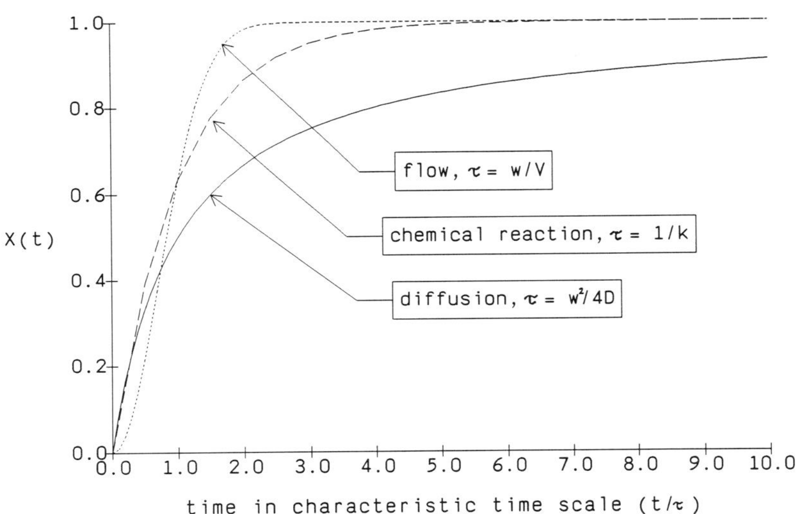

FIG. 1. Characteristic fluorescence recovery or correlation decay for diffusion, flow or drift, and chemical reaction. The system contains only a single fluorescent component and is supposed to be planar and infinite in extent. To mimic the recovery behavior usually observed in FPR measurements, we have plotted $X(t) = 1 - \Delta f(t)/(QP\bar{c}\kappa/2) = 1 - G(t)/G(0)$ versus t. To ensure that $\Delta f(t)/QP\bar{c}\kappa/2) = G(t)/G/(0)$ it is necessary to choose $\kappa \ll 1$. In this figure we have set $\tau_D = \tau_V = \tau_{\text{chem}} = 1$ where τ_D and τ_V are the characteristic times for diffusion and drift or flow defined in the text, and τ_{chem} is a characteristic chemical relaxation time which governs chemical relaxation according to $\Delta c(t) = \Delta c(0) \exp(-t/\tau_{\text{chem}})$. The solid, dotted, and dashed lines depict diffusion, flow or drift, and chemical relaxation, respectively. This figure illustrates the slow rate of diffusional recovery, especially after long periods of time.

as $X(t) = 1 - \Delta F(t)/(QP\bar{c}\kappa/2) = 1 - G(t)/G(0)$. Validity of the second equality requires that $\kappa \ll 1$. As κ increases, the apparent size of the photobleached region increases even at fixed w (cf. Axelrod *et al.*, 1976). Figure 2 illustrates the behavior of $F(t)$ for several values of κ.

When chemical reactions are present the analysis can become considerably more complicated. Because the observation system is open to transport of the fluorescence molecules, the reaction and transport terms are typically coupled in complex ways. Even simple reactions can generate relatively complex expressions which can sometimes be simplified to illustrate the characteristic contributions of transport and reaction kinetics (cf. Elson and Magde, 1974; Elson, 1986). Systems involving many components may be amenable only to numerical methods (e.g., Icenogle and Elson, 1983a). When chemical processes are separable, they typically conribute terms to the fluorescence recovery or correlation decay which depend exponentially on time (cf. Elson and Magde, 1974; Elson, 1986; and Fig. 1). The dependence of the characteristic times for diffusion, flow, and chemical reactions on w is illustrated in Table I.

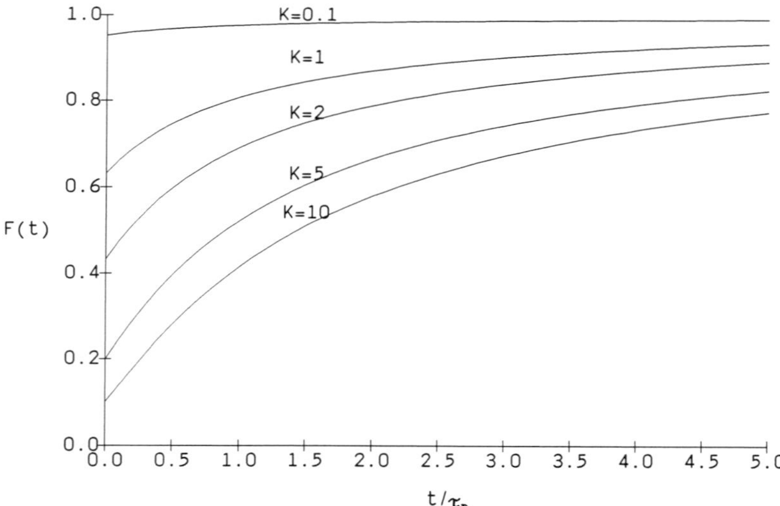

FIG. 2. Fluorescence recovery for different extents of photobleaching. $F(t)$ is plotted for different values of κ, the parameter which describes the extent of photobleaching. The initial value of the fluorescence is $F(0) = (QP\bar{c}/\kappa)[1 - \exp(-\kappa)]$. The curves are normalized by setting $F(\infty) = QP\bar{c} = 1$.

TABLE I

DIMENSIONAL CHARACTERISTICS OF MEASURED PROCESSES[a]

Process	Characteristic rate	Characteristic time
Diffusion	$D \propto \langle(\Delta x)^2\rangle/t$	$\tau_D \propto w^2/D$
Flow or drift	$V \propto \Delta x/t$	$\tau_V \propto w/V$
Chemical reaction	$k \propto 1/t$	$\tau_{\text{chem}} \propto w^0/k = 1/k$

[a] The distance Δx is traversed in time t. Diffusion is characterized in terms of the mean square distance traversed in time t, where $\langle\cdots\rangle$ represents an average over many particle displacements. For spot photobleaching measurements w is the $\exp(-2)$ radius of the excitation intensity profile; D, V, and k represent a diffusion coefficient, flow or drift velocity, and chemical rate constant, respectively.

In FPR measurements on cells, it is frequently found that the fluorescence does not recover to its initial prephotobleach level. The fraction of the fluorescence intensity which does recover is denoted R_f and is operationally defined as $R_f = [F(\infty) - F(0)]/[F(<0) - F(0)]$ where $F(\infty)$ is the fluorescence intensity measured at so long a time after the bleaching pulse that recovery has attained its maximum extent; $F(0)$ is the intensity immediately after the photobleaching pulse (defined as $t = 0$); and $F(<0)$ is the intensity prior to the photobleaching pulse (defined as $t < 0$). The nonrecoverable fraction $(1 - R_f)$ is usually attributed to the presence of fluorophores which are immobile on the time scale of the measurement. Immobile molecules which have been photobleached cannot be replaced by other corresponding unbleached immobile molecules from outside the observation region. It is important to realize, however, that fluorophores which appear immobile on a short time scale may have a measurable diffusion coefficient on a longer time scale. Hence an FPR measurement can yield both the effective diffusion coefficient of the mobile fluorophores and the fraction of fluorophores which are apparently immobile on the experimental time scale. There is, however, another possible cause for nonrecovery of a frction of the fluorescence. If the sample space available to the observed fluorophores is of a size comparable to that of the observation area, a nonrecoverable fraction could result from net depletion of the fluorophores by the photobleaching pulse, even if all the fluorophores are mobile. Then, as will be discussed below, the nonrecoverable fraction provides a measure of the area or volume of the sample space available to the freely mobile fluorophores.

III. Technical Issues

A. Practical Analysis of Gaussian Spot Data

Most commonly, the cylindrical Gaussian intensity profile introduced above is used for excitation in FPR and FCS experiments. In the simplest experiments there is only a single diffusing or drifting component which is confined to a plane coincident with the focal plane of the microscope objective. The latter condition can be fulfilled for measurements of molecules confined to planar membranes. More generally, there can be several components, each with different diffusion or drift velocities, and the molecules might be free to move in three dimensions as in a bulk system such as cytoplasm.

1. Analysis of Simple Recovery of Fluorophores Confined to a Plane

We consider first the simplest spot photobleaching measurements of diffusion of a single component. Experimental apparatus and procedures for carrying out these measurements and for analyzing records of fluorescence versus time based on Eq. (1b) or some variation thereof to yield optimal estimates of the diffusion coefficient are described in detail in Chapter 10 by Wolf (this volume) and elsewhere (Petersen *et al.*, 1986a). Similar methods can be used to fit recovery data for flow processes using Eq. (2b) or (3b). To determine diffusion coefficients and drift velocities from FCS and FPR measurements, it is necessary to know the size and shape of the observation region defined by $I(\mathbf{r})$. One approach, discussed by Schneider and Webb (1981), is to deduce the beam profile from the intensity record acquired by translating a knife edge or a small fluorescent bead through the beam at constant known velocity. An alternative is to use a video image method. The microscope is focused on a thin uniform layer of fluorophores to provide a fluorescent image of the observation region defined by $I(\mathbf{r})$. As long as the layer of fluorophores is uniform and the camera is not saturated by the highest levels of intensity, the emitted fluorescence intensity at each point $\mathbf{r}$ should be proportional to $I(\mathbf{r})$. A digital video record of this fluorescence distribution is acquired using a sufficiently sensitive camera and a standard digital imaging device and methods such as these are discussed in Chapters 1 and 2 by Young, and Jeričević *et al.*, this volume. The distance scale in the video image can be easily ascertained from the image of a microscopic object of well-known size, e.g., a grating. (The distance scale in both the x- and y-directions should be checked). Then the digital image can be analyzed to yield a plot of the intensity versus position, e.g., along a line through the center of the region. This can be fitted to the appropriate Gaussian function to determine w. Lines of various orientations can be analyzed or a contour plot of the entire intensity profile over the focal plane can be used to assess the circular symmetry of the region. Advantages of this approach are its simplicity, flexibility, ease of assessing angular symmetry, and its reliance on digital video equipment, which may be more readily available than the scanning stages required for other methods. Another advantage is that the entire distribution of intensity in two dimensions can be mapped and, if this deviates from the standard Gaussian profile, can be used in the analysis of the recovery curve [cf. Eqs. (2) and (3) in Elson, 1985], although this will rarely be necessary. For small spots comparable to λ, the wavelength of the excitation light, it is also necessary to account for broadening of the spot image by diffraction. The relation between the radius of the spot illuminating the system, w_0, and the radius of its image, w_i, is given approximately by $w_i^2 \simeq w_0^2 [1 + (\varepsilon/w_0)^2]$, where $\varepsilon = 1.22\lambda/2\text{NA}$ and NA is the numerical aperature of the objective.

2. Multicomponent Recovery

A complex biological system may contain labeled molecules with different diffusion coefficients. Hence a single diffusion coefficient may be insufficient to characterize the fluorescence recovery curve. In practice, the analysis of FPR measurements on even the simplest systems typically takes account of the possibility of an immobile as well as a mobile component. Hence not only τ_D (for determining the diffusion coefficient of the mobile molecules) but also $F(\infty)$ (for determining the mobile fraction) are derived from the records of $F(t)$ (Wolf, Chapter 10, this volume; Petersen *et al.*, 1986a). If species with different measurable diffusion coefficients are present, the situation is more complex. The recovery for a single component has the time course indicated in Eq. (1b). The slow increase of this function with time for $t \gg \tau_D$ impairs the resolution of components which diffuse more slowly. It is therefore useful to assess the sensitivity of the measurements to the presence of molecules with different diffusion coefficients. Figure 3 displays curves calculated for equal contributions of two fully mobile species, one with $\tau_D(1) = \tau_{\text{ref}}$, the other ranging from $\tau_D(2) = 0.1\tau_{\text{ref}}$ to $10\tau_{\text{ref}}$. The results depend on the range of time over which the recovery is measured. In Fig. 3 the recovery is measured out to $t = 3\tau_{\text{ref}}$. Figure 3 demonstrates that for $\tau_D(2) > \tau_D(1)$ a good fit is obtained to the recovery function for a single mobile component with the $\tau_D = (1.3\text{–}1.5)\tau_{\text{ref}}$ and an apparent immobile fraction which increases to 71% for $\tau_D(2) = 10\tau_D(1)$. Evidently, this apparent immobile fraction results from

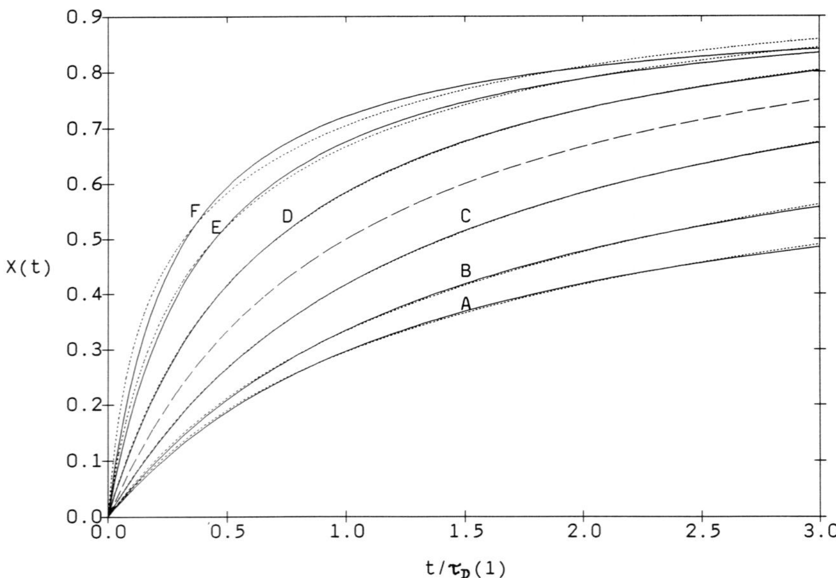

FIG. 3. Fluorescence recovery of two components with different diffusion coefficients. The ordinate measures $X(t)$ calculated as described for Fig. 1. The dotted curves are calculated for a system containing equal amounts of two components confined to an infinite plane. One component has $\tau_D(1) = \tau_{\text{ref}} = 1$. The recovery is followed out to $t = 3\tau_{\text{ref}}$. The solid curves represent the best fit to a single component recovery. The fit provides an estimate of the apparent (single component) diffusion time τ_D(app) and mobile fraction, R_f(app). The curves are described as follows:

Curve	$\tau_D(2)/\tau_D(1)$	τ_D(app)	R_f(app)
A	10.0	1.40	0.712
B	5.0	1.51	0.838
C	2.0	1.33	0.972
D	0.5	0.68	0.984
E	0.2	0.40	0.945
F	0.1	0.27	0.917

The dashed curve represents simple diffusional recovery for a single component.

the failure of the measurement to extend sufficiently into the time range over which the slow component contributes a major portion of its recovery. Conversely, for $\tau_D(2) < \tau_D(1)$, the apparent immobile fractions in the single component recovery are below 10%, but the apparent τ_D ranges from $0.27\tau_{\text{ref}}$ to $0.68\tau_{\text{ref}}$ as $\tau_D(2)/\tau_D(1)$ ranges from 0.1 to 0.5. Thus, for $\tau_D(2) < \tau_D(1)$, almost the full recovery of both components is registered, but the substantial contribution of the faster process has a large influence on the τ_D calculated, assuming only one mobile component. It is clear that, in studies of complex biological systems, it is useful to measure the fluorescence recovery over a range of times to determine whether components with different rates are present. Resolution of multicomponent recovery will, however, be limited by the accuracy of the data and the contributions both of systematic and random errors, especially shot noise in measurements of samples with low fluorescence intensity (cf. Petersen *et al.*, 1986a). Introduction of more than a single diffusional recovery process should be justified by a convincing demonstration of a systematic dependence of recovery rate and extent on measurement period.

A useful approximate approach to analyzing FPR data is provided by fitting the reciprocal of the fluorescence displacement $\Delta F(t) = -[F(t) - F(\infty)]$ as a function of time (Yguerabide *et al.*, 1982; Van Zoelen *et al.*, 1983). This approach is based on the close approximation of $\Delta F(t) \propto \Sigma_{n=1}(-\kappa)^n\{n![1 + n(1 + 2t/\tau_D)]\}^{-1}$ by the function $[1 + t/t_{1/2}]^{-1}$ $\Sigma_{n=1}(-\kappa)^n/(n + 1)!$ where $t_{1/2}$ is the half-time for recovery that depends on κ, the extent of photobleaching. This approximation is valid over the ranges $0 < t < 10\tau_D$ and $0 < \kappa < 6$ (about 85% bleached). Hence over this range a plot of $1/\Delta F(t)$ versus t should give a straight line for a single component recovery with intercept/slope $= t_{1/2}$. If a flow process contributes to the recovery, the reciprocal plot will have positive curvature and if there is a second, slower diffusing component, the plot will have negative curvature.

3. Diffusion in Three Dimensions

In spot FPR and FCS experiments the excitation light is typically focused on the sample through an objective lens. The light beam has a minimum radius and maximum intensity at the plane of focus. Because of the convergence of a focused beam, its diameter increases with increasing distance above or below the focal plane, and its intensity decreases correspondingly (cf. Schneider and Webb, 1981). This behavior can be represented approximately as $I(z) = I_0/[1 + (z/z_0)^2]$ where z is the distance from the focal plane and z_0 is the characteristic distance at which $I(z_0) = I_0/2$. Also $z_0 = n\pi w_0^2/\lambda$, where n is the refractive index of the medium containing the fluorophores, w_0 is the radius of the beam in the focal plane, and λ is the wavelength of the exciting light. This nonuniformity of the excitation beam shape can influence FPR measurements of diffusion rate in three-dimensional samples.

The simplest analysis of photobleaching measurements accounts only for the motion of fluorophores in the focal plane perpendicular to the axis of the beam of excitation light (the z-axis). Except when diffusing fluorophores are confined to a planar membrane, however, they will be able to move parallel to the beam axis. If the intensity of the beam varies along its axis, the emission intensity excited from a fluorophore will vary as the fluorophore moves in the z-direction. This can add an additional rate process to the fluorescence recovery or correlation decay which will depend on z_0. For simplicity suppose that the beam profile is Gaussian also in the z-direction so that $I(r) = I_0 \exp[-2(x^2 + y^2)/w^2 - 2z^2/(z_0')^2]$ where $z_0' = z_0/(\ln 2)^{1/2}$. Then, for small degrees of photobleaching (small κ), $G(t) \propto \Delta F(t) \propto [1 + t/\tau_D]^{-1}[1 + t/\tau_z]^{-1/2}$, where $\tau_D = w^2/4D$ and $\tau_z = (z_0')^2/4D$. From this it can be seen that if z_0' is substantially greater than w, the diffusion along the z-direction would contribute little to the time dependence of fluorescence recovery or correlation decay. This is because fluorescence recovery or correlation decay due to diffusion in the x-y plane would be much faster, and so would dominate the observed time behavior. In general, diffusion along the z-axis will not contribute significantly to the observed time behavior in FCS and FPR measurements. This can be seen from the expression for z_0 given above. For the highest magnification objectives $w \approx 2\lambda$. Hence $z_0'/w \approx 7$ and $\tau_z/\tau_D \approx 50$. For lower-power objectives these ratios will be even larger.

A more substantial contribution to the recovery behavior results from the variation of w with z. The variation can be approximated as $w(z) = w(0)[1 + (z/z_0)^2]^{1/2}$. Hence as the distance $|z|$ from the focal plane increases, $w(z)$ also increases. The characteristic diffusion time at $|z|$ increases according to $\tau_D(z) = w(z)^2/4D$. Hence the farther from the focal plane, the longer the time for diffusional recovery or correlation decay. To analyze this effect quantitatively, it is necessary to weight correctly the contribution of fluorophores at various z-levels. Ignoring this effect by analyzing recovery data using only the minimum spot size on the focal plane, w_0, would cause an underestimation of D. For measurements of diffusion in a three-dimensional sample using a 40× air objective, interpretation of the recovery curves for small levels of bleaching in terms of an effective w, $w_{\text{eff}} = \sqrt{2}w_0$ gives a good estimation of D. Hence ignoring this effect by interpreting the data in terms of w_0 would cause a 2-fold underestimation of D. For larger extent of bleaching, additional corrections would be necessary to account for the effective decrease in k with increasing distance from the focal plane.

Additional complications result in the interpretation of FCS amplitudes, which are important in FCS measurements of aggregation in three-dimensional samples as discussed below. These matters and a more complete analysis of FPR and FCS measurements of diffusion in three-dimensional systems will be deferred to a later publication (Qian and Elson, 1988b).

4. Diffusion in a Limited Area

The conventional analysis of FCS and FPR measurements assumes that the spatial domain available for diffusion is unlimited (Elson and Magde, 1974; Axelrod, *et al.*, 1976). This assumption is invalid for systems which are comparable in size to the illumination region. The apparent diffusion rate in both FCS and FPR measurements and the apparent mobile fraction in FPR measurements can be influenced by restricting the system size and shape.

The effect on apparent diffusion rate can be illustrated by considering an FCS or FPR measurement of a single fully mobile component with diffusion coefficient D confined to a planar rectangular region bounded by $\pm L_x$ and $\pm L_y$ on the x- and y-axes. As above, the excitation intensity profile is a Gaussian centered at the origin of coordinates with an exp(−2) radius of w. This problem is readily solved using standard Fourier transform methods (Qian and Elson, 1988a). Figure 4 displays results calculated for a square system area. For these calculations $w = 1$ μm so that the beam area is approximately 3.14 μm^2. The figure demonstrates that the apparent diffusion rate increases relative to that for an infinite area as the sample diminishes. This increase in apparent diffusion rate results from the elimination from the recovery process in the small area of the slow contributions due to diffusion from remote points which contribute to recovery in a large area. Hence, for a relatively compact region such as a square, limitation of the available area makes the diffusion appear faster than it would in an infinite space. The shape of the system area is also important, however. This is illustrated in Fig. 5 which presents recovery curves calculated for rectangular regions with the same area but different ratios of L_x to L_y. In elongated regions the recovery process includes contributions from remote points. Therefore the apparent diffusion rate appears slower than for compact regions with the same area. As long as the system area is substantially larger than the beam area and its shape is not too elongated, the effect of the finite size of the system is inconsequential. The effect of the area restriction becomes significant for square systems when the ratio of the system and beam areas is on the order of 40 : 1. Note, however, that this corresponds to a ratio of linear dimensions of only ≈6 : 1.

Restriction of the system area can also influence the fraction of fluorescence recovery and therefore the apparent mobile fraction. If the area of the spot defined by the excitation beam were comparable to the system area, the photobleaching pulse could eliminate a significant fraction of the fluorophores in the system. Hence the fractional recovery would be diminished even if all the fluorophores were rapidly mobile. The fractional

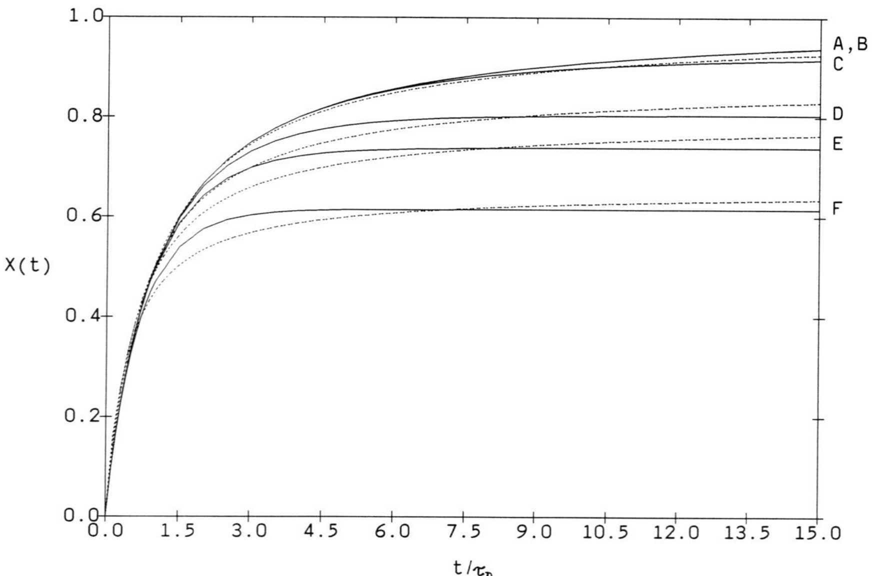

FIG. 4. Dependence of fluorescence recovery on system size. The fluorescence recovery for small κ was calculated supposing that the system was confined to a planar square (cf. Qian and Elson, 1988a), taking account of the nonrecoverable fraction due to depletion of the limited amount of fluorophore in small areas. The beam radius $w = 1$, so that the beam area is π. The calculated recovery curves (solid lines) were fitted to the best single component recovery curves (dotted lines) to provide an estimate for τ_D(app) and R_f(app) as in Fig. 3. The curves are described as follows:

Curve	System area	τ_D(app)	R_f(app)
A	∞	1.00	1.00
B	400	1.00	1.00
C	40	0.96	0.99
D	16	0.74	0.87
E	12	0.62	0.80
F	8	0.45	0.65

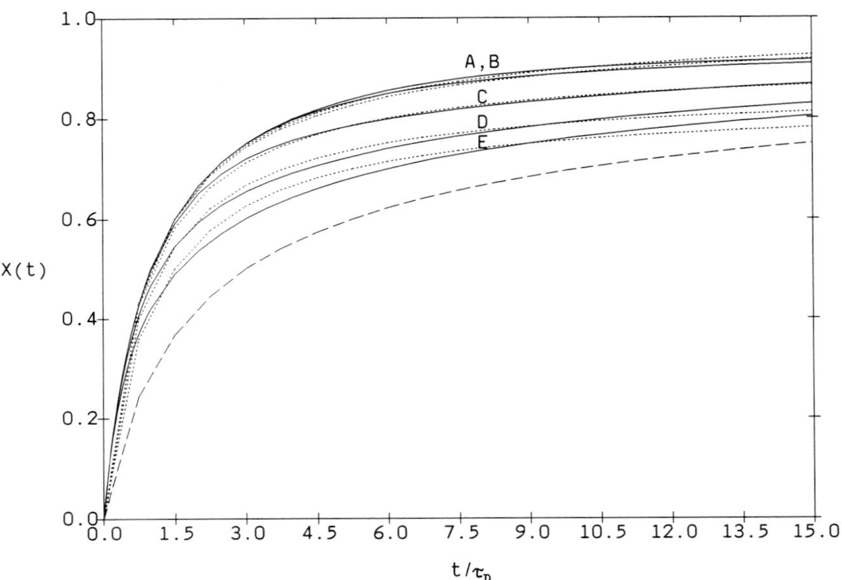

FIG. 5. Dependence of fluorescence recovery on system shape. The fluorescence recovery for small κ was calculated supposing that the system was confined to a planar rectangle of dimensions $2L_x$ by $2L_y$, with the area held constant at 40 square units (cf. Qian and Elson, 1988a). The nonrecoverable fraction due to depletion of the limited amount of fluorophore in small areas was taken into account. The beam radius $w = 1$, so that the beam area is π. The calculated recovery curves (solid lines) were fitted to the best single component recovery curves (dotted lines) to provide an estimate for τ_D(app) and R_f(app) as in Fig. 3. The curves are described as follows:

Curve	L_x	L_y	τ_D(app)	R_f(app)
A	3.16	3.16	0.96	0.99
B	2.50	4.00	0.94	0.97
C	1.67	6.00	0.85	0.91
D	1.25	8.00	0.86	0.86
E	1.00	10.00	0.99	0.83

The dashed curve is for diffusion in a one-dimensional space.

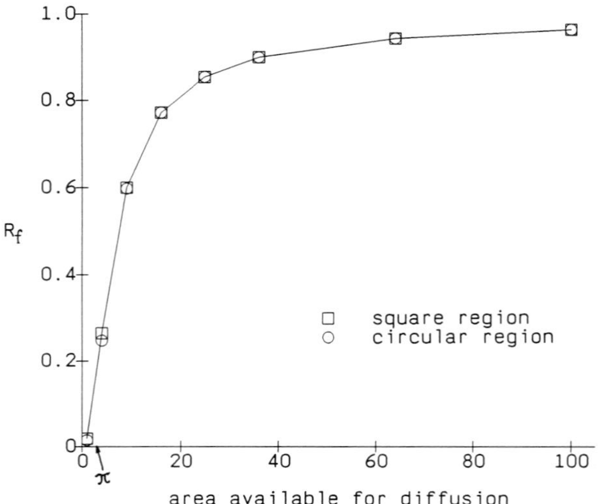

FIG. 6. A comparison of the recoverable fluorescence from small closed regions. The apparent recoverable fraction was calculated as described by Qian and Elson (1988a) for square (□) and circular (○) systems. The beam radius is $w = 1$ unit, and so the beam area is π. Assume $\kappa = 2$.

recovery would decrease to a greater extent both as the ratio of the bleached area to the total sample area and as the extent of photobleaching, measured by κ, increased. Figure 6 displays the dependence of the fractional fluorescence recovery on the area of square and circular systems assuming all fluorophores are mobile (Qian and Elson, 1988a). For this calculation $\kappa = 2$, corresponding to a fractional photobleaching of $1 - F(0)/F(<0) = 0.57$ of the initial fluorescence. This figure demonstrates that confinement of the system area can cause a 10% decrease in fractional recovery when the linear dimensions of the system and beam are in the ratio of ~3 : 1. For smaller values of κ, the effect of finite system area will be smaller. The figure also demonstrates that the effects on fractional recovery of square and circular spot shapes are essentially the same, except for very small system areas comparable in size to the photobleached spot.

In some instances the dimensions of the area available for fluorophore diffusion may not be immediately apparent. This might be true, for example, of a subregion such as a membrane domain which is embedded in a larger region. If the subregion were closed to the transport of fluorophores across its boundaries, it would behave as an independent small system in FCS and FPR measurements. A useful diagnostic criterion for the presence of small closed systems can be provided by repeatedly photobleaching the same spot. Consider first a system which is large compared to the excitation spot in which a fraction ϕ of the fluorophores are mobile and $1 - \phi$ are immobile. Because of the large size of the system, each photobleaching pulse eliminates a negligible fraction of the total fluorophores. After the first photobleaching pulse $R_f = \phi$. Subsequent photobleaching pulses further reduce the number of immobile fluorophores in the observation spot, but do not significantly alter the quantity of mobile fluorophores that recover. Hence relative to the initial prebleach fluorescence the final fluorescence, $F(\infty)$, approaches a constant nonzero value as the immobile components are more and more completely eliminated. Correspondingly, the mobile fraction, calculated with respect to the prebleach fluorescence $F(<0)$ appropriate for each successive photobleach, increases with successive photobleaches toward unity. This is in contrast to the response in a small closed area in which all the fluorophores are mobile. Then both $F(<0)$ and $F(\infty)$ go to zero as the fluorophores in the area are depleted by successive photobleaches (cf. Angelides *et al.*, 1988).

B. Different Excitation Beam Profiles

The most common approach to FPR measurements uses the cylindrical Gaussian beam profile introduced above and integrates the intensity from the entire illuminated region. A variation of this approach uses a central Gaussian profile for photobleaching and then scans the monitoring beam to detect the local fluorescence at different positions in and around the photobleached region (Koppel, 1979). Another variation on this approach is, first, to photobleach a spot using a focussed beam and then to follow the fate of the spot using quantitative digital video image analysis (Kapitza *et al.*, 1985). These approaches have the advantage of allowing detection of anisotropic diffusion.

Another useful approach is to photobleach a periodic pattern onto the system by placing an appropriate mask in a real image plane of the microscope (Smith and McConnell, 1978). The periodic distribution of the regions of high and low fluorophore concentration simplifies the analysis of the recovery process by providing directly a Fourier decomposition of the fluorophore diffusion. A square two-dimensional periodic pattern is useful for detecting different rates of diffusion along the two orthogonal directions of the pattern (Smith *et al.*, 1979). The precision of the pattern photobleaching method has been extended by modulating the acquisition of the fluorescence signal. In one approach, for example, a periodic pattern of stripes is photobleached onto the system, and then a striped pattern of excitation intensity with the same periodicity is scanned across the system (Lanni and Ware, 1982). As the excitation pattern passes into and out of phase with the bleached pattern, the measured fluorescence intensity has an AC component with frequency determined by the spacing and scan velocity of the pattern and a DC level determined by the extent of photobleaching. For a system with a single diffusing component, the first order decay rate of the AC component yields the diffusion coefficient. A similar approach uses the interference of two crossed laser beams to produce the periodic excitation profile for photobleaching and scanning of fluorescence (Davoust *et al.*, 1982). A third adaptation of this approach has been developed to measure simultaneously the diffusion coefficients of rapidly and slowly diffusing components (Koppel and Sheetz, 1983).

One advantage of the pattern photobleaching approach is its simplification of the analysis due to the direct measurement of the Fourier components of diffusion (cf. Koppel, 1981). The pattern method also discriminates against drifts both of the position of the sample relative to the optical axis of the measuring system and of the DC levels of fluorescence as might occur due to inadvertent photobleaching during the monitoring phase of the measurement. Another advantage is that the pattern method measures fluorescence recovery from a wide area of the system and so tends to average out spatial heterogeneity of diffusion rates. In cells, however, heterogeneity of diffusion rates may result from variations in cell structure. Hence the ability of spot photobleaching to measure separately the diffusion rates from different regions of a cell can provide a correlation of diffusion with specific structural features (e.g., Axelrod *et al.,* 1976). Thus because of its experimental simplicity and, more important, its capability for spatial discrimination spot, photobleaching is still most often used in measurements on cells.

C. Video Methods

The ready availability of intensified video cameras that can acquire images of very weakly fluorescent specimens and of devices for rapid digitization, processing, and analysis of video images has provided a new dimension to photobleaching methods. The simplest approach has been to use a laser beam to photobleach a spot onto a system as a marker for the motion of a fluorescence labeled component. This, for example, has been used to observe (Wang, 1985; Felder, 1984) or to measure (Amato and Taylor, 1985; Kreis *et al.,* 1982) the systematic motion of labeled actin, microinjected into living cells and incorporated into microfilaments. Kapitza *et al.,* (1985) determined the distribution of fluorescence around a photobleached spot at different times after the photobleaching pulse in a study of the isotropy of diffusion of lipid probes in synthetic membranes and cell surface membranes. A further extension of this work could fit the observed fluorescence intensity contours to the expected distribution of fluorescence as a function of time and position calculated for systematic drift as well as diffusion (Elson, 1985). The adaptation of video methods to the acquisition and analysis of photobleaching data can provide a powerful supplement to the more conventional approaches. In addition to the possibilities for observing anisotropic transport, this approach allows discrimination against spatial drift of the system and provides an immediate correlation of the transport and dissipation of photobleached spots with microscopically observable cell structures.

D. Sensitivity of Results to the Mechanism of Photobleaching

The simplest analysis of FPR measurements supposes that each fluorescent component is eliminated from the system by an irreversible first order reaction, the rate of which is proportional to the photobleaching excitation intensity (Axelrod *et al.*, 1976). This model could be inadequate for several reasons. The photolysis reaction might not be a single step process which is first order in the fluorescent component. Even if the bleaching mechanism is more complex, however, as long as it occurs rapidly on the time scale of the diffusional recovery and the extent of bleaching is small, then there should be little effect on the interpretation of the experiments. For more complex mechanisms the bleach parameter κ will have different interpretations, but this parameter is determined empirically in practice and its interpretation in terms of photobleaching kinetics is rarely important. Any complex bleaching mechanism, including higher order processes, behaves as a first order reaction in the limit of low extent of bleaching. For larger extents of bleaching, corrections to the simple theory must be derived for each bleaching mechanism. Reversibility of the photolysis could be more serious if the fluorescence recovery due to this process overlapped the recovery due to the transport and reaction processes to be studied. The potential contributions of reversibility to the fluorescence recovery should be directly checked in the experimental system, for example, by determining whether fluorescence recovery can be detected from immobilized fluorophores. In principle, the kinetics of photobleaching might also be proportional to a higher power of the excitation intensity. But this is extremely unlikely at the photobleaching intensities commonly used. Moreover, this complication would affect only the interpretation of κ and so would be inconsequential.

A potentially more substantial complication could result from the failure to assume that the photobleaching reaction for each component of a multicomponent system is independent of the others. This assumption is certainly incorrect for systems in which several fluorophores are present on a single kinetic unit (i.e., diffusable component), as occurs in multiple binding reactions and in polymers. In these systems there can be important differences between FCS and FPR measurements which can be understood qualitatively as follows. When each kinetic unit bears only a single fluorophore, a single photolysis event renders that component nonfluorescent. When many fluorophores are present on the same macromolecule, they are likely to be photobleached independently of each other. Then each photobleaching reaction will reduce the fluorescence of the macromolecule by only a fraction of its initial value. Diffusion of the kinetic unit out of the observation region plays the same role in an FCS measurement that photolysis does in an FPR experiment. For singly labeled molecules, both diffusion and photobleaching eliminate the molecule from observation. In an FCS measurement, diffusion of a multiply labeled macromolecule out of the observation region also eliminates its entire complement of fluorophores, whereas photobleaching eliminates only a fraction. A consequence of this difference is that FCS measurements are much more sensitive than FPR measurements to fluorophore aggregation (see below). The effects of the photobleaching mechanism in general and for relatively simple multicomponent systems have been discussed (Elson, 1985). For a more complex system, the multiple binding of the fluorescent ligand, ethidium bromide, to DNA, experimental FCS and FPR measurements have been analyzed in detail in terms of a plausible model for the photobleaching mechanism (Icenogle and Elson, 1983a,b).

IV. Measurement of Aggregation of Fluorophores by FCS

Methods for measuring fluorophore aggregation state have been developed to take advantage of the sensitivity of FCS to this property. Although the rate of translational diffusion depends on the size of the diffusant, this dependence is relatively weak. Hence FCS and FPR measurements of diffusion do not provide a sensitive indicator of aggregation. The amplitude of the FCS correlation function, $G(0)$, provides a much more sensitive indicator (Weissman *et al.*, 1976; Magde *et al.*, 1978). It may readily be shown that, for a multicomponent system, $G(0) \propto \Sigma_i(\varepsilon_i q_i)^2 \overline{c}_i$ where ε_i, q_i, and $\overline{c}_i$ are the absorption coefficient, fluorescence quantum yield, and the equilibrium concentration of the i'th component, respectively (Magde *et al.*, 1978). Consider an aggregation or polymerization reaction, and suppose that the i'th component bears i fluorophores and, for simplicity, that the aggregation process does not affect their absorption or emission properties. The q_i is the same for all i: $q_i = q$, and $\varepsilon_i = i\varepsilon$, where q and ε are the quantum yield and absorption coefficient of the individual fluorophores. Suppose that the system has only two states, monomers and 100-mers. Then it is readily shown that aggregation of only 1% of the monomers would double $G(0)$ (Magde *et al.*, 1978). Similarly, dimerization of all of the fluorophores would double $G(0)$. Under favorable conditions a doubling of $G(0)$ should be readily observable. FPR measurements would be unable to detect either of these processes with typical signal-to-noise ratios.

This approach has been further developed and applied to characterize aggregation on cell surfaces by Petersen *et al.* (1986a,b). The fluorescence labeled cell is translated at uniform velocity under a focused laser beam and the fluorescence emitted from each illuminated spot is recorded and $G(0)$ is calculated. Because of the uniform scanning rate, the resulting temporal fluorescence autocorrelation is equivalent to a spatial fluorescence autocorrelation function. Hence the fluctuations in the concentration of fluorophores in different spots on the cell surface are obtained. These data then provide the weight average aggregation state of the fluorophores. Additional information about the distribution of aggregate sizes may be obtained from computation of higher order correlation functions (Palmer and Thompson, 1987; Qian and Elson, 1988c). Characterization of aggregation by FCS can be useful in studies of cell biological problems, for example, in measurinng the polymerization of cytoskeletal filaments. The measurements can be difficult, however, due to requirements for long periods of signal acquisition and sample stability.

V. Interaction Systems

Interactions of a molecule with immobile or slowly moving structures can retard its motion. Hence measurement of diffusion rates can help to characterize molecular interactions. As pointed out above, if an interaction causes a substantial change in fluorescence properties, then its kinetics might be observable directly in an FCS or FPR measurement (Magde *et al.*, 1972; Icenogle and Elson, 1983a,b). Even in the absence of this sort of specific fluorescence change, important information is available from the measurements (e.g., Elson and Reidler, 1979). This is most readily illustrated in terms of a simple bimolecular combination reaction of the form

$$A + B \underset{k_b}{\overset{k_f}{\rightleftharpoons}} C$$

where A is a slowly moving nonfluorescent macromolecule with diffusion coefficient D_A, B is a more rapidly diffusing fluorescent molecule with diffusion coefficient D_B so that $D_B \gg D_A$; and k_f and k_b are the forward and reverse rate constants. Since A is much larger than B, we shall also suppose for simplicity that $D_C = D_A$. The fluorophore diffuses freely as component B at a rate which is determined by its size and shape and the viscosity of the medium. The mobility of the fluorophore is, however, retarded by the interaction of B with A to form C. The retardation is greater the stronger the interaction (i.e., the larger the equilibrium constant $k = k_f/k_b = \overline{c}_C/(\overline{c}_A\overline{c}_B)$ and the smaller the D_C.

The relative rates of reaction and diffusion determine the effect of the interaction of A and B on the experimental measurements. If the reaction is fast, each B will be in both the free and bound states many times during the time for diffusional recovery ("fast exchange" limit) and so will appear to move at a single intermediate rate. If the reaction is slow, each B for the most part will be either free or bound during the entire diffusional recovery time and so there will be two components, one of high and one of low mobility with diffusion coefficients D_B and D_C, respectively ("slow exchange limit"). To assess this quantitatively, the appropriate characteristic time for diffusion is that for component B: $\tau_D(B) = w^2/4D_B$, while that for reaction is the inverse of the relaxation rate obtained from linearized chemical kinetics: $\tau_{chem} = [k_f(\overline{c}_A + \overline{c}_B) + k_b]^{-1}$. Therefore single-component behavior will appear when $\tau_{chem} \ll \tau_D(B)$; two components, one fast and one slow, will appear when $\tau_{chem} \gg \tau_D(B)$. When D_C is so small that there is negligible diffusional recovery of C during the measurement period, the fluorophore will appear to be effectively immobile. This rationalizes the interpretation of a nonrecoverable fraction of fluorescence in terms of an immobile fraction of fluorophores (as long as the system is large enough, as discussed above).

In the fast exchange limit the diffusion coefficient of the single effective component, D_e, is $D_e = X_B D_B + X_C D_C$, where X_B and X_C are the equilibrium mole fractions of B and C:

$$X_C = \overline{c}_C/(\overline{c}_B + \overline{c}_C) = K\overline{c}_A\,/(1 + K\overline{c}_A)$$

$$X_B = 1 - X_B = \overline{c}_B/(\overline{c}_B + \overline{c}_C) = 1/(1 + K\overline{c}_A)$$

Therefore, in the fast exchange limit, K can be determined by measurements of D_e at various values of $\bar{c}_A$. Similarly determining the variation of recovery rate with $\bar{c}_A$ is the most definitive way to test this model for the effect of interactions on diffusion measurements. Another approach to distinguishing the effects of diffusion and interaction is by their different dependence on w: $\tau_D(B) \approx w^{-2}$; τ_{chem} is independent of w(cf. Table I). Hence increasing w, e.g., by using a lower-magnification objective, could cause the system to pass from the slow to the fast exchange limit so that a fast and a slow component would become merged into a single diffusive component. Although this is a powerful approach for analyzing mechanism in principle, it is limited in practice by the relatively narrow range over which w can be varied.

Applications of FCS and FPR even to simple reaction systems have been infrequent. Studies of complex reactions have been even rarer. One system, the multiple binding of ethidium bromide to DNA, has been studied in some detail (Icenogle and Elson, 1983a,b) and illustrates the complexity of the analysis required. The multiple binding system is formally similar to a polymerization reaction. Hence, the studies of the ethidium–DNA interaction provide a prototype for an analysis of the polymerization of cytoskeletal filaments. Application of FCS as discussed above to characterize the weight average and possibly even the distribution of the degrees of polymerization could also provide important information.

VI. Summary

The theoretical basis and experimental implementation of FCS and FPR measurements are now well established. Because of the requirements for system stability and long data acquisition times FCS is relatively rarely used. But FCS can provide unique information, especially about extents of aggregation or polymerization and therefore is a useful supplement to FPR for certain applications. FPR measurements are now carried out routinely in many laboratories in a variety of formats using different beam profiles, optical systems, and analytical schemes. A particular version may be better adapted to a specific application. The spot photobleaching approach, however, seems simplest and most versatile for cellular studies and is now most often used. Important experimental considerations in setting up a spot photobleaching instrument are discussed in detail in Chapter 10 by Wolf (this volume) and elsewhere (Petersen *et al.*, 1986a).

In interpreting FPR measurements it is also important to take into account the possibility of systematic errors from a number of sources. In

Chapter 10 in this volume, Wolf discusses many factors that must be properly controlled in carrying out FPR measurements. Additional consideration of some of these points is presented by Petersen *et al.* (1986a). One potentially troublesome type of error arises from the possibility that chemical reactions initiated by the photobleaching pulse or during the measurement of recovery could significantly perturb the system. Evidence from a variety of sources [summarized, for example, in Petersen *et al.* (1986a)] indicates that photobleaching fluorophores can induce chemical cross-linking of cellular proteins under some conditions. But measurements in a number of different systems have domonstrated that, even if these types of reactions occur in FPR measurements, nevertheless they do not perturb the measured mobilities. If possible, however, this point should be checked for each new system because variations in structure or environmental conditions could enhance the chemical cross-linking reactions mediated by photogenerated free radicals.

In practice, the principal difficulty in carryig out FPR measurements on cells is frequently the low intensity of the fluorescent signal which can be obtained from specifically labeled cell surface ligands or microinjected components. This low intensity results from the typically low capacity of an individual cell for the specifically labeled macromolecule. Even in the absence of systematic errors, low emission intensity will reduce the accuracy of measurements due to shot noise. This is an important practical limitation on measuring accuracy. Low measurement accuracy severely limits the extent to which the data can be interpreted mechanistically. Precision can be improved by averaging many recovery experiments. Measurements of mobile fraction, however, are valid only for the first bleach of a particular spot in the system. Furthermore, averages of single measurements from different spots on the same or different cells are distorted by regional and cellular variability (cf. Petersen *et al.,* 1986a). One way to improve accuracy would be to associate more fluorophores with each ligand or microinjected component as in the development of phycobiliprotein labels (Glazer and Stryer, 1984). Alternatively, the emission intensity is limited not only by the number of fluorophores in the observed region, but also by the intensity of the excitation beam. The emission intensity could therefore be increased in proportion to the excitation intensity. A severe limit on this, however, is the requirement that there be little or no photobleaching during the measurement of recovery. Hence, in many instances, it would be advantageous to have more photostable fluorophores so that the excitation (and therefore emission) intensity could be increased without unwanted photobleaching. It can be expected that the laser will have sufficient intensity to photobleach even these more stable fluorophores when desired. This development eventually will be limited by the ability of the cell to withstand the higher intensities used for photobleaching and monitoring fluorescence. Hence, in many instances, there are probably fundamental limitations of the ultimate accuracy of FPR and FCS measurements on living cells. It seems likely, however, that further improvements in fluorophores stability could substantially enhance measurement accuracy.

Acknowledgments

Recent work on FCS and FPR carried out in the authors' laboratory was supported by NIH grants GM 38838 and GM 30299.

References

Amato, P. A., and Taylor, D. L. (1985). *J. Cell Biol.* **103,** 1074–1084.
Angelides, K. J., Elmer, L. W., Loftus, D., and Elson, E. (1988). *J. Cell Biol.* **106,** 1911–1925.
Axelrod, D. (1983) *J. Membr. Biol.* **75,** 1–10.
Axelrod, D., Koppel, D. E., Schlessinger, J., Elson, E., and Webb, W. W. (1976). *Biophys. J.* **16,** 1055–1069.
Devoust, J., Devaux, P. F., and Leger, L. (1982). *EMBO J.* **1,** 1233–1238.
Edidin, M., Zagyansky, Y., and Lardner, T. J. (1976). *Science* **191,** 466–468.
Elson, E. L. (1985). *Annu. Rev. Phys. Chem.* **36,** 379–406.
Elson, E. L. (1986). *In* "Optical Methods in Cell Physiology" (P. De Weer and B. M. Salzberg, eds.), pp. 367–383. Wiley (Interscience), New York.
Elson, E. L., and Magde, D. (1974). *Biopolymers* **13,** 1–27.
Elson, E. L., and Reidler, J. A. (1979). *J. Supramol. Struct.* **12,** 61–67.
Elson, E. L., and Webb, W. W. (1975). *Annu. Rev. Biophys. Bioeng.* **4,** 311–334.
Felder, S. (1984). Ph. D. thesis submitted to Washington University, St. Louis, MO.
Glazer, A. N., and Stryer, L. (1984). *Trends Biochem. Sci.* **9,** 423–427.
Hackenbrock, C. R., Chazotte, B., and Gupte, S. S. (1986). *J. Bioenerg.* **18,** 331–368.
Icenogle, R. D., and Elson, E. L. (1983a). *Biopolymers* **22,** 1919–1948.
Icenogle, R. D., and Elson, E. L. (1983b). *Biopolymers* **22,** 1949–1966.
Jacobson, K., and Wojcieszyn, J. (1984). *Proc. Natl. Acad. Sci. U.S.A.* **81,** 6747–6751.
Jacobson, K., Derzko, Z., Wu, E.-S., Hou, Y., and Poste, G. (1976). *J. Supramol. Struct.* **5,** 565–576.
Kapitza, H. G., McGregor, G., and Jacobson, K. A. (1985). *Proc. Natl. Acad. Sci. U.S.A.* **82,** 4122–4126.
Keizer, J., Ramirez, J., and Peacock-Lopez, E. (1985). *Biophys. J.* **47,** 79–88.
Koppel, D. E. (1979). *Biophys. J.* **28,** 281–292.
Koppel, D. E. (1981). *J. Supramol. Struct.* **17,** 61–67.
Koppel, D. E., and Sheetz, M. P. (1983). *Biophys. J.* **43,** 175–181.
Koppel, D. E., Axelrod, D., Schlessinger, J., Elson, E. L., and Webb, W. W. (1976). *Biophys, J.* **16,** 1315–1329.
Kreis, T. E., Geiger, B., and Schlessinger, J. (1982). *Cell* **29,** 835–845.
Lanni, F., and Ware, B. R. (1982). *Rev. Sci. Instrum.* **53,** 905–908.
Luby-Phelps, K., Taylor, D. L., and Lanni, F. (1986). *J. Cell Biol.* **102,** 2015–2022.
McCloskey, M. A., and Poo, M.-M. (1986). *J. Cell Biol.* **102,** 88–96.
Magde, D., Elson, E. L., and Webb, W. W. (1972). *Phys. Rev. Lett.* **29,** 705–708.
Magde, D., Elson, E. L., and Webb, W. W. (1974). *Biopolymers* **13,** 29–61.

Magde, D., Webb, W. W., and Elson, E. L. (1978). *Biopolymers* **17,** 361–376.
Palmer, A. G., and Thompson, N. L. (1987). *Biophys. J.* **52,** 257–270.
Peters, R. (1981). *Cell Biol. Int. Rep.* **5,** 733–760.
Petersen. N. O. (1986). *Biophys. J.* **49,** 809–815.
Petersen, N. O., Felder, S., and Elson, E. L. (1986a). *Handb. Exp. Immunol.* **1,** 24.1–24.23.
Petersen, N. O., Johnson, D. C., and Schlesinger, M. J. (1986b). *Biophys. J.* **49,** 817–820.
Poo, M.-M. (1982). *Nature (London)* **295,** 332–334.
Qian, H., and Elson, E. L. (1988a). *J. Cell Biol.* **106,** 1921–1923.
Qian, H., and Elson, E. L. (1988b). *Proc.* SPIE **909,** in press.
Qian, H., and Elson, E. L. (1988c). In preparation.
Schlessinger, J. (1983). *Biopolymers* **22,** 347–353.
Schneider, M. B., and Webb, W. W. (1981). *App. Opt.* **20.** 1382–1388.
Smith, B. A., and McConnell, H. M. (1978). *Proc. Natl. Acad. Sci. U.S.A.* **75,** 2759–2763.
Smith, B. A., Clark, W. R., and McConnell, H. M. (1979). *Proc. Natl. Acad. Sci. U.S.A.* **75,** 5641–5644.
Van Zoelen, E. J. J., Tertoolen, L. G. J., and De Laat, S. W. (1983). *Biophys. J.* **42,** 103–108.
Wang, Y.-L. (1985). *J. Cell Biol.* **101,** 597–602.
Weissman, M., Schindler, H., and Feher, G. (1976). *Proc. Natl. Acad. Sci. U.S.A.* **73,** 2776–2780.
Yguerabide, J., Schmidt, J. A., and Yguerabide, E. E. (1982). *Biophys. J.* **40,** 69–75.

Chapter 12

Fluorescence Polarization Microscopy

DANIEL AXELROD

Biophysics Research Division and Department of Physics
University of Michigan
Ann Arbor, Michigan 48109

I. Introduction

Fluorescence polarization techniques have long been used to measure orientational distributions and rotational diffusion of molecules in solution or suspension. Most frequently, a commercial spectrofluorimeter is used for such studies. Fluorescence polarization through a microscope offers the possibility of analogous studies on the surface or interior of very small subvolumes of cells. However, fluorescence polarization with a microscope presents some unique experimental problems and opportunities not generally present with a spectrofluorimeter. This chapter emphasizes those special features, first in theory and then followed by practical suggestions.

The chief experimental problem is that the microscope objective often has a rather high aperture and can thereby "look around" the sample, effectively integrating fluorescence over a range of emission solid angles. A second problem is that the ideal perfectly polarized excitation illumination may suffer some depolarization as it is concentrated by that same objective (in the standard epiillumination system to which we limit the discussion). A third problem is that the nearby substrate (often a glass coverslip or tissue culture polystyrene) may influence the polarization of rays emitted at high incidence angles but still captured by the objective.

An important opportunity presented by the microscope is that highly oriented samples, such as elongated fluorophores aligned with the acyl chains of membrane phospholipids, can be viewed easily. By contrast, a typical sample in a spectrofluorimeter is randomly oriented. This chapter emphasizes the polarization from fluorophores in membranes oriented flat on the microscope stage. Another opportunity is that the excitation light can be concentrated to a degree sufficient to remove a substantial portion of the fluorophores from the ground state, leaving behind an orientation-dependent population of fluorophores, the time dependence of which reflects the rotational rate of the fluorophore. Also, of course, the microscope allows study of fluorescence polarization from very small areas or volumes.

Except as noted, this chapter is confined to discussions of quantitative steady-state fluorescence, i.e., fluorescence changes much slower than the fluorescence lifetime. Fluorescence lifetime measurements, either by time-resolved or phase-modulation methods, are certainly possible through a microscope, and many of the experimental considerations from steady-state configurations will still apply. Standard epiillumination is assumed, either from an ionized gas arc or a continuous-wave laser. We also neglect effects on polarization of lens aberrations and angle-dependent transmission of rays through microscope optics. The first effect is difficult to discuss in general; the second effect is somewhat ameliorated by lens coatings, which tend to increase transmission and reduce reflections.

II. Theory

This section discusses special features encountered in fluorescence polarization as seen through a microscope: high-aperture objectives, oriented samples with nonzero angles between absorption and emission dipoles, molecular rotation in oriented samples, and methods for characterizing the orientational distribution.

A. Definitions

We first set up a coordinate system. The optical axis of the microscope is called the x-axis. The median direction of the collected emission propagates in the positive x-direction. Because of the cylindrical symmetry of the objective, the excitation polarization can be along any transverse direction (i.e., in the plane of the microscope stage). In practice, this polarization will be "front-to-back" or "side-to-side" as seen from the front of the microscope; whichever is chosen, the excitation polarization direction is called the z-axis. The remaining orthogonal direction in the microscope stage plane is the y-axis, with its sense selected to form a right-handed coordinate system.

A polarizer placed in the path of the emission light "downstream" from the objective and epiillumination dichroic mirror/barrier filter combination is oriented in either of two possible directions: along z ($\|$ to the excitation polarization) or along y ($\perp$ to the excitation polarization). The observed fluorescence in each of these two modes is denoted by $\mathcal{F}_{\|}$ and $\mathcal{F}_{\perp}$. These two measurements will in general differ even in a randomly oriented sample because the polarized incident light selects an anisotropic distribution of fluorophores for excitation. However, $\mathcal{F}_{\|}$ and $\mathcal{F}_{\perp}$ tend to become equal if the excited fluorophores rotationally diffuse very rapidly before they emit.

The fluorescence anisotropy r, the conventional variable describing fluorescence polarization, is defined as

$$r = \frac{\mathcal{F}_{\|} - \mathcal{F}_{\perp}}{\mathcal{F}_{\|} + 2\mathcal{F}_{\perp}} \tag{1}$$

The peculiar normalization represented by the denominator is chosen because, for the common three-dimensional case of fluorophores symmetrically distributed around the z-axis and small observation apertures, it is proportional to the number of excited fluorophores at any given time (and thereby, to the total emitted fluorescence) and is invariant as these symmetrically distributed fluorophores undergo isotropic rotational diffusion. (Because of this normalization feature, the time dependence of the decay of anisotropy from a bulk solution or suspension after a flash excitation contains only terms involving the rotational diffusion coefficients and not the fluorescence lifetime.)

Since rotation during the fluorescence lifetime disorders the distribution of fluorophores, the r of a homogeneous sample is highest immediately after a flash excitation, or in steady state, in samples with zero rotational motion. The maximum r in these circumstance is called r_0.

The numerical aperture A of a microscope objective is defined as:

$$A \equiv n \sin \sigma \tag{2}$$

where n is the index of refraction of the medium in which the sample is imbedded, and σ is the cone half-angle subtended by the objective as viewed from the sample. By Snell's law, in which $n \sin \sigma$ is unaltered as a ray passes through a planar interface such as a microscope coverslip, n can alternatively represent the index of refraction of the objective's immersion medium and σ the angle between the optical axis and the most extreme ray that enters the objective. Given the numerical aperture of the objective and the intended immersion medium, σ can be readily calculated.

At any given time t, the distribution of *excited* fluorophores will be a function $\rho(\Omega, t)$ of a set of orientational angles symbolically denoted by Ω.

B. High-Aperture Observation of Polarized Emission

In a small aperture limit, $\mathcal{F}_{\parallel}$ is proportional to the square of the projection that each emission dipole casts on the z-axis only, integrated over the excited emission dipole distribution $\rho(\Omega, t)$ in the sample. Similarly, $\mathcal{F}_{\perp}$ depends on the projection on the y-axis only. With a finite aperture, however, both the $\parallel$ and $\perp$ modes can "see" the projections of the emission dipoles on each of the other two axes. The total fluorescence $\mathcal{F}_{\parallel,\perp}$ then becomes

$$\mathcal{F}_{\perp} = K_a F_x + K_c F_y + K_b F_z \tag{3}$$

$$\mathcal{F}_{\parallel} = K_a F_x + K_b F_y + K_c F_z \tag{4}$$

where $F_{x,y,z}$ are the fluorescence intensities that would be seen through a polarizer oriented along x, y, or z, respectively, by a small aperture detector positioned on any axis orthogonal to the particular polarizer orientation, and $K_{a,b,c}$ are weighting factors that depend on the actual numerical aperture (Axelrod, 1979).

$F_{x,y,z}$ can be written in general form as follows (with multiplicative constant factors involving absorption, emission, and collection efficiencies suppressed):

$$F_x(t) = \int (\hat{\mu} \cdot \hat{x})^2 \rho(\Omega, t)\, d\Omega \tag{5}$$

$$F_y(t) = \int (\hat{\mu} \cdot \hat{y})^2 \rho(\Omega, t)\, d\Omega \tag{6}$$

$$F_z(t) = \int (\hat{\mu} \cdot \hat{z})^2 \rho(\Omega, t)\, d\Omega \tag{7}$$

where $\hat{\mu}$ is a unit vector in the direction of the emission dipole moment and $\hat{x}$, $\hat{y}$, or $\hat{z}$ is a unit vector along the x, y, or z-axes.

$K_{a,b,c}$ can be written in exact form as follows:

$$K_a = (1/6)(2 - 3\cos\sigma + \cos^3\sigma)(1 - \cos\sigma)^{-1} \tag{8}$$

$$K_b = (1/24)(1 - 3\cos\sigma + 3\cos^2\sigma - \cos^3\sigma)(1 - \cos\sigma)^{-1} \tag{9}$$

$$K_c = (1/8)(5 - 3\cos\sigma - \cos^2\sigma - \cos^3\sigma)(1 - \cos\sigma)^{-1} \tag{10}$$

These parameters are normalized so that as we approach the small aperture limit $\sigma \rightarrow 0$, then $K_c \rightarrow 1$ and $K_{a,b} \rightarrow 0$. Nonzero factors K_a and K_b lead to the mixing in of fluorescence intensities from dipole moment components not normally observed at small aperture. For small σ, $K_c \gg K_{a,b}$ as expected. Note that the mixing of $F_{x,y,z}$ described in Eqs. (3) and (4) implies that $\mathcal{F}_{\parallel} + 2\mathcal{F}_{\perp}$ in the denominator of r is no longer the total emitted fluorescence. Certain convenient features of r, such as its above-mentioned independence from the fluorescence lifetime in time-resolved experiments, are thereby abolished by high apertures.

Since $K_{a,b,c}$ depend only on σ, the high-aperture effects are here described as functions of ratio A/n [see Eq. (2)].

By combining Eqs (1) and (3), anisotropy r can be rewritten:

$$r = \frac{(F_z - F_y)(K_c - K_b)}{(F_z + 2F_y)(K_c + K_b) + (F_z - F_y)K_c + 3F_xK_a} \tag{11}$$

Evaluation of Eq. (11) clearly depends on knowing not just the numerical aperture but also the actual orientation distribution of the sample (so that the integrals for $F_{x,y,z}$ may be performed). The effect of high apertures also depends in general on the intramolecular angle between absorption and emission dipoles and on the amount of diffusional rotation of the fluorophore during the lifetime of the excited state, since both of these parameters affect the orientational distribution of excited fluorophores. In general, the correction factor needed in r for numerical aperture is not constant, but depends on the exact distributional case. The following sections discuss some distributions likely to be encountered.

1. Three-Dimensional Isotropic Frozen Samples: Anisotrophy versus Angle Between Dipoles

In three-dimensional frozen isotropic orientation distributions, the expressions for $F_{x,y,z}$ (with all common factors suppressed) are

$$F_x = 1 + \sin^2 \alpha \tag{12}$$

$$F_y = 1 + \sin^2 \alpha \tag{13}$$

$$F_z = 3 - 2 \sin^2 \alpha \tag{14}$$

When these are installed into Eq. (11), one can generate curves for r_0 versus numerical aperture and versus angle α as displayed in Fig. 1. Again, the effect of high aperture on the r_0 in samples with nonzero angle α is not described by a simple multiplicative factor.

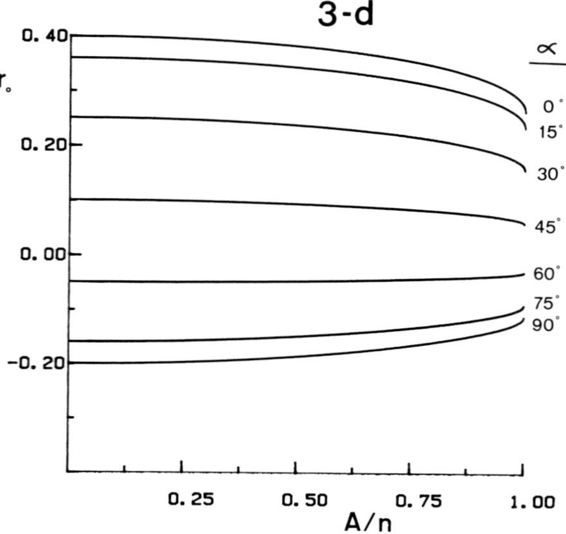

FIG. 1. Fluorescence anisotropy r_0 versus A/n for three-dimensional frozen isotropic distributions with various angles α between the absorption and emission dipoles. The r_0 equals zero for all apertures at the "magic angle" α where $\cos^2 \alpha = 1/3$.

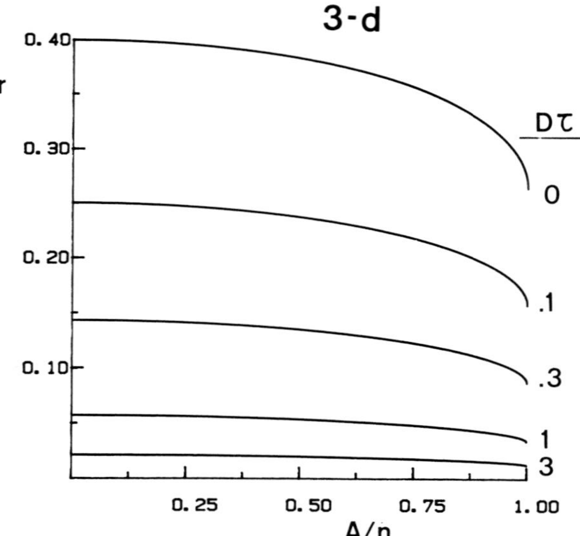

FIG. 2. Fluorescence anisotropy r versus A/n for three-dimensional isotropic distributions with various rotation rates $D\tau$ and fixed angle $\alpha = 0$.

2. Three-Dimensional Isotropic Diffusing Samples: Anisotrophy versus Diffusion Coefficient

Here we assume that angle $\alpha = 0$, but that the three-dimensional isotropic sample undergoes three-dimensional isotropic rotational diffusion with coefficient D and fluorescence lifetime τ. $F_{x,y,z}$ (with common factors suppressed) are

$$F_x = 1 - \frac{2}{5}(6D\tau + 1)^{-1} \tag{15}$$

$$F_y = 1 - \frac{2}{5}(6D\tau + 1)^{-1} \tag{16}$$

$$F_z = 1 + \frac{4}{5}(6D\tau + 1)^{-1} \tag{17}$$

Figure 2 shows r versus numerical aperture for various products $D\tau$.

3. Two-Dimensional Oriented Samples: Anisotrophy versus Angle of Emission Dipole

A fluorophore on a membrane (or on a protein in a membrane) has only one symmetry feature in general: an azimuthally isotropic distribution around the membrane normal. Otherwise, several other parameters affecting r could take on any physically meaningful value: the three independent angles describing the relative orientations among the absorption dipole, the emission dipole, and the plane of the membrane; the orientation of the membrane itself with respect to the optical axis; and the ratio of the characteristic rotation time to the fluorescence lifetime. A complete and general set of solutions is presented in Axelrod (1979) for $F_{x,y,z}$ (there denoted as J_1, J_2, J_3).

Here we show the effect of numerical aperture on a special common case, a reasonable model of how diacyl carbocyanine dyes situate in the membrane: the membrane oriented with its normal along the optical axis (i.e., the membrane lying in the plane of the microscope stage, the y-z plane); the absorption dipole at any angle with respect to the membrane surface; the emission dipole at some variable angle θ with respect to the plane of the membrane and in the plane formed by the normal to the membrane and the absorption dipole; and with a variable ratio of characteristic rotation time (the reciprocal of the rotational diffusion coefficient D) to the fluorescence lifetime τ. In this case,

$$F_x = \sin^2 \theta \tag{18}$$

$$F_y = \frac{1}{2} \cos^2 \theta \left[1 - \frac{1}{2(4D\tau + 1)} \right] \tag{19}$$

$$F_z = \frac{1}{2} \cos^2 \theta \left[1 + \frac{1}{2(4D\tau + 1)} \right] \tag{20}$$

Figure 3 displays results for r_0 versus numerical aperture for various angles θ.

In general, the more the fluorophore dipoles point along the optical axis (the x-axis), the stronger the effect of finite apertures. For example, 1,6-diphenylhexatriene (DPH), a commonly used membrane probe with a dipole approximately transverse to the membrane (and thereby along the x-axis for a membrane in the plane of the microscope stage), the correction is large even for the lowest apertures. Conversely, for diacyl carbocyanine dyes, another popular class of membrane probes with dipoles approximately parallel to the membrane, even high-aperture objectives can be used with little need for correction. In fact, of all the distributions (two- or

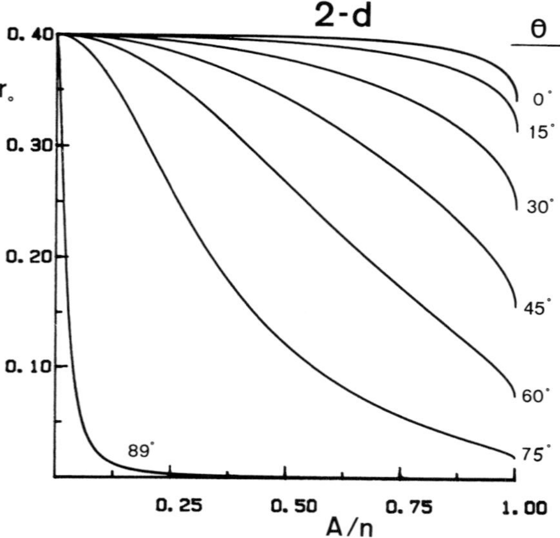

FIG. 3. Fluorescence anisotropy r_0 versus A/n for two-dimensional frozen isotropic distributions with emission dipoles out of the plane by the angles θ shown. Note that the cases with the largest out-of-plane angles are extremely sensitive to numerical aperture.

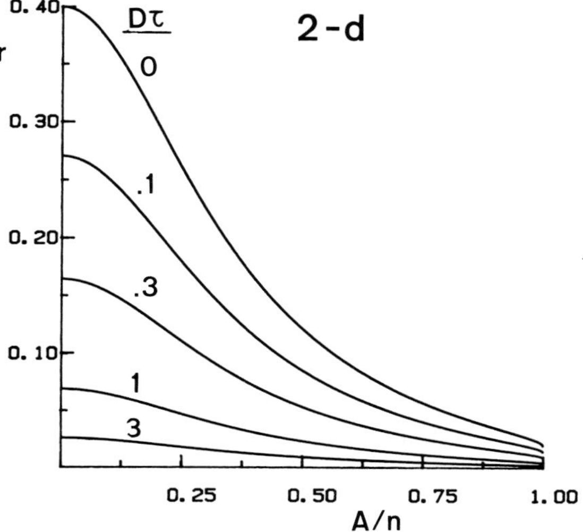

FIG. 4. Fluorescence anisotropy r versus A/n for two-dimensional isotropic distributions undergoing various rates of rotational diffusion $D\tau$, for angle $\theta = 75°$.

three-dimensional) considered here, that particular one shows the least numerical aperture dependence, because it has the smallest F_x.

Clearly, one cannot simply multiply a measured anisotropy by some correction factor in all cases to recover the small aperture value for r, because the correction factor depends on the angular distribution of fluorophores.

4. Two-Dimensional Oriented Samples: Anisotrophy versus Diffusion Coefficient

Figure 4 shows the effect of varying rotation rate $D\tau$ on the numerical aperture dependence of r, where Eqs. (18)–(20) are used with angle θ chosen equal to 75°.

C. High Angle of Convergence Polarized Excitation

To this point, we have discussed the emission polarization *assuming* that the excitation is perfectly polarized. However, since the last optical element in the excitation path is not a polarizer operating on a plane wave, but rather, a high-aperture objective with a high angle of ray convergence, the purity of the polarization is not necessarily maintained.

For the general case where the excitation beam incident on the back of the objective is a coherent, collimated, purely polarized laser beam with a Gaussian intensity profile (i.e., the structure of the standard TEM_{00} mode), the complete solution for the electric field vectors near the focus is given by Yoshida and Asakura (1974). The polarization near the focus is a function of the lateral position from the optical axis, the longitudinal distance from the focal plane, the angle of convergence of the laser beam rays, and the degree to which the incident beam is "cropped" by the objective's aperture. Here, we discuss a case of common interest: the polarization exactly in the focal plane.

We use the same coordinate system as in the above sections, in which x is along the optical axis and z is along the direction of the laser beam's pure polarization just before it enters the objective. (Note that Yoshida and Asakura's system uses these two axes with their roles reversed from those here.) The complex vectorial electric field $\mathbf{E}(r)$ in the plane of the microscope stage at the focal plane is

$$\mathbf{E}(r) = (E_x\hat{x} + E_y\hat{y} + E_z\hat{z})\, e^{-i\omega t} \tag{21}$$

with x, y and z components,

$$E_x(r) = 2AI_1(r) \cos \phi \tag{22}$$

$$E_y(r) = -iAI_2(r) \sin 2\phi \tag{23}$$

$$E_z(r) = -iA[I_0(r) + I_2(r) \cos 2\phi] \tag{24}$$

where (r, ϕ) are the angular coordinates of the observation point in the y-z focal plane (with ϕ measured from the polarization direction, the z-axis) and $I_{0,1,2}(r)$ are real integrals over Bessel functions of the first kind J_n as follows:

$$I_0(r) = \int_0^\sigma \exp(-\sin^2 \theta/\tan^2 \gamma) \cos^{1/2} \theta \sin \theta\, (1 + \cos \theta) J_0(kr \sin \theta)\, d\theta \tag{25}$$

$$I_1(r) = \int_0^\sigma \exp(-\sin^2 \theta/\tan^2 \gamma) \cos^{1/2} \theta \sin^2 \theta\, J_1(kr \sin \theta)\, d\theta \tag{26}$$

$$I_2(r) = \int_0^\sigma \exp(-\sin^2 \theta/\tan^2 \gamma) \cos^{1/2} \theta \sin \theta\, (1 - \cos \theta) J_2(kr \sin \theta)\, d\theta \tag{27}$$

The constants used in Eqs. (22)–(27) are as follows: $k \equiv 2\pi/\lambda$, where $\lambda = \lambda_0/n$ is the actual wavelength in the immersion medium. $A \equiv (\pi f/\lambda)A_0$, where A_0 is the field amplitude at the center of the laser beam before it enters the objective, f is the focal length of the objective, and

$\tan \gamma \equiv w_0/f$ is the ratio of the e^{-1} radius w_0 of the incident laser beam electric field strength before it enters the objective to the focal length f. Angle γ is thereby the angle of convergence of the laser beam as it emerges from the objective, as followed by the ray following the $1/e^2$ intensity radius. This angle can be measured directly by measuring the spread of the beam at two positions along the optical axis, or by measuring w_0 directly and calculating f from the geometrical optics formulas: $(M + 1)/I = 1/f$ where I, the image distance, is the "tube length" (usually ~160 mm) and M is the magnification of the objective. σ is maximal cone half-angle subtended at the sample by the objective's numerical aperture.

The fact that I_1 and I_2 are nonzero leads to the depolarization. The relative amplitudes of the I integrals clearly depend on two variable: σ, which describes the cropping of the beam by a limited objective aperture; and γ, which describes the degree of ray convergence. Figure 5 schematically shows the off-optical axis polarization at selected positions in the focal plane. As can be deduced immediately from the phase-factor in

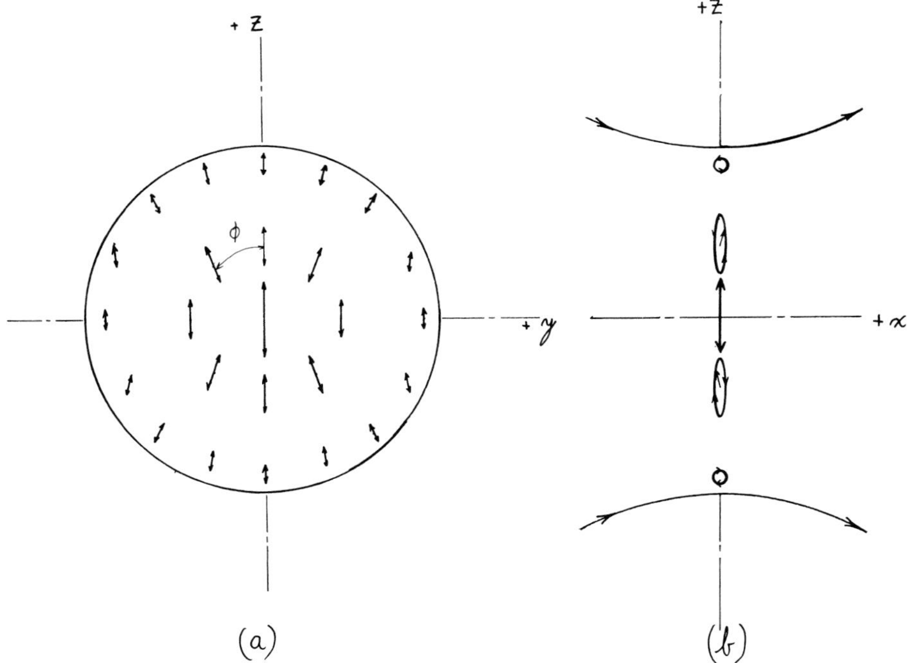

FIG. 5. Schematic drawing of the electric field vectors at selected points in the focal plane: (a) as projected into the (y-z) focal plane; and (b) as projected in the x-z plane. The components shown are not to scale: the $E_{x,y}$ are exaggerated for pictorial clarity.

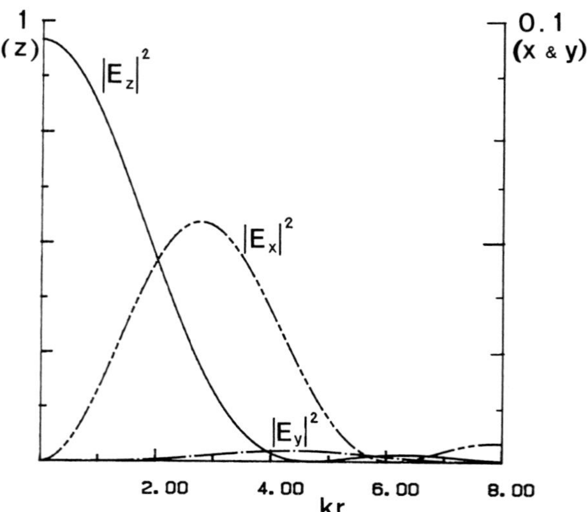

FIG. 6. The local intensities $|E_{x,y,z}|^2$ as a function of effective radius kr as observed along the $\phi = 45°$ diameter in the focal plane for a typical objective. Note that the scale of the x- and y-polarized powers has been expanded by a factor of 10 for pictorial clarity. Otherwise, the absolute units for intensity are arbitrary. For this calculation, the following optics are assumed: a NA = 1.25 objective, $n = 1.52$ immersion oil, a $1/e^2$ intensity convergence half-angle of 75°. At that convergence angle, considerable cropping of the Gaussian tails occurs.

Eqs (22)–(24) the polarization is linear but at an angle to the z-axis as projected into the focal plane, but elliptical as projected into the x-z plane or x-y plane. The only location in the beam profile that retains the original pure z-polarization is along the x-axis diameter line.

Figure 6 shows the local intensities $|E_{x,y,z}|^2$ as a function of r as observed along the $\phi = 45°$ diameter in the focal plane for a typical objective. In general, the depolarization arising from E_x (the component along the optical axis) is considerable larger than that from E_y.

How severe is the loss of pure z-polarization when averaged over the illuminated area of a focused beam? It is this average power that will give rise to the fluorescence excited by the whole focused spot. The average power $P_{x,y,z}$ is simply $|E_{x,y,z}|^2$, respectively, integrated over all r and ϕ in cylindrical coordinates. As can be seen from Eqs. (22)–(27), the average power depends strongly on the angle of convergence γ of the beam. Figure 7 shows how $P_{x,y,z}$ varies as a function of γ (which can be experimentally varied by adjusting w_0). Note that the average depolarization terms $P_{x,y}$ are negligible for small γ as expected, then rise rapidly, approaching an asymptote for larger angles.

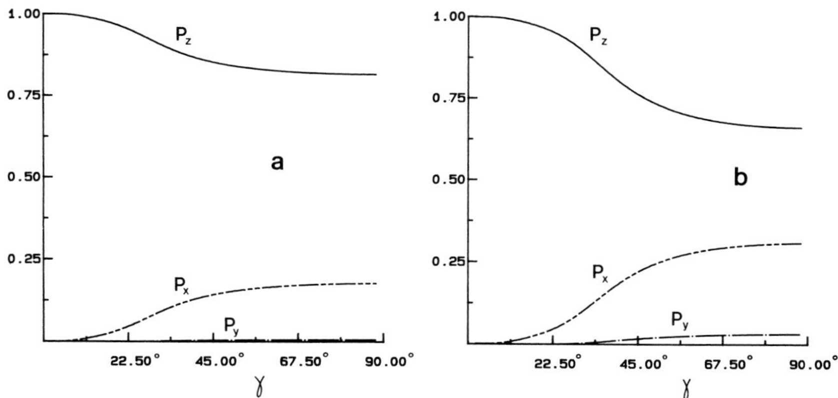

FIG. 7. The integrated intensity $P_{x,y,z}$ (i.e., the power) over the focal plane for each of the polarizations versus $1/e^2$ convergence angle γ. The ordinate scale is expressed as a fraction of the total laser power $P_x + P_y + P_z$. For this calculation, we assume two different aperture angles as depicted: (a) $\sigma = 55.3°$, the aperture angle for a NA = 1.25 objective in $n = 1.52$ immersion oil; and (b) $\sigma = 89.3°$. The latter σ is unrealistic but comparison of the curves illustrates how the cropping effect in (a) reduces the depolarization for large w_0 and γ. Both sets of curves are almost superimposable for $\gamma < 25°$. As long as $\gamma \ll \sigma$, these curves are applicable to any objective lens.

The flattening of the curves in Fig. 7 at large γ angles (corresponding to large w_0) occurs where the interior walls of the objective (subtending an angle σ) tend to "crop" the tails of the incident beam. At very large γ, only the central portion of the incident Gaussian transmits through the objective. Diffraction rings develop as expected, but since the angular spread of the rays becomes limited by the aperture, the average polarization varies little with further increases in γ. The only effect of changing numerical aperture on the shape of the curve in Fig. 7 will be to alter the angle γ at which the flattening takes place. Fig. 7a and b compare the effects of two different apertures. It is clear from Fig. 7 that the depolarization produced even by moderately convergent focusing of the excitation light with any objective can significantly alter measured fluorescence emission polarizations. On the other hand, an illumination beam from each point in a conventional arc source is designed to be completely defocused at the sample (Kohler illumination) and, in that case, high angle of convergence depolarization effects may not be important.

D. Deducing Orientational Distributions

A measured polarization anisotropy ratio r, even after correction for the depolarizing effects of wide-angle excitation and emission, is not an unambiguous indicator of orientation distribution. However, because the sample's orientation can be changed easily and the excitation polarization can be rotated, several independent measurements of anisotropy (termed "linear dichroism" of fluorescence excitation) can be made on a single sample. That set of results, combined with any knowledge of symmetry in the sample, can be used to unambiguously deduce some features of the orientational distribution. A general treatment of this subject has been presented by Burghardt (1984), in which the orientational distribution is ultimately described by an expansion in a complete set of standard orthonormal angular functions. In practice, certain samples may require this general model-independent analysis, but many others have only a limited set of reasonable models for orientational distributions. The predictions of each for anisotropy r can then be tested and perhaps all but one reasonable model thereby excluded.

Figure 8 shows an example of how orientational distribution can be deduced by varying the relative orientation of the sample, the excitation polarization, and the emission polarizer. In that case, 3,3′-dioctade-

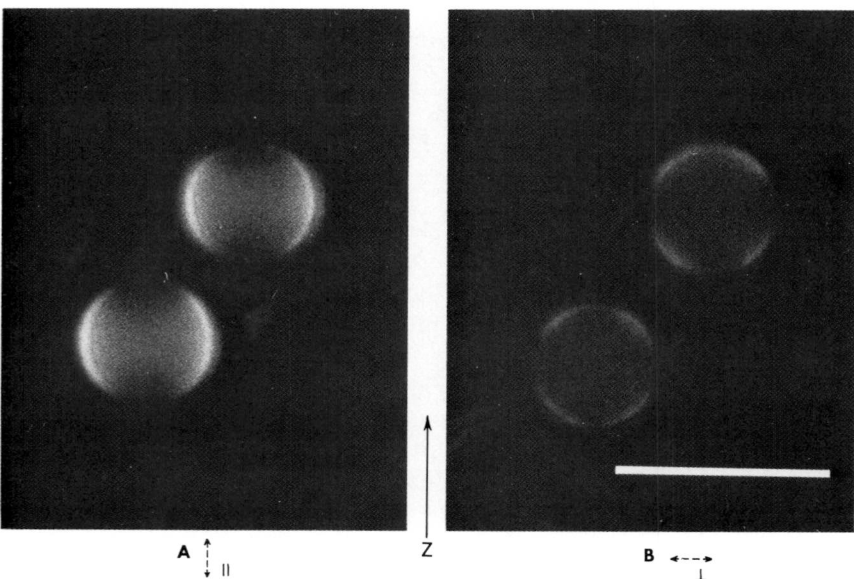

FIG. 8. DiI-labeled erythrocyte ghost, viewed by a 100 × 1.25 NA oil-immersion objective in a Leitz Diavert microscope under argon laser illumination, with the emission polarizer in the (A) $\parallel$ and (B) $\perp$ orientations and the excitation polarization along z as denoted by the arrow. Space bar = 10 μm.

cylindocarbocyanine (diI) labeling of an erythrocyte ghost, the sample orientation is varied simply by viewing different regions of the spherical surface. Even a qualitative interpretation of the photographs shows that the dipole moment must be oriented roughly parallel to the membrane; any other thermodynamically reasonable distribution will not produce the effect observed.

E. Polarized Fluorescence Photobleaching Recovery

In conventional nonmicroscopic fluorescence polarization spectroscopy, samples are often isotropic solutions or colloids with rapid rotational rates. By using time-resolved anisotropy techniques (either the decay of anisotropy after a picosecond or nanosecond excitation flash, or the phase relationship between a sine-wave-modulated excitation at tens or hundreds of megahertz and the resulting sine-wave fluorescence emission intensity), the rotational diffusion coefficient can be deduced. For an explanatory review of this subject, see Lakowicz (1985). These techniques basically measure how much the anisotropic orientational distribution of fluorophores excited by polarized light becomes more isotropic during the nanosecond-scale lifetime of the fluorescence.

In biological samples likely to be viewed under a microscope, the rotational rates are often somewhat slower, too slow to be seen during the fluorescence lifetime. One solution to this problem is to lengthen the lifetime by switching to phosphorescence instead of fluorescence. Phosphorescence, usually achieved by the extrinsic probe eosin isothiocyanate covalently attached to the cell component under study, has a lifetime on the order of hundreds of microseconds. Phosphorescence anisotropy decay requires the sample to be deoxygenated to lengthen the triplet state lifetime and increase its population; such deoxygenation may be incompatible with certain cell biological samples.

Another solution is to combine microscope polarization techniques with fluorescence photobleaching recovery (FPR) techniques (see Chapter 10 by Wolf, and Chapter 11 by Elson and Qian, this volume). In the combination, called "polarized FPR," a brief flash of polarized laser light creates an anisotropic distribution of unbleached fluorophores which relaxes by rotational diffusion, leading to a time-dependent postbleach fluorescence (Velez and Axelrod, 1988). Since it does not depend on excited triplet states, polarized FPR is applicable to any time scale from seconds to microseconds, and does not require sample deoxygenation. The optical theory and practice of polarized FPR is very similar to that of "polarized fluorescence depletion" (Johnson and Garland, 1982; Yoshida and Barisas, 1986). In fact, the only essential difference is that polarized fluorescence depletion uses deoxygenated samples to enhance temporary trapping of excited molecules in the triplet state, which is a special form of photobleaching.

Polarized FPR does contain one complication. At fast (microsecond) time scales, a partial postbleach recovery independent of molecular orientation is observed which tends to obscure rotational effects. The effect of this "reversible photobleaching" can be overcome by performing two successive experiments on the same sample: one in which the bleach ("pump") pulse is polarized parallel to the observation ("probe") beam; and the other in which the pump and probe beams are orthogonal. The fluorescence in these two modes, as observed through a polarizer oriented parallel to the probe beam polarization, are denoted by $F_{\parallel}(t)$ and $F_{\perp}(t)$, respectively. To eliminate the effect of reversible photobleaching, a sample ratio is taken:

$$R(t) = [F(-) - F_{\perp}(t)]/[F(-) - F_{\parallel}(t)] \tag{23}$$

where $F(-)$ is the prebleach fluorescence. A more familiar-looking ratio reminiscent of fluorescence anisotropy is $r_{\mathrm{b}}(t)$ (where the subscript stands for "bleaching"):

$$r_{\mathrm{b}}(t) \equiv \frac{1 - R(t)}{1 + 2R(t)} \tag{28}$$

Anisotropy $r_{\mathrm{b}}(t)$ is, in theory, a monotonically decreasing function of t, representing the loss of anisotropicity by rotational diffusion. Its exact form depends, of course, on the model chosen for rotational distribution and diffusion. Below we present the results for two common cases.

1. Two-Dimensional Diffusion around x-Axis

In this case, the fluorophores are free to rotate around an axis parallel to the optical axis of the microscope, e.g., as for a fluorophore-tagged component in a membrane lying in the plane of the microscope stage. Then (see Velez and Axelrod, 1988)

$$r_{\mathrm{b}}(t) = \frac{2b\, e^{-4Dt}}{3a - b\, e^{-4Dt} - 3c\, e^{-16Dt}} \tag{29}$$

where D is the rotational diffusion coefficient and a, b, and c are complicated functions of the duration and intensity of bleaching, the extent of possible fast wobble of the fluorophore around its covalent bond of attachment, and the mean angle between the fluorophore dipole and the membrane surface. In general, however, c is considerably smaller than b, so that to a fair approximation, Eq. (29) can be written in terms of a single exponential.

2. Three-Dimensional Diffusion

In this case, fluorophores are oriented isotropically and are free to rotate isotropically in any direction, e.g., as for liposomes or spherical proteins in solution or cells in suspension. Then

$$r_b(t) = \frac{3b\, e^{-6Dt} - \frac{5}{3} c\, e^{-20Dt}}{3a - \frac{14}{3} c\, e^{-20Dt}} \tag{30}$$

where a, b, and c are complicated functions of bleaching and wobble, but different from those in Eq. (29). Again, c is generally smaller than b, so that Eq. (30) can be written approximately in terms of a single exponential.

III. Experiment

Discussion of the optical setup for polarized microscopy can be conveniently divided between the excitation path and the emission path.

A. Excitation Path

For an arc-lamp source, a film polarizer is placed between the last lens of the source and the field diaphragm plane of the epiilluminator. Because the beam cannot be collimated without losing too much intensity, use of an arc lamp somewhat limits the manipulations available in the excitation path.

For a laser source (most often a continuous-wave argon laser of 3 W "all-lines" power), the excitation beam intensity and polarization can be manipulated by a variety of mechano- or electrooptical devices. For experiments involving polarized FPR, the beam intensity must be modulated. This is conveniently done with an acoustooptic modulator (AOM), a commercially available device that splits the incident laser beam into several diffraction orders. The relative intensity of the orders can be affected by an electrical signal (supplied by a logic circuit or a computer) such that the first order is ordinarily quite dim (for the "probe" beam) but can be made transiently brighter by at least 5000-fold (for the bleaching "pump" beam).

The laser beam itself is highly polarized (usually vertically) by internal Brewster windows in the laser cavity. However, the AOM introduces a slight depolarization. This may be eliminated by passing the first order from the AOM through a crystal polarizer. This is usually a commercially available Glan–Thompson prism rated to withstand the high intensity of a laser beam. (For this reason, film polarizers are not adequate.)

For experiments in which the excitation polarization must be switched rapidly (such as during the bleaching pulse in the "⊥ mode" of polarized FPR), the first-order beam is then passed through a quarter-wave plate, the axis of which is permanently rotated at 45° to the laser polarization, so that the beam emerges approximately circularly polarized (approximately, because the quarter-wave plate is designed for exact operation at a particular wavelength only). The circularly polarized light then enters a commercially available transverse Pockels cell. This electrooptical device can alter the relative phase of the horizontal and vertical components of incident polarization, based on the level of an applied high-voltage signal. This applied signal, generated by a fast DC amplifier driven by a logic circuit or computer, can be adjusted such that the emerging light can be varied quickly between vertical and horizontal polarizations.

The "purity" of the emerging beam's polarization is not high: approximately 10% of the intensity may be found in the "wrong" polarization, orthogonal to the one desired. This is because the Pockels cell does not polarize a laser beam uniformly across the beam's width. Although a diameter line in the direction of the desired polarization may be purely polarized, other regions of the beam cross-section emerge with a slight elliptical polarization. Nevertheless, the impurity may still be acceptable for some purposes.

After passing through any focusing lenses installed to control the size of the illuminated region at the field diaphragm (and hence the sample) plane, the beam reflects at the epiillumination dichroic mirror and passes through the objective. Since the beam has seen several optical elements which affect its intensity in a polarization-dependent manner (imparting an effective birefringence), the polarization should be checked as it would appear on the sample. This may be done by temporarily placing a photodiode (for example, a PIN photodiode with a flat front face) directly in the path of the laser beam as it diverges from the sample plane. (Be sure to use immersion fluid and a coverslip if so required by the particular objective.) A rotatable film polarizer should be temporarily positioned between the objective and the photodiode. By rotating the film polarizer, the actual purity of the final polarization may be determined.

If the excitation polarization is meant to be varied (say, by a Pockels cell as described) for use in linear dichroism or polarized FPR, it is essential to determine if the intensity and polarization purity in the two orthogonal directions of the microscope stage are identical. If the two orthogonal polarizations are not identical, then the brighter of the two polarizations should be reduced in intensity. This can be done easily by placing a glass microscope slide somewhere "upstream" in the laser path (before it enters the microscope), at an angle that will transmit different relative intensities of the vertical and horizontal beam intensities. (Recall that transmission and reflection at a dielectric interface is polarization dependent according to the Fresnel coefficients.) By experimentally choosing the proper direction and angle for the slide orientation, the polarization dependence of the final intensity at the sample may be made to disappear.

B. Emission Path

In an epiillumination system, the fluorescence emission passes through the sample substrate, the objective, the dichroic mirror, and a barrier filter before it is incident upon a film polarizer placed at a position fixed by the microscope manufacturer. The film polarizer mechanism may be a slider holding two filters for fluorescence in the two orthogonal polarizations $\mathcal{F}_{\parallel}$ and $\mathcal{F}_{\perp}$, or it may be a single rotatable filter. From that polarizer plane, the emission then traverses through several other mirrors and lenses, an image plane diaphragm (which allows observation of small selective regions on the sample and exclusion of the surroundings), and finally falls upon the photomultiplier photocathode.

This optical path likely introduces a birefringence (i.e., a greater transmittivity for one polarization over the orthogonal one) which must be measured in order to compensate for its effect on $\mathcal{F}_{\parallel}$ and $\mathcal{F}_{\perp}$. The intrinsic birefringence of the microscope can be measured directly using a light source that is completely unpolarized with the light directed along the optical axis toward the objective. A tungsten lamp behind a diffusing screen is a suitable source here. If the objective requires immersion fluid or coverslip, they should be used for this experiment. By interposing a rotatable film polarizer between the unpolarized source and the objective, the transmittivity through the microscope of the two orthogonal polarizations of equal intensity can be measured by the photomultiplier.

The sample substrate should be as birefringence free as possible. In general, glass or fused silica is better than plastic. If plastic (as in tissue culture dishes) must be used, the orientation of the dish and location under observation should be noted. After the fluorescence experiment is finished, the dish alone (with sample removed) should be checked for birefringence at the same orientation and location, by the method discussed in the preceding paragraph.

Care should be taken to ensure that the polarizing filter(s) does not induce a slight optical shift of the image when changing between $\parallel$ and $\perp$ polarizations. If any such a shift is noted by visual inspection of the images in the eyepiece against a graticule image, then the polarizers should be replaced by a better-matched set or the rotatable polarizer remounted. On a heterogeneous sample such as a biological cell, a slight relative shift can introduce a large error in fluorescence anisotropy measurements.

The photomultiplier itself should, in general, be as sensitive as possible. High sensitivity minimizes the amount of photobleaching that occurs while a statistically adequate reading is made. Photomultipliers with gallium–arsenide photocathodes are presently the most sensitive: two popular models for photon counting are the RCA C31034A and the Hamamatsu R946-02. High sensitivity often goes with high dark-count, so photomultiplier cooling is beneficial.

ACKNOWLEDGMENTS

This work was supported by USPHS NIH grant NS 14565.

REFERENCES

Axelrod, D. (1979). *Biophys. J.* **26,** 557–574.
Burghardt, T. P. (1984). *Biopolymers* **23,** 2383–2406.
Johnson, P., and Garland, P. B. (1982). *Biochem. J.* **203,** 313–321.
Lakowicz, J. R. (1985). "Principles of Fluorescence Spectroscopy." Plenum, New York.
Velez, M., and Axelrod. D. (1988). *Biophys. J.* **53,** 575–591.
Yoshida, A., and Asakura, T. (1974). *Optik* **41,** 281–292.
Yoshida, T. M., and Barisas, B. G. (1986). *Biophys. J.* **50,** 41–53.

Chapter 13

Fluorescence Microscopy in Three Dimensions

DAVID A. AGARD,* YASUSHI HIRAOKA,* PETER SHAW,† AND JOHN W. SEDAT*

**Howard Hughes Medical Institute and Department of Biochemistry*
University of California, San Francisco
San Francisco, California 94143

†John Innes Institute
Colney Lane, Norwich NR4 7UH, England
United Kingdom

I. Introduction

Light microscopy is unique in its ability to allow the examination of biological specimens in a hydrated state in living samples or under conditions that closely approximate the living state. As discussed elsewhere in this volume and in Volume 29, fluorescence microscopy adds to the fundamental power of light microscopy the ability to examine the spatial distribution of specific cellular components. The availability of fluorescently-labeled antibodies for protein localization, dyes such as Hoechst 33258 or DAPI to specifically label DNA (Arndt-Jovin and Jovin, Chapter 16, this volume) as well as pH- (Waggoner, 1986) and Ca^{2+}- (Grynkiewicz *et al.,* 1985; see Tsien, Chapter 5, this volume) dependent fluorophores allows details of cellular organization and function to be probed with a precision previously unknown.

Fundamental issues often concern the spatial relationships and interactions of cellular components. In this regard it must be remembered that cells are intrinsically three-dimensional structures. Unfortunately, the high-numerical aperture (NA) objective lenses required for high-resolution spatial analysis and optimal light gathering power have a very limited depth of focus (typically less than 0.4 μm for a 63× 1.4NA oil lens). This is too narrow to image an entire cell clearly, yet too wide to give good optical sectioning. Thus any image of the specimen is contaminated with out-of-focus information from focal planes above and below the current focus setting. Often this problem can be so severe as to preclude the desired analysis. Although the situation is somewhat better with interference microscopic methods such as differential interference contrast which can optically minimize out-of-focus contributions, the lack of specific labels renders these methods less useful.

Fortunately, the development of relatively inexpensive computers and digital image acquisition systems has now made possible the three-dimensional reconstruction of images taken from the optical microscope. Three-dimensional data are collected by recording a set of images taken at sequential focal planes throughout the specimen. By analogy with physical sectioning techniques, this method is called optical sectioning. Most of the common light microscopic imaging methods can be used with this technique. Each recorded image represents the sum of in-focus information from the focal plane and out-of-focus information from the remainder of the sample. Much of the out-of-focus information can be computationally removed. Out-of-focus removal can be accomplished implicitly as in confocal microscopic methods (see Chapter 14 by Brakenhoff *et al.,* this volume) or explicitly in optical section microscopy. For optical section microscopy, digital image processing methods are used to remove the out-of-focus contributions. In practice, both methods require computer manipulation of the data, and sufficient advancements have been made in the image processing that similar results can be obtained with minimal computational treatment.

In many cases in cell biology, it is not necessary to fully reconstruct the specimen in three dimensions. For relatively thin specimens or where only a representative view is desired, it may be sufficient to provide a single "cleaned-up" image. This can be accomplished directly via confocal microscopy or by collecting data from three image planes, one above and one below the desired focal plane and using the nearest-neighbor processing schemes described below. In other cases it may be desirable to examine the entire specimen either as an in-focus projection or as an in-focus stereo pair. With either confocal or conventional optics, this will require that the sample be optically sectioned and that all the data be integrated into the final mono or stereo image. For conventional microscopy, we call this synthesis of a projeted image (or stereo pair) from a set of individual images "synthetic projection microscopy."

The choice between approaches will ultimately depend on such issues as photobleaching, time resolution, choice of available excitation wavelengths, existing equipment, etc. As discussed below, the nature of three-dimensional imaging dictates that the generation of stereoscopic images or fully in-focus synthetic projections can be accomplished with considerably less photobleaching using conventional optical sectioning microscopy rather than using confocal methods. This may be critically important in the analysis of living cells. Using either approach, it is now possible to analyze cellular architecture of intact specimens in three dimensions.

In this chapter, we briefly discuss the theory of image formation within the microscope and the use of optical section microscopy coupled with image processing methods to generate in-focus projected images, stereo pairs, and full three-dimensional representations. Throughout, we will attempt to concentrate on practical issues such as data collection, computational methods for out-of-focus removal, and ways of displaying the results. Finally, we will consider the impact of technological advancements as they relate to image quality and real-time analysis.

II. Image Formation in Three Dimensions

Although one normally thinks of a light microscope as a two-dimensional imaging device, it can be used to record data from three-dimensional specimens. It is important to remember that in two dimensions and especially in three dimensions, all microscopic imaging systems are imperfect, hence they distort the image being observed in a characteristic manner. Although the high-quality, multielement objective lenses found in modern microscopes have been designed to minimize image distortion, residual chromatic and spherical aberrations and the finite numerical aperture all lead to errors in recording images from two-dimensional specimens (Inoué, 1986). Not surprisingly, because of limited depth of focus, three-dimensional specimens are more drastically affected.

Perhaps the most straightforward way to understand what happens in an optical system is to consider how an idealized point object would appear. The imperfect optical properties will cause the point to spread out and take on a characteristic shape which is called the Point Spread Function or PSF. In practice, the microscope's PSF is determined almost exclusively by the objective lens, making it the most critical imaging component in the microscope. Once the microscope's PSF is known, the way in which an arbitrary image is distorted can be predicted from the following imaging equation describing the convolution of the "true" image with the PSF:

$$o(x,y) = \sum\sum_{uv}\sum i(u,v)\, s(u - x, v - y) \tag{1}$$

where o is the observed image, i the "true" image and s the PSF or smearing function. For a general discussion of convolutions, see the relevant sections in Bracewell (1965) or Casteleman (1979). As considered previously (Agard, 1984; Castleman, 1979) this treatment can be readily extended to three dimensions:

$$o(x,y,z) = \sum\sum_{uvw}\sum i(u,v,w)\, s(u - x, v - y, w - z) \tag{2}$$

where the z index now refers to the focus direction. That is, at any point (x, y) in any focal plane z, the observed image is sum of contributions from the entire specimen volume weighted by the smearing function s. This three-dimensional PSF can be derived directly from a set of two-dimensional PSFs taken at different distances from the focal plane. Equation (2) can be written in a more compact form using $*$ to signify the convolution operation of Eqs. (1) and (2): $o = i * s$. It is often convenient to recast Eqs. (1) and (2) in terms of their Fourier transforms:

$$O(X,Y) = I(X,Y)\, S(X,Y)$$

and

$$O(X,Y,Z) = I(X,Y,Z)\, S(X,Y,Z) \tag{3}$$

where capitals are used to refer to the Fourier transforms and the convolution is now simplified as a multiplication. The Fourier transform of the PSF is called the CTF or contrast transfer function. Hopkins (1955),

Stokseth (1969), and Castleman (1979) have all described more-or-less approximate formulae for calculating the two-dimensional CTF of an objective lens as a function of focus $S(X, Y, \Delta z)$. Previously, we have used the approximation given by Castleman (Agard and Sedat, 1984; Gruenbaum *et al.,* 1984). The full three-dimensional CTF $S(X, Y, Z)$ can be calculated from the set of two-dimensional CTFs by Fourier transformation:

$$S(X, Y, Z) = \sum S(X, Y, \Delta z)\ e^{2\pi i \Delta z Z} \quad (4)$$

This Fourier transformation can be accomplished numerically (Agard, 1984) or analytically (Erhardt *et al.,* 1985).

Unfortunately, many assumptions of ideality go into the calculation of the two-dimensional CTFs. The inaccuracy of these assumptions becomes especially pronounced for high-numerical aperture (high-NA) lenses and becomes greatly exaggerated when calculating three-dimensionsal CTFs. Another approach to obtaining the three-dimensional CTF is by direct experimental measurement using very small ($<0.1\ \mu$m diameter, smaller than the resolution limit) fluorescent beads to act as point sources. Recently, the use of very high quality charge coupled devices (CCDs, Hiraoka *et al.,* 1987; and Aikens *et al.,* volume 29) has provided the necessary data quality to permit the experimental determination of three-dimensional CTFs of several high-numerical aperture objective lenses under epifluorescence conditions (Hiraoka *et al.,* 1988). A set of point images taken at different levels of defocus is shown in Fig. 1. It should be apparent that unlike ideal lenses, the PSFs of real lenses are not circularly symmetric and can show varying amounts of residual spherical aberration (visualized as lack of symmetry above and below the focal plane). The two-dimensional PSFs of Fig. 1 can be radially averaged and Fourier transformed to yield a set of radially symmetric two-dimensional CTFs which can then be Fourier transformed [Eq. (4)] to yield the three-dimensional CTF (Fig. 2) which can be compared to the theoretical CTF for the same lens. It is important to note that there is a significant difference in behavior of the two CTFs, especially at low spatial frequencies. Much of the difference results from the fact that the high-numerical objective lens serves as a high-NA condenser and provides a partial confocal behavior to the imaging process. Thus we feel it is extremely important to use the experimental CTF wherever possible. As there is probably less difference between similar high-NA lenses than there is between experimental and theoretical functions, it is reasonable to simply interpolate from experimental curves to correct for changes expected to result from changing NA and wavelength.

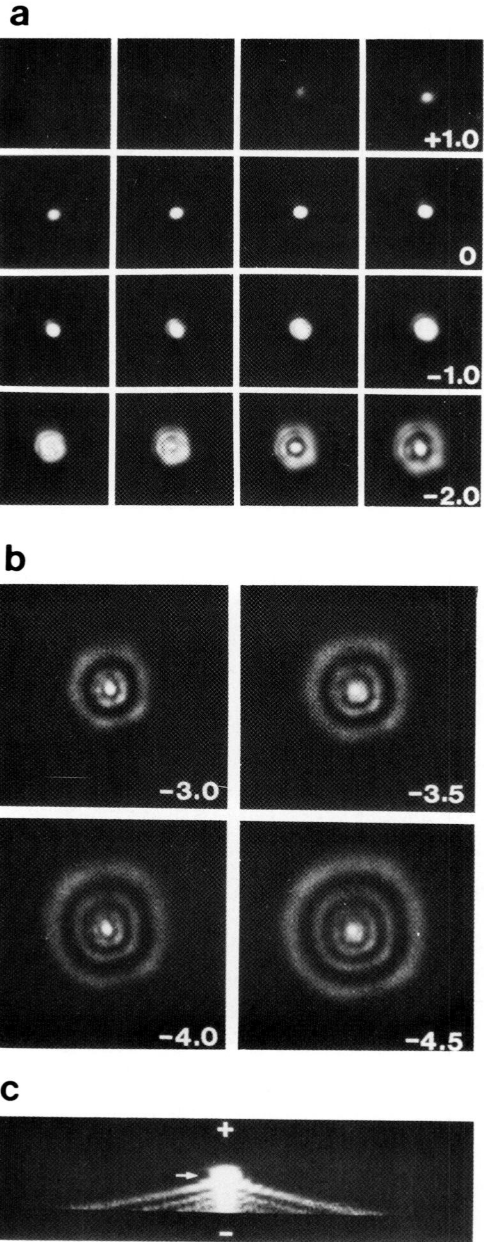

FIG. 1. Through-focus behavior of point images. Fluorescent microspheres (0.1 μm diameter; Pandex Laboratories, Inc.) were examined using a Leitz 1.4NA 63× planapo oil immersion lens with immersion oil recommended by Leitz. The serial focal series was taken with the CD in 0.25-μm focal steps. (a) The point images from +1.75 to −2.0 μm are shown every 0.25-μm from left to right and top to bottom; −3.0, −3.5, −4.0, and −4,5 μm are shown in b. The *x-z* edge view of the three-dimensional stack of 31 point images from +3.0 to −4.5 μm at 0.25-μm intervals is shown in c; the in-focus plane is marked by an arrow. All images are displayed on a logarithmic scale to facilitate viewing. From Hiraoka *et al.* (1987).

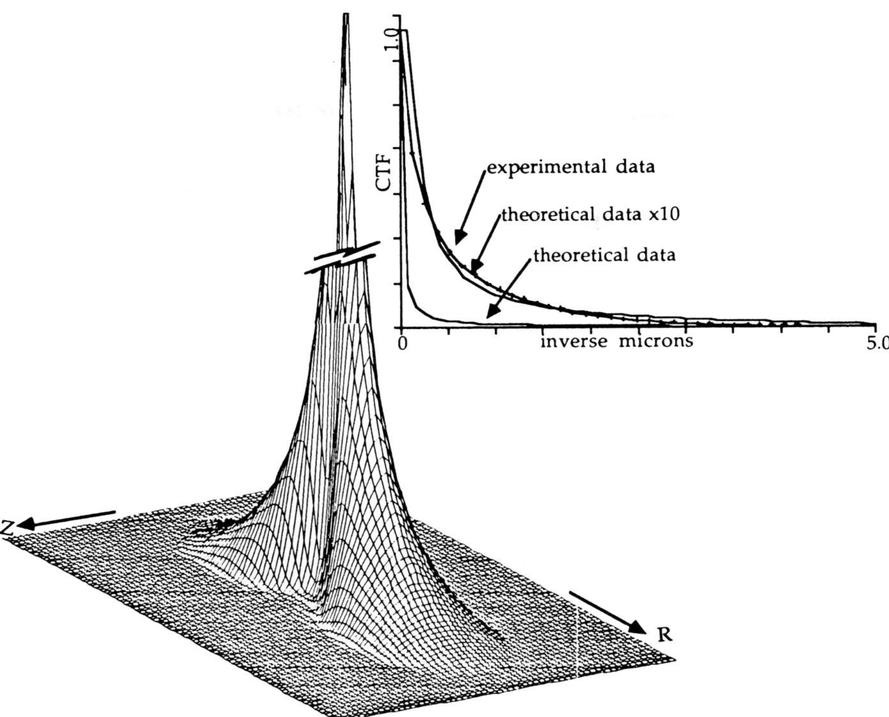

FIG. 2. Experimental three-dimensional CTF from an epifluorescent microscope. Optical section data from individual fluorescent microbeads were collected at 0.25-μm intervals using a Zeiss 1.2NA coverslipless water immersion lens. Data from several beads within each section image were averaged, three-dimension-Fourier transformed, and cyclindrically averaged in Fourier space. The resultant radially symmetric three-dimensional CTF is shown (R is the radial term corresponding to the X-Y plane, and Z denotes the focal direction). The inset shows $Z = 0$ lines for the experimental data and from theoretical calculations for the same lens. Note the significantly different falloffs. Even when scaled up 10-fold, the theoretical curve still fails to match the experimental data at low resolution.

III. Removal of Out-of-Focus Information

Given accurate knowledge of the three-dimensional PSF or CTF, it should be possible to remove the out-of-focus information that contaminates each of the observed images. However, examination of the three-dimensional CTF indicates that there is a region of Fourier space along the Z-axis (focus direction) where little or no data are observed. Because of this, it is not possible to make a perfect restoration; the very-low-resolution Z-axis terms are mostly severely compromised in any restoration scheme. Because of the optical properties of confocal microscopy, the amount of missing data is greatly reduced; however, resolution is still substantially reduced in the Z-direction. Fortunately, most biological imaging problems involve localizing spatially distinct objects; this reduces the importance of the very-low-resolution Z terms. In general, most specimens can be accurately reconstructed.

Not surprisingly, doing a better job with the deblurring requires more work. Because not all problems require ultimate three-dimensional resolution, we have developed three processing schemes providing three levels of deblurring: nearest neighbor, three-dimensional inverse filtering, and constrained-iterative three-dimensional deconvolution. It is important to note that even the simplest of the methods produces a substantial improvement and is suitable for all but the most demanding applications, with results roughly comparable to the MRC-Lasersharp confocal microscope.

A. Nearest-Neighbor Deblurring

The simplest scheme for deblurring optical section data sets uses image data from a single focal plane above and a single focal plane below to correct the central focal plane. As signals from the adjacent planes will contribute most strongly to the central plane, their effects are most important to consider. This was the rationale first used by Castleman (1979) and later by us (Agard, 1984) to justify using only nearest neighbors:

$$o_j = i_j * s_0 + i_{j-1} * s_{-1} + i_{j+1} * s_{+1} + \cdots \tag{5}$$

where the subscripts on i and o refer to the z plane numbers and the subscripts on s refer to the number of interplane spacings (Δz) away from focus for the two-dimensional PSF, s. Thus if we know the $i_{j\pm1}$, we could then use our knowledge of s_0 and $s_{\pm1}$ to calculate i_j. By assuming for the purposes of subtraction that $i_{j\pm1} \simeq o_{j+1}$, we can then calculate an approximation to i_j:

$$i_j = [o_j - c(o_{j-1} + o_{j+1}) * s_1] * g \tag{6}$$

where c is an adjustable constant, the negligible difference between s_{-1} and s_{+1} is ignored, and g is the filter function corresponding to the inverse of s_0. Although in Castleman's original analysis $c = 0.5$, we found that this overcorrected the image, as judged by holes in the image corresponding to strong regions on the adjacent planes. Our best results were obtained with

$c \simeq 0.45$ (Gruenbaum *et al.*, 1984). In fact, the results with this method were far better than expected even with interplanar spacings of $\leq 0.25\ \mu$m. Further analysis suggests a reason for the favorable behavior as well as an improvement. Let's consider explicitly all terms for the triplet of sections that contain i_j up to $i_{j\pm 2}$:

$$o_j = i_j * s_0 + i_{j-1} * s_1 + i_{j+1} * s_1 + i_{j-2} * s_2 + i_{j+2} * s_2 \cdots$$

$$o_{j-1} = i_j * s_1 + i_{j-1} * s_0 + i_{j+1} * s_2 + i_{j-2} * s_1 + i_{j+2} * s_3 \cdots$$

$$o_{j+1} = i_j * s_1 + i_{j-1} * s_2 + i_{j+1} * s_0 + i_{j-2} * s_3 + i_{j+2} * s_1 \cdots$$

Now if we blur $o_{j\pm 1}$ by the difference between s_0 and s_1, s_Δ, then all s_0 terms become exactly equivalent to s_1. Furthermore, for $\Delta z < 1\ \mu$m, $s_1 * s_\Delta \simeq s_2$ and $s_2 * s_\Delta \simeq s_3$. Summing the blurred adjacent sections and gathering terms:

$$\begin{aligned}(o_{j-1} + o_{j+1}) * s_\Delta &= 2i_j * s_1 * s_\Delta + (i_{j-1} + i_{j+1}) * (s_1 + s_2 * s_\Delta) \\ &\quad + (i_{j-2} + i_{j+2}) * (s_2 + s_3 * s_\Delta) + \cdots \\ o_j - c(o_{j-1} + o_{j+1}) * s_\Delta &= i_j * (s_0 - 2cs_1 * s_\Delta) \\ &\quad + (i_{j-1} + i_{j+1}) * (s_1 - c(s_1 + s_2 * s_\Delta)) \\ &\quad + (i_{j-2} + i_{j+2}) * (s_2 - c(s_2 + s_3 * s_\Delta)) + \cdots \end{aligned} \tag{7}$$

At low spatial frequencies, all of these filters contribute nearly equally, thus to best remove the low frequencies, c should be set to 0.5. Unfortunately, with $c = 0.5$, all of the low spatial frequencies from i_j are also removed. As discussed in Agard (1984), $c \simeq 0.45$ seems optimal in practice. At high spatial frequencies, the difference between nonequivalent blurring functions (e.g., s_1 and $s_2 * s_\Delta$) becomes significant and only 1/2 of the desired correction is made. Furthermore, Eq. (7) suggests that the inverse filter function, g, should be the inverse of $s_0 - 2cs_1 * s_\Delta$ and not simply of s_0 as given in Eq. (6). Another possible improvement is to effectively make c a function of frequency so that it starts at a value of 0.45 at low frequencies and increases to a value of 0.9 at high spatial frequencies. It is somewhat simpler to consider this filter in Fourier space; thus a reasonable function might be $S_0/(S_0 + S_1 S_\Delta)$, where the capital S denotes the CTF instead of the PSF. Putting this all together, we can now rewrite Eq. (6) as

$$i_j = [o_j - c(o_{j-1} + o_{j+1}) * f] * g \tag{8}$$

where c is again a constant of about 0.45 and the Fourier transform of the new filter function f is given by

$$F = S_0 S_\Delta/(S_0 + S_1 S_\Delta) \tag{9}$$

The choice for the sharpening function g in practice is not very critical. With the use of the more complex filter function F, above, it is reasonable to again use an inverse filter based on s_0. For such inverse filtering situations, it is standard practice to use a Wiener inverse filter to minimize the effects of noise that can dominate at high spatial frequencies. The Fourier transform of the inverse filter is then defined as

$$G = S_0/(S_\Delta + \alpha) \tag{10}$$

with α being an empirical constant that depends on the noise level in the data. A value of 0.5 will lead to negligible enhancement (suitable for very noisy data), while a small value such as 0.01 should work well on rather noise-free data.

The real space analogs of the filter functions defined in Eqs. (9) and (10) can be approximated as small matrices, in the range of 5×5 to 11×11, to allow the computations required by Eq. (8) to be done using generally available image processing hardware. Thus, it should be possible to do the all of the required calculations in essentially real time.

B. Three-Dimensional Inverse Filtering

A more accurate approach to removal of the out-of-focus information requires that the full contributions of *all* of the observed data be utilized. The most direct way to utilize the entire data set is to perform a three-dimensional inverse filter operation. Erhardt used this approach with a theoretical CTF (Erhardt *et al.*, 1985). As discussed above, we find that the differences between experimental and theoretical CTFs are significant; hence we get best results using a Wiener filter and the experimental CTF:

$$I(X, Y, Z) \simeq O(X, Y, Z)/[O(X, Y, Z)^2 + \alpha] \tag{11}$$

where α is a constant whose choice is based on the signal-to-noise ratio of the data and, since no other enhancements are being performed, should be in the range of 0.001 to 0.1.

Although performing three-dimensional filter or inverse filter operations is quite commonplace, the practical implementation with discrete Fourier transforms such as the FFT deserves a few comments. The FFT is cyclical, that is, the sampling in Fourier space is such that all axial dimensions should be thought of as being infinitely replicated. As a consequence, left and right, top and bottom, etc., image borders are immediately adjacent to one another, creating discontinuities. The importance of these discontinuities is directly related to their magnitude *and* the three-dimensional PSF. In the x-y plane it is often possible to choose a region that encompasses the object so that only background deviations are involved. When this is not possible, borders can be smoothly rolled down to a background value using for example, a Gaussian weighting function. The three-dimensional light microscope CTF is such that errors arising from *x-y* discontinuities are weighted toward higher frequencies, producing mainly local perturbations at the edges. Furthermore, procedures like rolling down to background are consistent with the in-plane CTF and hence are quite effective. Unfortunately, the same is not true in the *z* (focal) direction. Because data along the *z*-axis in the three-dimensional Fourier space are unobserved in the microscopy experiment, it is imperative that the values from the top of the stack and the bottom of the stack make a very smooth transition from one to the other. We have found that expanding the *z*-axis by 1.5 to 2× with data arranged as a weighted pixel-by-pixel linear combination of the top and bottom sections works quite well. The need to store and transform the expanded data can be essentially eliminated by arranging that the three-dimensional FFT be done first as a set of two-dimensional FFTs based on the observed data, followed by *z*-axis one-dimensional transforms. Before each *z*-axis line is transformed, it is expanded by linearly interpolating between the first and last values. When performing a three-dimensional filter, the expanded values are transformed in *z*, filtered, inverse transformed in *z*, and only those sections corresponding to those originally observed are inverse transformed in the *x-y* plane. The result is reasonable behavior with minimal computational or storage overhead. Furthermore, as radix-2 FFTs are more efficient, *z* expansion to the next power-of-2 provides a computationally efficient way to deal with data collected as an arbitrary number of sections.

C. Constrained Iterative Deconvolution

In terms of signal recovery in the presence of noise, the Wiener inverse filter provides a nearly optimal linear restoration. One drawback of all such linear methods is that negative ripples build up around strong features which can obscure or distort nearby weaker features. Nonlinear restoration methods that constrain the result to be positive can eliminate this problem while using positivity and other *a priori* information to provide a better restoration. The drawback is that these methods are invariably iterative and hence quite computationally expensive in three dimensions. For some time, we have have been working on constrained iterative deconvolution schemes for data processing problems in one dimension (Agard *et al.*, 1981; Thomas and Agard, 1985), two dimensions (Agard and Sedat, 1980), and three dimensions (Agard, 1984). We have recently made changes that significantly improve performance for three-dimensional optical sectioning data. The method is stable, and shows substantially accelerated convergence compared to previous methods. Convergence is reached in 5–10 cycles (instead of 20–50) and since most of the improvement occurs in the first 5 cycles, iterations can be terminated early with little degradation.

The strategy is to develop a positively constrained solution, g, that, when convolved with the known smearing function s, will regenerate the observed data o. The pixel-by-pixel differences between the convolved guess and the observed data are used to update the guess. Two different update schemes can be used: an additive method initially developed by van Cittert and later modified by Jannson (Jannson *et al.*, 1970) and by us (Agard *et al.*, 1981), and a multiplicative method developed by Gold (1964):

$$\begin{aligned} &\text{(a)} \quad o^k = i^k * s \\ &\text{(b)} \quad i^{k+1} = i^k + \gamma(o)(o - o^k) \\ &\text{(c)} \quad \text{if } i^{k+1} \leq 0 \text{ then } i^{k+1} = 0 \qquad (12) \\ &\text{(d)} \quad k = k + 1 \\ &\qquad \gamma = 1 - [o^k - A]^2/A^2 \end{aligned}$$

where A is a constant set to the maximum value of $o/2$. For the Gold method, the update equation on line b is changed to

$$\text{(b}') \quad i^{k+1} = i^k(o/o^k)$$

In three dimensions these schemes suffer from rather slow convergence, although the Gold method is faster. The problem stems from inadequate correction of the high-frequency components. This can be seen by the following argument performed only with the additive method for simplicity. At each cycle we desire to correct our current guess with a function, δ, so that $o \simeq (g_n + \delta) * s$ or in Fourier space: $O \simeq (G_n + \delta)S$. Rearranging, we get

$$\delta \simeq (O - G_nS)/S \qquad (13)$$

From this we can see that a better approximation to the update is to use an inverse filtered version of the difference between the observed data and the convolved guess. In practice, we use a Wiener filter to minimize effects of noise and only perform the inverse filtered update for the first two cycles. After that, we switch to the modified Van Cittert update described above [Eq. (12)]. At each cycle, the new guess is corrected to maintain positivity and any other desired real-space constraints, such as spatial boundedness. Every five cycles or so, the guess is smoothed with a Gaussian filter to minimize the buildup of noise. The result is an extremely powerful reconstruction scheme that avoids negative ripples that plague linear methods, minimizes noise buildup, is exceedingly stable, and converges quite rapidly.

D. Combination of Multiple Views

All of the discussions so far have assumed that a single set of optical sections was recorded in the imaging experiment. The microscope's three-dimensional CTF indicates that the resolution is substantially greater in the x-y direction than it is in z. Resolution in the focal direction is further complicated by the missing data near the z-axis. As discussed previously (Agard, 1984), this suggests that a significant improvement in imaging could be obtained by combining two optical section data sets collected at 90° to one another. In fact, by collecting several data sets at different tilt angles, it should be possible to reconstruct a three-dimensional image with uniform resolution in all directions; limited only by the in-plane resolution of the objective lens.

There are two practical problems with this approach. First, it is necessary to be able to tilt the specimen to reasonable angles. Although it is not possible to use a slide (the maximum tilt is $\leq \pm 15°$), we have had significant success by mounting the specimen on the *outside* of a glass fiber using poly-L-lysine as an adhesive. Using a Zeiss 1.2NA water immersion lens it has been possible to achieve tilts of $\geq \pm 45°$.

A second, very significant problem is the need to properly align the data sets to better than 500 Å before they can be merged. In principle this can be accomplished by designing an ultraprecise eucentric tilt stage or by computational methods. Since it is unlikely that the requisite physical alignment can be obtained, the computational approach must be considered. Although, the methodology is rather involved, we have recently developed all of the necessary computer programs and have successfully merged several three-dimensional data sets. Although still preliminary, it does appear that substantial resolution improvements in the z-direction can be obtained. A full description of the approach will be published elsewhere.

IV. Synthetic Projection and Stereographic Imaging

For many problems of biological interest, reconstructing the entire three-dimensional fluorescence distribution within the sample is either unnecessary or irrelevant. This is often the case when dealing with tissue culture cells or other specimens that are only a few microns thick. Such samples are sufficiently three-dimensional so that out-of-focus effects can obscure relevant information, but all that may be desired is localization in the *x-y* plane or perhaps calculation of a stereo pair. Either of these two types of calculations can be performed exceedingly efficiently using optical section microscopy. In fact, the optical behavior of the microscope is such that the necessary calculations can be performed more rapidly and with considerably less photobleaching than is possible with confocal microscopy. Although the details of this method will be published elsewhere, an outline of the procedure will be given here. We call the technique synthetic projection microscopy because it synthesizes an in-focus projected image of the object from a small number of optical sections. A simple variation is then used to generate stereo pairs.

Consider a three-dimensional object $i(x, y, z)$ whose projection we wish to calculate. Let i_j refer to the set of optical sections taken at intervals of approximately the depth of focus of the objective lens being used. In practice values between 0.5 and 1.0 μm are reasonable. Then the desired projected image distribution i_p is given by: $i_p = \Sigma i_j/N$. Using the Fourier projection theorem (Bracewell, 1965), the desired two-dimensional projection i_p can be understood to arise only from the $Z = 0$ central section of the object's three-dimensional Fourier transform, $I(X, Y, Z)$:

$$i_p = \sum\sum I(X, Y, O)\, e^{2\pi(Xx+Yy)} \tag{14}$$

By a similar argument, the relationship between the projection of the observed data and its Fourier transform is given by

$$o_p = \sum\sum O(X, Y, O)\, e^{2\pi(Xx+Yy)} \tag{15}$$

Furthermore, we know that the relationship between $I(X, Y, Z)$ and $O(X, Y, Z)$ is specified by the three-dimensional CTF, $S(X, Y, Z)$ as

$$O(X, Y, Z) = I(X, Y, Z)\, S(X, Y, Z) \tag{16}$$

Combining Eq. (14)–(16) gives the following result:

$$o_p = \sum\sum I(X, Y, O)\, S(X, Y, O)\, e^{2\pi(Xx+Yy)}$$

which, by the convolution theorem, can be returned to real space:

$$o_p = i_p * s_{Z0} \tag{17}$$

where s_{Z0} is the two-dimensional Fourier transform of the $Z = 0$ plane of the three-dimensional CTF. The desired in-focus projected image can then be obtained by deconvolving the observed o_p with s_{Z0}. This can be done either by a Weiner filter or by a two-dimensional constrained iterative deconvolution. As discussed earlier, the inverse filter can be recast as a local matrix convolution problem, allowing real time processing.

Generation of stereo pairs is accomplished during the summation of the observed frames. Two sums are accumulated for each new frame. The old trick of sliding the new image an integral number of pixels left to generate the left image and an equivalent number to the right for the right image works very well. The pixel displacement correlates with the stereo viewing angle and should generally be rather small. After the two separate images are accumulated, they are then deblurred as described for mono images.

V. Experimental Considerations

A. Lens Choice

Perhaps the most important microscopic considerations for three-dimensional imaging are the choice of the objective lens and its proper use. For high-resolution analysis, the use of a high-NA (NA $\geq$ 1.2) lens is crucial. High-quality lenses can now be obtained with NAs between 1.2 and 1.4 over the magnification range of 40× to 100×. An extremely important consideration for three-dimensional imaging that is generally neglected concerns residual spherical aberration in the objective lens. Immersion objective lenses are designed to image specimens located immediately against the underside of a defined thickness coverslip. Spherical aberration increases dramatically if this condition is not met. Furthermore, all of the high-NA oil lenses that we have tested show residual spherical aberrations. Small amounts of spherical aberration have rather little effect on two-dimensional imaging, but will substantially degrade three-dimensional images. A simple way of checking for spherical aberration is to image a small bright object equal distances above and below the plane of optimal focus. As pointed out by Shinya Inoué, spherical aberration manifests itself as a lack of symmetry in the out-of-focus images. Ideally, one would use a lens with a correction collar and adjust it for minimum asymmetry. Unfortunately, none of the high-NA oil lenses have correction collars. We have taken two approaches in dealing with this problem: (1) when possible we have used a Zeiss 1.2NA coverslipless water immersion lens, thereby avoiding the use of a coverslip entirely, or (2) we have experimentally found the correct index of refraction immersion oil to properly compensate for the spherical aberration with the oil lenses being used in the laboratory. For example, the proper index of refraction oil for our Leitz 1.4NA planapo lens is $\eta_D^{25°C} = 1.518$ instead of 1.515.

B. Data Collection

Beyond objective lens considerations, there are a few aspects of data collection that warrant discussion. The first is that it is particularly useful to automate the collection of the through-focus microscopic images. To allow high-resolution three-dimensional imaging it is useful to be able to step in focus ≤ 0.1 μm. This is readily accomplished by attaching a microstepper motor to the fine-focus knob of the microscope, either directly or through an electromagnetic clutch (Mathog *et al.*, 1985). Microstepper motors are available from several manufacturers and have between 10,000 and 50,000 steps per revolution; complete computer-controllable systems using an RS-232 interface are also available (for example, from Compumoter Inc., Petaluma, CA).

Data acquisition is best performed using a high-quality image detector such as a cooled CCD camera (Photometrics Ltd., Tucson, Az). As discussed by *Aikens et al.,* in Volume 29 of this series and Hiraoka *et al.* (1987), the combination of high sensitivity, large dynamic range, photometric linearity, and geometric precision makes the CCD a nearly ideal imaging device. In many ways, three-dimensional imaging relies on being able to interpret the *differences* between adjacent images. Thus, all of these properties are important. If three-dimensional resolution is not a crucial factor, it is quite reasonable to use a SIT camera and image processor to digitally average for 64–256 frames to minimize noise. However, it should be emphasized that the results obtainable with the CCD are truly remarkable.

C. Photobleaching

In recording three-dimensional data it is common to record anywhere from 10 to 64 separate images from a given sample. This requirement simultaneously places demands on the sensitivity of the detector and on sample preparation. Of particular importance is the use of antioxidants (Giloh and Sedat, 1982; Bock *et al.*, 1985) to minimize photobleaching of the fluorophore.

VI. Display

Once three-dimensional images are collected and deblurred, it is necessary to be able to examine this vast amount of data in order to extract the relevant biological information. Unfortunately, we have found the use of contouring, either as three-dimensional cage contours (FRODO, Jones,

1982) or as solid surface representations, inappropriate for examining specimens. In many ways, this is the most difficult aspect of three-dimensional microscopic imaging.

A common way to examine three-dimensional image data is to consider it as a set of slices that can be displayed one after the other as a digital-loop movie. This can be accomplished using either capabilities of the display hardware on the image processor or using a frame-accurate video recording device, probably the best of which is the optical video disk recorder (Panasonic) which permits full random access control on recording and playback.

Although this approach is by far the easiest to accomplish, it is the least satisfactory. Unfortunately, human perception is not really geared to remembering shapes when moving forward and backward. As users of any macromolecular display system know, the trick to perceiving three dimensions is to use a rotating, projected view of the object. Thus, in generating the digital-loop movies, a set of rotated projected views is substituted for the translational set normally used. We have found this approach to be far superior, and use it routinely.

The process of calculating rotated projected views can be greatly speeded up by combining the two operations. Consider a two-dimensional slice of the three-dimensional image taken perpendicular to the desired rotation axis: for example, the x-z-plane for tilts about the y-axis. Then the desired coordinate transformation is given by

$$x' = \cos\alpha(x - x_c) - \sin\alpha(z - z_c) + x'_c$$
$$z' = \sin\alpha(x - x_c) + \cos\alpha(z - z_c) + z'_c$$

Instead of rotating all the data and then projecting it down the z' axis, we loop through the desired x' range and the original z solving for x as we go.

$$x = [(x' - x'_c) + \sin\alpha(z - z_c)]/\cos\alpha + x_c \qquad (18)$$

The value to be added into the projection vector at x' is then given by one-dimensional interpolation:

(a) loop on x' in output projection vector
(b) loop on z in image
(c) calculate x in image that corresponds to current x', z
(d) linearly interpolate in image to get value and add to value in projection vector at x'

By saving one loop, this approach can speed calculations by as much as 30-fold. For $|\alpha| \geq 45°$, because of the $\cos\alpha$ in the denominator, it is advisable to loop on x', x and solve for z. Rotations about other axes can likewise be efficiently accomplished.

Sets of rotated stereo pairs can be used to enhance the realism of the display. Both Tektronix (Portland, OR) and StereoGraphics (San Rafael, CA) have developed electrooptic polarizing screens that can be mounted on the front of the video display monitor. The left and right images are alternately displayed on the CRT, and the polarizing screen will switch from left to right circular polarization in synchrony with the images. Viewers need only wear left and right circular polarized "sun glasses" to see stereo. When combined with rotation, the effect is quite dramatic; more importantly, it is possible to extract significantly more biological information from the images.

VII. Experimental Results for Imaging Chromosomes

Much of our laboratories' effort has been devoted to elucidating the detailed high-resolution structure of eukaryotic chromosomes and their spatial organization within the nucleus. Three-dimensional high-voltage and intermediate-voltage electron microscopic tomographic approaches are being used to analyze how the fundamental 110 Å nucleosomal fibers are organized into higher-order structures within mitotic and interphase chromosomes (Belmont *et al.*, 1987). For some time we have been using *Drosophila melanogaster* polytene chromosomes as a model interphase system for studying the relationship between three-dimensional spatial organization within the nucleus and biological function (Agard and Sedat, 1984; Mathog *et al.*, 1984; Gruenbaum *et al.*, 1984; Hochstrasser *et al.*, 1986). The earliest studies were done with images recorded on film and then digitized and aligned before deblurring. The process was greatly simplified with the use of direct video data acquisition. Before and after comparisons of a single section from a data set of 24 focal planes (1.2 μm spacing) of a 4′,6-diamidino-2phenylindole (DAPI)-stained salivary gland polytene nucleus is shown in Fig. 3. The images were deblurred with the nearest-neighbor scheme of Eq. (6).

The results of these studies as well as the development of a growing interest in chromosome dynamical behavior during the cell cycle have led us to attempt to analyze the three-dimensional organization of diploid chromosomes as a function of the cell cycle. The comparative difficulty associated with examining a diploid nucleus 5 μm in diameter is demonstrated in Fig. 3. By comparison with the polytene studies, an increase in three-dimensional resolution of nearly an order of magnitude was required. Several factors were crucial in reaching the necessary resolution; foremost were the use of a cooled CCD to collect very-high-quality image

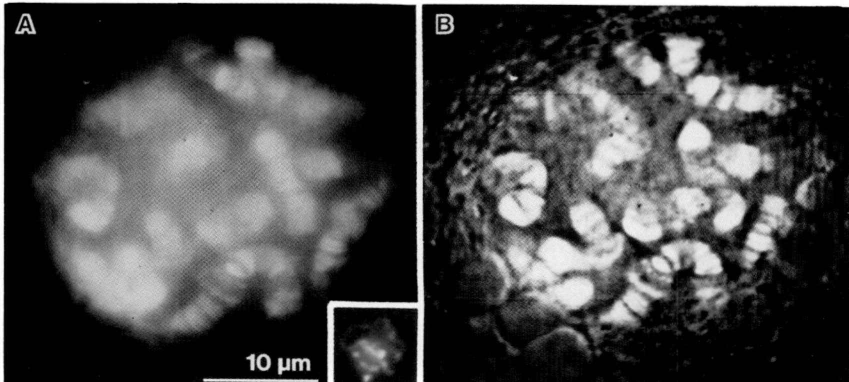

FIG. 3. Low-resolution polytene chromosome images before and after processing. *Drosophila* salivary glands were stained with the DNA-specific dye DAPI as described in Mathog *et al.* (1984). Twenty four images were recorded at 1.2-μm focal intervals using a SIT camera and each image was digitally averaged for 256 video frames. The nearest-neighbor algorithm [Eq. (6)] was used to process the images. A representative before section is shown in A and the corresponding after processing image is shown in B. Puffed regions indicating intense transcriptional activity are clearly seen in B. The inset in A shows a *Drosophila* embryonic diploid nucleus at the same magnification for comparison.

data and the use of the experimental CTF (which required the CCD to be determined) in the image processing. The substantial improvement in image quality, which was made possible primarily through the use of the CCD camera, and secondarily through the use of a finer interplanar spacing, is shown in Fig. 4. Processing again made use of the same simple nearest-neighbor scheme used in Fig. 3 which can be implemented on essentially any image processing system.

High-resolution three-dimensional images of diploid chromosomes from *Drosophila melanogaster* early embryos are shown as several optical sections before (Fig. 5A) and after (Fig. 5B) image processing. A stereo pair computed from the entire nucleus is shown in Fig. 6. Stereo pairs calculated from another diploid *Drosophila* nucleus are shown in Fig. 7A. The sister and homologous chromatids are clearly separated; the bright regions represent clusters of centric heterochromatin. The three-dimensional paths of each of the chromatids have been traced using our interactive modeling software (Mathog, 1985). This is the first step in our analysis of the dynamical behavior of diploid chromosomes throughout the cell cycle.

Once a model describing the three-dimensional path of a chromatid has been built, it is then possible to use that model as a map for topologically

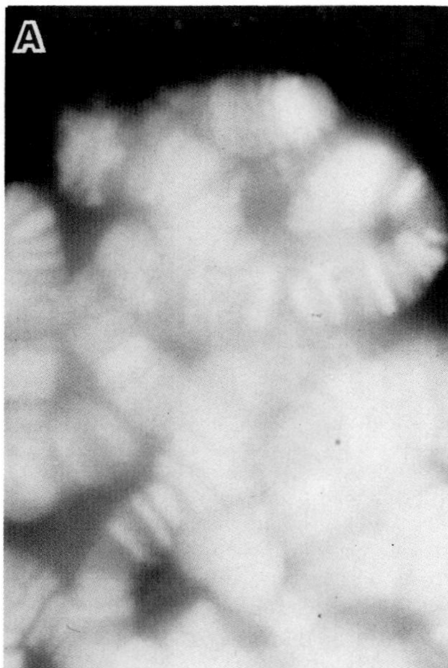

FIG. 4. High-resolution polytene chromosome images before and after processing. Polytene chromosomes were stained with DAPI and imaged on our cooled CCD system (Hiraoka *et al.*, 1987) with an interplanar spacing of 0.25 μm. Data were processed as in Fig. 3. Note the significantly higher resolution possible using the CCD to collect better data and finer interplanar spacings.

unfolding the chromosome. This is extremely useful to us because it allows us to uniquely identify each chromatid using its characteristic staining pattern. An example of this is shown in Fig. 8.

VIII. Summary

The combination of the specificity provided by fluorescence microscopy and the ability to quantitatively analyze specimens in three dimensions allows the fundamental organization of cells to be probed as never before. Key features in this emergent technology have been the development of a wide variety of fluorescent dyes or fluorescently labeled probes to provide the requisite specificity. High-quality, cooled charge-coupled devices have recently become available. Functioning as nearly ideal imagers or "electronic film," they are more sensitive than photomultipliers and provide extraordinarily accurate direct digital readout from the microscope.

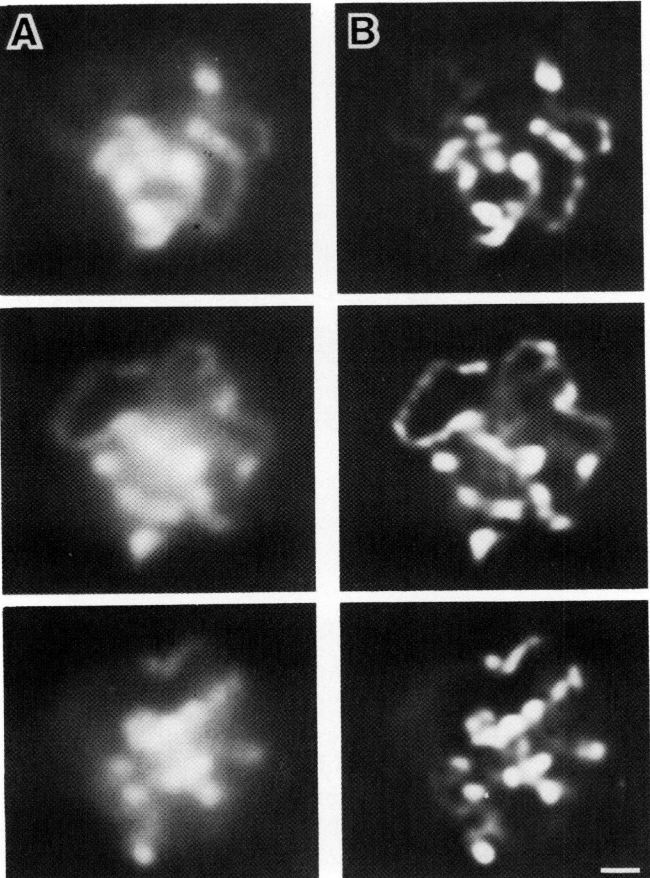

FIG. 5. Embryonic diploid nucleus before and after the removal of out-of-focus image information. *D. melanogaster* embryos stained with 0.1 μg/ml DAPI were attached a glass slide and mounted in buffer A without a coverslip. Using a Zeiss coverslipless water lens (63×/NA = 1.2), optical section images were taken on the CCD at 0.25-μm focal intervals. Optical section images were processed by the iterative constrained three-dimensional deconvolution method [Eq. (12)] with the experimental CTF determined for the water lens. This figure shows the same nucleus shown in the inset of Fig. 3 before (A) and after (B) processing; each section in the figure is separated by 1 μm in the z-direction. Bar represents 1 μm.

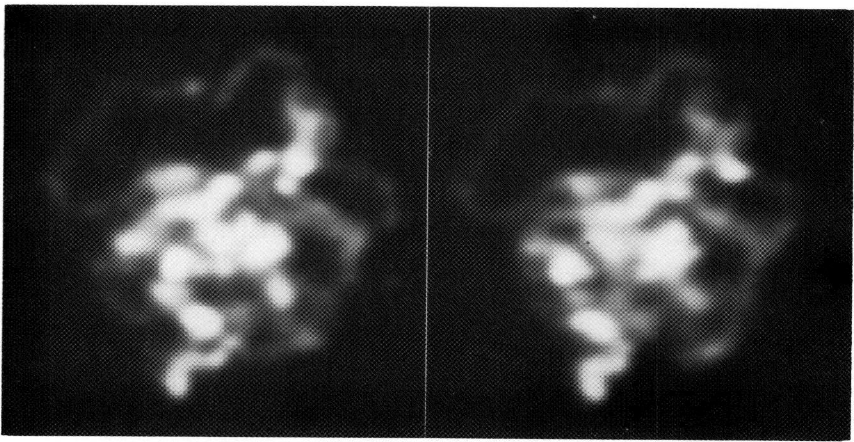

FIG. 6. Stereo image of diploid embryonic chromosomes. A stereo projection was constructed from the optical sections shown in Fig. 5 after deblurring using the rotated projection method [Eq. (18)].

Not only is this precision crucial for accurate quantitative imaging such as that required for the ratioing necessary to determine intracellular ion concentrations, but it also opens the way for sophisticated image processing. It is important to realize that image processing isn't simply a means to improve image aesthetics, but can directly provide new, biologically important information. The impact of modern video microscopy techniques (Allen, 1985; Inoué, 1986) attests to the fact that many biologically relevant phenomena take place at the limits of conventional microscopy. Image processing can be used to substantially enhance the resolution and contrast obtainable in two dimensions, enabling the invisible to be seen and quantitated.

Cells are intrinsically three-dimensional. This can simply be a nuisance because of limited depth of focus of the microscope or it could be a fundamental aspect of the problem being studied. In either case, image processing techniques can be used to rapidly provide the desired representation of the data. In this chapter we have discussed the nature of image

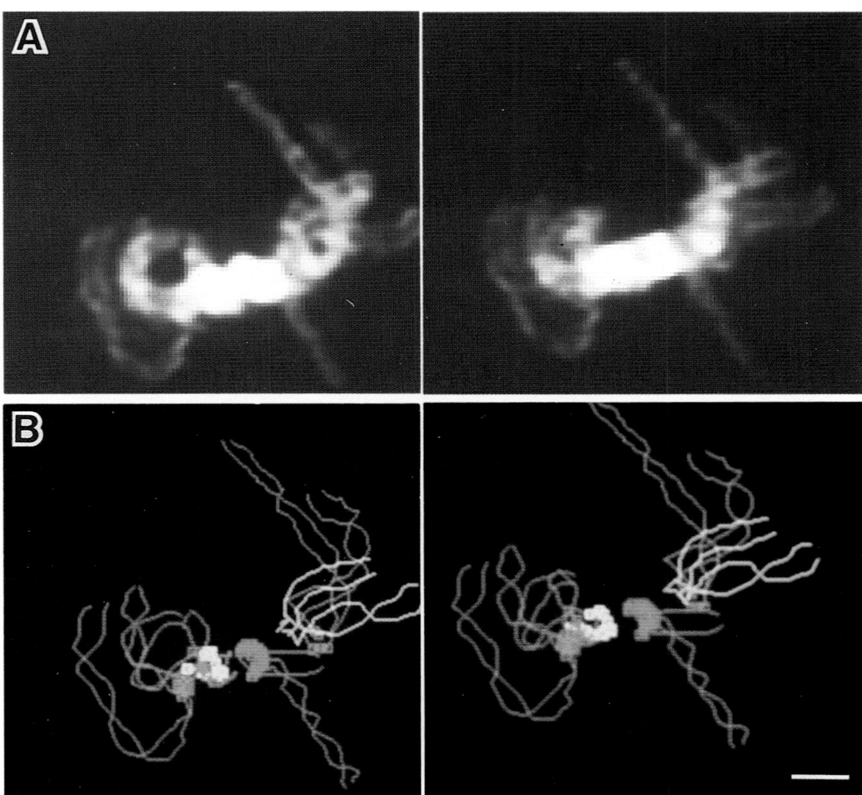

Fig. 7. Stereo model of the 3D spatial arrangement of diploid chromosomes. The stereo image (A) was constructed from optical section images taken with 0.1 μm step size in z. Out-of-focus information was removed as in Fig. 5. Chromosome paths were traced in the stack of optical sections using our interactive modeling software (Mathog, 1985) and are shown in stereo in (B). Each line represents a sister chromatid; the central bright area is the centric heterochromatin cluster. Bar represents 1 μm.

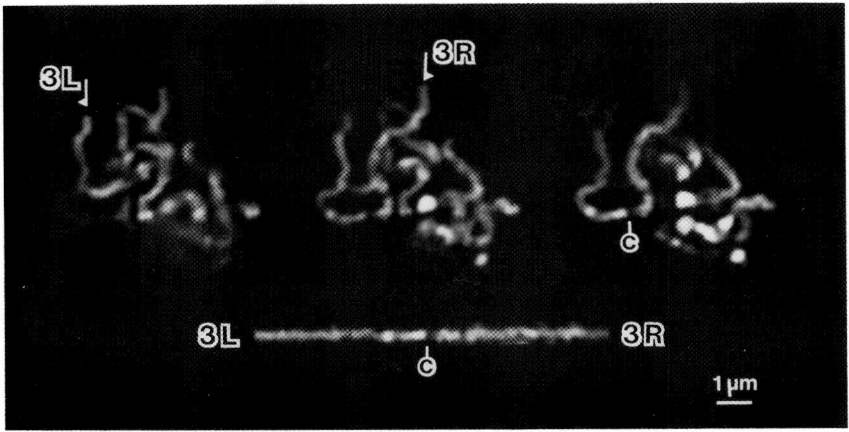

FIG. 8. Cytological assignment of diploid chromosomes. Gently squashed embryos were imaged at 0.5-μm focal intervals, using a Leitz oil lens (63×/NA = 1.4). Immersion oil with the refractive index of 1.518 (Cargille Laboratory, Inc.) was used to minimize spherical aberration. Chromosome paths were traced and then the traced chromosomes were computationally straightened using the modeled chromosome path. This figure demonstrates the results of straightening the third chromosome. c, Centromere.

formation in three dimensions and dealt with several means to remove contaminating out-of-focus information. The most straightforward of these methods uses only information from adjacent focal planes to correct the central one. This approach can be readily applied to virtually any problem and with most commonly available image processing hardware to provide a substantially deblurred image in almost real time. In addition to covering more sophisticated algorithms where the utmost in three-dimensional imaging is required, we have developed a method for extremely rapidly and accurately producing an in-focus, high-resolution "synthetic projection" image from a thick specimen. This is equivalent to that produced by a microscope having the impossible combination of a high-NA objective lens and an infinite depth of focus. A variation on this method allows efficient calculation of stereo pairs.

Because these synthetic projection methods are efficient both computationally and in their use of the fluorescent light emitted from the sample, they are ideally suited for examining living cells or when using fluorophores that are very sensitive to photobleaching. Currently, real-time three-dimensional imaging is limited equally by the necessary integration time required to form a high-quality CCD image (approximately 0.5 second) and the rather slow readout rates available with the high-precision Photometics CCD camera controllers (approximately 1 second for a 512 × 512 area at

12 bits). High-quality stereo imaging can be accomplished with about 10 views. By comparison, computation time is negligible. Thus it should now be possible to generate complete stereo pairs at a rate of 1 every 20 seconds or so. CCD readout rates can be expected to substantially increase in the next year, and the use of more intense light sources or by compromising somewhat on image quality by tolerating slightly noisier images, lower resolutions, or reduced stereo depth should reduce the time per stereo pair to about 2–5 seconds. This should be sufficiently fast to allow complex biological dynamical phenomena to be studied in three dimensions in living cells.

Although our studies have focused on the three-dimensional organization of chromosomes within the nucleus, it should be apparent that the methods being used are quite general and are suitable for many problems in structural cell biology. In our own laboratories, we have begun collaborations with colleagues at UCSF to study polarization phenomena in epithelial cells and spatial aspects of regulated transport through the Golgi. With the development of processing, display, and modeling software, it is now possible to quantitatively determine, analyze, and compare three-dimensional distributions of cellular components.

Acknowledgments

We would like to thank Andrew Belmont for critical reading of the manuscript. This work was supported in part by NIH GM25101-09 and NIH GM32803-03 to J.W.S. and NIH GM31627 to D.A.A., and more recently by Howard Hughes Medical Institute (J.W.S. and D.A.A.). D.A.A. was also supported by a grant from the NSF Presidential Young Investigator Programs. Y.H. was supported by Damon Runyon-Walter Winchell Cancer Fund Fellowship DRG903.

References

Agard, D. A. (1984). *Annu. Rev. Biophys. Bioeng.* **13,** 191.

Agard, D. A., and Sedat, J. W., (1980) *Proc. Soc. Photo-Opt. Instrum. Eng.* **264,** 110.

Agard, D. A., and Sedat, J. W. (1984) *Nature* (*London*) **302,** 676.

Agard, D. A., Steinberg, R. A., and Stroud, R. M. (1981). *Anal. Biochem.* **111,** 257.

Allen, R. D. (1985). *Annu. Rev. Biophys. Chem.* **14,** 265.

Belmont, A. S., Sedat, J. W., and Agard, D. A. (1987). *J. Cell Biol.* **105,** 77.

Bock, G., Hilchenback, K., Schoenstein, K, and Larch, G. (1985). *J. Histchem. Cytochem.* **33,** 699.

Bracewell, R. (1965). "The Fourier Transform and Its Applications," pp. 108–12. McGraw-Hill, New York.

Castleman, K. R. (1979). "Digital Image Processing." Prentice-Hall, Englewood Cliffs, New Jersey.

Erhardt, A., Zinser, G., Komitowski, D., and Bille, J. (1985). *Appl. Opt.* **24,** 194.

Giloh, H., and Sedat, J. W. (1982). *Science* **217,** 1252.

Gold, R. (1964). AEC Research and development report. ANL 6984, Argonne National Laboratory, Argonne, Illinois.

Gruenbaum, Y., Hochstrasser, M., Mathog, D., Saumweber, H., Agard, D. A. and Sedat, J. W. (1984). *J. Cell Sci. Suppl.* **1,** 223.

Grynkiewicz, G., Poonie, M., and Tsien, R. Y. (1985). *J. Biol. Chem.* **260,** 3440.

Hiraoka, Y., Sedat, J. W., and Agard, D. A. (1987). *Science* **238,** 36.

Hochstrasser, M., Mathog, D., Gruenbaum, Y., Saumweber, H., and Sedat, J. W. (1986). *J. Cell Biol.* **102,** 112.

Hopkins, H. H. (1955). *Proc. R. Soc.* (*London*) *Ses. A* **231,** 91.

Inoue, S. (1986). *In* "Video Microscopy," Plenum, New York.

Jannson, P. A., Hunt, R. M., and Plyler, E. K. (1970). *J. Opt. Soc. Am.* **60,** 596.

Jones, T. A. (1982). *In* "Computational Crystallography" (D. Sayre, ed.), pp. 303–317. Oxford Univ. Press, London.

Mathog, D. (1985). *J. Microsc.* **137,** 253.

Mathog, D., Hochstrasser, M., Gruenbaum, Y., Saumweber, H., and Sedat, J. W. (1984). *Nature* (*London*) **308,** 414.

Mathog, D., Hochstrasser, M., and Sedat, J. W. (1985). *J. Microsc.* **137,** 241.

Stokseth, P. A. (1969). *J. Opt. Soc. Am.* **59,** 1314.

Thomas, G. J., Jr., and Agard, D. A. (1984). *Biophys. J.* **46,** 763–768.

Waggoner, A. S., (1986). *In* "Applications of Fluorescence in the Biomedical Sciences" (D. L. Taylor, A. S. Waggoner, R. F. Murphy, F. Lanni, and R. R. Birge, eds.) pp. 3–28. Liss, New York.

Williams, D. A., Fogarty, K. E., Tsien, R. Y., and Fay, F. S. (1985). *Nature* (*London*) **318,**558.

Chapter 14

Three-Dimensional Confocal Fluorescence Microscopy

G. J. BRAKENHOFF, E. A. VAN SPRONSEN,
H. T. M. VAN DER VOORT, AND N. NANNINGA

Department of Electron Microscopy and Molecular Cytology
University of Amsterdam
1018 TV Amsterdam, The Netherlands

I. Introduction

Biological material is organized in four dimensions: three spatial ones and a temporal one. The ideal microscope should be able to make accessible the information about this organization with the highest possible resolution, while leaving the organism as much as is feasible in its original state. Light microscopy is able to visualize biological objects in their natural watery condition and, in principle, during their temporal development. In this aspect, it has a great advantage over electron microscopy, where specimens can, in practice, only be observed in a dehydrated dead condition. In addition, preparation steps for EM may effect the morphology of the biological structure. For instance, in bacteria, shrinkages of up to 50% during fixation, dehydration, and embedding have been observed. (Woldringh *et al.,* 1977). With respect to resolution, confocal scanning microscopy takes a middle position in between both techniques. It has, as a specific advantage an excellent capability for visualizing directly the three-dimensional structure of biological objects.

The idea for confocal imaging in light microscopy was conceived in the late 1970s. Sheppard (1977) showed theoretically that in confocal scanning microscopy a fundamental improvement in imaging, as compared to normal light microscopy, could be expected. That this expectation could be realized in practice at high numerical apertures (NA) (where this improvement really counts) was demonstrated by Brakenhoff *et al.* (1979) in transmission confocal microscopy. Resolutions down to 140 nm have been observed in UV in confocal transmission microscopy (Brakenhoff *et al.*, 1980). The confocal principle can also be applied to imaging in fluorescence (Cox *et al.*, 1982; Brakenhoff *et al.*, 1985; Wijnaendts van Resandt *et al.*, 1985; Carlsson *et al.*, 1985). In this case, we have, in addition to the improved imaging, the optical sectioning property by which the contributions from out-of-focus areas in the specimen are effectively suppressed. In normal microscopy, these contributions lead to a strong reduction in the available image contrast.

This sectioning effect and the serial way in which the data become available make a confocal microscope very suitable for coupling to a computer system. We obtain in fact a three-dimensional microscope where each data point as collected represents the quantity of the specific contrast parameter used at a certain point in space. Agard and Sedat (1983) have developed deconvolution techniques for eliminating the out-of-focus information from conventional fluorescence microscopy. Confocal microscopy can deliver directly "clear" optical sections without the use of time-consuming image reconstruction algorithms. In addition, image processing can also be used to enhance the confocal images.

Below, after an explanation of the technique and an imaging test, we demonstrate using a number of biological examples the usefulness of confocal imaging to biology. We use computer-generated stereoscopic images for the visualization of the three-dimensional biological information. Finally, we discuss a number of aspects of confocal imaging, especially the advantages and drawbacks of the various scanning approaches.

II. Confocal Microscopy

A. Principle and Theory

The basic idea of confocal microscopy as illustrated in Fig. 1 is that one and the same point in the object is optimally illuminated by a point light

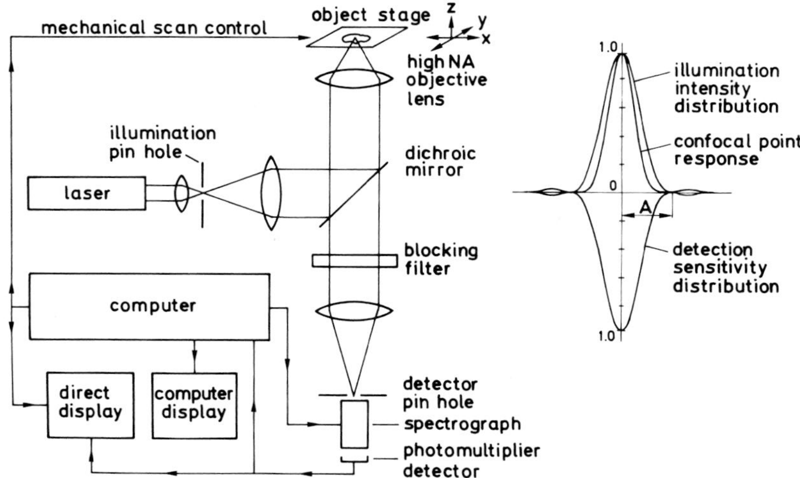

FIG. 1. Schematic diagram of the confocal scanning fluorescence microscope. Light originating from the laser-illuminated pinhole is focussed on a certain point in the object. The same point is subsequently imaged on the detector pinhole. The term confocal relates to the fact that the image of the illumination pinhole and the back-projection of the detection pinhole have a common focus in the object. The specimen is scanned mechanically through this confocal point. A direct image is displayed on a cathode-ray tube, modulated by the detected intensity and runing in synchrony with the scan. At the same time, the image data are digitized and stored in the computer system. The optical arrangement dichroic beam splitter/blocking filter is as commonly used in fluorescence microscopy. The spectrograph enables one to select a certain band from the fluorescence radiation for image formation. The bandwidth as well as the wavelength selection of the spectrograph can be set by the computer system.

source as well as imaged on a point detector. A number of the properties of the imaging can be readily understood from a treatment of the imaging of a point object, under the assumption of diffraction-limited optics. For the full theory for a general object, we would like to refer to treatments based on Fourier optics (Sheppard and Choudhury, 1977; Wilson and Sheppard, 1984). Figure 1 shows the optical arrangement for confocal microscopy in fluorescence (or reflection when the dichroic beam splitter is replaced by a normal one). As is well-known, the lateral intensity distribution in the object from a point light source imaged by diffraction limited optics is given by the distribution in the Airy pattern (Born and Wolf, 1975).

$$I_1(u) = \text{const}\ (J_1(u)/u)^2 \tag{1}$$

with $J_1(u)$ the first-order Bessel function. The optical units u, v, w are related to the transverse coordinates x, y, and z, respectively, by the relation

$$u = n2\pi x \sin \alpha/\lambda; \qquad v = n2\pi y \sin \alpha/\lambda; \qquad w = n2\pi z \sin^2 \alpha/\lambda \quad (2)$$

$n \sin \alpha$ is the numerical aperture (NA) of the optics employed, λ the wavelength of light, and n the refractive index. The intensity distribution $I_a(w)$ on the optical axis is given by

$$I_a(w) = \text{const}(\sin w/4)^2/(w/4)^2 \quad (3)$$

Confocal conditions are realized when the detection pinhole is positioned in such a way as to make its backprojected image in object space exactly coincide with the focal point of the illumination distribution. The functional dependence in this backprojected image actually indicates the degree to which the various parts of the illuminated object contribute to the detected signal, we will call this the detection sensitivity distribution. If the same lens is used for illumination and detection purposes, then both distributions, the illumination and detection, will be identical. The detected signal from one point object located at a certain position in the object space will be the product of the intensity and detection sensitivity distribution in this location. Hence, if the lateral and axial detection sensitivy distributions are defined by $D_1(u)$ and $D_a(w)$, respectively, then the confocal response $C_1(u)$ and $C_a(w)$ to a point object in object space can be written as

$$C_1(u) = I_1(u)\, D_1(u) \text{ and } C_a(w) = I_a(w) D_a(w) \quad (4)$$

This result is also found from a Fourier optical treatment of the case of a point object (Sheppard and Choudhury, 1977). A number of aspects of confocal imaging can be realized immediately from this simple approach.

1. A fundamental improvement in imaging with respect to conventional microscopy. In normal microscopy the point response is given by $D_1(u)$. Due to multiplication, a narrowing of the response (see Fig. 1) by a factor of 1.38 results, as can be calculated. This improvement in linear resolution takes place in both directions transverse to the optical axis as well as along the optical axis, so that finally one image point in the confocal laser scanning microscope (CSLM) will correspond to a volume element in the specimen that is smaller by a factor of $1.389^3 = 2.64$ than in conventional microscopy. As a result, there is a proportional increase in the amount of information which can be extracted from the specimen.

2. Due to the multiplication, the sidelobe intensity, which is typically 1% in conventional microscopy, is reduced to 0.01%. As the influence of scattering on various lens surfaces in the optical chain is also effectively suppressed by the presence of the pinhole, a greatly improved dynamic

range of the system results, from 100 to 1 in conventional microscopy to 10,000 to 1 in confocal.

3. In fluorescence confocal microscopy, the lateral confocal response is given by

$$C_1(u) = \text{const}\ I_1(u)\ D_1(\lambda_1 u/\lambda_2) \tag{5}$$

and the corresponding expression for the axial response, where λ_1 is the wavelength of the excitation radiation and λ_2 of the longer fluorescence wavelength. This means that the point resolution in fluorescence, due to broadening of the detection response, will be somewhat reduced (for instance, 7% for $\lambda_2 = 1.25\lambda_1$) with respect to the case $\lambda_1 = \lambda_2$.

4. In practice, in confocal microscopy, pinholes of finite dimensions always have to be used for illumination and detection. Due to the high intensity of laser light sources in general, the illumination pinhole can be made sufficiently small so that it in fact functions as a point light source. However, on the detection side, especially in fluorescence confocal microscopy, larger pinholes will often have to be used in order to collect sufficient light from weakly fluorescent objects. The effective detection sensitivity distribution then becomes a convolution of the point sensitivity distribution with the aperture function of the finite size pinhole. This will result in an effective widening of the detection sensitivity distribution and negatively affect the resolution of the system. To quantify these effects, we would like to use the quantity A as the radius of the first dark ring in the Airy pattern (see Fig. 1). Incidentally, A is very close to the Rayleigh distance for two-point resolution in incoherent conventional microscopy. By integrating $D_1(u)$ over the pinhole area for various radii, we calculated the width at half-intensity of the point response in fluorescence as a function of detector pinhole size (r_p) for the case $\lambda_2 = 1.25\lambda$. The result is plotted in Fig. 2 related to the point response width in conventional fluorescence microscopy at λ_2. We see that for $r_p < 0.4A$ the confocal resolution is close to optimal. For larger r_p the resolution is gradually reduced to a value which is still about 20% above the conventional value. This is because, in that case, the response is determined by the width of the shorter wave illumination distribution, while in conventional microscopy the longer wavelength of the fluorescence radiation determines the resolution. We found a comparable effect for the axial resolution dependence (Van der Voort *et al.*, 1987).

For general objects in either transmission or reflection confocal microscopy, coherence will play a role and the effects of this cannot be predicted in this simple model. However, in confocal fluorescence microscopy, due to the incoherence of fluorescence emission, the detected intensity from

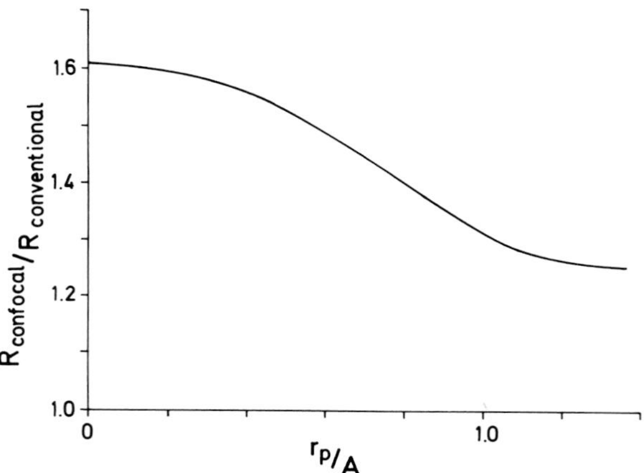

FIG. 2. Resolution gain in confocal fluorescence microscopy over conventional fluorescence microscopy. Plotted is the dependence on detection pinhole radius r_p for the case $\lambda_{fluorescence} = 1.25\lambda_{excitation}$. A is the radius of the first dark ring in the Airy pattern.

the various object elements will be a *linear* sum of the local fluorescence distribution weighted with the confocal fluorescence point response as derived above. We can therefore use this response also to describe the imaging of general objects and expect to be able to image such objects with resolutions as indicated in Fig. 2. Wilson and Sheppard (1984) found from a Fourier decription of the imaging an improvement of the two-point resolution by a factor of 1.3 in incoherent confocal fluorescence microscopy over conventional incoherent microscopy at $\lambda_1 = \lambda_2$. At $\lambda_2 = 1.25\lambda_1$ this factor would become $1.25 \times 1.3 = 1.62$, in agreement with the result shown in Fig. 2.

Implicit in the confocal principle, but deserving specific attention, is the sectioning property. In normal microscopy *all* the fluorescence radiation generated over the depth of the specimen reaches the image plane. The in-focus layer is imaged sharply, while the out-of-focus layers create a contrast-reducing background. In confocal microscopy, due to the combined effect of the focussed illumination and the effect of the detection pinhole, only the radiation generated in the specimen layer imaged reaches the detector efficiently, thus preventing the contribution of out-of-focus areas to the image. A good illustration of this property is that one is unable to find the axial position of a thin uniform fluorescent layer in normal microscopy because, independent of height, the same amount of radiation reaches the image plane. In confocal microscopy the position of the layer can be determined with the resolution as defined by $C_a(w)$.

B. Instrumentation

1. Optics and Scanning

The imaging in a confocal microscope takes place point by point. There are three ways to build up an image in such a system: (1) By moving the specimen with respect to the confocal spot (on-axis) (Brakenhoff *et al.*, 1979). (2) By moving the confocal spot over the specimen with the help of a tilting-mirror system (off-axis imaging) (Carlsson *et al.*, 1985, White *et al.*, 1987). (3) By imaging an array of illuminated pinholes on the object and arranging the detection in such a way that the outcoming radiation passes a corresponding array of pinholes. The holes are located on a rotating disk in a Nipkov arrangement (off-axis imaging) (Petran *et al.*, 1985). With the term on-axis we indicate that in the image formation only the area of the image field of the confocal lens *on* the optical axis is used. In off-axis scanning the whole field of the confocal lens is utilized.

For reasons discussed below, we have opted for scanning the specimen mechanically with respect to the confocal spot (Marsman *et al.*, 1983). We will describe the apparatus that is presently in use in our institute. See Wilson (1980) for an alternative approach to on-axis scanning. In order to really profit from the improved imaging characteristics, we used confocal optics with highest possible NA: immersion optics with NA = 1.3. The optical layout is given in Fig. 1. Figure 3 shows a cross-section of the confocal optics/scanning unit. The most critical element for a satisfactory image formation is the mechanical scan movement of the object with respect to the confocal optics. The stability of the object positioning and repeatability of the scan should be appreciably better than the expected optical resolutions in the range of 100 to 200 nm. In the approach indicated in Fig. 3, we were able to realize a relative stability of 10 to 20 nm over time, sufficient for focussing and image acquisition. The scan format is conventional, a fast line scan frequency (range 50–150 Hz) with the image repetition frequency depending on the number of lines per frame. Typical values are 64 or 128 lines per frame or subframe during searching and focussing, and 2048 lines for direct recording on film. For input into the computer, we mostly used 256 lines per frame. The system is equipped with a Krypton-Ion laser providing a choice of excitation wavelengths between 330 and 600 nm.

2. Instrument Control and Computer System

For the operation and control of the microscope, we use two computer systems. The first, a relatively slow 8-bit one, takes, via a number of

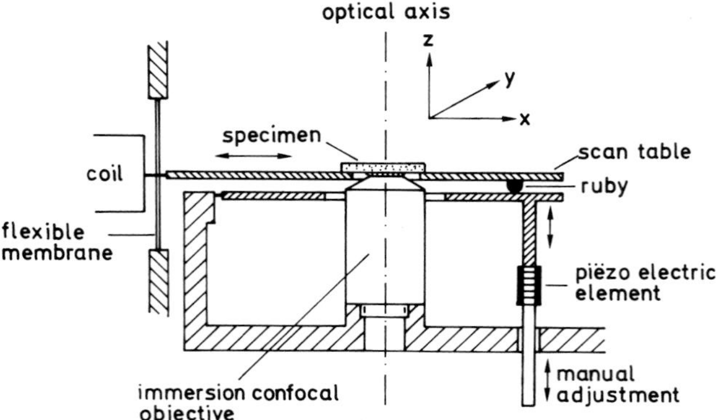

FIG. 3. Confocal optics/scanning unit. The specimen is placed on the scantable and is scanned above the confocal immersion lens. The height or z-position of the scan table is determined by the skating of the ruby attached to the table over a hard metal polished surface. The movement in the x- (shown) and y-directions (not shown) is guided by membranes which are flexible in the scan direction. The lens is a high-numerical aperture lens with NA = 1.3. The hard metal reference plate on which the ruby skates can be positioned in the z-direction by a combination of coarse manual adjustment and a piezoelectric actuator for the fine control. A special feature of the inverse optical arrangement is that the specimen is easily accessible for experimentation during CSLM imaging.

interfaces, care of the various household functions like scanning, scan magnification, specimen height control, optical alignment control, photomultiplier setting, spectrograph wavelength and bandwidth setting, The second, a 16-bit system with 68000 processor, is used during image acquisition for data collection and storage. It also generates at that stage from the contents of the frame store a continuous 60-Hz repetition frequency viewing image. This image is independent of the slower mechanical scan movement of the microscope.

A very useful routine developed for the instrument acquires automatically, after having obtained from the operator the necessary parameters, a series of images at various heights in the object (optical sections) and stores these on a hard disk. These sections are the starting material for the generation of stereoscopic images, etc. (see below). Image acquisition parameters are at present as follows: with a 50-Hz line scan the data from a section digitized on a 256 × 256 pixel grid are collected in 5 seconds. For the applications shown below, we mostly collected the data from 6 to 16 sections, therefore requiring 0.5 to 1.5 minutes. These values can be easily upgraded, 100 to 150 Hz are practical scan frequencies, with the data collected on a 512 × 512 raster or higher.

3. Image Processing Facilities

We use the same 68000 processor for off-line processing of the microscope data for various applications. Available are various two dimensional image processing routines like shifting, stretching, magnification, rotation, Gaussian and median filtering and a local contrast enhancement filter for contrast manipulation. In addition to this, a 68020 based multiprocessor system is available for more computation-intensive tasks like 3-D image processing and analysis.

To present to the user the three-dimensional information obtained in the confocal microscope, we have written a stereoscopic image generation routine. Due to the stability of the scan motion, all the image elements in a series of optical sections, as acquired by the scan program, are aligned with respect to each other well within the resolution limit, so that the data can be used without preprocessing. The algorithm we employ operates as follows: let the pixel values in a stack of images be represented by $I_{i,j,k}$ with $i, j: 0 \ldots 255$ and $k: 0 \ldots 15$. An image can then be generated by a routine in which to a pixel $S_{i,j}(s_1, s_2)$ the maximum value of the set of pixels $I_{i+_s1k,\ j+s2k}$ (k running from $0 \ldots 15$) is assigned. Two images, $R = S(s_1, s_2)$ and $L = S(s_3, s_4)$, can then be viewed as a stereoscopic image pair. By a suitable choice of the values of s_1 to s_4, the object can be examined from any direction with any desired stereoscopic viewing angle. Running on the multiprocessor system, the algorithm is capable of generating a stereoscopic pair from a $256^2 \times 16$ data set in less than 3 seconds. For a large class of objects, the maximum routine gives good results. One can choose also alternatives like adding the set of pixel values or taking the minimum. For the results presented below we mostly chose s_1, s_2, $s_3 = 0$, $s_4 = 1$ or -1.

III. Confocal Imaging Results

A. Three-Dimensional Fluorescence Point Response

We have determined this response with the help of a 90-nm fluorescent bead (Fig. 4). The data for this response were collected by the through-focus routine described above. The section spacing was 0.15 μm and the lateral sample spacing was 22 nm. The noise in the original data (bleaching puts a limit to the intensity of the excitation radiation) was reduced by suitable filtering (based on Fourier transforms, where only frequencies within a bandwidth of about three times the highest expected optical frequency were passed). From Fig. 4 it can be seen that, in our present

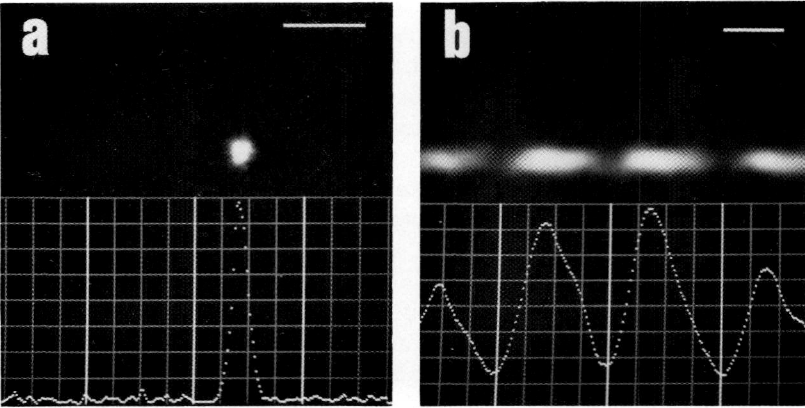

FIG. 4. Three-dimensional confocal imaging of a 90-nm fluorescent bead (excitation $\lambda = 482$ nm, detection > 510 nm). (a) An *x-y* image through the center of the bead is shown. (b) a number of *x-z* cross-sections through the bead taken at distances in the *y*-direction of 90 nm are shown. The apparent width of the fluorescence response on this approximate point object is 220 nm lateral and 0.9 μm axial. (Both pictures: bar = 1 μm.)

system, we have a lateral point response width of about 220 nm and 0.9 μm in the axial direcction. From Eq. (5) we calculate, in a NA = 1.3 system with a point detector at the excitation and detection wavelengths used, a lateral response width of 175 nm and, along the axial direction, of 800 nm. The differences with the observed values can be partly accounted for by the finite size of the pinhole used ($R = 0.5$, see above) and partly by the fact that the theoretical expectation is based on relations derived in the paraxial approach. At high NA, the expected response will then be about 10% broader (Wilson and Sheppard, 1984).

B. Biological Applications

1. Optical Sectioning of Chloroplasts

Plant chloroplasts contain several photosynthetic pigments that show bright autofluorescence while remaining relatively insensitive to photobleaching. For this reason, we used this material to demonstrate several features of the CSLM. The images were made of living cells, thus avoiding fixation techniques which often make the interpretation of microscopic pictures difficult. The chloroplast of most green algae differ from the ones found in higher plants in the sense that their thylakoids are not stacked in a regular pattern. These chloroplasts, however, may exhibit very intricate forms. While being a simple disk-shaped structure in higher plants, green algae chloroplasts may be spiral, bell or star shaped, or they may form irregular networks. It is this complicated three-dimensional organization that makes them difficult to visualize in conventional microscopy and, at the same time, particularly amenable to making optical sections in confocal microscopy.

Chlamydomonas is a single-celled organism which contains one large chloroplast. Its shape has been revealed previously by making serial sections for electron microscopy (Schötz *et al.*, 1972). By illuminating a specimen of *Chlamydomonas eugametos* with 483 nm laser light, we recorded a bright red fluorescence. In Fig. 5 such a chloroplast is sectioned optically by means of the CSLM. Successive optical sections were made 800 nm apart and, as predicted by theory for a confocal microscope, there is no apparent overlap between them. There are no grana-like structures in *Chlamydomonas* and so the bell-shaped chloroplast shows a more or less equal fluorescence throughout. About halfway through this series of sections, a large pyrenoid occurs. Because this is the place where starch accumulates, it does not exhibit fluorescence.

2. Low-Magnification Scanning Microscopy

Closterium cells consist mainly of a large chloroplast, split in two halves by the nucleus. In Fig. 6 a fluorescent image, as seen in a conventional fluorescence microscope, is presented. The nucleus has been made visible by staining the cell with a 20 μM solution of the DNA-specific dye 4′6-diamidino-2-phenylindole (DAPI) (Coleman *et al.*, 1981). This is a so-called vital stain and we observed that cells continue to grow and divide normally after this treatment. Because of the length and the thickness of this organism, it was impossible to photograph it properly with a 100×, high-NA lens in an ordinary microscope, and a 40×/0.70 lens had to be used.

In Fig. 7 the same species is visible in a CSLM picture. Making low-magnification pictures with a 100×/1.3 lens in an on-axis scanning microscope is a matter of increasing the scan amplitude of the table and part of the difference in image quality between Fig. 6 and Fig. 7 must be attributed to this effect.

Figure 8 shows a detail of a chloroplast after "zooming in" (that is, decreasing the scan amplitude) on the image shown in Fig. 7.

3. Dual-Wavelength Recording

In Fig. 9, a nucleus from a single *Closterium monoliferum* cell is visualized. The DAPI-stained DNA is visible in the form of a very large

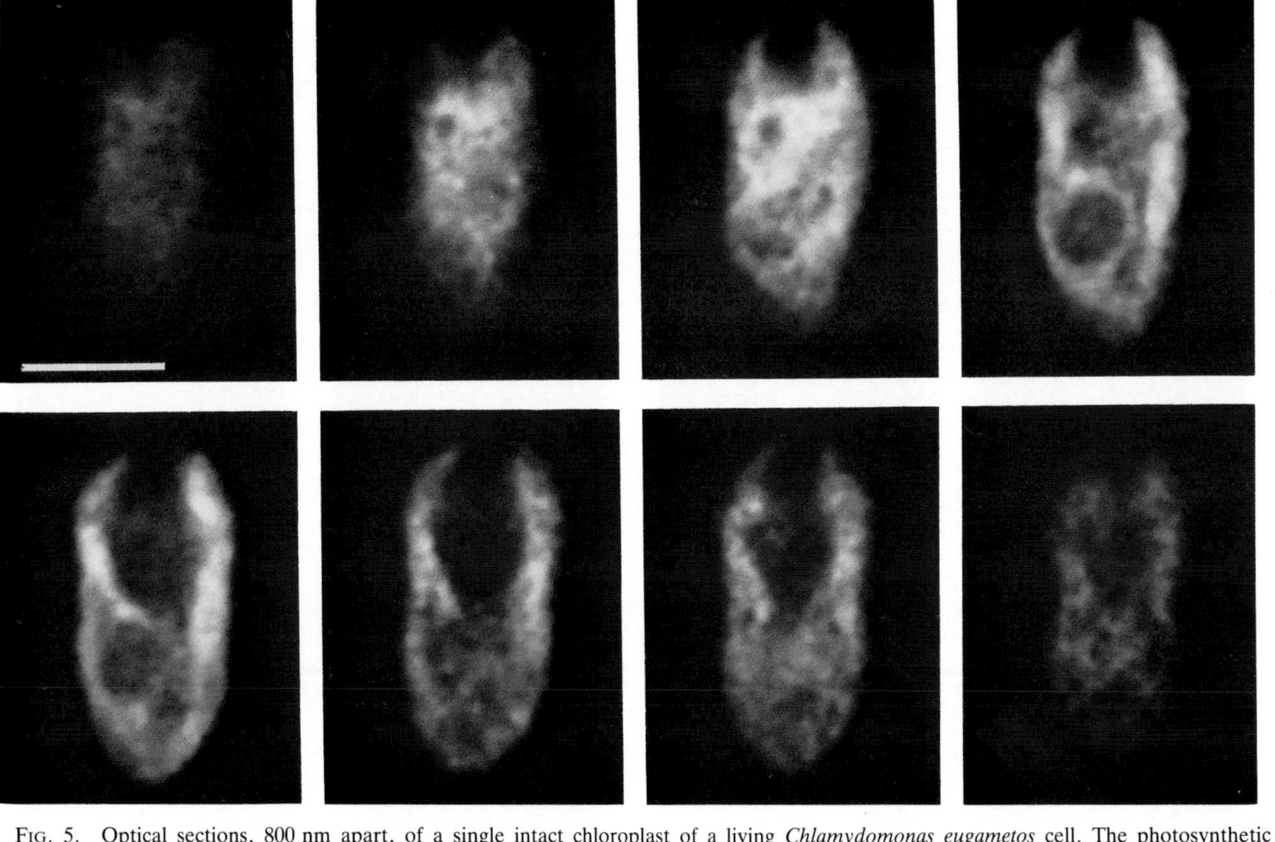

FIG. 5. Optical sections, 800 nm apart, of a single intact chloroplast of a living *Chlamydomonas eugametos* cell. The photosynthetic pigments in this bell-shaped structure show a bright emission peak in the red with illumination at 483 nm and a 520 nm blocking filter. The dark spot in the lower left part, about halfway through the series, is the pyrenoid. (Bar = 5 μm)

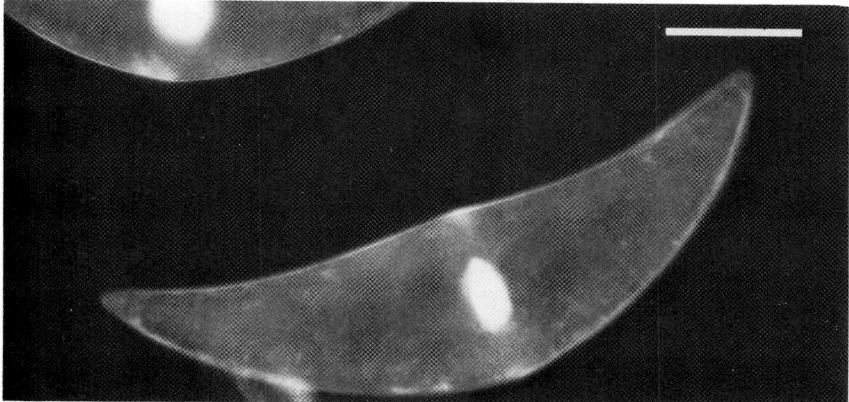

FIG. 6. Fluorescence image of the green alga *Closterium moniliferum,* photographed in a conventional microscope. The cell's nucleus is stained with the DNA-specific dye DAPI and the violet light used to excite it is also used to excite the cell's photosynthetic pigments. These are contained in the large chloroplast that makes up the major part of the cell body. Because of the thickness of this specimen, much of its structure is blurred by out-of-focus fluorescence. (Bar = 50 μm.)

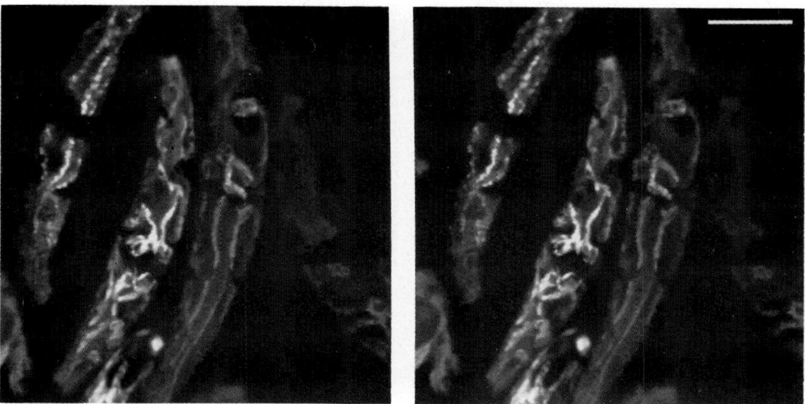

FIG. 7. Stereoscopic fluorescence CSLM image of the same species as in Fig. 6, excited with 413 nm laser light. This picture consists of eight sections, superimposed during stereo processing, each displaying only infocus information and only the color to which the spectrograph was set (in this case red). (Bar = 50 μm.)

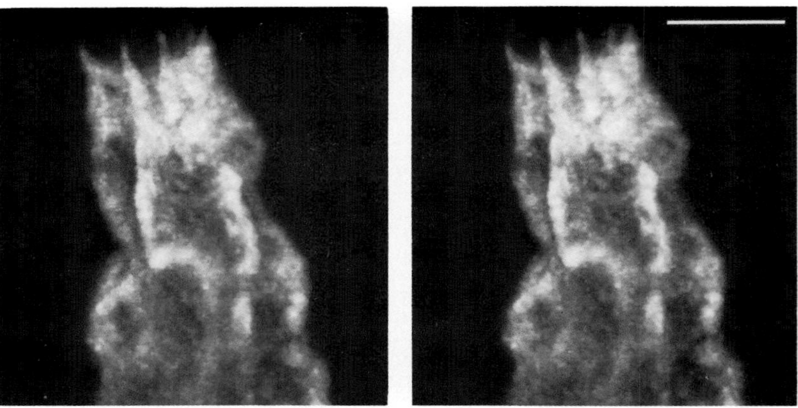

FIG. 8. A detail of the chloroplast in Fig. 7 made by "zooming in" on the cell by reducing the amplitude of the scan table of the CSLM. (Bar = 10 μm.)

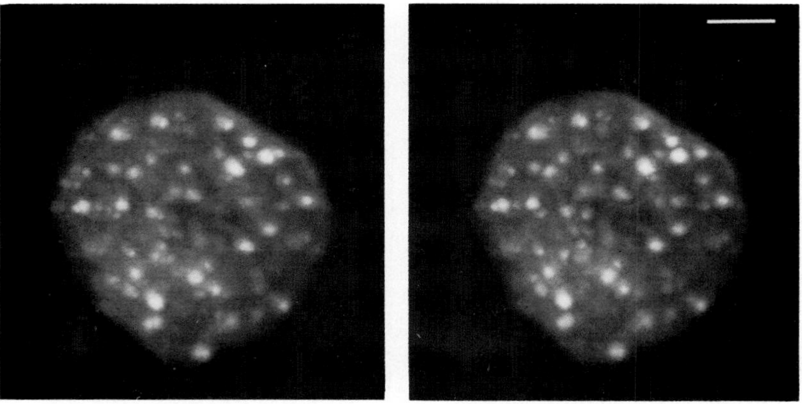

FIG. 9. The nucleus of *Closterium moniliferum,* which fills the gap between the chloroplast halves (clearly visible in Fig. 6 and 7), is stained with a DNA-specific dye (DAPI). The spectrograph was set to blue emission so as to select the dye's signal. The three-dimensional distribution of a large number of small chromosomes can easily be discerned in this picture. (Bar = 5 μm.)

number of small chromosomes. This picture was made with the spectrograph tuned to the emission waveband of DAPI, which is in the blue, whereas the chloroplast autofluorescence (Figs. 7 and 8) is in the red. In the picture made with a conventional fluorescence microscope (Fig. 6) a long pass filter was used to block the violet excitation light from the fluorescence signal. Therefore, both blue and red fluorescence is visible in the same image. Because the spectrograph in the CSLM actually performs as a variable bandpass filter, fluorescent images can easily be split into their constituent colours (in this particular case blue for the nucleus and red for the chloroplast) and be processed independently.

4. Analysis of Three-Dimensional Structures in Lymphocytes

There is a continuing debate about whether the distribution of chromosomes inside nuclei is entirely random, or may be subjet to some sort of organization. A significant problem that has to be overcome before answering this question is the poor visibility of individual chromosomes outside the metaphase. Obviously, the solution to this problem is closely coupled to the quality of imaging technology. Here we present an approach to this problem in which the specific advantages of the CSLM are used in combination with a powerful new hybridization procedure to reveal aspects of the three-dimensional organization of chromosomes in lymphocytes.

Human peripheral lymphocyte interphase nuclei, isolated from fresh venous blood, were hybridized *in situ* with a biotinylated chromosome-specific DNA probe. The hybridization procedure was entirely performed in suspension in order to preserve morphology and allow three-dimensional analysis. Bound probe was visualized using fluoresceinated avidin (bright spots in Figs. 10 and 11), while all the DNA was counter-stained with propidium iodide (grey). In Fig. 10 a probe specific for the long arm of chromosome Y was used. In these male cells only one hybridization spot is clearly visible, representing the target region of the probe. Particularly apparent is the localization of this region near the nuclear surface. This kind of observation is hard to make in conventional fluorescence microscopy. In Fig. 11 a probe for chromosome 1 can be seen. In most cells two hybridization spots are visible, representing the centromeres of both chromosome 1 homologs. In addition, in some nuclei so called minor hybridization spots are visible, representing centromeres of other chromosomes. Also shown in this picture, the localization of the DNA targets seems to be related to the nuclear membrane.

5. High-Resolution Confocal Microscopy of Yeast Cells

Actin in its polymerized form (F-actin) plays an important role as an essential element of the cytoskeleton of eukaryotic cells. The availability of phalloidin (Wulf *et al.*, 1979) offers a means of visualizing this element in

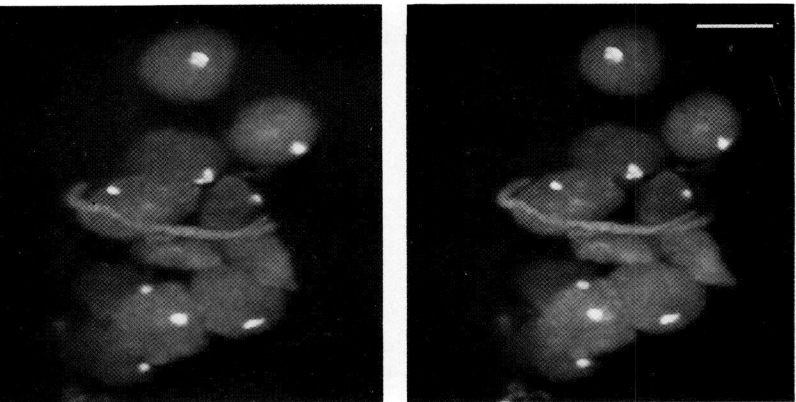

FIG. 10. Interphase human lymphocyte nuclei in which one arm of the Y chromosome is stained with a DNA probe. One bright fluorescent spot per nucleus is visible in these male cells against a grey background caused by staining all of the nucleus's DNA with propidium iodide. (Bar = 10 μm.)

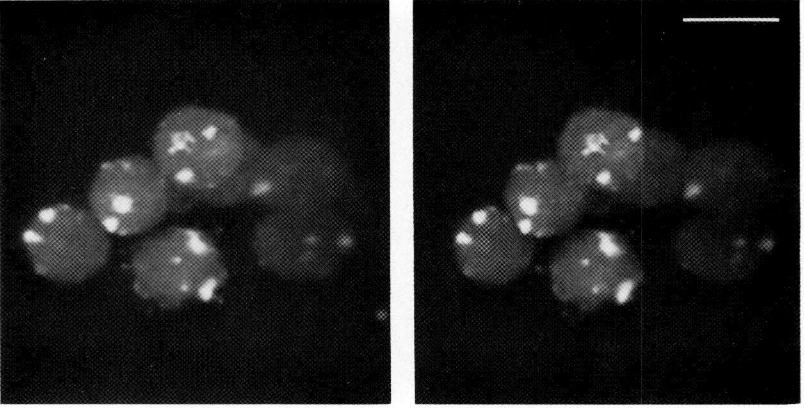

FIG. 11. Lymphocytes from the same individual as in Fig. 10, now with both chromosomes 1 stained at their centromeres (specimen and photograph by H. van Dekken). (Bar = 10 μm.)

the fluorescence microscope. This mushroom poison, which very specifically attaches to F-actin, can be fluoresceinated without any notable interference with its normal functioning and has recently become commercially available. Because it is a relatively small molecule (MW$\approx$1250), it passes cell walls after only a mild fixation procedure. We used this

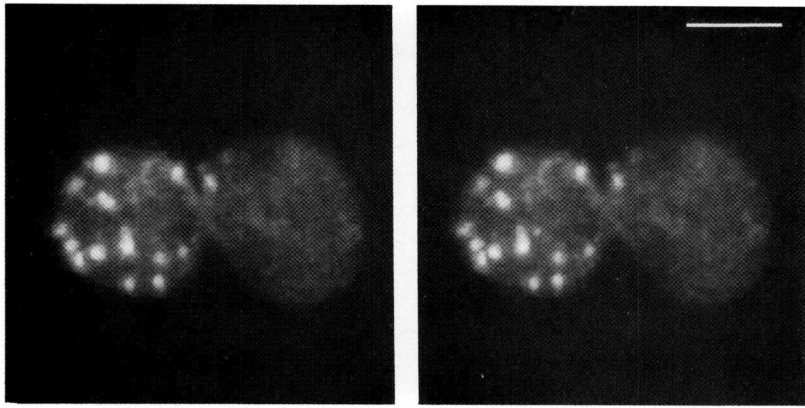

FIG. 12. Actin visualized in the yeast *Saccharomyces cerevisiae* by staining with fluoresceinated phalloidin, a mushroom poison. Actin is mainly present near the cell wall of the growing bud. (Bar = 2 μm.)

staining technique to study the distribution of actin inside the cells of the budding yeast *Saccharomyces cerevisiae* (Fig. 12). In this organism, the occurence of actin is rather different from that seen in most other eukaryotes. Instead of fiberlike structures, yeast actin forms patchlike structures, which seem to be located near the cell wall (Adams and Pringle, 1984; Kilmartin and Adams, 1984). However, because of their apparent size of about 300 nm, the three-dimensional organization of these patches is hard to visualize by means of high-magnification conventional microscopy. In the confocal image presented here, the three-dimensional distribution of actin patches near the cell wall and their prevalence in places where deposition of new cell wall material is likely to occur (that is the growing buds), strongly suggests a role for actin in the growth processes of *S. cerevisiae.*

IV. Discussion

A. Scanning Systems

In the previous sections, we have shown the results obtained in our on-axis system where we have the specimen moving with respect to the confocal spot located on the optical axis. The advantage of such an approach from the optical point of view is twofold: first, one uses for imaging that part of the lens for which it is optimally corrected, or vice versa, one can use lenses which are only well corrected on-axis. Second, the area that can be imaged (field of view) is determined by the mechanical scan amplitude and can thus be much larger than the optically determined field of view. For instance, Fig. 7 could not have been obtained with NA = 1.3 immersion optics in off-axis confocal systems. An additional virtue of an on-axis system, in view of the subsequent image processing, is that the imaging conditions for each point of the three-dimensional image are exactly identical.

In off-axis confocal microscopy, the confocal spot is scanned over the specimen optically with the help of a scanning mirror system. The main advantages of this approach are first, that the specimen can remain stationary, and second, that image acquisition times can be reduced due to the higher scan rates possible with optical scanning. In the Nipkov-type off-axis confocal microscopy, the image repetition frequency can be increased up to TV frequencies (Petran *et al.,* 1985). However, a penalty has to be paid in this system because of the fact that the sizes of both the illumination and the detection pinholes cannot be made very small for technical and light-efficiency reasons. This means that the effective resolution and sectioning capabilities of such a system will be negatively affected.

The desired signal-to-noise ratio in the image is often the limiting factor in the image acquisition time. If that is the case, then the integration times needed in conventional fluorescence microscopy (or for that matter Nipkov-type confocal microscopy) become equivalent to the time it takes to build up a scanning image. Per image pixel the amount of radiation collected is identical. In "whole-frame" microscopy, the signal is collected parallel by all the pixels at low light levels; in scanning microscopy it is collected pixel by pixel at the much higher illumination levels made possible by the high intensity laser light sources used. For rapidly (0.1 second) changing phenomena a scanning confocal microscope seems not to be the first choice. Video-enhanced microscopy (Allen and Allen, 1982) is more appropriate for this purpose (see Inoué, Chapter 3, this volume). However, it is possible in scanning microscopy to reach image repetition frequencies of about 5 Hz by reducing the number of lines per image. By reducing the size of the scan field, we can still have a high-resolution image over a limited area.

With the respect to the mechanical scanning of the object, we have not experienced this as a limiting factor in our confoal work. The specimens examined are in general rather small, and if a specimen is embedded in a medium of comparable specific gravity, as is often the case, then no scan-induced deformation would be expected, nor did we observe such a phenomenon in our objects. For practical reasons and to be able to examine heavy objects, it nevertheless seems desirable to leave the object stationary and still have on-axis confocal imaging. This is possible by moving the confocal optics with respect to the object in a controlled way (Hamilton, 1986). We recently constructed a system with scanning high-NA optics which produced images of comparable quality to those obtained in the object scanning system (Brakenhoff, 1988). To summarize this section: the future user should weigh the arguments mentioned above against his needs and make his choice, balancing the optical and practical merits of on-axis scanning systems against those of the off-axis approach.

B. Final Remarks

We have presented the various aspects of confocal fluorescence microscopy. As demonstrated, the combination of a high transverse resolution with optical sectioning results in very-high-quality images of biological specimens. The instrument is versatile, can handle multiple fluorescing labels, and is relatively easy to use. When coupled to a computer, the following parallel can be drawn: where in normal microscopy we make a two-dimensional image on a photographic plate, a charge-coupled device, etc., we can project in confocal microscopy a three-dimensional image of the structure in a three-dimensional computer memory array. For weakly absorbing objects, these data represent directly the distribution of fluorescent material over the sampled volume of the specimen.

The instrument is suitable for the study of phenomena changing on a not-too-rapid time scale. For instance, cell development in cultures can be followed with three-dimensional image acquisition in 0.5 to 1 minute, depending on circumstances, and in two dimensions at considerably faster rates. The gain is especially notable with respect to electron microscopy, where it can take months (sectioning, photographing, aligning images, digitizing images etc.) before the three-dimensional structure of a specimen can be studied.

We have developed various representation methods, among which are stereoscopic images, to make the three-dimensional structure visually accessible, but we are of the opinion that the optimal use of the three-dimensional confocal data will be made when they form the basis for structural analysis of the biological objects studied. In that stage, the property of equal imaging conditions for all data points in on-axis confocal systems may be very important. The presented biological examples demonstrate that the instrument can be fruitfully applied to many biological problems and we expect that this technique, in any of its indicated forms, will be extensively used for the elucidation of the two- and three-dimensional structure of biological objects in the future.

Acknowledgments

We thank H. van Dekken (TNO, Rijswijk) for the lymphocyte specimen and A. Tomson and R. Demets (Department of Plant Physiology, University of Amsterdam) for the *Chlamydomonas* specimen. This work was supported by Stichting voor de Technische Wetenschappen (STW) and by the Foundation for Fundamental Biological Research (BION), which is subsidized by the Netherlands Organization for the Advancement of Pure Research (ZWO).

References

Adams, A. E. M., and Pringle, J. R. (1984). *J. Cell Biol.* **98,** 934–945.

Agard, D A., and Sedat J. W. (1983). *Nature (London)* **302,** 676–681.

Allen, R. D., and Allen, N. S. (1982). *J. Microsc.* **129,** 3–17.

Born, M., and Wolf, E. (1975). "Principles of Optics." Pergamon, Oxford.

Brakenhoff, G. J. (1988). To be published.

Brakenhoff, G. J., Blom, P., and Barends, P. (1979). *J. Microsc.* **117,** 219–232.

Brakenhoff, G. J., Binnerts, J. S., and Woldringh, C. L. (1980). *In* "Scanned Image Microscopy" (E. A. Ash, ed.), pp. 183–201. Academic Press, London.

Brakenhoff, G. J., Van der Voort, H. T. M., Van Spronsen, E. A., Linnemans, W. A .M., and Nanninga, N. (1985). *Nature (London)* **317,** 748–749.

Carlsson, K., Danielson, P. E., Lenz, R., Liljeborg, A., Majlöf, L., and Aslund, N. (1985). *Opt. Lett.* **10,** 53–55.

Coleman, A. W., Maguire, M. J., and Coleman J. R. (1981). *J. Histochem. Cytochem.* **29,** 959–968.

Cox, I. J., Sheppard, C. J. R., and Wilson, T (1982). *Optik* **66,** 391–396.

Hamilton, D. K., and Wilson, T. (1986). *J. Phys. E. Sci. Instrum.* **19,** 52–54.

Johnson, G. D., and de C. Noqueira Araujo, G. M. (1981). *J. Immunol. Methods* **43,** 349–350.

Kilmartin, J. V., and Adams, A. E. M. (1984). *J. Cell Biol.* **98,** 922–933.

Marsman, H. J. B., Stricher, R., Wijnaendts van Resandt, R. W., and Brakenhoff, G. J. (1983). *Rev. Sci. Instrum.* **54,** 1047–1052.

Petran, M., Hadravsky, N., and Boyde, A. (1985). *Scanning* **7,** 97–106.

Ruch, F. (1970). *In* "Introduction to Quantitative Cytochemistry" 431–450 (G. L. Wied and G. F. Bar, eds.), Vol. 1, pp. 431–450. Academic Press, New York.

Schötz, F., Barthelt, H. Arnold, C. G., and Schimmer, O. (1972). *Protoplasma* **75,** 229–254.

Sheppard, C. J. R., and Choudhury, A. (1977). *Optica* **24,** 1051–1073.

Van der Voort, H. T. M., Brakenhoff, G. J., and Janssen G. C. A. M. (1987). Optik **78,** 48–53.

White, J. G., Amos, W. B., and Fordham, M. (1987). *J. Cell Biol.* **105,** 41–48.

Wijnaendts van Resandt, R. W., Marsman, H. J. B., Kaplan, R., Davoust, J., Stelzer, E. H. K., and Stricker, R. (1985). *J. Microsc.* **138,** 29–34.

Wilson, T. (1980). *Appl. Phys.* **22,** 119–128.

Wilson, T., and Sheppard, C. J. R. (1984). "Theory and Practice of Scanning Optical Microscopy." Academic Press, New York.

Woldringh, C. L., de Jong, M. A., v.d. Berg, W., and Koppes, L. (1977). *J. Bacteriol.* **131,** 270–279.

Wulf, E., Deboben, A., Bautz, F. A., Faulstich, H., and Wieland, T. (1979). *Proc. Natl. Acad. Sci. U.S.A.* **76,** 4498–4502.

Chapter 15

Emission of Fluorescence at an Interface

DANIEL AXELROD AND EDWARD H. HELLEN

Biophysics Research Division and Department of Physics
University of Michigan
Ann Arbor, Michigan 48109

I. Introduction

In many of the experiments involving fluorescence in cell biology, the fluorophores are located near a surface. Usually this surface is an aqueous buffer/glass or plastic interface upon which cells grow. Occasionally, the interface may have a thin coating on it, such as a synthetic polymer, a metal, or a lipid bilayer. The presence of the surface affects several aspects of the fluorescence, including its intensity, angular distribution, quantum yield, and lifetime.

The effects referred to here are not "quantum chemical," i.e., effects upon the orbitals or states of the fluorophore in the presence of any static fields associated with the surface. Rather, the effect is "classical optical," i.e., effects on the electromagetic field generated by a classical oscillating dipole in the presence of an interface between any media with dissimilar refractive indices. Of course, both types of effect may be present simultaneously in a given system. But the quantum-chemical effects vary with the detailed chemistry of each system, whereas the classical-optical effects are more universal. Occasionally, a change in the emission properties of a fluorophore at a surface may be attributed to the former when in fact the latter is responsible.

This chapter deals only with the classical-optical effects. It emphasizes the emission properties as they might be observed through a microscope, with particular attention to the bare-glass/water and metal film-coated glass/water interfaces. The results suggest some practical experiments that take advantage of the special optical effects at surfaces. These experiments include deducing the relative concentration of fluorophore as a function of distance from the surface, quenching unwanted "background" fluorescence from fluorophores nonspecifically adsorbed to a substrate; and optimizing collection of fluorescence by a microscope objective.

II. Theory

A. Description of the Model

Surface optical effects can be calculated at various levels of approximation. The simplest (and least accurate) approach is to model the fluorophore as an oscillating electric dipole of fixed amplitude generating only rays of propagating light (the "far field"). The rays (which are actually symbols for propagating plane waves) interact with the surface according to Snell's Law and the law of reflection. Their uniformly spaced wave fronts (within each uniform refractive index medium) extend as semi-infinite planes. This approach considers the interference between such rays of propagating light directly emitted from the dipole and light rays reflecting off the interface. The results are valid only for distances z' from the surface of greater than the light wavelength (~500 nm).

A better approximation must consider the so-called "near field." The mathematical form of a dipole radiation pattern cannot be expressed simply as a superposition of plane waves/rays propagating in different directions with direction-dependent amplitudes. Rather, it is necessary to suppose that some of the wave fronts do not extend infinitely far from the dipole but instead exponentially decay. A whole set of such exponentially decaying fields exists with a continuous range of decay constants. When a fluorophore (say, in water) is near a higher refractive index surface (say, glass), each of these exponentially decaying near-field components can interact with the surface and ultimately become propagating waves in the glass at its own unique angle θ (with respect to the normal). Angle θ is always greater than the critical angle θ_c for total internal reflection (see Axelrod, this volume). This conversion of exponentially decaying waves from the near field of the dipole in water into supercritical angle propagating waves in the glass can be significant for fluorophores within about one wavelength of the surface. At a metal surface, consideration of the near field is even more important, because the metal converts its electromagnetic energy into heat. Figure 1 is a schematic diagram of the near field interacting with a surface.

Another feature of the simplest model that needs modification is the assumption of a fixed-dipole amplitude. Because of the efficient capture of nonpropagating near fields by a surface, a fixed dipole emits more power the closer it moves to a surface. But in steady-state fluorescence, the emitted power can only be as large as the (constant) absorbed power (or less, if the intrinsic quantum yield of the isolated fluorophore is less than 100%). Therefore, the fluorophore must be modeled as a constant *power* (and variable *amplitude*) dipole.

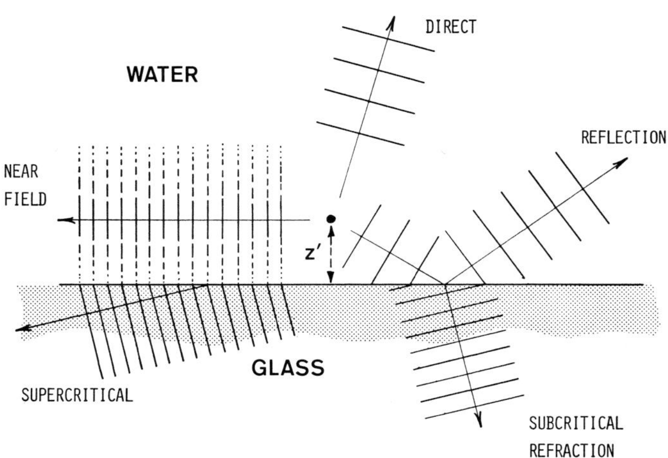

FIG. 1. Schematic drawing of the interaction of dipole radiation with a dielectric interface, showing the direct far field, the reflected far field, the subcritical refracted far field, the near field, and the supercritical light arising from the near field.

The two above features which modify the simplest theory extend the range of distance z' between the fluorophore and the surface over which the results remain valid from a minimum of several hundred nanometers without the modifications, to less than 10 nanometers with them. Those two features are incorporated into the results displayed here. Other refinements, not included here, involve consideration of energy transfer to electron-hole pairs (for metals only at $z' < 10$ nm) and nonhomogeneous atomic field effects ($z' < 0.25$ nm).

B. Mathematical and Physical Origin of the Effects

Details of the calculations discussed here, along with numerous references to prior work are given by Hellen and Axelrod (1987). This section briefly summarizes the mathematical approach and emphasizes the physical meaning.

The model here consists of a medium 3 (refractive index n_3, e.g., glass), and sandwiched layer of thickness t (n_2, e.g., polymer, metal, lipid, or more of medium 3) and a medium 1 (n_1, e,g., water). The $z = 0$ origin is at the $n_1 : n_2$ interface and the dipole resides in medium 1 at $z = +z'$.

In general, an electric field $\mathbf{E}(\mathbf{r})$ of an isolated, fixed-amplitude dipole (i.e., no surfaces nearby) can be expanded as an integral over plane waves (with sinusoidal time dependence suppressed) as follows:

$$\mathbf{E}(\mathbf{r}) = \int d\mathbf{k} \exp(i\mathbf{k} \cdot \mathbf{r})\, \mathbf{E}(\mathbf{k}) \tag{1}$$

where wave vector $\mathbf{k} = k_x\hat{x} + k_y\hat{y} + k_z\hat{z}$, and the integration extends over all $\mathbf{k}$, subject only to the restriction that the frequency of the light ω remain fixed. $\mathbf{E}(\mathbf{k})$ can be determined directly via Maxwell's equations, or from the known form of dipole radiation $\mathbf{E}(\mathbf{r})$ via Eq. (1). Since the magnitude of $\mathbf{k}$ is

$$|\mathbf{k}| = (k_x^2 + k_y^2 + k_z^2)^{1/2} = \omega n_1/c \tag{2}$$

the integration in Eq. (1) requires only that $|\mathbf{k}|$ be held fixed. Propagating (sinusoidal) waves are described by k_x^2, k_y^2, and k_z^2 all positive. However, if we allow $(k_x^2 + k_y^2) > |\mathbf{k}|^2$, then $k_z^2 < 0$. Those "plane waves" have an imaginary k_z and correspond to exponentially decaying waves in either direction along the z-axis, starting from the position of the dipole. This set of plane waves is the dipole's "near field." The amplitude and phase of each wave exp(**ikr**), whether propagating or exponentially decaying, is given by $\mathbf{E}(\mathbf{k})$.

The near-field wave fronts are all parallel to the z-axis, exponentially decaying in amplitude in either direction along the z-axis starting from the dipole position, and traveling radially outward parallel to the x-y plane.

The apparent wavelength of each exponentially decaying wave is shorter than that of the propagating waves, corresponding to the large radial **k**-vector amplitude given by

$$k_r = (k_x^2 + k_y^2)^{1/2}. \tag{3}$$

Up to this point, the choice of x-, y-, and z-axes is quite arbitrary. However, the placement of the $n_1 : n_2$ interface as the x-y plane assigns a fixed meaning. In that case, any waves traveling toward the interface can interact with it, by both reflecting and refracting. The amplitudes of the reflected and refracted plane waves are given by Fresnel coefficients, appropriately modified for a three-layer system. The directions of each are given by Snell's Law of refraction and the law of reflection. The Fresnel coefficients are functions of the dielectric constants $\varepsilon_i(= n_i)$ in the three media. The dielectric constant is simply the square of the refractive index. For a dielectric, ε is positive and real. But for a metal, ε is complex, with a negative real part and a positive imaginary part.

The electric field $\mathbf{E_1}(\mathbf{r})$ observed at point $\mathbf{r}$ in medium 1 is then the superposition of the direct field [calculated from Eq. (1) for waves traveling away from the surface in the $+z$ direction] and reflected waves (calculated as described above), integrated over all allowed **k** vectors. The electric field $\mathbf{E_3}(\mathbf{r})$ observed in medium 3 is an integral of the refracted waves, also integrated over all allowed **k**. Note that Snell's law demands that the spacing between successive wave fronts as projected on a dielectric boundary must be the same on both sides of the boundary. This requirement is the same as the statement that k_r be continuous across the boundary. Then the exponentially decaying near field waves, with their short wave front spacing, will refract only into supercritical angles in medium 2.

For fluorophores close to the surface, exponentially decaying waves with a wide range of decay constants ik_z will extend to the surface, giving rise to a wide band of supercritical refraction angles. But for fluorophores somewhat farther away, only those exponentially decaying waves with the longest decay distances (i.e., the smallest ik_z) will reach the surface, giving rise to a rather narrow band of supercritical emission angles extending only slightly above the critical angle. Therefore, by viewing only supercritical angle emission at a fixed angle θ, one will detect only fluorophores near the surface. The higher the θ, the closer to the surface are the detected fluorophores.

One *might* conclude that: (1) at any particular supercritical observation angle θ in the glass, the observed intensity will decrease exponentially with z' as a fluorophore is pulled away from the surface; and (2) the total emission into all other angular ranges is unaffected by moving the fluorophore. However, neither of these conclusions is correct. Recall that a fluorophore must be modeled as a fixed-power rather than fixed-amplitude dipole. That means that any increase in emited power into any one set of directions, e.g., supercritical angles into the glass, will be at the expense of emitted power into other directions. Furthermore, the intensity emitted into medium 1 is determined partly by the phase-dependent interference between direct and reflected plane waves, which is also a function of z'. The intensity at any angle must be normalized by a function describing the total power $P_T(z')$ released by the dipole (including any lost into heat in a dissipative medium 2). In general, $P_T(z')$ can be calculated from

$$P_T(z') = (\omega/2)\, Im(\boldsymbol{\mu} \cdot \mathbf{E}) \tag{4}$$

Physically, this formula describes the power dissipated by a harmonic oscillator (the dipole with moment $\boldsymbol{\mu}$) as it is driven by the force felt at its own location from its own emitted and reflected electric field. P_T is calculable given all the refractive indices and Fresnel coefficients of the layered model (Hellen and Axelrod, 1987; Ford and Weber, 1984).

Incorrect conclusion (1) above is sometimes said to derive from the "reciprocity principle," which states that light waves in any optical system all could be reversed in direction without altering any paths or intensities and remain consistent with physical reality (because Maxwell's equations are invariant under time reversal). Applying this principle here, one notes that an evanescent wave set up by a supercritical ray undergoing total internal reflection (see Axelrod, chapter 9, this volume) can excite a dipole with a power that exponentially decays with z'. Then (by the reciprocity principle) an excited dipole should lead to a supercritical emitted beam intensity that also decays exponentially with z'. Although this prediction would be true *if* the fluorophore were a fixed-amplitude dipole in both cases, it cannot be modeled as such in the latter case.

The radiated intensity $\hat{S}(\mathbf{r}, z')$ from a fluorophore near a surface per unit of adsorbed power can be derived from the Poynting vector magnitude. In terms of $\mathbf{E_1(r)}$ and $\mathbf{E_3(r)}$, it is

$$\hat{S}_i(\mathbf{r}, z') = \frac{cn_i|\mathbf{E}_i(\mathbf{r}, z')|^2}{8\pi P_T(z')} \tag{5}$$

where c is the speed of light in vacuum. If we multiply $\hat{S}$ by the input power to the fluorophore's absorption dipole, then (assuming a 100% quantum yield) we get the intensity $I(\mathbf{r}, z')$ radiated from the fluorophore:

$$I_i(\mathbf{r}, z') = |\boldsymbol{\mu} \cdot \mathbf{E}^{\text{ex}}|^2\, \hat{S}_i(\mathbf{r}, z') \tag{6}$$

Two additional feature can be incorporated into Eqs. (3)–(6): the dipole orientation distribution and the concentration distribution in systems consisting of many dipoles. The orientation of the dipole with respect to the surface, described by angles $\Omega' \equiv (\theta', \phi')$, affects $\mathbf{E}_i$ and all the other measurables derived from it. The details are provided in Hellen and Axelrod (1987). Consider a concentration distribution of dipoles in both orientation and distance from the surface specified by $C(\theta', \phi', z')$. Since the dipoles all oscillate incoherently with respect to one another, the integrated intensity $\mathcal{I}$ due to this distribution is simply

$$\mathcal{I}_i(\mathbf{r}) = K \int d\Omega' \, dz' \, C(\Omega', z') \, I_i(\mathbf{r}, \Omega', z') \tag{7}$$

The total fluorescence power $\mathcal{F}$ collected from the fluorophore distribution by a microscope objective centered on the normal line at a distance r is an integral of $\mathcal{I}$ over the objective's aperture which subtends a solid angle Ω:

$$\mathcal{F}_i(r) = \int r^2 \, d\Omega \, \mathcal{I}_i(\mathbf{r}) \tag{8}$$

Returning to the case of a single dipole, we find another parameter to be useful: the fluorescence collection efficiency Q. Parameter Q is the fraction of the total energy dissipated by a fixed-amplitude dipole that is collected by a microscope objective centered on the normal at distance r.

$$Q(z', \theta') = \frac{\text{Collected power}}{\text{Total dissipated power}} = \int r^2 \, d\Omega \, \hat{S}_i(\mathbf{r}, z', \theta') \tag{9}$$

III. Graphical Results

Rather than displaying the rather complicated explicit forms for $\hat{S}$, $\mathcal{I}$, and $\mathcal{F}$ in general terms of n_i, t, Ω^1, Ω, and C, we show here graphical results for certain specific configurations. The qualitative features of these results will be relevant for most other configurations. We specialize in two particular interfaces: bare glass/water, where the intermediate layer is just an extension of medium 3 (i.e., $n_2 = n_3$); and glass/aluminum film (22 nm thick)/water.

A. Radiated Intensity from Excited Dipoles versus Observation Angle

Figure 2 shows the radiated intensity $\hat{S}$ as a function of the observation angle θ for a dipole 80 nm from the surface. For simplicity, the azimuthal angle of observation ϕ is averaged. (This is equivalent to assuming that the excited dipole distribution is azimuthally symmetric about the surface normal.)

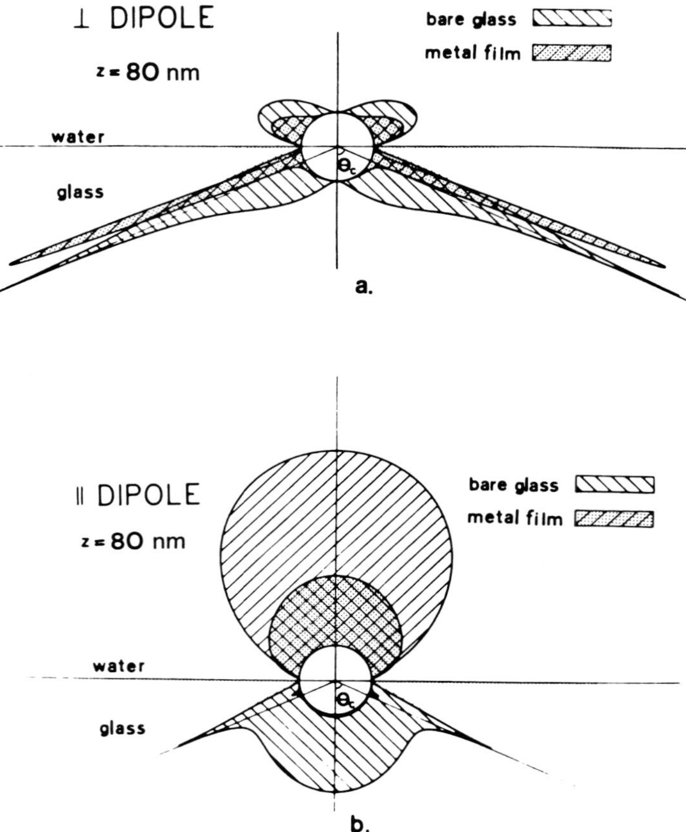

FIG. 2. Normalized intensities $\hat{S}$, averaged over azimuthal angle ϕ, for a fluorophore with dipole either (a) perpendicular or (b) parallel to the surface, as a function of observation angle θ. A distance z' of 80 nm, and for the metal film case, a 22-nm thick coating of aluminum, are assumed. At each angle, the radial distance from the center circle (with the fluorophore at the center) is proportional to the radiated power. The following constants are assumed: $\lambda_0 = 520$ nm; $\varepsilon_1 = 1.77$; $\varepsilon_2 = -32.5 + 8.4i$ (for the metal film case only); and $\varepsilon_3 = 2.13$.

In the bare-glass case, note that a rather strong peak of intensity is drawn into the glass, maximal at exactly $\theta = \theta_c$, but with significant intensity into supercritical angles. The effect is especially pronounced for dipoles oriented perpendicular to the surface, but is present for any position or orientation distribution.

In the aluminum film case, a peak of intensity directed into the glass is again presented, but here centered in an extremely narrow band at some

$\theta = \theta_p > \theta_c$. Angle θ_p is called the "surface plasmon" angle and arises from near-field waves from the dipole whose radial **k**-vector magnitude k_r is exactly matched to resonant electronic vibrations which can propagate on the metal surface (Weber and Eagan, 1979) and then reemit light into the hollow cone pattern depicted. Note that dipoles perpendicular to the metal surface can furnish energy to the surface plasmons quite effectively, leading to an apparent transmission of light through a virtually opaque metal film. However, dipoles parallel to the surface are very unsuccessful at coupling with surface plasmons and almost all the radiated emission appears in the water.

B. Supercritical Radiated Intensity into Glass versus Distance from Surface

Figure 3 shows the intensity that would be detected in the glass at a particular supercritical angle. Only the results for perpendicular dipoles are shown (so that averaging over azimuthal ϕ is unnecessary). The results for parallel dipoles are qualitatively similar except that the metal film case would be very much reduced in overall intensity.

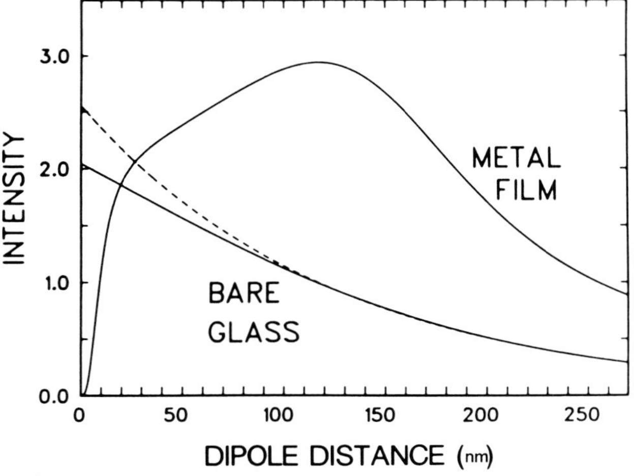

FIG. 3. Intensity emitted at supercritical angle $\theta = 70°$ into the glass versus z' for fluorophores oriented perpendicularly to the surface. The solid lines show the normalized intensity $\hat{S}$ for fixed-power dipoles with and without a 22-nm aluminum film on the glass. The dashed line shows the exponentially decaying intensity that would be obtained if the fluorophore was (incorrectly) modeled as a fixed-amplitude dipole. All the intensities are independent of ϕ for perpendicular dipoles. The dielectric constants and λ_0 are as in Fig. 2.

In the bare-glass case, note that the decay is not exponential, as otherwise would be expected if the "reciprocity principle" had been misapplied here. Nevertheless, by viewing only supercritical angles, one can selectively observe only those fluorophores within several hundred nanometers of the surface, even if the excitation (rather than the emission) is not surface-selective at all.

In the metal film case, the intensity is virtually zero for distances less than 5 nm. This quenching effect occurs at all angles, not just supercritical ones. The excitation energy is almost entirely converted into heat in the metal film. At larger distances, the dipole near field couples with surface plasmons which emit into the glass centered around $\theta = \theta_p$. At even larger distances, the near field is too weak to interact with the surface, and the supercritical intensity drops toward zero.

C. Total Power and Lifetime from Fixed-Amplitude Dipoles versus Distance

As previously discussed, the total power (including heat) emitted in the systems described is equal to the constant input power for actual fluorophores. But the total power P_T emitted by fixed-*amplitude* dipoles is still important to calculate, both because it appears in the denominator of Eq. (5) and because it is inversely proportional to the fluorescence lifetime (for systems with a 100% intrinsic quantum yield).

Figure 4 shows P_T for both parallel and perpendicular dipoles in the bare-glass case. Note that the power increases slightly for perpendicular dipoles very close to the surface. This corresponds to an approximately 10% decrease in the fluorescence lifetime of fluorophores. This effect should be taken into account when measuring fluorescence lifetimes near dielectric surfaces. For parallel dipoles, P_T exhibits only slight undulations.

Figure 4 also shows P_T for the aluminum film case. There is a dramatic increase in P_T (actually approaching infinity) for both dipole orientations at small distances. Virtually all of that energy is converted into heat in the metal, thereby accounting for the strong fluorescence quenching on metal surfaces. For dipoles oriented parallel, an additional factor further promotes quenching: when the dipole is very near the surface ($z' \ll \lambda$), its oppositely charged mirror image in the metal virtually cancels out the emitted electric field by simple wave interference.

D. Collection Efficiency

The collection efficiency Q measures the fraction of the total power emitted by a fluorophore that can be gathered as light by the microscope

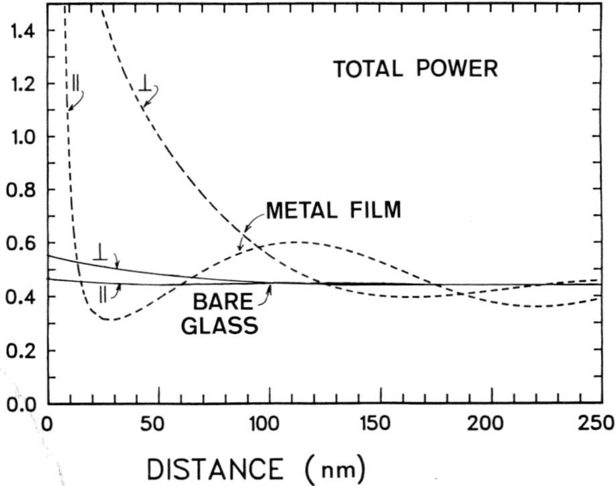

FIG. 4. Total power dissipated by fixed-amplitude dipoles oriented parallel and perpendicular to the interface for bare glass (solid lines) and 22-nm aluminum film (dashed lines). Dielectric constants and λ_0 are as in Fig. 2.

objective. Figure 5 shows Q for both parallel and perpendicular dipole orientations, for an objective positioned either to look through the water or to look through the glass substrate.

Figure 5a shows the bare glass case. It shows that viewing through the glass substrate is better than viewing through the water, at least for fluorophores very near the substrate surface. Around 60% of the emitted energy can be captured by a numerical aperture 1.4 objective by viewing through the glass substrate; only around 30% can be captured by viewing through the water. Much of this advantage is due to the ability of the high aperture to gather the emitted peak centered at $\theta = \theta_c$ (see Fig. 2). For smaller-aperture objectives, the relative advantage of viewing through the substrate diminishes. Clearly, a 1.4 aperture objective is much better than a 1.3 aperture objective because of its special ability to gather the $\theta = \theta_c$ peak.

Figure 5b and c (with expanded abscissa scale) show the aluminum film case. In the example shown, the surface plasmon peak is gathered (but just barely) by the objective; if it were not, the emission into the glass would be much less. But even with this high aperture, it is still more efficient to view the fluorophores through the water for most distances outside the strong quenching region of $z' < 10$ nm. For large z' distances, viewing through the water is very efficient. This is simply because the metal surface acts like a mirror for far-field propagating light emitted by the dipole.

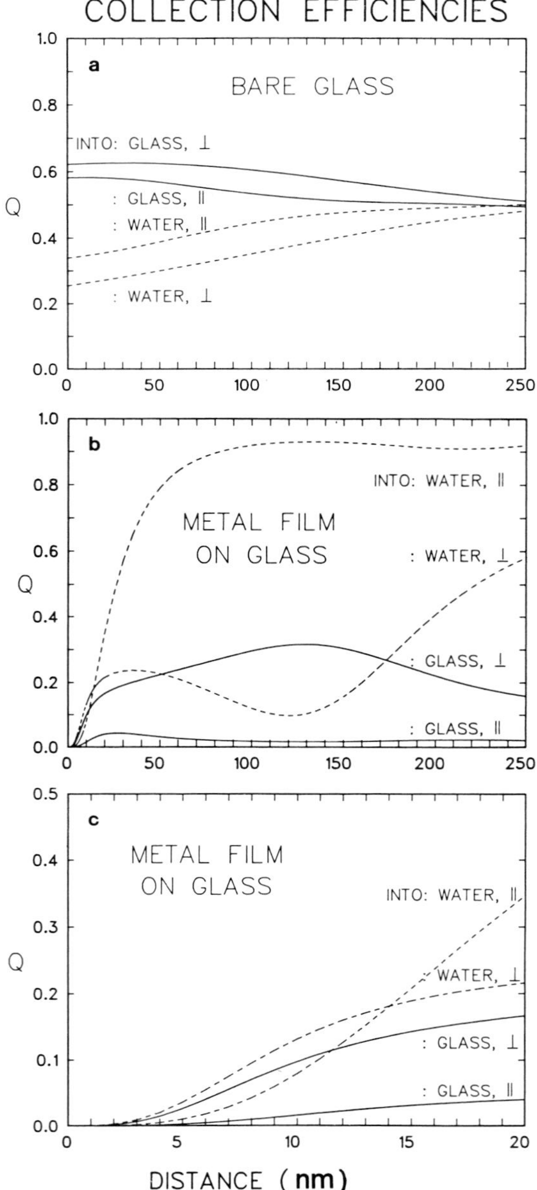

FIG. 5. Collection efficiencies Q for an objective with a numerical aperture of 1.4 versus z^{i}. Two positions of the objective are indicated: collection of light emitted into the water (dahsed lines) and into the glass (solid lines). Two dipole orientations relative to the surface are indicated: parallel and perpendicular. The dielectric constants and λ_0 are as in Fig. 2. (a) Bare glass; (b) 22-nm aluminum film on glass; (c) expanded abscissa scale of b, showing the quenching effect in more detail.

IV. Consequences for Experiments

A. Polarization

A complete treatment shows the polarizations of the emitted fields (Hellen and Axelrod, 1987). The graphs for intensity displayed in this chapter assume that no polarizing analyzer is used. The polarization field of each emitting dipole is always in the same plane as that dipole's orientation at the instant of emission and perpendicular to the direction of emitted light propagation. The polarization from a distribution of excited dipoles spread over different orientations will be different from that observed from an identical distribution observed in the absence of an interface, because the emission intensity pattern at an interface depends upon dipole orientation. Moreover, the polarization will depend upon the dipole distance from the surface. These facts are clear from Fig. 4, which shows that the collected power versus distance curves are differently oriented dipoles. Furthermore, the observed polarization will depend upon what range of angles θ is observed.

B. Fluorescence Efficiency and Lifetime

The preceding treatment assumed, for simplicity, that the quantum yield of the isolated dipole (i.e., at $z^1 = \infty$) was 100%. Here we assign it a more general value of q_0. The following definitions are useful:

$$p = P_T(z')/P_T(\infty)$$
$$f = \frac{\int d\Omega\ S(z', \Omega)}{\int d\Omega\ S(\infty, \Omega)} \tag{10}$$

Both p and f are ratios of a power emitted at position z' relative to that for an isolated dipole; p refers to total power (light plus heat) whereas f refers to radiated power only, derived by integrating the fixed-amplitude dipole radiated intensity S [given by Eq. (5) without the P_T normalization in the denominator] over 4π steradians.

We seek an expression for the actual quantum efficiency observed for a fluorophore at distance z', which we denote as q. One can show that

$$\frac{1}{q} = \frac{1}{f}\left[\frac{1}{q_0} + p - 1\right] \tag{11}$$

Given a "natural" (i.e., no radiationless decay) fluorescence lifetime τ_0 for an isolated fluorophore, one can show that the actual observed lifetime for a real fluorophore near an interface is

$$\tau = q\tau_0/f \tag{12}$$

Note that this expression differs from the more familiar $\tau = q\tau_0$ applicable to systems in which the rate of power emission P_T is constant and alterations in the fluorescence arise only from changes in the radiationless decay rate. Near a surface, however, P_T itself can change by three possible routes: (1) the radiated power rate can change due to interference between the dipole's reflected and directly emitted fields; (2) the near field of the dipole, which normally carries away no energy, may be converted into a radiating field in the denser medium by interaction with the surface; and (3) the surface may be a dissipative medium, such as a real metal, thereby converting the dipole near field into heat. In most cases with $z' < 10$ nm, effects 2 and 3 combine to increase the total dissipated power P_T, hence decreasing the lifetime for fluorophores close to the surface. Note that the degree of lifetime shortening does depend on the orientation of the dipole.

It is possible in principle for effect 1 to exert an opposing effect by tending to decrease P_T through destructive interference between the reflected and direct fields. However, for most materials likely to be encountered, either dielectrics or metals, the net result will be a lifetime decrease, not an increase, for small z'.

For bare glass, the lifetime decrease is only slight (~10% for $q_0 = 1$ and 5% for $q_0 = 0.5$); for metal-coated glass, the effect is dramatic. Particularly on metals, the expected decrease in lifetime may help protect the fluorophore against photobleaching that arises from excited-state chemical reactions involving diffusional collisions.

C. Concentration versus Distance from the Surface

The dependence of fluorophore concentration C on distance z' can serve as a means of measuring the distance of a sheet of fluorophore (e.g., a membrane) from the surface or as a measure of the conformation and thermodynamics of adsorption of large and pliable polymers at the surface (Ausserre *et al.*, 1985). The dependence of the intensity $\mathcal{I}_3$ versus θ emitted into the glass at supercritical angles ($\theta > \theta_c$) can provide information about $C(z')$. Eqs. (5)–(7) give

$$\mathcal{I}_3(\mathbf{r}) = K\frac{cn_3}{8\pi}\int d\Omega'\, dz'\, C(\Omega', z')|\boldsymbol{\mu}^{\mathbf{a}} \cdot \mathbf{E}^{\mathbf{ex}}(z')|^2 \frac{|\mathbf{E}_3(\mathbf{r}, \Omega', z')|^2}{P_T(\Omega', z')} \tag{13}$$

Consider first the adsorption term involving $\mathbf{E}^{\mathbf{ex}}$. If epiillumination is used, that term is not a function of z'. If total internal reflection (TIR) is used (see Axelrod, Chapter 9, this volume), that term is a simple decaying exponential in z'. Now consider that $\mathbf{E}_3$ term. According to Hellen and Axelrod (1987), that term is a function of θ and dipole orientation, denoted here as $g(\Omega', \theta)$, multiplied by a simple decaying exponential in z'. Therefore, Eq. (13) becomes

$$\mathcal{I}_3(\mathbf{r}) = J \int d\Omega' \, dz' \, C(\Omega', z') \frac{\exp[-\alpha(\Omega') \, g(\Omega', \theta) z']}{P_{\mathrm{T}}(\Omega', z')} \tag{14}$$

where α and J are factors that do not depend on θ or z'. Equation (14) *would* be a Laplace transform of C except for two features: the generally nontrivial integration over Ω'; and the nonconstant, nonexponential dependence of P_{T} on z'. Nevertheless, α, g, and P_{T} are known functions, in principle, and some reasonable simplifying assumption about the orientations of the dipoles (e.g., all aligned) can be made. Then, by measuring $\mathcal{I}_3$ versus θ, one can use Eq. (14) to calculate $C(z')$. The situation can be simplified further by the additional approximation that P_{T} is constant in z', which according to Fig. 3 is correct within 10% for bare glass/water interfaces. In that case, Eq. (14) is in the form of a Laplace transform, and inverse Laplace transforming an experimental measurement of $\mathcal{I}_3$ versus $\alpha g(\theta)$ will yield $C(z')$.

If excitation by the evanescent wave of TIR is used, the observed intensity from a single fluorophore does *not* decrease exponentially as it is moved away from the surface, despite the exponential decay of the excitation with z'. This nonexponentiality in the emitted intensity exists regardless of the observation angle θ (sub- or supercritical), aperture, or direction (through the glass or water), due to the presence of P_{T} in the denominator of any integrands for $\mathcal{I}_1$ or $\mathcal{I}_3$. Nevertheless, exponentiality is a fair approximation (to with 10%) for a bare glass/water surface.

D. Collection System Design

The significant anisotropy in the emission pattern from a fluorophore near a surface (Fig. 2) suggests how to maximize the collection of emitted light. When viewing through the glass, it is clearly desirable to use an objective with an numerical aperture (NA) high enough to collect the sharp peaks in the pattern.

For a bare glass/water interface, this criterion is

$$\mathrm{NA} > n_1 = 1.33 \tag{15}$$

For a glass/metal film/water interface, this criterion is

$$\mathrm{NA} > \mathrm{Re}\left[\left(\frac{\varepsilon_1 \varepsilon_2}{\varepsilon_1 + \varepsilon_2}\right)^{1/2}\right] \tag{16}$$

where the ε_i ($= n_i^2$) are the dielectric constants. This condition predicts that NA = 1.4 would capture the surface plasmon peak from an aluminum-film surface but not from a silver-film surface. Therefore, since objectives with aperture higher than 1.4 are rather rare, an alumimum film is a better choice.

E. Selective Detection of Adsorbed Fluorophores

One bare glass, the supercritical angle emission that occurs only from fluorophores near the glass suggests a method of selective detection of such fluorophores even in the presence of excited fluorophores farther out in the solution. An objective with a high numerical aperture (e.g., 1.4) can be masked to exclude any emission at less than the critical angle into the glass, thereby excluding all light from distant fluorophores (Fig. 6). This approach avoids the necessity of selective excitation near the surface, as is done by evanescent wave excitation in total internal reflection fluorescence (see Axelrod, Chapter 9, this volume); epiillumination would achieve the same surface-selective effect.

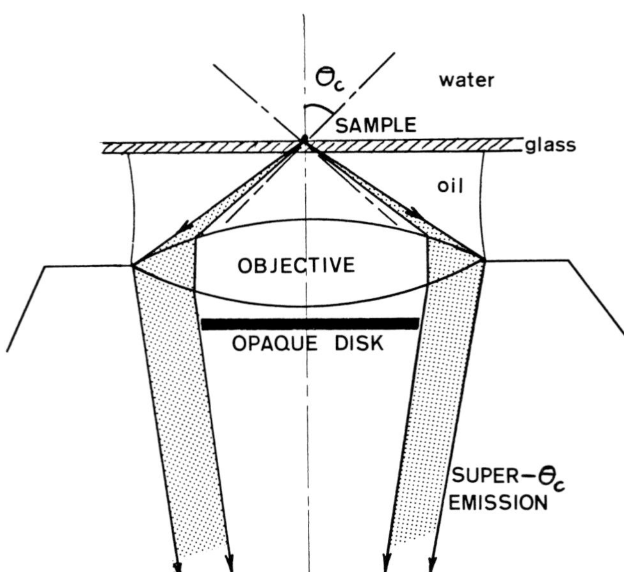

FIG. 6. Schematic diagram of a high-numerical aperture objective modified to accept only supercritical light emitted through a glass coverslip.

F. Selective Surface Quenching

On a metal coated surface, the highly effective and highly z'-dependent quenching could be used to distinguish between fluorophores close to the surface (e.g., $z' < 2$ nm) and those farther out (e.g., $z' < 7$ nm). For example, a metal-coated surface could be coated with an artificial, reconstituted, or flattened biological phospholipid bilayer membrane. Fluorescence from the more distal bilayer half would not be quenched nearly so strongly as that from the proximal half. This possibility of selectively detecting fluorophores in only one-half of a bilayer may find application to studies of membrane asymmetry, transmembrane transport, and lipid "flip-flop." This selective quenching can be used quite generally, since an aluminum-coated surface can be chemically treated with organosilanes to derivatize it with a wide range of functionalities (see Axelrod, Chapter 9, this volume).

Metal coatings are subject to heating more than dielectric coatings because, for most incidence angles, a significant portion of the incident light can be adsorbed. At high but accessible incident focused laser intensities, microscopic boiling in the water can be seen. As a general rule, the incident laser intensity should be reduced by at least two orders of magnitude from this point.

Semiconductor surfaces also quench nearby fluorescence. This effect has been applied experimentally (Nakache *et al.*, 1985) but not yet treated theoretically.

V. Conclusions

Consideration of the behavior of light emitted by a fluorophore near a surface complicates the accurate interpretation of fluoroscence data. Polarization, intensity, lifetime, and calibration of distances from the surface are all affected.

On the other hand, some of the surface-induced optical phenomena can be put to experimental advantage, including systems for selective surface fluorescence detection, selective surface fluorescence quenching, and optimal collection of fluorescence power.

Acknowledgement

The authors are indebted to Drs. N. L. Thomson, W. L. Weber, and G. W. Ford for useful discussions on this topic. This work was supported by USPHS NIH grant NS 14565.

References

Ausserre, D., Hervet, H., and Rondelez, F. (1985). *Phys. Rev. Lett.* **54,** 1948–1951.

Ford, G. W., and Weber, W. H. (1984). *Phys. Rep.* **113,** 195–287.

Hellen, E. H., and Axelrod, D. (1987). *J. Opt. Soc. Am.* **4,** 337–350.

Nakache, M., Schrieber, A. B., Gaub, H., and McConnell, H. M. (1985). *Nature (London)* **317,** 75–77.

Chapter 16

Fluorescence Labeling and Microscopy of DNA

DONNA J. ARNDT-JOVIN AND THOMAS M. JOVIN

Department of Molecular Biology
Max Planck Institute for Biophysical Chemistry
D-3400 Göttingen, Federal Republic of Germany

I. Introduction

Volume 29 and this volume deal with the *in vivo* labeling of macromolecules and other cellular constituents and their measurement by fluorescence microscopy. An inherent pitfall in this approach is that either the

METHODS IN CELL BIOLOGY, VOL. 30

labeling and/or observation technique may perturb the cell such that the information one derives is misleading or outright erroneous. The problem certainly applies to the determination of DNA content and cell cycle progression in living cells. Therefore, as we discuss the presently available techniques for *in vivo* measurements, an implicit requirement is that all experiments must be carefully controlled in the sense of demonstrating that the conditions for labeling and measurement do not impair cell viability. In some cases, this condition cannot be met and there is no satisfactory alternative but to use techniques based on fixation and labeling protocols that allow one to "take a snapshot in time" of the DNA metabolism in the living cell. Such experiments will be alluded to in this chapter as well. Our emphasis will be on the use of fluorescence digital imaging microscopy (FDIM) for measuring DNA, its metabolism, and its structure.

II. Labeling Strategies

A. Labeling Goals

The selection of a specific DNA probe is dictated by the labeling strategy best suited to the goal of the microscope study. The nucleus of the living cell is a complex structure directing the activity of the entire cell. In some instances the experimental requirement is simply a *qualitative identification of the nucleus,* e.g., as for the determination of ploidy, and cell type. Morphometric information such as shape, area, and volume can be derived readily for this purpose. In other cases, the positional information is of essence, such as in the identification of contact points between organelles and the nuclear membrane. Many DNA probes are suitable for the identification of the nucleus. However, only a relatively few compounds can be used for the *quantitative measurement of DNA content,* particularly *in vivo*. A third experimental objective may be to establish the *state of the DNA* in the nucleus, that is, condensation, helical structure, strandedness, topological distribution, and accessibility. In this case, other probes and methods are generally required, some of which can only be used with fixed cells.

B. Labeling Methods

In Table I we define a number of labeling methods and include representative examples of their application to DNA measurements, some of which are discussed in greater detail below. It should be emphasized that the techniques are not mutually exclusive. That is, several types of DNA

TABLE I

LABELING METHODS FOR DNA IN CELLS

Method	Mechanism(s)	Examples
Passive	Uptake of a drug by the cells. Binding to DNA leads to a fluorescence signal	Ethidium; BBI-342
Active	Uptake of a drug and metabolism or modification to a fluorescent, active agent which forms a complex or adduct with DNA	Carcinogens such as benzo[α]pyrene; photorelease of "caged" compounds
Photo	Photoactivation of nonreactive precursor molecules after cellular uptake. Photoinduced modification of DNA	Psoralen addition to DNA; thymine dimer formation
Induced	Quenching or activation of DNA fluorescence elicited by another drug. Example: fluorescence resonance energy transfer	Enhanced metaphase chromosome banding
Primary	Modification of the bases in DNA, allowing them to be specifically detected by fluorescence techniques	BrdUrd substitution for dT; AAF[a] or cis-platinum addition to bases
Hybridization probes	Labeling of DNA or RNA probes directly by chemical, immunological, or biochemical means	Biotinylated DNA or RNA; Hg^{2+} addition; rhodamine end addition
Structure specific	Use of drugs, immunological or reactive probes which sense helical nature or handedness, condensation state, or accessibility	Antibodies to Z-, B-, ss-DNA; drugs which sense ss-DNA; Protein footprinting reagents
Indirect	*A posteriori* fluorescence detection after *in vivo* labeling	Immunofluorescence detection of BrdUrd incorporated into DNA
Microinjection	Introduction of a probe by microinjection into the cell with a fine capillary	Can be applied to impermeable drugs or to immunofluorescent probes

[a] BrdUrd, 5-bromodeoxyuridine; AAF, acetylaminofluorene.

measurements may be derived from a single probe or, conversely, a single technique may be applicable to a variety of probes for different types of measurements. Although some of the listed probes have been applied only for *a posteriori* fluorescence labeling, i.e., with fixed cells, they could just as well be adapted for *in vivo* measurements by use of microinjection techniques.

C. DNA-Specific Dyes and Drugs

Several classes of DNA-binding dyes or drugs exist. (We will use the generic term "DNA drug".) As mentioned above, many of these compounds inhibit normal DNA synthesis, a property that in many instances led to their discovery (Kersten and Kersten, 1974). The utility of such drugs in the suppression of tumor growth or induction of desirable mutations in microorganisms has provided the impetus for much of the chemical, physical, and structural characterization of the various complexes with DNA (for a review, see Waring, 1981). It is a fortuitous circumstance that useful fluorescence phenomena are often associated with binding. With the advent of techniques for systematic molecular design, one can expect a proliferation of new compounds specifically targeted to the quantitative determinations of DNA via some form of light emission process.

In Table II are classified the modes of interaction of drugs with DNA. Many of these binding parameters have been established by Müller and colleagues (1975), and substantiated by crystallographic studies. The general chemical structures of the most commonly used noncovalent DNA drug families are shown in Fig. 1, while Table III summarizes some of the relevant fluorescence spectral properties, i.e., the excitation and emission maxima.

TABLE II

MODES OF INTERACTION OF DRUGS WITH DNA

Noncovalent
- a. Intercalation
- b. Groove binding: major, minor
- c. "Outside" electrostatic binding, e.g., to the phosphate groups

Covalent
- a. Single-point adduct formation
- b. Cross-linking: intrastrand, interstrand
- c. Reaction with strand breakage
- d. New chemical linkages introduced during *in vitro* synthesis

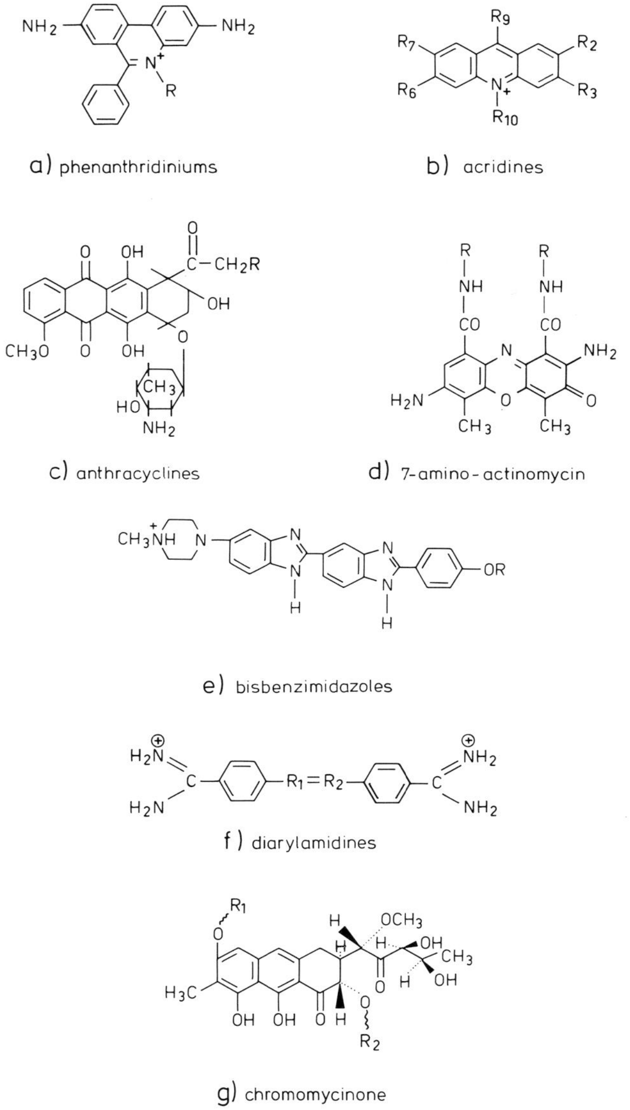

a) phenanthridiniums

b) acridines

c) anthracyclines

d) 7-amino-actinomycin

e) bisbenzimidazoles

f) diarylamidines

g) chromomycinone

FIG. 1. Chemical structure of commonly used noncovalent DNA drug families. The heterocyclic parent structure is shown. (*a*) ethidium, R = C_2H_5; propidium, R = $(CH_2)_3N(CH_3)_2$; homoethidium, R = $(CH_2)_8(NH_2)_2X$. (*b*) R = H unless otherwise noted; acridine orange, R_3 and R_6 = $N(CH_3)_2$; acriflavine, R_3 and R_6 = NH_2, R_{10} = CH_3; proflavin, R_3 and $R_6 = NH_2$; quinacrine, $R_3 = Cl$, $R_7 = OCH_3$, $R_9 = NHCH(CH_3)(CH_2)_3N(C_2H_5)_2$. (*c*) Daunomycin, R = H; adriamycin, R = OH. (*d*) R = cyclic pentapeptide. (*e*) These drugs have been referred to in the literature as "Hoechst-(number designation)." We prefer the generic abbreviation of the trivial chemical name, BBI, rather than the explicit reference to a firm. BBI-342, R = C_2H_5; BBI-258, R = H. (*f*) hydroxystilbamidine, R_1 and R_2 = CH; berenile, R_1 = N_2H, R_2 = N. (*g*) Mithramycin, R_1 = β-olivosyl (1,3)-β-olivosyl, $C_{12}H_{21}O_6$; R_2 = β-mycarosyl (1,4)α-oliosyl (1,3)β-olivosyl, $C_{18}H_{33}O_9$.

TABLE III

FLUORESCENCE EXCITATION AND EMISSION MAXIMA OF COMMON NONCOVALENT BINDING DNA DRUGS

Compound	λ_{max} excitation (nm)		λ_{max} emission (nm)	
	Drug alone	DNA complex	Drug alone	DNA complex[a]
Ethidium bromide[bc]	480	525	620	605
Propidium iodide[c]	495	530	639	615
Acridine orange[d]	495	502, 450[e]	535	538, 640[e]
Proflavin[f]	444	460	516	492
Quinacrine[f]	424	455	516	495
BBI-342[g]	339	352	510	465
DAPI[g]	340	360	456	456
7-Aminoactinomycin D[c]	502	555	672	655
Mithramycin (+ Mg^{2+})[h]	425	425	575	580
Hydroxystilbamidine[i]	345	360	450, 600	450, 600
Tris(4,7-diphenylphenanthroline) ruthenium (II)[j]	470	470	620	630

[a] Underlined values denote enhanced emission in the bound state.
[b] From Le-Pecq and Paoletti (1967).
[c] From Crissman *et al.* (1979).
[d] From Rigler and Ehrenberg (1973).
[e] Bound to single-stranded DNA or RNA.
[f] From Pachman and Rigler (1972).
[g] From this laboratory.
[h] From Hayasaka and Inoue, (1969); Crissman *et al.* (1979).
[i] From Festy and Duane (1973).
[j] From Barton *et al.* (1984).

III. Measurements

A. Passive and Active DNA Labeling for DNA Content and Cell Cycle Analysis

1. NONCOVALENT DNA DRUGS

Of drugs that intercalate between the base pairs of B-DNA, the phenanthridine derivatives (Fig. 1a) such as ethidium bromide (EtBr) or propidium iodide (PI) and the acridines (Fig. 1b) such as acridine orange, proflavin, and quinacrine are used most commonly. These drugs bind to RNA as well as to DNA. The cationic forms also bind electrostatically to other macromolecules such as proteins, polysaccharides, or glycosaminoglycans, and to membranes. A variety of conditions have been employed to suppress the electrostatic interactions with other cellular constituents in fixed cells (Darzynkiewicz, 1979). EtBr and PI bind to DNA of any base composition and may insert from either the major or the minor groove (Neidle and Berman, 1983) with little or no sequence preference (Nelson and Tinoco, 1984). The binding constant of some dimeric derivatives of the intercalators (e.g., ethidium homodimer, Gaugain *et al.*, 1978; Markovits *et al.*, 1983) is several orders of magnitude greater than for the monomer drug, thereby providing an increased sensitivity in DNA measurements.

R. Haugland (private communication) has synthesized a new series of intercalating phenanthridine derivatives that are termed "caged" ethidiums. These nonfluorescent compounds are taken up by cells and photochemically activated by irradiation to a fluorescent species that then intercalates into DNA. Thus, individual cells or regions of a nucleus may be selectively observed and measured. Homoethidium is a membrane-permeable nonfluorescent derivative that can be cleaved by cellular peroxidases to ethidium and accumulates in the nucleus.

The binding equilibria for intercalation into double-stranded DNA and RNA tend to differ, and provide a basis for selective observation of a particular nucleic acid complex. For example, the emission wavelength of the acridine orange complex with double-stranded DNA peaks in the green (near 530 nm), whereas the complex with single-stranded nucleic acid emits in the red, i.e., at >580 nm. (The basis for these emission properties is discussed in Rigler and Ehrenberg, 1973; Darzynkiewicz, 1979; and Kapuscinski *et al.* 1982). As we discuss in the section on fluorescence detection (see Section V, B), the two types of nucleic acid complexes may be better distinguished using phase-modulation fluorescence detection. However, in order to obtain accurate measurement of the DNA content of cells with any of these drugs, it is generally necessary to carry out a prior digestion of RNA.

Some of the earliest quantitative electronic imaging in the microscope was performed by Caspersson (1950) who measured DNA absorption of acid-hydrolyzed, Feulgen-reacted cells. Caspersson's contribution to quantitative DNA measurements in the microscope extend to the present era as well (Caspersson, 1979). The Feulgen reaction can be performed with a fluorescence drug such as acriflavine to form the Schiff base derivative, a procedure that still constitutes one of the most reliable measurements of cellular DNA content.

The differential banding pattern on metaphase chromosomes produced by acridine derivatives, such as proflavin and quinacrine, presumably reflects the superposition of many complex processes: (1) fluorescence enhancement and quenching by A-T rich and G-C rich DNA sequences, respectively; (2) photophysical interactions such as resonance energy transfer; (3) excluded binding due to chromosomal proteins (Arndt-Jovin *et al.,* 1979; Darzynkiewicz and Kapuscinski,1988; Schröter *et al.,* 1985; reviewed in Latt, 1976).

The anthracycline (Fig. 1c) drugs daunomycin and adriamycin are intercalators (Wang, 1987) that are stabilized in their interaction with DNA by specific hydrogen bonding in both the major and minor grooves. Over 500 derivatives of the anthracyclines have been synthesized, most of them for purposes of tumour therapy. The unwinding of the helix (induced by all intercalators) and particularly the interaction of the amino sugar with the minor groove are thought to prevent the function of both DNA and RNA polymerases. Whether any of the derivatives that are ineffective as antitumour agents could be used for quantitating DNA *in vivo* without major pertubation has not been systematically studied.

The phenoxazome ring of the antibiotic actinomycin D intercalates between G-C base pairs while the two cyclic peptides bind in the minor groove (Sobell *et al.,* 1977). A fluorescent derivative has been synthesized, 7-aminoactinomycin D (Fig. 1d, Table III), that has the attractive property (for fluorescence microscopy) of being excited maximally in the bound state at 555 nm and emitting deep in the red (Gill *et al.,* 1975). Furthermore, although the drug inhibits RNA synthesis and affects metabolism, it permeates living cells so that some types of viable cell measurements can be accomplished.

Most of the nonintercalating DNA-binding drugs interact with the minor groove of double-stranded B-DNA as was shown for the nonfluorescent pyrrole-amide chains like neotropsin and distamycin (Makarov *et al.,* 1979; Kopka *et al.,* 1985). Fluorescent synthetic compounds with a similar sequence preference and geometry are the bis-benzimidazole dyes (Fig. 1e) such as BBI-342 (Mikhailov *et al.,* 1981; Zimmer and Wähnert, 1986). Recent X-ray crystallographic studies (Kopka *et al.,* 1987; Pjura *et al.,* 1987) demonstrate how the drugs stabilize the B double helix and why they do not bind to single-stranded DNA or RNA. The A-T-rich tracts that form the binding site have been shown in biophysical studies (reviewed in Zimmer and Wähnert, 1986) and by footprinting analysis of natural DNAs to consist preferentially of a series of five A-T base pairs, of which at least three are preferentially nonalternating (Harshman and Dervan, 1985).

Diarylamidines (Fig. 1f) are minor groove-binding A-T-specific fluorescent drugs that interact with thymidine residues by hydrogen bonding somewhat differently (Festy, 1979; Zimmer and Wähnert, 1986) than the BBI compounds. Their antibacterial activity may result from interference with RNA and DNA synthesis, as in the case of netropsin and related drugs, but detailed studies with mammalian cells have not been carried out.

Although chromomycinone (Fig. 1g) derivatives with oligosaccharide side chains, e.g., mithramycin, also bind in the minor groove of B-DNA, they are specific for G-C base pairs. These drugs show high specificity for double-stranded DNA and do not form complexes with single-stranded DNA nor RNA. However, they do not permeate the plasma membrane of living cells and require high concentrations of Mg^{2+} ions for binding, properties that have limited their utility to fixed cell preparations.

The nonfluorescent triphenylmethane derivative, methyl green, provides an unusual example of a drug binding exclusively in the major groove (Nordén and Tjerneld, 1977), a mechanism that is otherwise employed extensively by DNA regulatory proteins, e.g., the repressor proteins cro and lambda (Giniger and Ptashne, 1987; Miller *et al.*, 1985).

2. Vital Staining and Toxicity

The pleomorphic nature of *in vivo* drug/cell interactions should be recognized. The toxicity of some drugs may not be due directly to the formation of complexes with DNA or RNA. Compounds with lipophilic character inhibit vital cellular systems via binding to membranes. For example, distamycin A blocks transformation in bacteria by binding cell membrane receptor sites required for the penetration of donor DNA rather than through direct action on the DNA itself (Mazza *et al.*, 1973). The intercalating DNA drugs cause mutation in DNA by interfering with normal replication of the chromosomes. Additionally, they cause changes in transcription patterns, i.e., gene expression, probably due to steric hindrance to the translocation of RNA polymerase along the double helix. Recent studies have shown that bacterial RNA polymerase pauses upon encountering an intercalated ethidium bromide molecule (R. Clegg, personal communication). Depending upon the thermodynamic and kinetic properties of the equilibria involving drug, DNA, RNA, and protein, the transcriptional complex may dissociate and thereby terminate transcription. Neither ethidium bromide nor propidium iodide cross the plasma membrane of living cells but they diffuse freely into damaged or dead cells. Thus, these drugs have been used extensively in flow cytometric studies to identify dead cells in populations being screened for other cellular parameters, thereby providing a "gate" with which to selectively extract the information from the living cohort. Since phase-contrast optics in the microscope can distinguish living from dead cells relatively easily, this application of the phenathridinium drugs is not usual in microscopy.

However, EtBr and PrI *can* be introduced into living cells by suspending them in physiological buffers containing dimethyl sulfoxide.

The acridines are more permeable to living cells but have the same effects on transcription and replication as mentioned above. They have the added disadvantage that they are not absolutely specific for nucleic acids and stain many other negatively charged macromolecules, as stated previously. Under certain conditions and with fixed cells they may be useful for determining DNA/RNA ratios (see Darzynkiewicz *et al.*, 1980) but the absolute quantitation of DNA is more reliable with the other drugs.

The bis-benzimidazole BBI-342 (Loewe and Urbanietz, 1974) is one of the few DNA-specific drugs that can penetrate the living cell without prior perturbation of the cell membrane by detergents or organic solvents (Arndt-Jovin and Jovin, 1977). As a consequence, it has been used extensively in flow cytometry to analyze and sort living cells according to DNA content. This dye does not interfere with replication per se (cells progress through the S phase) but it appears to inhibit chromosome condensation and thus, normal mitosis. The extended minor groove binding site shown in crystallographic studies (Kopka *et al.*, 1987) probably accounts for the fact that higher order chromatin condensation is more difficult in the presence of the drug. Nonetheless, since BBI-342 can traverse the cell membrane, the drug can also be removed from the chromatin to a large extent so that many cell lines retain full viability (Arndt-Jovin and Jovin, 1977).

Clearly, the permeability of the plasma membrane or the accessibility of the chromatin may vary between cell types. Some years ago, it was reported that lymphocytes could be distinguished on the basis of staining with BBI-342 at low concentrations (Loken, 1980), i.e., below that required for saturation of the DNA. Recently, Krishan (1987) has shown that the variable staining of several cell lines with BBI-342 is due to a rapid efflux pump that reduces intracellular dye concentrations. This pump appears to be the same as that responsible for multiple drug resistance in some cell lines and can be inhibited by blockers such as phenothiazines or Ca^{2+} channel blockers. These drugs allow the stoichiometric staining of hitherto refractory cells with BBI-342 and result in *in vivo* DNA distributions identical to those for fixed cells.

During the last 10 years, since its introduction as a dye for sorting viable cells on the basis of DNA content, BBI-342 has been used as a vital stain for many established lines as well as a variety of primary cells from many different hosts. As can be seen in Table III, it is excited maximally near 350 nm, usually by a mercury lamp or argon ion laser. High intensities of UV irradiation are, in general, damaging to cells since a number of cellular constituents excite at this wavelength and deleterious photoaddition products can be produced. Thus the clonability of cells following drug exposure is related to the time of exposure, the concentration of the drug, and the level of UV radiation. Each cell line or cell type must be checked extensively in order to optimize these parameters before BBI-342 can be used as a vital stain (Fried *et al.,* 1982; Pallavicini *et al.,* 1979). However, acceptable conditions can be found in most cases, as Van Zant and Fry (1983) have shown, even for primary bone marrow precursor populations in the mouse, in which case viable progenitor populations were sorted. Although BBI dyes can be used directly for vital staining in the fluorescence microscope, attention must be paid to the excitation intensities, as this discussion has pointed out. In flow analyzers or sorters the cells are exposed to UV irradiation for only microseconds (which in some cases is long enough for saturation to be achieved, see Bartholdi *et al.,* 1983) rather than milliseconds to minutes as in the microscope. In developing a new method for measurements of fluorescence resonancce energy transfer (FRET, see Section III,C) in the microscope (Jovin and Arndt-Jovin, 1988) we determined the bleaching kinetics of a number of commonly used DNA drugs such as BBI-342 using a 50 W mercury arc source and a 63×, 1.25NA oil immersion objective. Total bleaching can be achieved with diploid cells in a matter of seconds and, in this case, exploited in quantitation of FRET.

3. Staining of Fixed Cells for DNA Content

As mentioned above, it is sometimes difficult to find conditions that allow specific DNA staining and yet maintain cell viability. In such cases, it may be preferable to use a specific DNA stain with fixed cells in order to determine DNA content under prescribed conditions. A number of excellent treatises are available concerning fixation and subsequent cell treatment for quantitative analysis of DNA and RNA in cells (Crissman *et al.,* 1979; Latt, 1979; Darzynkiewicz, 1979). Most of the DNA drugs mentioned above can yield a quantitative estimation of DNA after RNase digestion of fixed cells; thus, the selection often is based on optimizing the excitation and emission wavelengths in relation to the simultaneous immunofluorescence determination of other cellular constituents such as cell surface receptors. All of the DNA drugs bind less to chromatin than to naked DNA but only some of them seem to be affected by the state of chromatin condensation. (For a discussion of this point and comparative DNA distributions, see Darzynkiewicz *et al.,* 1984.) If RNase digestion is not desired, DNA can be specifically and quantitatively stained with either BBI-342, mithramycin (Crissman and Tobey, 1974), or DAPI (4′,6-diamidino-2-phenylindole) without contribution from RNA species. The derivative BBI-258, though not as permeable to the cell membrane in living cells as BBI-342, can be used interchangeably with BBI-342 for fixed cell studies. DAPI has spectral properties similar to BBI-342, but is not as photosensitive. Spectral studies with DAPI show a larger enchancement of fluorescence when bound to A-T base pairs than when bound to G-C base pairs, but the molecular nature of its interaction with DNA has not been established (Cavatorta *et al.,* 1985).

B. DNA Synthesis Measurements Using Primary Labeling and Immunofluorescent Probes

The position in the cell cycle (i.e., DNA content) can be measured using vital dyes such as BBI-342 as discussed above. However, other questions concerning the DNA metabolism of cells are better probed by indirect methods. The availability of monoclonal antibodies to BrdUrd has provided an exciting tool for the complex analysis of the spatial as well as temporal distribution of DNA synthesis in cells and organisms. Briefly, the technique consists of incorporation of BrdUrd in place of thymidine into living cells for a limited period of the cell cycle (as short as 1–2 minutes of incorporation can be monitored). The cells are subsequently fixed, the DNA denatured or depurinated to reveal single-stranded regions containing the br^5U base, and a monoclonal antibody against the modified base is bound and visualized using normal immunocytochemical reactions (direct or indirect fluorescent antibody labeling). Gonchoroff *et al.,* 1985) have isolated a specific monoclonal antibody that recognizes the br^5U in double-stranded DNA. The reader is referred to a special issue of *Cytometry* **6**, No. 6 (1986) for an overview of the technique and applications. Although hundreds of papers have been published using this technique, the exploitation of the spatial information through quantitative fluorescence microscopy has been meager to date.

Clear differences in the spatial distribution of immunofluorescence patterns in cells pulse-labeled with BrdUrd can be demonstrated and quantitated with a CCD camera (Fig. 2) [see photographic data of Nakamura *et al.,* (1986) as well]. We have observed that the "horseshoe or circular patterns" that are visible in some nuclei are replication sites occurring on condensed heterochromatin as can be seen from the immunofluorescence and companion DNA-staining images (Fig. 3). In mouse cells, satellite DNA and presumably other heterochromatin remain in a number of large aggregates within the nuclei throughout most of interphase. We have shown by DNA staining of both living and fixed cells that these heterochromatin aggregates are not generated by the immunofluorescence staining

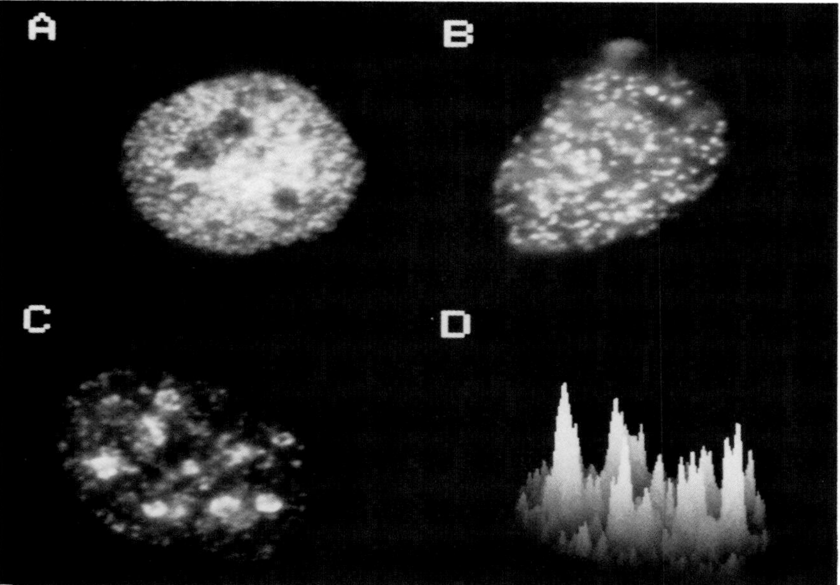

FIG. 2. Digital images of immunofluorescence patterns of newly replicated DNA in BrdUrd pulse-labeled mouse fibroblasts. Mouse 3T3B cells growing on 10-well Multitest slides (Flow Labs) were synchronized by serum starvation followed by pulse-labeling with BrdUrd for 5 or 20 minutes at different times after release into medium containing 10% fetal calf serum. Visualization after fixation and DNA denaturation with acid was made by staining with 10 μg/ml of a monoclonal antibody to BrdUrd (Traincard *et al.*, 1983) followed by rhodamine-labeled goat anti-mouse IgG and a counterstain of 5 μM BBI-342. The slides were mounted in 25% glycerol buffered at pH 8 containing 10% polyvinyl alcohol (Mowiol 4-88) to reduce photobleaching. (A) A nucleus in mid S phase. DNA synthesis is proceeding at the maximum rate with large numbers of replicons simultaneously incorporating BrdUrd throughout the nucleus. The nucleoli are obvious as negative areas within the fluorescent nucleus. (B) A replication pattern showing replication sites near the nuclear membrane in earlier S phase than (A). (C) Replication of condensed heterochromatin as observed late in S phase. (D) A three-dimensional representation of the immunofluorescence intensitites in the optical section of panel (C) generated with INCOS2 (image processing system from Signum, Munich). Microscopy: Neoflua 63×, 1.25 NA, oil immersion objective; Optovar 1.25×; excitation source, 50-W mercury arc lamp; excitation at 546 nm using an 8-nm HW (bandwidth at half height) bandpass filter and a 580 dichroic filter; emission >590 nm using a long pass filter. Camera: thermoelectrically cooled CCD detector (CSF-Thomson 384 × 576 pixels), model CH220 Photometrics (Tucson, AZ). The camera has a dark noise of 0.6 counts/second, a dynamic range of 0–16,000 counts (14-bit ADC), and a linearity of >99%. Flat field corrections for inhomogeneity in the excitation and detection fields were made by division with an image generated from a fluorescent uranium glass. Fluorescence images were acquired with an integration time of 10 second, during which time no perceptible photobleaching could be detected.

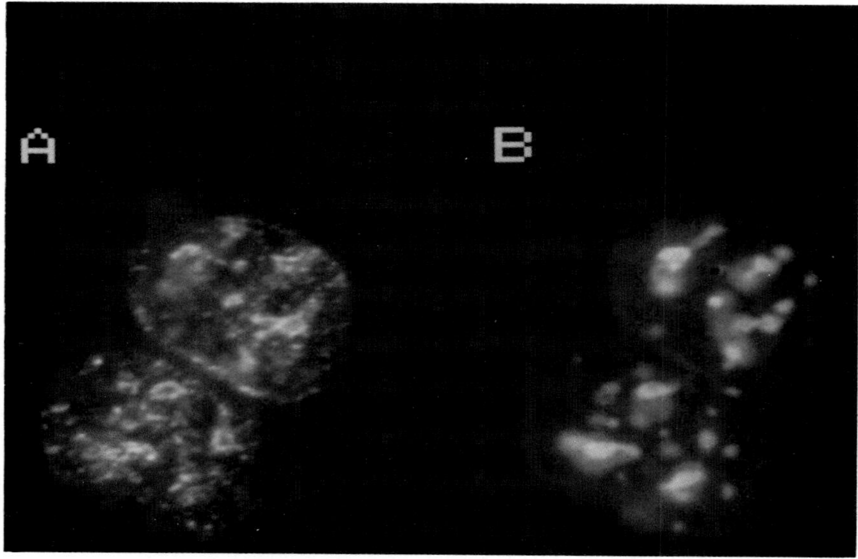

FIG. 3. Correlation of a replication pattern in BrdUrd pulse-labeled nuclei with the regions of condensed heterochromatin. (A) Digital immunofluorescence image of mouse fibroblasts (conditions as in Fig. 2) labeled with BrdUrd in late S phase. (B) DNA fluorescence digital image of the same cell as panel (A) counterstained with 5 μM BBI-342. Strong fluorescence intensity of A-T-rich condensed chromatin correlates with the position of the rings in (A). Staining with the G-C-specific drug mithramycin shows that the condensed chromatin contains such regions as well (data not shown). Instrumentation as in Fig. 2. Excitation of BBI-342 was at 365 nm (12 nm HW with a 395 nm dichroic filter) and the emission was collected with a 418-nm long pass filter. The lamp was attenuated with a 1 OD neutral density filter and the integration time was 2 seconds for BBI-342 measurements.

conditions (data not shown, Arndt-Jovin and Jovin, 1988). Thus, dual-labeling FDIM can distinguish structures below the normal resolution of the light microscope and provides a guide to the spatial arrangement of the replication process. Clearly, higher-resolution studies are needed to determine details of the patterns and the temporal as well as spatial progress of replication. The confocal laser scanning microscope (CLSM) (see Brakenhoff, Chapter 14, this volume) is an ideal tool for such investigations. Because of the high sensitivity of the BrdUrd method and short labeling times required, an almost dynamic picture of DNA synthesis can be achieved with a timed series of "snapshots" of the fixed cells. Additionally, the antibody technique can be coupled to measurements of other cellular parameters that are not destroyed by the fixation techniques.

C. Energy Transfer: Selective DNA Observation

The measurement of FRET (Förster, 1948, see review by Stryer *et al.*, 1982) between a *donor* and a suitable *acceptor* chromophore in the microscope provides spatial resolution about 2 orders of magnitude greater than that defined by diffraction optics. Thus, one can distinguish between aggregation or association states of molecules in different locations within a cell. In particular, FRET can be used to study specific loci on chromosomes or processes such as condensation of DNA and chromatin. Prior to the introduction of FDIM, *qualitative* demonstrations of FRET in the microscope were used to study drug binding to metaphase (Latt *et al.*, 1977, 1980) and interphase (Hamori, *et al.*, 1980) chromosomes.

Latt *et al.*, (1977) showed that the A-T specificity of the BBI dyes could be exploited to study replication in BrdUrd-substituted DNA. The fluorescence of dye binding near to the Br-substituted base is reduced, possibly by FRET. The degree of quenching serves as a measure of the extent of replication. Other drugs show diminished, or in some cases enhanced, fluorescence when bound to BrdUrd-containing DNA. However, the magnitude of quenching is greatest for the BBI dyes, a feature which combined with their DNA specificity makes them particularly suitable. An important application of this quenching phenomenon is the determination of sister chromatic exchanges (SCE) in metaphase spreads (Latt, 1977).

Sahar and Latt (1980) have utilized combinations of fluorescent and nonfluorescent DNA drugs and FRET to enhance metaphase chromosome banding or to selectively distinguish between unbanded metaphase chromosomes in suspension (as for flow analysis and sorting). Other methods for highlighting particular chromosomes or chromosome loci will be discussed below under hybridization probes.

More recently, we have developed a quantitative method for high-resolution FRET measurements in the microscope that is termed "photobleaching FRET digital imaging microscopy"(pbFRET-DIM) (Jovin and Arndt-Jovin, 1988), using a high-resolution solid-state charged-coupled device (CCD) sensor. We have compared this new technique with a more conventional method of determining energy transfer based on an image of sensitized emission and find a number of advantages. PbFRET-DIM avoids (1) the problem of image registration, (2) low accuracy in the calculation of low transfer efficiency, (3) distortion of data due to differential bleaching, (4) the necessity to determine spectral overlap factors, and (5) the necessity to use a fluorescent acceptor. We have applied pbFRET-DIM to the study of DNA bands in polytene chromosomes (Jovin and

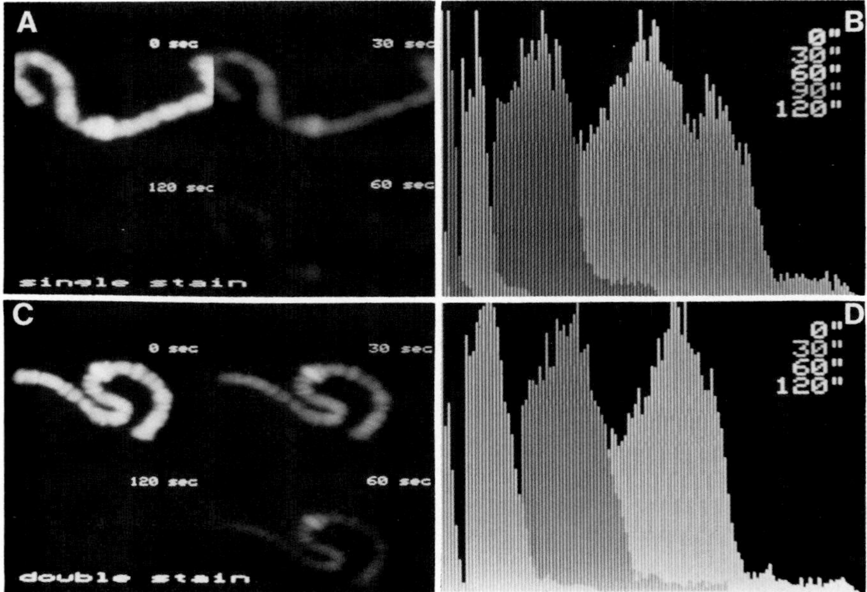

FIG. 4. FRET between BBI-342 and ethidium bromide bound to polytene chromosomes. Squashed *Chironomus thummi* salivary gland nuclei were fixed in acetic acid: ethanol, 1:3 (v/v) washed in buffered Ringers, stained, washed in Ringers, and mounted in 50% glycerol–Ringers. Objective: Neofluar 63×, 1.25 NA oil immersion. Excitation: 50-W mercury lamp with electronic shutter. Full lamp intensity was used for bleaching but was attenuated with a 2 OD neutral density filter for monitoring images. Filters: excitation, 365 nm, 12 nm HW bandpass with a 395 nm dichroic filter, emission, 418 long pass filter plus a 450 nm, 20 nm HW bandpass filter. Camera: Photometrics CCD as in Fig. 2. (A) A series of images of a polytene insect chromosome, stained with 10 μM BBI-342 only, each taken for 2 seconds in the monitoring mode (1% the excitation intensity) during the course of photobleaching at times 0, 30, 60, and 120 seconds. (B) The corresponding gray-level histograms for the images in (A) as well as for an image taken after 90 seconds of photobleaching. (C) A series of images of a chromosome, simultaneously stained with 10 μM BBI-342 and 50 μM ethidium bromide, taken in the monitoring mode during photobleaching at times 0, 30, 60, and 120 seconds. (D) The corresponding gray-level histograms for the images in (C).

Arndt-Jovin, 1988). Figure 4 shows images acquired during pbFRET-DIM of these chromosomes stained with DNA drugs of different specificities (BBI-342 alone or together with ethidium bromide as acceptor). The kinetics of the photobleaching in both cases are plotted in Fig. 5. From these we can determine an efficiency of energy transfer.

An unusual *in vivo* application of energy transfer is the selective killing of newly replicated cells using high-intensity UV excitation of BrdUrd-

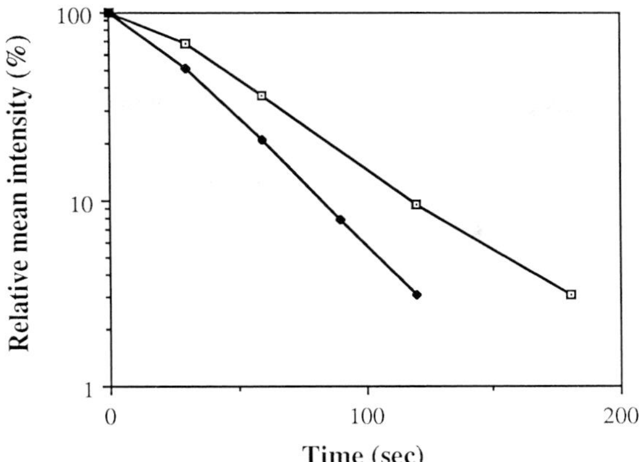

FIG. 5. Determination of the efficiency of FRET between DNA dyes using pbFRET-DIM. Photobleaching kinetics of BBI-342 (donor) bound to polytene chromosomes in the presence or absence of ethidium bromide (acceptor). The $\log_{10}$ of the means of the histograms of the monitoring images (panels B and D in Fig. 4) are ploted versus time for the single (◆) and double (□) labeled chromosomes. The energy transfer efficiency $E(i)$ for each image point is given (Jovin and Arndt-Jovin, 1988) by the equation $E(i) = 1 - [\tau_{b1}]/[\tau'_{b1}(i)]$ in which the numerator of the ratio is the bleaching time (reciprocal rate) for the donor alone and the denominator is the bleaching time for the donor in the presence of acceptor. In the present example, we calculate a mean FRET efficiency for the entire image of ~35%. Since the donor, BBI-342, has a sequence preference but the acceptor does not, we see a rather constant image of the donor distribution during bleaching. With another dye pair consisting of an acceptor that is base-pair specific and a donor that distributes indiscriminately, FRET results in enhanced contrast and differential banding of the chromosomes.

labeled cells stained with BBI dyes (Stetten *et al.*, 1976). In this case, the excitation energy is transferred from the dye to the BrdUrd, presumably causing the release of a free radical that causes mutation and death in these cells.

FRET measurements might be used to localize more precisely proteins bound to chromatin. That is, energy transfer between DNA-bound drugs and fluorescently labeled proteins, introduced by microinjection, would allow one to selectively observe the chromatin-bound proteins.

D. Nuclear versus Mitochondrial DNA

Although mitochondrial DNA makes up as much as 10% of the DNA content in yeast cells, the large mammalian genome (~3×10^9 bp) and a smaller mitochondrial DNA (17,600 bp) result in the contribution of mitochondrial DNA to the total DNA fluorescence of <0.1%. This amount is at least an order of magnitude lower than the best coefficient of variation attained for cellular DNA content, even in flow sorters. The small size of the mitochrondrial genome and the limited dynamic range of most imaging systems means that, in most cases, the mitochondrial DNA contribution will be overlooked in fluorescence microscopy. However, one can exploit the spatial resolution of the microscope and quantitative, sensitive detectors to restrict DNA measurements to specific areas (e.g., nucleus or cytoplasm). The *localization* of mitochrondria can be accomplished using specific drugs such as rhodamine 123 (see Chen, Volume 29, this series).

E. DNA Structure and Topology

The structure of chromatin in the nucleus has been shown to correlate with cellular state or activity and can be influenced by external perturbations on the cell. The regions of protein-free double-stranded DNA, double-stranded RNA, or DNA–RNA hybrids can be probed *in vivo* by photocross linking with psoralens (for a review of the methods, derivatives, and applications, see Cimino *et al.,* 1985). Such measurements can give *in vivo* information on nucleosome phasing in active or inactive chromatin, replication intermediates (D-loop formation), RNA transcriptional complexes (R-loop formation), and on the presence of cruciform structures. To date, the analysis of the psoralen adducts has been conducted by electron microscopy or restriction enzyme digestion and gel electrophoresis of radiolabeled derivatives. The very low fluorescence quantum yields (0.01) of psoralen adducts and their emission wavelengths in the region of the spectrum where cellular autofluorescence is most pronounced would be expected to make direct fluorescence observation difficult. However, with specific antibodies to the adducts, visualization and quantitation of the distribution of immunofluorescence should be possible by imaging techniques.

Other photoactivation or photoaddition reactions *in vivo* have been used to characterize protein-binding patterns on DNA and single-stranded regions in the DNA (Becker and Wang, 1984). Even though the analysis of the *in vivo* photo reactions is performed *a posteriori*, the reactions themselves are rapid [the time window is that of the excited state (10^{-9}–10^{-6})] and may be made site specific by focusing the light source. These reactions provide a tool to look at kinetic and structural changes in chromatin coupled to physiological stimuli. A caveat should be made about the interpretation of structural probes and the necessity of adequate controls in all experiments. The mere presence of an external probe may distort an existing chromatin structure or even induce a new one because of the dynamic nature of DNA and chromatin.

Barton investigated the DNA-binding properties of the fluorescent chiral metal (ruthenium or cobalt) complexes of the tris(phenanthroline) ring (described by Nordén and Tjerneld, 1976). She found that some enantiomorphs were capable of distinguishing different structures along the DNA duplex, including the chirality of the helix itself (Barton, 1986; Barton *et al.*, 1984). The enantiospecific cobalt diphenylphenanthroline complexes have been shown (Barton and Raphael, 1984) to cause oxidative cleavage of the sugar phosphate backbone upon irradiation preserving the recognition properties of the compounds.

DNA–RNA hybrids can be visualized and quantitated by fluorescent antibodies (Stollar, 1986) after cell fixation but microinjection techniques might make *in vivo* measurements possible.

The polymorphic nature of double-stranded DNA becomes more evident as we create more sophisticated probes of DNA structure. The most dramatic manifestation of a change in helical structure is the transition from the right-handed B- to the left-handed Z-DNA conformation. Immunological probes to both helical forms have been raised (Stollar, 1986; Jovin *et al.*, 1983) and used for *in vivo* and *in vitro* probing of various organisms by FDIM (Robert-Nicoud *et al.*, 1988; Arndt-Jovin *et al.*, 1985; Arndt-Jovin and Jovin, 1985). Sequence-dependent structural variations within the B-DNA family can also be detected by immunological means (Dieckmann and Zarling, 1987).

Einck and Bustin (1983) showed that microinjected fluorescent antibodies to histone and some nonhistone proteins localize to the nucleus *in vivo* and are capable of inhibiting transcription. A variety of studies using microinjection of antibodies to topoisomerases or polymerases or of directly labeled nonhistone proteins can be conceived, whereby their localization on the chromatin could be studied quantitatively with FDIM.

Three-dimensional fluorescent imaging techniques (see Agard *et al.*, chapter 13, and Brakenhoff *et al.*, chapter 14, this volume) open the entire area of DNA topology to investigation. Several important questions in molecular genetics concern the localization of genes and/or gene products (RNA or protein) in cells or developing organisms. In this case, nucleic acid hybridization probes can be used to detect specific regions of the DNA or specific RNA. That is, prelabeled DNA or RNA sequences are hybridized to the cellular nucleic acids, followed by detection with immunofluorescence and/or by amplification schemes based on avidin (streptavidin) or enzymes. The earliest probes were directly labeled RNAs (Bauman *et al.*, 1980). More recently, the sensitivity of DNA and RNA probes reacted with AAF (Landegent *et al.*, 1984), Hg^{2+}, and biotin (Hopman *et al.*, 1986) has been extended such that 5-kb single-copy genes are detected. The localization of genes or gene families by hybridization probes requires fixation of cells. Thus, correlations of chromosome topology with functional states (cell cycle position, metabolism, hormonal stimulation, etc.) will be assessed by probing *a posteriori* (Table I) with probes localized to specific genetic loci. One aspect of the hybridization technology that could be conceivably applied *in vivo* is the localization and quantitation of mRNA by use of labeled anti-sense RNA or DNA.

As early as 1980, fluorescence correlation spectroscopy was used to probe the state of chromatin in the cell nucleus (Sorscher *et al.*, 1980) by observing fluctuations in the fluorescence amplitude of ethidium bromide bound to monkey cell nuclei. Differences between resting and cycling nuclei were detected and interpreted as indication of changes in the compaction of chromatin.

IV. Measurement Techniques

A. Excitation Sources

As seen in Table III, the excitation wavelengths of the various DNA-specific fluorophores extend from the UV to the red. Thus, versatile light sources are desirable. Most fluorescence microscopes rely on 50–100 W mercury or xenon arc lamps for illumination and linepass or bandpass excitation filters combined with epiillumination and dichroic filters to block the excitation light from entering the detectors. Increasingly, lasers have been employed, particularly for dynamic diffusion measurements such as fluorescence photobleaching recovery (FPR) (see Wolf, this chapter 10, this volume; see also discussion below). Of particular note is the recent proliferation of laser scanning fluorescence microscopes (Brakenhoff *et al.*, 1979; Wilson and Sheppard, 1984; Wijnaendts van Resandt *et al.*, 1985; Wilke, 1984; White *et al.*, 1987; see Brakenhoff *et al.*, chapter 14, this volume). We are aware of at least six commercial instruments on the market at present. Multiple laser sources on these instruments provide excitation wavelengths from 356 to 665 nm. They can be coupled to dye lasers, frequency doubled, or operated in a pulsed mode (e.g., by mode locking or incorporating opto-modulators), thereby providing an arbitrary selection of spectrum, temporal profiles, polarization state, and energy.

With any of the above sources, electronic shutters and attenuating filters or polarizers are usually employed in order to limit the power and duration of the excitation source. These two factors are of paramount importance in the case of quantitative microscopy and/or vital staining. Bleaching of the fluorophore is usually to be avoided and photodamage to the preparation by the excitation light itself or generation of photoreactive by-products of the fluorophore must be controlled. Simple repetitive quantitative measurements of the fluorescence intensity of a preparation serve to demonstrate the maximum acceptable excitation irradiance and dose before bleaching ensues. Oxygen scavangers reduce bleaching, but are usually not compatible with vital staining. Although bleaching may not be occurring, photodamage may still pose a problem if nonfluorescent cellular constituents absorb strongly at the excitation wavelength. Outgrowth experiments can be used to determine the long-term effects on cell viability.

B. Quantitative Detection

There are two aspects to the problem of quantitation in FDIM (Arndt-Jovin and Jovin, 1985). The first relates to the selection of a sensitive detector. Chapters by Spring and Lowry and Aikens *et al.*, Volume 29, discuss in detail the types of detectors that may be used for quantitative microscopy using conventional light sources. Fluorescence is usually of low light intensity such that intensified video cameras or high-resolution, cooled CCD arrays are required. The problems of linearity and dynamic range of individual devices are discussed elsewhere. Our experience suggests that high-performance CCDs are, in most respects, superior to video detectors, except when high-speed repetitive measurements of large areas are required. In the particular case of DNA measurement, if the total DNA content of individual cells is the issue, eg., to determine cell cycle kinetics, then an efficient low-power objective with a large object field is preferable. The corresponding detector must have a high dynamic range, good linearity, and high spatial resolution in order to make precise measurements involving a small number of pixels for any one object. Such criteria are met by the above-mentioned CCD arrays.

The second consideration concerns the nature of the specimen. From the experience with flow sorters using a number of DNA-specific drugs, we know that diploid cell lines have a coefficient of variation (CV) in DNA content of 1.5–4%. The measurement of DNA in the microscope by absorption in systems with flying spot microphotometers (see Goldstein, 1986 for a discussion of errors) or CCD arrays (Carlson *et al.*, 1984) can give CVs comparable to flow cytometers (<5%). Systems for quantitation of DNA by fluorescence microscopy have been less successful, probably due to the lack of linearity of many TV-based sensors. We have recently addressed this problem using a CCD camera and a low-magnification 16×, 0.6 NA oil immersion lens. A typical field of growing mammalian fibroblasts will have 70–100 nuclei under these conditions. The set of images shown in Fig. 6 was taken after a 10-minute fixation in −20°C methanol,

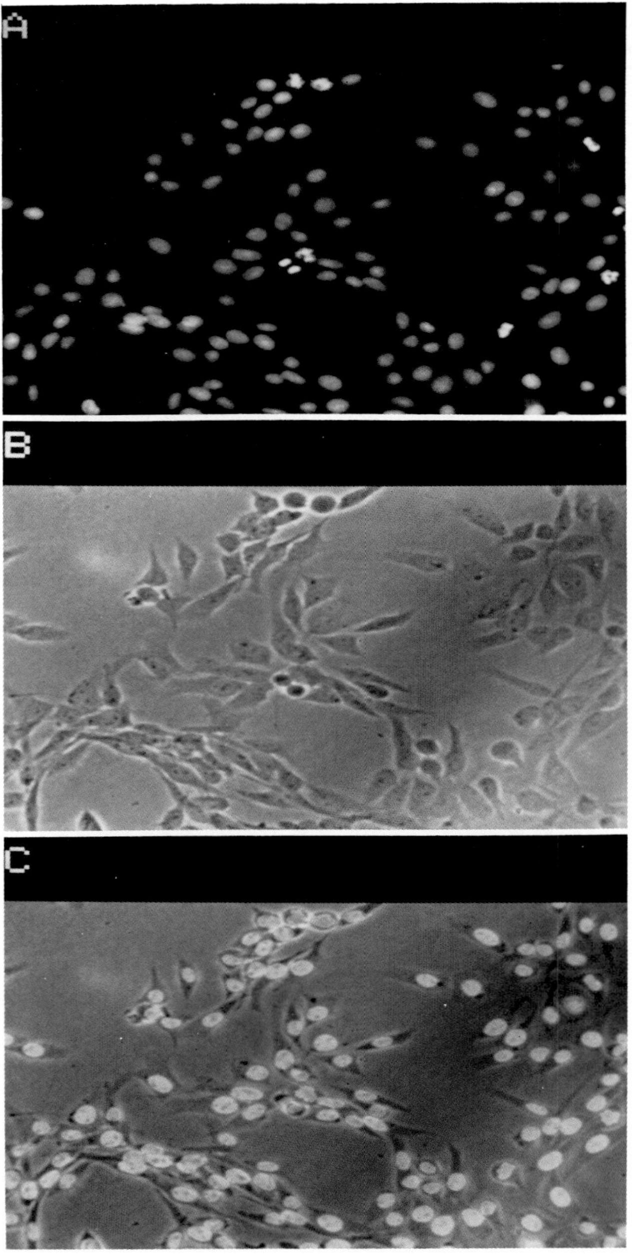
A
B
C

staining with mithramycin-Mg^{2+}, and rapid measurement. The fluorescence field without image processing is shown in Fig. 6A, the corresponding phase contrast image in Fig. 6B and an overlay of the DNA mask used for area determinations on the phase contrast image in Fig. 6C. Image processing is carried out to remove border objects and overlapping nuclei, and to combine the data from several fields. The evaluation of four fields (350 cells) for DNA content and nuclear size is shown in Fig. 7. Despite the statistical uncertainty imposed by the low number of cells, it appears that CVs comparable to those of absorption-based DIM systems can be obtained with this CCD FDIM. Similar results can be achieved *in vivo* using the dye BBI-342 on this cell line (data not shown).

As already mentioned, a new generation of light microscopes is coming into commercialization and laboratory use based on scanning with laser beams. Of greatest interest and potential impact is the *confocal* implementation, i.e., the confocal laser scanning microscope (CLSM). Brakenhoff reported the first CLSM with the capability of fluorescence detection (1979). In the present commercial instruments a focused laser beam is scanned over the object and the emission or transmission measured with a photomultiplier detector. Confocal optics allows an optical section to be made through transparent but thick objects (>500 μm), the information being derived from a relatively small z-axis focal element ($\sim$0.7–0.8 μm) with diffraction-limited resolution in the x-y plane (as low as 0.2 μm dependent inversely upon the excitation wavelength and the confocal pinhole diameter). The point-by-point illumination of the object ensures that background fluorescence and scatter, the vexing problems of conventional light microscopes, are largely avoided. The photomultiplier detector system provides both linearity and high sensitivity for quantitative measurements. The large increase in apparent resolution of CLSMs due to increased contrast enhancement and effective suppression of out-of-focus contributions is illustrated in Fig. 8, in which we

---

FIG. 6. Digital images of chinese hamster fibroblasts, stained specifically for double-stranded DNA. Log phase cells growing on 10-well Multitest slides were fixed for 10 minutes in $-20°C$ methanol, washed in saline buffer, stained with 50 μM mithramycin in saline containing 15mM $MgCl_2$, and measured after 5 minutes in the same solution containing 25% glycerol. (A) Fluorescence digital image corrected for bias, flat field, and background with camera software before transfer to the image processing system. (B) Digital phase-contrast image after illumination field and bias correction only. (C) Superposition of the binary mask from the fluorescent objects on the phase-contrast image processing to remove overlapping nuclei and border objects. Objective: Plan Neofluar 16×, phase 2, 0.5 NA oil immersion. Excitation of mithramycin was at 436 nm (8 nm HW with a 460 nm dichroic filter) and emission was collected >470 nm. The excitation source was attenuated with a 1 OD neutral density filter and fluorescence images acquired with an integration time of 1 second. CCD camera as in Fig. 2.

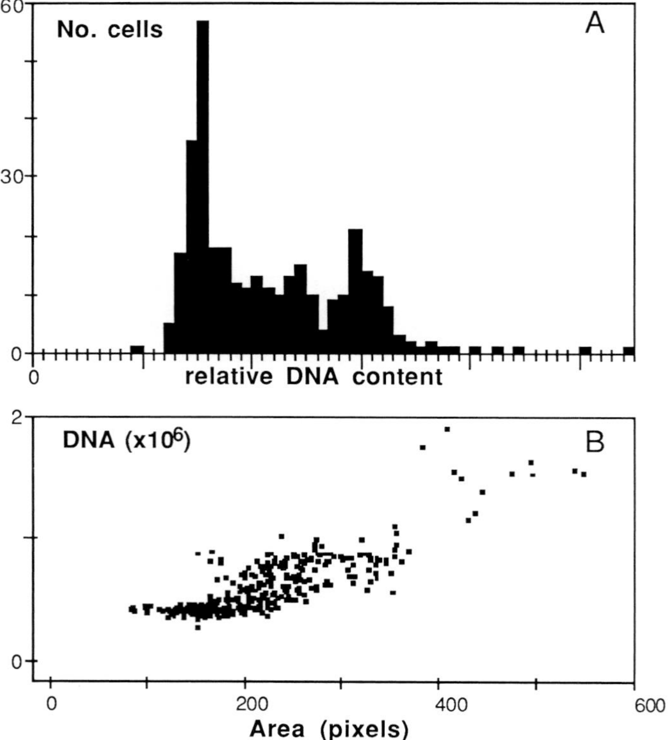

FIG. 7. DNA distributions generated from FDIM of attached chinese hamster cells stained with mithramycin. One of the original fluorescence images is shown in Fig. 6. The images were corrected for bias, background, and field inhomogeneity using Photometrics firmware. The 14-bit digitized data were transferred to a DEC μ VaxII and processed with the TIPS software package (TPD Image Processing System from the Institute of Applied Physics, TNO, and the Applied Physics Department of the Technical University of Delft). The elimination of nuclei on the image borders, the segmentation of remaining nuclei and the generation of a binary mask were effected by automatic routines. Any overlapping nucleic were eliminated interactively, and those remaining were labeled and analyzed by TIPS routines. The integrated intensity (proportional to DNA content) and area (in pixel units) for each nucleus were transferred to an Apple Macintosh computer for generation of the histograms and scatter plots shown in this figure using the commercial programs Statworks and Cricket Graph. (A) The maximal DNA content plotted was 1.5×10^6 (in units of counts from the Photometrics camera) and the histogram was resolved into 55 subpopulations. Despite the poor statistical significance due to the small number of cells analyzed (350), a distribution typical for proliferating cells is perceived with modal values in the expected relationship. One can infer an inherent CV for the DNA determination of ~4%. Note that a DNA content of 10^6 integrated counts (in our units) would correspond to a signal-to-noise ratio from counting statistics alone of 10^3. (B) Scatterplot of DNA content versus nuclear area (as a projection on the image plane). Clearly, the latter is correlated positively with increasing DNA content, as perceived further in (C) in the form of a three-dimensional histogram with a resolution of 30 (DNA) $\times$ 15 (area).

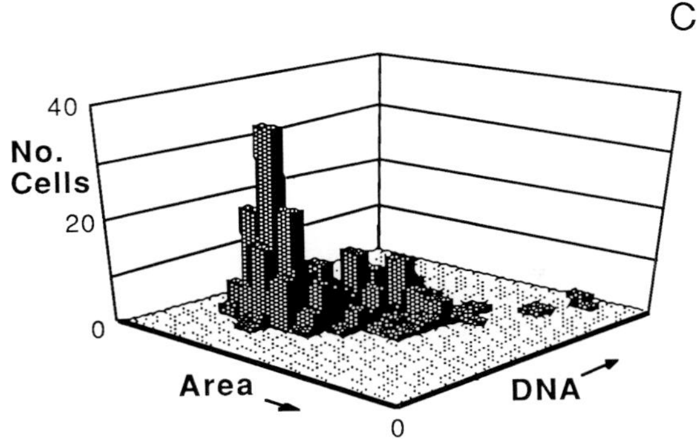

FIG. 7. *(continued)*

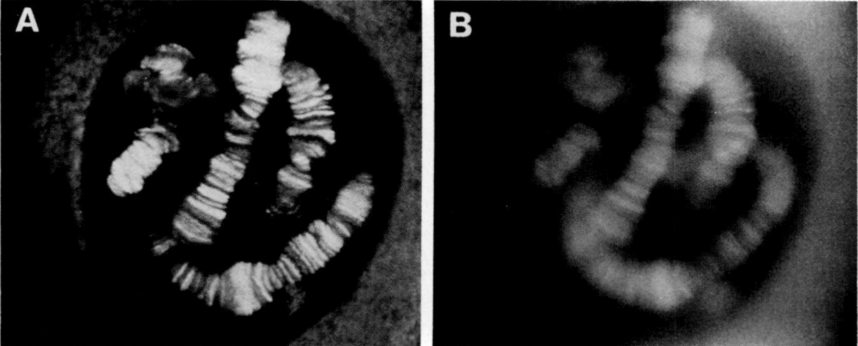

FIG. 8. Laser scanning microscope images of a whole mount salivary gland from *Chironomus thummi* stained *in vivo* with acridine orange. A mid-plane section of a single nucleus, object field 110 μm, in (A) confocal scanning mode and (B) nonconfocal mode. The four individual chromosomes are 10 μm in diameter. Images (512 × 512 pixels) were taken with a modified Carl Zeiss (Oberkochen) confocal laser scanning microscope LSM-44 equipped with three laser sources. Objective: Neofluar 40×, 0.75 NA; Excitation: 488 nm, 200 mW attenuated to (A) 1% or (B) 0.1%. The emission was collected through an aperture of 20 μm in the confocal mode. The images were acquired in 32 seconds by 16× line average scanning.

compare the same fluorescence micrograph viewed by an LSM with and without confocal optics. The whole salivary gland from an insect, *Chironomus thummi,* was stained *in vivo* with the drug acridine orange and viewed directly with the CLSM. The polytene chromosomes in this nucleus are 200 μm long and a clear delineation of the DNA bands of 0.5–2 μm thickness is visible in the confocal mode operation. The dramatic enhancement obtained in the confocal mode is obvious.

V. New Directions in FDIM

Most fluorescent microscopy is carried out with steady-state illumination and broadband detection of spectral emission and does not utilize time-resolved measurements. Fluorescence and other emission processes involve a number of photophysical phenomena that have not been exploited to a significant degree in microscopy but are realizable with present-day technology. The objectives outlined in this chapter with respect to DNA determinations would greatly benefit from the systematic application of these techniques.

A. Time-Resolved Measurements

A time-resolved fluorescence measurement in the microscope developed in the 1970s, fluorescence photobleaching recovery (FPR) uses a short (millisecond) high-intensity laser pulse to photobleach a spot in a fluorescent ligand field in a cell or membrane and then a low-intensity continuous laser illumination to monitor the recovery of the fluorescence by diffusion processes. The rates and extents of translational diffusion can be correlated with structural and functional states of cellular membranes and other organelles (for a review see Jovin and Vaz, 1988; see Wolf, Chapter 10, and Elson and Qian, Chapter 11, this volume). Jacobson and co-workers have extended FPR to whole-cell imaging with video cameras (Kapitza *et al.*, 1985) With the advent of the CCD technology it can be anticipated that more FPR measurements will be made in conjunction with imaging. In addition, the coupling of gated intensifiers to CCD detectors provides the following very general capabilities: time resolution down to a few microsecond, detection synchronized to an excitation pulse ("boxcar detection"), and quantum conversion at the intensified photocathode. The rotational motions of cellular macromolecules can also be assessed in the microscope by related techniques involving polarized fluorescence and phosphorescence emission (Yoshida and Barisas, 1986; Garland and Birmingham, 1986).

A more fundamental aspect of fluorescence is the lifetime of the fluorophore, which is generally in the range of 1–10 nanoseconds. The environment as well as competing physical processes can alter the fluorescence lifetime and provide us with physical information about the cellular milieu in which the fluorophore is localized. Since many DNA drugs exhibit altered quantum yields (and thus lifetimes) upon binding, time-resolved measurements could provide significant increases in sensitivity and selectivity. Such information, with the additional spatial resolution that the microscope provides, could distinguish, for example, between

DNA, RNA, or membrane binding of a probe. Few papers have dealt with attempts to measure these quantities in the microscope (Andreoni *et al.*, 1979; Arndt-Jovin *et al.*, 1979; Bottiroli *et al.*, 1980). Using a pulsed laser excitation and delayed digital (CCD) acquisition, we have recently measured delayed fluorescence and phosphorescence images.

The sensitivity of many FDIM measurements, especially using high-magnification objectives, is limited by scatter (both Rayleigh and Raman) and autofluorescence of the cells (Aubin, 1979) and immersion oils. Flurophores with long-wavelength excitation and emission maxima can overcome this problem to some extent but the gain may be offset by loss in detection sensitivity in the case of photomultiplier but not solid-state detectors. Another strategy is to select fluorophores (such as the lanthanides) or phosphorescent probes that have lifetimes in the micro- to millisecond range. By using pulsed light sources and gated detectors, these emissions can be resolved from the perturbing signals.

B. Phase-Resolved Measurements

Electro-optical or acousto-optical couplers can modulate an excitation laser source at a defined frequency or set of frequencies. By selecting a phase shift in the detector corresponding to a certain lifetime, a pronounced selectivity for particular fluorophore can be achieved (Gratton *et al.*, 1984). It follows that scanning the phase of the emission should permit the resolution of a number of components, each of which could be used individually or in combination to generate useful images. Thus, to continue the example cited in the previous section, one could visualize a drug bound to double-stranded DNA, to single-stranded nucleic acid(s), or associated with proteins and membranes. Although the components for a microscope phase detection system are commercially available, such an instrument has not yet been reported.

C. Spectrally Resolved Measurements

With a relatively stationary object in the microscope, the ability to alternate excitation and emission wavelengths by use of selective filters is easily achieved, e.g., using motor-driven filter wheels, and has been dealt with by several authors (Bright *et al.*, 1987). In many studies it is important to quantitate and correlate several cellular functions simultaneously (see Waggoner *et al.*, Chapter 17, and Bright *et al.*, Chapter 6, this volume). This is particularly true with the measurement of DNA content, since many cellular processes are dependent on the position in the cell cycle. Obviously, multiple fluorescent probes with spectrally separated excitation or emission properties provide the means to make such measurements. Some imaging systems can combine ratio images in real time whereas others store the images for later processing. If the spectral changes are associated with fast functional changes, then such a real-time system is desirable though not essential. The shift in excitation wavelength between the free and bound form of a fluorescent probe has been exploited in the microscope to measure Ca^{2+} by the dyes fura-2 or quin 2 (Tsien, Chapter 5, this volume) using ratio imaging. The emission spectra of other fluorescence probes show a remarkable sensitivity to local environment and thereby to the state of association of the macromolecule to which they are conjugated (Marriott *et al.*, 1988). It can be anticipated that microscope systems will increasingly exploit the entire range of spectral information. Several systems have been reported in which a spectrum analyzer has been coupled to the emission detector (Wampler, 1986). *In vivo* imaging with such a system is cumbersomely slow and bandpass filter systems with multiple detectors are technologically superior.

D. Spatially Resolved Measurements

Fluorescence is a particularly useful phenomenon for probing the state of association and the environment of molecules. In particular, FRET provides information about molecular interaction at distances of <0.01 μm since the efficiency varies inverse with the sixth power of the separation. As discussed in Section III,C, the measurement of FRET in the microscope provides an enormous spatial resolution and the ability to distinguish between aggregation or association states of molecules in different locations within a cell. In addition, the determinations of fluorescence polarization (Weber, 1986) can provide complementary information since rotational depolarization is also influenced by the local environment (see Axelrod, Chapter 12, this volume).

E. Other Prospects

Bio(chemi)luminescence may form the basis for even higher resolution and sensitivity in FDIM measurements. It should be possible in the near future to use antibody- or biotin-coupled enzymes such as luciferase as an alternative to peroxidases or photophatases in conjunction with DNA hybridization probes, and thereby increase the sensitivity by orders of magnitude.

Systematic design of probes coupled to molecular graphics and force-field calculations should provide greater specificity and selectively for

DNA, its structure, and metabolic state. It seems likely that the spectral repertoire of probes will also expand into the red, where autofluorescence of living cells is not such as problem.

FDIM is a quantitative descriptor of cellular structures and processes. The method has become accessible to the average biologist because of advances in hardware and software used for data processing. The latter will become faster and more efficient in the future with the use of special-purpose processors and graphics engines, developments that will be especially important for dealing with the mass of correlated data of nuclei generated in three dimensions from CLSMs.

Acknowledgments

We acknowledge the generosity of Dr. R. Aikens in supplying the camera system used in this study. We thank our colleagues, Michel Robert-Nicoud for assistance in the preparation of Fig. 8 and Robert Clegg and Gerard Marriott for helpful discussions.

References

Andreoni, A., Sacchi, C. A., and Svelto, O. (1979). *In* "Chemical and Biochemical Applications of Lasers" (B. Moore, ed), Vol. 4, pp. 1–12. Academic Press, New York.

Arndt-Jovin, D. J., and Jovin, T. M. (1977). *J. Histochem. Cytochem.* **25,** 585–589.

Arndt-Jovin, D. J. and Jovin, T. M. (1985). *Science* **230,** 247–256.

Arndt-Jovin, D. J., Latt, S. A., Striker, G., and Jovin, T. M. (1979). *J. Histochem. Cytochem.* **27,** 87–95.

Arndt-Jovin, D. J., Robert-Nicoud, M., Baurschmidt, P., and Jovin, T. M. (1985). *J. Cell Biol.* **101,** 1422–1433.

Arndt-Jovin, D. J., and Jovin, T. N. (1988). In preparation.

Aubin, J. E. (1979). *J. Histochem. Cytochem.* **27,** 36–43.

Bartholdi, M. F., Sinclari, D. C., and Cram, L. S. (1984). *Cytometry* **3,** 395–401.

Barton, J. K. (1986). *Science* **233,** 727–34.

Barton, J. K., and Raphael, A. L. (1984). *J. Am. Chem. Soc.* **106,** 2466–2468.

Barton, J. K., Basile, L. A., Danishefsky. A., and Alexandrescu, A. (1984). *Proc. Natl. Acad. Sci. U.S.A.* **81,** 1961–1965.

Bauman, J. G. J., Wiegant, J., Borst, P., and van Duijn, P. (1980). *Exp. Cell Res.* **128,** 485–490.

Becker, M. M., and Wang, J. C. (1984). *Nature (London)* **309,** 682–687.

Bottiroli, G., Cionini, P. G., Docchio, F., and Sacchi, C. A. (1980). *In* "Lasers: Photomedicine and Photobiology" (R. Pratesi and C. A. Sacchi, eds.), pp. 175–179. Springer-Verlag, Berlin.

Brakenhoff, G. J. (1979). *J. Microsc.* **117,** 233–242.

Brakenhoff, G. J., Blom, P., and Barends, P. (1979). *J. Microsc.* **117,** 219–232.

Brakenhoff, G. J., van der Voort, H. T. M., van Spronsen, E. A., and Nanninga, N. (1987). *Anno N. Y. Acad. Sci.* **483,** 405–415.

Bright, G. R., Rogowska, J., Fisher, G. W., and Taylor, D. L. (1987). *Biotechniques* **5,** 556–563.

Bucana, C., Saiki, I., and Nayar, R. (1986). *J. Histochem. Cytochem.* **34,** 1109–1115.

Carlson, L., Auer, G., Kudynowski, J., Lindblom, R., and Zetterberg, A. (1984). *Cytometry* **4,** 319–326.

Caspersson, T. (1950). *Exp. Cell Res.* **1,** 595–598.
Caspersson, T. (1979). *Acta Histochem.* **20,** 15–28.
Cavatorta, P., Masotti, L., and Szabo, A. G. (1985). *Biophys. Chem.* **22,** 11–16.
Cimino, G. D., Gamper, H. B., Isaacs, S. T., and Hearst, J. E. (1985). *Annu. Rev. Biochem.* **54,** 1151–1193.
Crissman, H. A., and Tobey, R. A. (1974). *Science* **184,** 1297–1298.
Crissman, H. A., Stevenson, A. P., Kissane, R. J., and Tobey, R. A. (1979). *In* "Flow Cytometry and Sorting" (M. R., Melamed, P. F. Mullaney, and M. L. Mendelsohn, eds.), pp. 243–262 Wiley, New York.
Darzynkiewicz, Z. (1979). *In* "Flow Cytometry and Sorting" (M. R., Melamed, P. F. Mullaney, and M. L. Mendelsohn, eds.), pp. 285–316 Wiley, New York.
Darzynkiewicz, Z., and Kapuscinski, J. (1988). *Cytometry* **9,** 7–18.
Darzynkiewicz, Z., Traganos, F., and Melamed, M. R. (1980). *Cytometry* **1,** 98–108.
Darzynkiewicz, Z., Traganos, F., Kapuscinski, J., Staiano-Coico, L., and Melamed, M. R. (1984). *Cytometry* **5,** 355–363.
Diekmann, S., and Zarling, D. A. (1987). *Nucleic Acids Res.* **15,** 6063–6074.
Einck, L., and Bustin, M. (1983). *Proc. Natl. Acad. Sci. U.S.A.* **80,** 6735–6739.
Festy, B. (1979) *In* "Antibiotics, Vol. 2. Mechanism of Action of Antieucaryotic and Antiviral Compounds" (J. W. Corcoran and F. E. Hahn, eds.), pp. 223–235. Springer-Verlag, Berlin.
Festy, B., and Duane, M. (1973). *Biochemistry* **12,** 4827–4834.
Förster, T. (1948). *Ann. Phys. Leipzig* **2,** 55–75.
Fried, J., Doblin, J., Takamoto, S., Perez, A., Hansen, H., and Clarkson, B. (1982). *Cytometry* **3,** 42–47.
Garland P. B., and Birmingham, J. J. (1986). *In* "Applications of Fluorescence in the Biomedical Sciences" (D. L., Taylor, A. S., Waggoner, R. F., Murphy, F., Lanni, and R. R., Birge, eds.), pp. 245–254. Liss, New York.
Gaugain, B., Barbet, J., Capelle, N., Roques, B. P., LePecq, J. -B., and LeBret, M. (1978). *Biochemistry* **17,** 5078–5088.
Gill, J. E., Jotz, M. M., Young, S. G., Modest, E. J., and Sengupta, S. K. (1975). *J. Histochem. Cytochem.* **23,** 793–799.
Giniger, E., and Ptashne, M. (1987). *Nature* (*London*) **330,** 670–672.
Goldstein, D. J. (1986). *Cytometry* **7,** 532–535.
Gonchoroff, N. J., Greipp, P. R., Kyle, R. A., and Katzmann, J. A. (1985). *Cytometry* **6,** 506–512.
Gratton, E., Jameson, D. M., and Hall, R. D. (1984). *Annu. Rev. Biophys. Bioeng.* **13,** 105–124.
Hamori, E., Arndt-Jovin, D. J., Grimwade, B. G., and Jovin, T. M. (1980). *Cytometry* **1,** 132–135.
Harshman, K. D., and Dervan, P. B. (1985). *Nucleic Acids Res.* **13,** 4825–4835.
Hayasaka, T., and Inoue, Y. (1969). *Biochemistry* **8,** 2342–2347.
Hopman, A. H. N., Wiegant, J., Raap, A. K., Landegent, J. E., van der Ploeg, M., and van Duijn, P. (1986). *Histochemistry* **85,** 1–4.
Jovin, T. M., and Arndt-Jovin, D. J. (1988). *In* "Microspectrofluorometry of Single Living Cells" (E., Kohen, J. S., Ploem, and J. G., Hirschberg, eds.). Academic Press, San Diego, in press.
Jovin, T. M., and Vaz, W. H. C. (1988) *In* "Methods in Enzymology." (R. Wu. ed.), Vol 155, Academic Press, Orlando, Florida, in press.
Jovin, T. M., van de Sande, J. H., Zarling, D. A., Arndt-Jovin, D. J., Eckstein, F., Füldner, H. H., Greider, C., Grieger, I., Hamori, E., Kalisch, B., McIntosh, L. P., and Robert-Nicoud, M. (1983). *Cold Spring Harbor Symp. Quant. Biol.* **47,** 143–154.

Kapitza, H. G., McGregor, G., and Jacobsen, K. A. (1985). *Proc. Natl. Acad. Sci. USA* **82,** 4122–4126.

Kapuscinski, J., Darzynkiewicz, Z., and Melamed, M. (1982). *Cytometry* **2,** 201–211.

Kelly, J. M., Murphy, M. J., McConnell, D. J., and OhUigin, C. (1985). *Nuclei Acids Res.* **13,** 167–184.

Kersten, H., and Kersten, W. (1974). "Inhibitors of Nucleic Acid Synthesis," pp. 1–184. Springer-Verlag, Heidelberg.

Kopka, M. L., Yoon, C. Goodsell, D., Pjura, P., and Dickerson, R. E. (1985). *J. Mol. Biol.* **183,** 553–563.

Kopka, M. L., Pjura, P. E., Goodsell, D. S., and Dickerson, R. E. (1987). *In* "Nucleic Acids and Molecular Biology 1" (F. Eckstein and D. M. J. Lilley, eds.), pp. 1–24. Springer-Verlag, Heidelberg.

Krishnan, A. (1987). *Cytometry* **8,** 642–645.

Landegent, J. E., Jansen in de Wal, N., Baan, R. A., Hoeijmakers, J .H. H., and van der Ploeg, M. (1984). *Exp. Cell Res.* **153,** 61–72.

Latt, S. A. (1976). *Annu. Rev. Biophys. Bioeng.* **5,** 1–38.

Latt, S. A. (1977). *Can. J. Genet. Cytol* **19,** 603–623.

Latt, S. A. (1979). *In* "Flow Cytometry and Sorting" (M. R., Melamed, P. F. Mullaney, and M. L. Mendelsohn, eds.), pp. 263–284. Wiley, New York.

Latt, S. A., Sahar, E., Eisenhard, M. E., and Juergens, L. A. (1980). *Cytometry* **1,** 2–11.

Latt. S. A., George, Y., and Gray, J. W. (1977). *J. Histochem. Cytochem.* **25,** 927–934.

LePecq, J. B., and Paoletti, C. (1967). *J. Mol. Biol.* **27,** 87–106.

Loewe, H., and Urbanietz, J. (1974). *Arzneim. Forsch.* **24,** 1927–1933.

Loken, M. R. (1980). *J. Histochem. Cytochem.* **28,** 36.

Makarov, V. L., Poletaev, A. I., Sveshnikov, P. G., Kondrat'eva, N. O., Pis'menskii, V. F., Doskocil, J., Koudelka, J., and Vol'kenshtein, M. V. (1979). *Mol. Biol. (Engl. Trans.)* **13,** 339–352.

Markovits, J., Ramstein, J., Roques, B. P., and LePecq, J.-B. (1983). *Biochemistry* **22,** 3231–3237.

Marriott, G., Zechel, K., and Jovin, T. M. (1988). *Biochemistry* **27,** in press.

Mazza, G., Galizzi, A., Minghetti, A., and Siccradi, A. (1973). *Antimicrob. Agents Chemother.* **3**, 384–391.

Mikhailov, M. V., Zasedatelev, A. S., Krylov, A. S., and Furskii, G. V. (1981). *Mol. Biol. (Engl. Trans.)* **15,** 541–553.

Miller, J., McLachlan, A. D., and Klug, A. (1985). *EMBO J.* **4,** 1609–1614.

Müller, W., and Crothers, D. M. (1975). *Eur. J. Biochem.* **54,** 267–277.

Müller, W., and Gautier, F. (1975). *Eur. J. Biochem.* **54,** 385–394.

Müller, W., Bünemann, H., and Dattagupta, N. (1975). *Eur. J. Biochem.* **54,** 279–291.

Nakamura, H., Morita, T., and Sato, C. (1986). *Exp. Cell Res.* **165,** 291–297.

Neidle, S., and Berman, H. (1983). *Prog. Biophys. Mol. Biol.* **41,** 43–66.

Nelson, J. W., and Tinoco, I. (1984). *Biopolymers* **23,** 213–233.

Nordén, B., and Tjerneld, F. (1976). *FEBS Lett.* **67,** 368–370.

Nordén, B., and Tjerneld, F. (1977). *Chem. Phys. Lett.* **50,** 508–512.

Pachman, U., and Rigler, R. (1972). *Exp. Cell Res.* **72,** 602–608.

Pallavincini, M. G., Lalande, M. E., Miller, R. G., and Hill, R. P. (1979). *Cancer Res.* **39,** 1891.

Pijura, P., Goodsell, D., Grezeskowiak, K., and Dickerson, R. E. (1987). *J. Mol. Biol.* **197,** 257–271.

Ramponi, R., and Rodgers, M. A. J. (1987). *Photochem. Photobiol.* **45,** 161–165.

Rigler, R., and Ehernberg, M. (1973). *Q. Rev. Biophys.* **6,** 139–199.

Robert-Nicoud, M., Arndt-Jovin, D. J., Schormann, T., and Jovin, T. M. (1988). *Eur. J. Cell Biol*, in press.
Sahar, E., and Latt, S. A. (1980). *Chromosoma* **79,** 1–28.
Schröter, H., Maier, G., Ponstingl, H., and Nordheim, A. (1985). *EMBO J.* **4,** 3867–3872.
Schultz, P. G., and Dervan, P. B. (1984). *J. Biomol. Struct. Dyn.* **1,** 1133–1145.
Sobell, H., Tsai, C. C., Jain, S. C., and Gilbert, S. (1977). *J. Mol. Biol.* **114,** 333–365.
Sorscher, S. M., Bartholomew, J. C., and Klein, M. P. (1980). *Biochim. Biophys. Acta* **610,** 28–46.
Stetten, G., Latt, S. A., and Davidson, R. L. (1976). *Somatic Cell Genet.* **2,** 285–290.
Stollar, B. D. (1986). *CRC Crit. Rev. Biochem.* **20,** 1–36.
Stryer, L., Thomas, D. D., and Meares, C. F. (1982). *Annu. Rev. Biophys. Bioeng.* **11,** 203–222.
Traincard, F., Ternynck, T., Danchin, A., and Avrameas, S. (1983). *Ann Immunol.* **134,** 399–405.
Van Zant, G., and Fry, C. G. (1983). *Cytometry* **4,** 40–46.
Wampler, J. E. (1986). *In* "Applications of Fluorescence in the Biomedical Sciences" (D. L. Taylor, A. S., Waggoner, R. F., Murphy, F., Lanni, R. R., Birge, eds.), pp. 301–320. Liss, New York.
Wang, A. H.-J. (1987). *In* "Nucleic Acids and Molecular Biology 1" (F. Eckstein and D. M. J. Lilley, eds.). pp 53–69. Springer-Verlag, Heidelberg.
Waring, M. J. (1981). *Annu. Rev. Biochem.* **50,** 159–192.
Weber, G. (1986). *In* "Applications of Fluorescence in the Biomedical Sciences" (D. L. Taylor, A. S., Waggoner, R. F., Murphy, F., Lanni, and R. R. Birge, eds.), pp. 601–616. Liss, New York.
White, J. G., Amos, W. B., and Fordham, M. (1987). *J. Cell Biol.* **105,** 41–48.
Wijnaendts van Resandt, R. W., Marsman, H. J. B., Kaplan, R., Davourt, J., Stelzer, E. H. K., and Stricker, R. (1985). *J. Microsc.* **138,** 29–34.
Wilke, V. (1984). *Scanning* **7,** 88–96.
Wilson, T. (1987). *Ann. N.Y. Acad. Sci.* **483,** 416–427.
Wilson, T., and Sheppard, C. (1984). "Theory and Practice of Scanning Optical Microscopy." pp. 1–213. Academic Press, London.
Yoshida, T. M., and Barisas, B. G. (1986). *Biophys. J.* **50,** 41–53.
Zimmer, C., and Wähnert, U. (1986). *Progr. Biophys. Molec. Biol.* **47,** 31–112.

Chapter 17

Multiple Spectral Parameter Imaging

A. WAGGONER, R. DEBIASIO, P. CONRAD, G. R. BRIGHT, L. ERNST, K. RYAN, M. NEDERLOF, AND D. TAYLOR

Department of Biological Sciences and
Center for Fluorescence Research in the Biomedical Sciences
Carnegie Mellon University
Pittsburgh, Pennsylvania 15213

I. Introduction

It is clear that cell functions such as cell division, endocytosis, and cell migration involve a complex temporal and spatial interplay of multiple organelles, macromolecules, ions, and metabolites. A goal of cell biology is to define the role of each cellular constituent and to define the molecular interplay of multiple constituents that are required for the completion of specific cell functions. Therefore, a technique is required that will permit the quantitation of multiple physiological parameters in time and space within the same cells.

Quantitative fluorescence microscopy, when combined with the increasing number of sensitive fluorescent and other luminescent probes now available, offers a powerful approach for defining the chemical and molecular dynamics within living cells in time and space. We have developed an approach that will now allow the analysis and correlation of up to five separate parameters based on the spectral isolation of distinct fluorescent probes (DeBiasio *et al.*, 1987). The individual parameters can include a combination of ratio measurements of various ions (see Tsien, Chapter 5, and Bright *et al.*, Chapter 6, this volume), the mobility and orientation of fluorescent analogs on or in cells (Wolf, Chapter 10, and Axelrod, Chapter 12, this volume, and several chapters in Volume 29), membrane potential (Gross and Loew, Chapter 7, this volume), the distribution and activity of selected organelles (e.g., Chen and Maxfield, Volume 29), and the organization of DNA (Arndt-Jovin and Jovin, Chapter 16, this volume), to name just a few. The key is to develop and to use probes that can be spectrally isolated. Furthermore, the use of a properly designed imaging workstation will allow the investigator to employ these methods in both two and three dimensions and to harness the power of video-enhanced contrast microscopy (Inoué, Chapter 3, this volume).

The goal of this chapter is to outline the requirements for performing multiple spectral parameter imaging on living cells. Of course, many of these considerations also apply to fixed preparations.

II. Practical Aspects of Multiple Spectral Parameter Imaging

Multiple spectral parameter imaging requires an integration of several technologies, including fluorescent probe chemistry, biochemistry, fluorescence spectroscopy, fluorescence microscopy, digital image processing, digital image analysis, and digital image display. The spectral characteristics of the probes and their interaction with cell components must be well understood. The probes must be properly incorporated into the cells with minimal cellular damage. Cells containing multiple fluorescent probes must be maintained in good physiological condition for a time equal to the duration of the experiment. Multiple images at different wavelengths of excitation and/or emission must be acquired and stored over a time period that is short compared to the rate of the cellular function. The images must be "restored," quantitative information must be extracted from each individual image, and the spatial and temporal correlations between different parameters (distinct images in the image set) must be determined. And finally, images and the tabular and graphical results of the analyses must be presented at high resolution on a graphics monitor and in equally high-quality hardcopy form, preferably ready for publication.

The following discussion separates the practical aspects into distinct categories.

A. Selection and Use of Fluorescent Probes

A fluorescent probe can be a single fluorophore or it can be a biological (e.g., fluorescent analog; see Wang, Volume 29) or nonbiological material (e.g., fluorescent dextran; see Luby-Phelps, Volume 29) that is labeled with one or more fluorophores. The fluorescence signal from the probe contains information about the location and amount of the probe in the specimen and, if the probe is sensitive to its local environment, it will provide relevant physiological information, such as pH, calcium concentration, etc.

The kinds of biological questions that can be approached by fluorescence methods is dictated by the availability of sensitive and selective probes. Measurement of a number of structural and functional parameters at one time from a single preparation requires independent detection of the signals from each probe. At present, this can most easily be done by using probes with sufficiently different absorption and emission spectra carefully matched to the excitation and emission filters of the microscope. For example, the spectra of five probes with distinct spectral properties are shown in Fig. 1. One could also imagine separate detection of the probes based on the different fluorescent and phosphorescent lifetimes. Then a combination of spectral and lifetime selections could extend the number of parameters even further.

Let us consider the types of fluorescent probes available and the kinds of biological information they provide. The discussion is divided into two types of experiments: fixed cell studies and analysis of living cells. The information of fixed cell probes is included since every live cell investigation can be ended by fixation and analysis with probes for fixed cells (see Gorbsky and Borisy, Volume 29).

1. Probes for Fixed Preparations

Antibodies, avidin, lectins, and nucleic acids can be labeled with a range of commercially available fluorescent tags and used in established procedures to measure the distribution and amount of antigens, biotin-labeled materials, carbohydrate moieties, and complementary DNA and RNA sequences. Fluorescence detection of genetic sequences is only in its infancy but it will have a large impact on biology and medicine. It is also

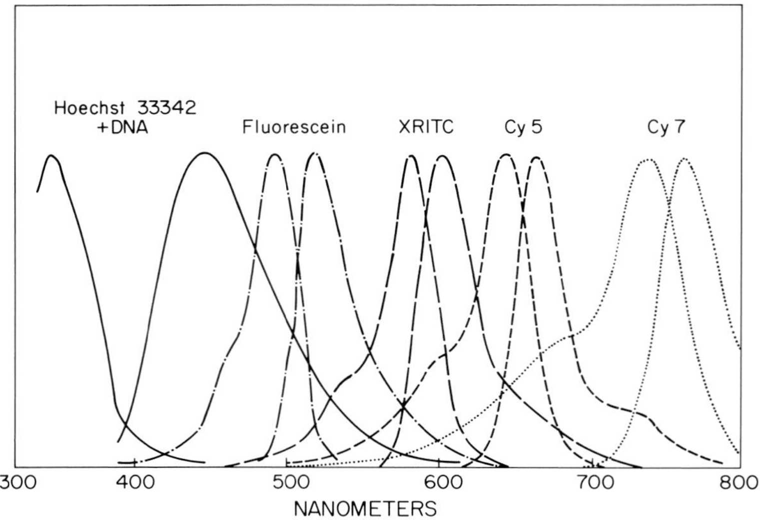

FIG. 1. Absorption and emission profiles of five spectrally distinct fluorescent probes. For each, the trace on the left is the absorption spectrum, while the trace on the right is the emission spectrum. Hoechst 33342 is a DNA content probe (see Table I). The other four probes are fluorescent labeling reagents. The membrane potential probe $diIC_1(5)$ (DeBiasio *et al.,* 1987) has spectra nearly identical to Cy5.

possible to incorporate fluorescent analogs of biological molecules into living cells and to obtain fluorescence images of the cells after they are fixed by a method that preserves morphology (see Gorbsky and Borisy, Volume 29 in this series).

Many of the reactive reagents available for tagging these materials are listed in Table I. (The structures of these molecules and of many other fluorescent probes discussed below can be found in Waggoner, 1986.) Until recently, most labeling reagents available absorb and emit light in the 350–600 nm region of the spectrum but new molecules (Cy5 and Cy7; Mujumdar *et al.,* 1988; Ernst *et al.,* 1988) have been synthesized that emit at longer wavelengths, even as far as the near infrared. Procedures have also been developed for attaching the fluorescent protein, allophycocyanin. (which emits at 680 nm) to proteins (Hardy *et al.,* 1983). The reagents can be attached to proteins through a variety of reactive groups (Haugland, 1983). Isothiocyanates are widely used to attach chromophores to neutral amino groups in buffers at pH 8.5–9.5. However, there are other alternatives for labeling amino groups that include succinimido active esters, imidoesters, anhydrides, chlorotriazinyl groups, aziridines, and so on. Iodoacetamido and maleimido groups on fluorochromes are commonly

TABLE I

SPECTROSCOPIC PROPERTIES OF SELECTED PROBES

Parameter	Probe[a]	Absorption Maximum[b]	Extinction Maximum[c]	Emission Maximum[b]	Quantum Yield	Measurement Conditions	References[d]
Covalent labeling reagents	Fluorescein-ITC, DCT, IA, MAL[e]	490	67	520	0.71	PH 7, PBS	1, W, MP
		494	76			pH 9.2, 0.1 *M* borate	2
	FITC-antibody	490	*	520	0.1–0.4[f]	pH 7, PBS	W
	TRITC-amines	554	85	573	0.28	pH 7, PBS	1, MP
	XRITC-amines	582	79	601	0.26	pH 7, PBS	W
	XRITC-antibody	580	*	604	0.08[g]	pH 7, PBS	W
	Texas Red-amines	596	85	620	0.51	pH 7, PBS	3, W, MP
	Texas Red-antibody	596	*	620	0.01[h]	pH 7, PBS	W
	Lissamine rhodamine sulfonamide	570		590	Medium		MP
	Eosin-ITC	522	83	540	0.19	PH 8, 0.5 *M* NaCl	4, 5
	Eosin-ITC-protein	525–528	~84	550		pH 8, 50 m *M* phosphate	
	Erythrosin-ITC, IA, MAL[e]	530	83	550	0.02	pH 8, 50 m *M* phosphate	5
	Erythrosin-ITC-protein	540	~84	560		pH 8, 50 m *M* phosphate	
	NBD-amine	478	24.6	520–550	0.36/0.21	Ethanol methanol	6, 7, 8
	NBD-S-CH_2CH_2OH	425	12.1	531	0.002	pH 7.5, 10% glycerol	
					0.015	80% glycerol	
	Dansyl-NH-CH_3	340 (pH 7.4, 0.1 *M* Tris)	3.4	578	0.068	Water	MP, 9
				539	0.5	Ethanol	
				508	0.41	Chloroform	

(*continued*)

TABLE I (*continued*)

Parameter	Probe[a]	Absorption Maximum[b]	Extinction Maximum[c]	Emission Maximum[b]	Quantum Yield	Measurement Conditions	References[d]
	Coumarin–phalloidin	387		470		Water	29
	Phycoerythrin-R	480–565	1960	578	0.68	pH 7, PBS	10
	Allophycocyanine	650	700	660	0.68	pH7, PBS	10
DNA–RNA content[i]	Hoechst 33342	340	120	450	0.83	+DNA (excess)	W
	DAPI	350		470		+DNA (excess)	W
	Ethidium Bromide	510	3.2	595		+DNA (excess)	11
	Propidium iodide	536	6.4	623	0.09	+DNA (excess)	W
	Acridine orange	480 440–470		520 650		+DNA +RNA	12, 13
	Pyronin Y	549–561 560–562 497	67–84 70–90 42	567–574 565–574 563	0.04–0.26 0.05–0.21 Low	+ds DNA[j] +ds RNA[j] +ss RNA	14, 15
	Thiazole orange	453	26	480	0.08	RNA	16
Membrane potential	diO-Cn-(3)	485	149	505	0.05	Methanol	17, W
	dil-Cn-(3)	548	126	567	0.07	Methanol	17, W
	dil-Cn-(5)	646	200	668	0.4	Methanol	17, W
	dil-Cn-(7)	740	240	770	0.28	Ethanol	W, 18

	diBA-Isopr-(3)	493	130	517	0.03	Methanol	17, W
	diBA-C_4-(5)	590	176	620		Ethanol	W
	Rhodamine 123	511	85	534	0.9	Ethanol	EK, 19
	WW 781 (XXV of Ref. 27)	614		638		Ethanol	27
	RH 160	526	50	685		Ethanol	27, 28
Lipid content and fluidity	Nile red	485 530		525 605		Heptane Acetone	20, 21
	Diphenylhexatriene (DPH)	330, 351, 370	77 (351 nm)	430		Hexane	MP
	dil-C_{18}-(3)	546	126	565	0.07	Methanol	W
	Dansyl-PE	335	4.5				
	NBD phosphatidylethanolamine	450	24[k]	530		Lipid	22
	Anthroyl stearate	361, 381	8.4, 7.5	446		Methanol	23
	Pyrene-sulfonamidoalkyls	350	30	380–400			MP
pH	6-Carboxyfluorescein	495 450		520		High pH Low pH	MP MP
	BCECF	505 460		530		High pH Low pH	MP MP
	SNARF-1 (pK_a = 7.5)	518–548 574		587 636		pH 5.5 pH 10.0	MP MP
	DCDHB[l]	340–360 340–360		500–580 420–440		High pH Low pH	24, MP

(*continued*)

TABLE I (*continued*)

Parameter	Probe[a]	Absorption Maximum[b]	Extinction Maximum	Emission Maximum[b]	Quantum Yield	Measurement Conditions	References[d]
Calcium	Fura-2	335	33	512–518	0.23	Low calcium	25
		360	27	505–510	0.49	High calcium	
	Indo-1	330	34	390–410	0.56	High calcium	25
		350	34	482–485	0.38	Low calcium	
Enzyme substrates	Rhodamine-di-arg-CBZ substrate	—	Low at 495 nm	532	0.09	HEPES pH 7.5+ 15% ethanol	26
	Product of rxn (rhodamine)	495	67	523	0.91	HEPES pH 7.5+ 15% ethanol	
	Coumarin-glucoside substrate	316	13	395		Acetate pH 5.5+ 1% Lubrol	W
	Product of rxn (hydroxycoumarin)	370	17	450		Glycine pH 10+ 1% Lubrol	W

[a] Abbreviations: ITC, Isothiocyanato-; DCT, Dichlorotrizinyl-; IA, Iodoacetamido-; MAL, Maleimido-; BCECF, 2',7'-bis-(2-carboxyethyl)-5(and 6)carboxyfluorescein; DCDHB, dicyano-dihydroxybenzene; DAPI, 4',6-diamidino-2-phenylindole; FITC, fluorescein isothiocyanate; PBS, phosphate-buffered saline.

[b] Measured in nanometers.

[c] Multiply value listed by 1000 to get liters/mol. cm. *, Suggest using the value for the amine adduct.

[d] EK, Eastman Kodak Chemical Catalog; MP, Molecular Probes, Inc, catalog or personal communication; W, Waggoner laboratory determination; 1, Haugland (1983); 2 Wilderspin, and Green (1983); 3 Titus *et al.* (1982); 4 Cherry *et al.* (1976); 5 Garland and Moore (1979); 6 Kenner, RA & Aboderin, AA: (1971); 7 Allen and Lowe (1973); 8 Bratcher *et al.* (1979); 9 Chen (1968); 10 Oi *et al.* (1982); 11 Pohl *et al.* (1972); 12 Kapuscinski *et al.* (1982); 13 Shapiro (1985); 14 Darzynkiewicz *et al.* (1987); 15 Kapuscinski and Darzynkiewicz (1987); 16 Lee *et al.* (1986); 17 Sims *et al.* (1974); 18 Duggan *et al.* (1983); 19 Kubin and Fletcher (1983); 20 Greenspan and Fowler (1985); 21 Sackett and Wolff (1987); 22 Struck *et al.* (1981); 23 Waggoner and Sryer (1970); 24 Valet *et al.* (1981); 25 Grynkiewicz *et al.* (1985); 26 Leytus *et al.* (1983); 27 London *et al.* (1986); 28 Grinvald *et al.* (1984); 29 Small *et al.* (1988); 30 Wolfbeis *et al.* (1983);

[e] Covalently bound to amino group (ITC, DCT) or sulfhydryl group (IA, MAL).

[f] Dye/antibody ratio of 2 to 5.

[g] Dye/antibody ratio of 2.5.

[h] Dye/antibody ratio of 1.2.

[i] See Arndt-Jovin and Jovin (Table III), Chapter 16, this volume, for additional DNA content probes.

[j] Base pair dependent.

[k] Value for NBD-ethanolamine in methanol which has an absolute maximum at 470 nm and an emission maximum at 550 nm (Barak and Yocum, (1981).

[l] See Tsien, Chapter 5, this volume, for additional details.

used for specific labeling of sulfhydryl residues and disulfide-bearing fluorophores will label sulfhydryl groups by disulfide exchange.

It is obvious that it is desirable to have the labeled material as brightly fluorescent as possible. That means that the extinction coefficient and the quantum yield should be large. The commonly used xanthene dyes (rhodamine and fluorescein, for example) have extinction coefficients around 70,000 liters/mol.cm and quantum yields in optimal buffers around 0.7. The respective values for cyanines are 200,000 and 0.3. The values for other labels are generally lower, except for the phycobiliproteins, some of which contain as many as 34 chromophores integrated into their protein structure, have extinction coefficients as high as 2×10^6, and quantum yields near 1.0 (Oi *et al.,* 1982; Kronick and Grossman, 1983). While the phycobiliproteins are extremely fluorescent and have found a wide use, particularly in flow cytometry as antibody labels, they are large and reduce the rate of binding of the labeled antibodies to their antigens.

Additional factors determine the final brightness of materials that have multiple fluorophores attached. If the fluorophores are physically close enough to behave as dimers or larger physical aggregates that are nonfluorescent, the overall fluorescence of the labeled species is reduced. Other nonaggregated fluorophores attached to the protein can be quenched by Forster energy transfer to nonfluorescent fluorophores that act as energy transfer sinks. Some labeling reagents, particularly the rhodamines, tend to become quenched when only a few are attached to an antibody. However, we believe it is possible to engineer the structure of new fluorophores to minimize this kind of quenching. These changes often improve the solubility of the labeled antibody.

Besides improving the brightness, water solubility, and wavelength range of labeling reagents, significant improvements need to be made in their photostability. Fluorescein, for example, can be excited about 100,000 times before it photobleaches (Mathies and Stryer, 1986) but fading is readily observed within seconds upon visual examination in a fluorescence microscope. Fortunately, low-light-level microscopy significantly reduces the rate of fading (see Spring and Lowy and Aikens *et al.,* Volume 29). Reagents are available to retard photobleaching but add unwelcome complexity and cannot be used in live cell experiments (Picciolo and Kaplan, 1984; Bock *et al.,* 1985).

DAPI, Hoechst 33342, and propidium have been used as fluorescent probes for visualization of DNA in cells. Although many flow cytometry studies have been published using these reagents to quantify total cell DNA for cell cycle and aneuploidy analyses (Muirhead *et al.,* 1985), little quantitative data has been accumulated with fluorescence microscopes. The difficulties of using vidicon detectors with appropriate linearity and background corrections undoubtedly account for the minimal number of microscope studies. Solid-state cameras should change this situation (see Aikens *et al.*, Volume 29, this series). It may also prove useful to borrow other probes from the flow cytometry field. For example, it should be possible to simultaneously determine cellular DNA and RNA content using acridine orange or pyronin Y, both of which fluoresce at longer wavelengths when bound to single-stranded RNA (Darzynkeiwicz *et al.*, 1987; Kapuscinski *et al.*, 1982; Shapiro, 1981).

2. Probes of Living Cells

Fluorescent tags listed in Table I can be covalently attached to purified biological molecules to form fluorescent analogs. If the analog has a function, the function must be shown to be intact before reincorporation into the cell (Wang *et al.*, 1982; Taylor *et al.*, 1986; Simon and Taylor, 1986; Wang, Volume 29, this series). Fluorescent lipid analogs (Table I) have been synthesized by a variety of investigators and used to study the compartmental movements of lipids within cells (Pagano, Volume 29 this series), the binding of LDL particles to their receptors (Barack and Webb, 1981; Webb and Gross, 1986), the translational mobility of lipids in membranes (Peters, 1981), and the microviscosity of lipid regions (Garland and Birmingham, 1986).

Certainly one of the main dangers in the use of fluorescent analogs is the perturbation of the function of the analog and other functions of the cell by the fluorophore on the analog. The perturbation can result from the physical size of the fluorophore, which may interfere with the normal binding or function of the analog, or it can result from the effects of photogenerated reactive oxygen products. One must also avoid changes in the analog that occur during photocross-linking of the fluorophore to adjacent structures during excitation of the specimen. Furthermore, evidence that labeled actin can become fragmented under strong illumination suggests that this effect could happen with other analogs (Simon *et al.*, 1988). These photodestructive effects need to be reduced in future fluorophores. Furthermore, the total dose of irradiance must be kept to a minimum (Taylor and Salmon, Volume 29).

Most fluorescent pH indicators that are used to quantify intracellular pH are derivatives of fluorescein. There are two approaches to these measurements. In the first method, membrane-soluble acetoxy derivatives of carboxylated fluoresceins are added to the medium bathing the cells. Once inside the cells, nonspecific esterases remove the acetoxy groups, leaving the negatively charged carboxyl groups on the probe, which prevent its escape from the cell. The disadvantage of this method is that acetoxy-bearing molecules, or more likely, the products of hydrolysis, are sometimes toxic to cells. In addition, acetoxy-modified probe molecules can diffuse from the cytoplasm into other intracellular compartments, where they may be trapped if esterase activity is present in the compartment. The result is a fluorescence signal that represents intracellular compartments as well as the cytoplasm (see Tsien Chapter 5, this volume).

The second method is to introduce membrane-impermeant pH indicators into cells by bulk loading or microinjection methods. These methods have the disadvantage that the plasma membrane is damaged during incorporation of the reagent. Surprisingly, little long-term damage seems to be done by these loading techniques (Bright *et al.*, 1987; Bright *et al.*, Chapter 6, this volume; see also McNeil, Volume 29). Fluorescein derivatives with multiple charges can be used in this method. But to ensure that no nonspecific interaction of the probe occurs between membrane and protein surfaces, some investigators have used fluorescein derivatives that are covalently linked to low-molecular-weight dextrans. It is essential that high-molecular-weight dextrans not be used because they are known to be compartmentalized in some regions of the cytoplasm (Luby-Phelps *et al.*, 1986; Luby-Phelps and Taylor, 1988; see also Luby-Phelps, Volume 29).

Calcium indicators (Table I) are used with single cells in an analogous way to pH indicators. The most common approach for incorporating the indicator into cells is the use of acetoxy derivatives (see Tsien, Chapter 5, this volume, for details). Bulk loading or microinjection of calcium indicators that are covalently attached to dextrans would provide a valuable alternative method, but these materials are not available at this time.

Cell and organelle transmembrane electrical potential is an important functional parameter that can often be measured with fluorescent probes. Rapid (millisecond) voltage changes of nerve and muscle cells are best detected with the "fast dyes" which insert deeply enough into the excitable membrane to sense the changing electric field within the lipid bilayer. Representative probes are shown in Table I. The potential-dependent light absorption and fluorescence changes occur by a variety of mechanisms. When the potential is changed, some charged cyanine and oxonol dyes electrophorese to a new chemical environment, e.g., membrane to water, with a resultant wavelength shift (George *et al.*, 1988). Others rotate in the changing electric field to a position where nonfluorescent dimers begin to form (Wolf and Waggoner, 1986). Yet others show electrochromic wavelength shifts due to potential-dependent shifts of electron density on the probe molecule (Gross and Loew, Chapter 7, this volume). Molecular mechanisms of potential sensitivity have been reviewed (Waggoner, 1985). Images of optical changes reflecting neural activity have been obtained for invertebrate ganglia, brains, and secretory organs. This work has been reviewed elsewhere (London *et al.*, 1986; Grinvald, 1984; Blasdel and Salama, 1986; Salzberg *et al.*, 1983). Images of potential-dependent spectral changes of probes on single membranes have been obtained (Gross *et al.*, 1986).

Slow dyes (Table I) are membrane-permeant charged dyes that redistribute across membranes according to the potential difference (Sims *et al.*, 1974). Fluorescence intensity changes occur when accumulated dye molecules adhere to intracellular membranes and proteins. But the major contribution to cell and organelle fluorescence viewed in a microscope is simply the quantity of dye within each compartment. In principle, the Nernst or Goldman–Hodgkin–Katz equations can be used to relate the quantity of dye (fluorescence level) in each compartment to a numerical value of the membrane potential. However, binding of dye to membranes and other intracellular components, as well as differences in quantum yields of free and bound dye, preclude such quantitation. Relative changes in membrane potential can be estimated, however, and it has been possible to use drugs and specific ion environments to set the cell membrane potential to known values and generate calibration curves (Freedman and Laris, 1988). Rhodamine 123 has been used for fluorescence microscopy studies of mitochondrial membrane potential in cultured cells (Johnson *et al.*, 1981; Weiss and Chen, 1984). Cyanine dyes, on the other hand, have been used for single cell membrane potential studies by flow cytometry (Simons *et al.*, 1988; Wilson *et al.*, 1988; Ransom and Cambier, 1988; Seligmann *et al.*, 1988). Negatively charged oxonol dyes, which tend to be excluded from mitochondria, may be superior for plasma membrane potential measurements (Seligmann *et al.*, 1988; Wilson *et al.*, 1988). Fluorescence imaging methods offer the exciting promise of being able to simultaneously measure plasma membrane and mitochondrial potentials of cultured cells that are stimulated with cytokines, chemotactic factors, drugs, toxins, and other perturbing agents.

Fluorogenic substrates become fluorescent or change their spectral properties as a result of the action of enzymes (Dolbeare, 1981, 1983). Most are substrates for hydrolase enzymes: peptidases, glycosidases, phosphatases, esterases, sulfatases, etc. Table I contains a few examples of the fluorophores that serve as a basis for constructing substrates. Naphthylamine and rhodamine can be linked to peptide groups that determine the peptidase specificity of the substrate (Leytus *et al.*, 1983). Fluorescein and hydroxycoumarins can be converted to esters, glycosides, phosphates, and sulfates (Koller and Wolfbeis, 1985). Most often these substrates are used to detect enzymes in homogeneous solutions. Cathepsin B substrates have been used in flow cytometry experiments to detect the arrival of pinocytosed material lysosomal compartments, and with a flow cytometer (Murphy and Roederer, 1986) and a fluorescence microscope (Dolbeare and Vanderlaan, 1979) to detect the presence of cathepsin B-containing cells in culture. Nevertheless, there are problems with existing substrates. Generally, the substrates are relatively membrane impermeant but the fluorescent products of the reaction are quite permeant and diffuse out of the cell before a good image can be obtained. Methods for attaching the substrate to other materials may be useful for localizing the product.

Endogenous fluorophores can also be useful probes. NADH levels, for example, can be measured as an indicator of the energy state of a cell (Thorell, 1983; Kohen *et al.,* 1983).

3. Future

The development of fluorescent and other luminescent probes is only in its infancy and major advances are expected in the next few years. Probes are the key to extracting biological information from cells by fluorometric, flow cytometric, and fluorescence imaging techniques. Chemists will synthesize probes for many important biological processes and structures. Improvements will also be made in the stability, brightness, and sensitivity of probes, while future probes will be designed to be less toxic or perturbing to the cellular environment. In addition, new classes of fluorophores will have to be developed. The effect of local molecular environment on the spectral properties of the probes will have to be determined, and ways of targeting the probes to the sites of interest will have to be developed.

B. Fluorescence Microscope

The basic information about fluorescence microscopy is presented in the chapters by Taylor and Salmon and Wampler and Kurz in Volume 29 of this series. The issues that are critically important for multiple spectral parameter imaging are discussed in more detail here. The system described is based on either a standard upright microscope or an inverted microscope, both operated in the epifluorescence mode.

1. Light Sources

The microscope should have at least two ports for white light sources, such as a xenon arc and a tungsten filament lamp. In addition, a laser might be used for specific wavelengths where an intense source is required (Spring and Smith, 1987; Bright *et al.,* Chapter 6, this volume). The light source must have a spectral output that coincides with at least a portion of the excitation spectra of the fluorescent probes selected for a particular experiment (see Section II, A above and Taylor and Salmon and Wampler and Kurz, Volume 29). The choice between a xenon arc and a tungsten filament lamp will depend on the requirements for the intensity of the light output at selected wavelengths. Having two distinct white light sources that can be separately shuttered allows the investigator to rapidly switch from one condition of illumination to another. The shutters should be fast in order to block the illumination of the specimen except during image acquisition.

It is imperative that the light sources exhibit a high degree of temporal and spatial stability. We have found that a 12-V tungsten–halogen filament lamp shows excellent stability when operated at constant current with a power supply (Bright *et al.,* 1987; Bright *et al.,* Chapter 6, this volume; Wampler and Kurz, Volume 29, this series). The arc lamps that we have evaluated exhibit a wide range of arc wander and temporal fluctuations in intensity. The stability of arc lamps must be determined before initiating quantitative studies (Wampler and Kurz, Volume 29, this series).

2. Objectives

Objectives are chosen for a specific application based on the combined characteristics of magnification, numerical aperture (NA), and percent transmission of excitation and emission wavelengths of light. The total magnification of the system should be kept to the minimum for the spatial information required, since the brightness of the image is proportional to the inverse of the square of the magnification (see Taylor and Salmon, Volume 29, this series). The signal-to-noise ratio of the image will be larger when the same number of photons from a fluorescent cell fall on a smaller area of the detector. Similarly, the NA of the objective should be maximized for any given magnification since the brightness of the image varies as the square of the NA (see Taylor and Salmon, Volume 29, this series). The high-NA objective will also yield the smallest depth of field. Therefore, a small depth of the specimen in the z-axis will be in focus. This is not a problem for thin cells such as many mammalian cells in culture. However, cells with a thickness much greater than the depth of field of the objective, must have a three-dimensional correction applied to remove out-of-focus light (see Taylor and Salmon, Volume 29; Wampler and Kurz, Volume 29; Jeričević *et al.,* Chapter 2, this volume; Young, Chapter 1, this volume; Agard *et al.,* Chapter 13, this volume). When a lower total magnification can be used, it is advantageous to use a high-magnification, high-NA objective and then to demagnify the image into the camera using a demagnifying lens (Ploem, 1986). This approach combines the light-gathering power of the high-NA objective with the increased signal-to-noise of a smaller total magnification.

Modern fluorescence objectives have been produced with glass that transmit light efficiently down to approx ~340 nm. Special quartz optics are required for applications at lower wavelengths, but we have found that quartz optics are not required for probes such as Hoechst and fura-2. Unfortunately, the microscope manufacturers have focused attention on the lower wavelengths, while probe chemists have been working on extending the usable spectrum to the far red and near infrared (see Section II, A above). We believe that the most critical spectral range for multiple spectral parameter imaging is between ~350 and ~900 nm. Therefore, the characteristics of objectives at the longer wavelengths must be determined. We have just initiated these evaluations and cannot describe the characteristics of objectives from the different companies. Ask the microscope manufacturer for the percent transmission and aberrations over the spectral range described above.

Most of our preliminary work has utilized a Zeiss 63× (1.2 NA) water immersion objective. This lens has very little spherical aberration and the image quality is excellent. The microscope manufacturers are now producing new lines of objectives and they should be evaluated. Be particularly careful with spherical aberration, which can be detected by taking a through-focus series of images of a fluorescent bead ~0.2 μm diameter (see Sisken, Chapter 4, this volume; Agard *et al.*, Chapter 13, this volume).

3. Spectral Selection

There are three properties of monochromators that must be considered: rate of changing wavelength; half bandwidth of the transmitted light; and the ability to modulate the intensity of the transmitted light. The ideal system would modulate the wavelength and the intensity of the light at a rate of change of at least 1–5 milliseconds for both the excitation and the emission light separately. This ideal has not yet been achieved, but Kurtz *et al.* (1987) and Westinghouse are developing acoustooptical tunable filters that have the potential to serve this purpose (see Wampler and Kurtz, Volume 29). In the meantime, it is possible to construct or to purchase interference filter wheels from a variety of vendors, including some of the microscope companies. We believe that this approach is the best solution for the immediate future compared to using expensive combinations of monochromators. Several companies make interference filters and dichroic mirrors with the required specifications for percent transmission, spectral half bandwidth, and low and high wavelength blocking (e.g., Omega Optical, Brattleboro, VT).

Filter wheels for modulating excitation light are simple to attach to the microscope between the light source and the microscope stand (see Wampler and Kurz, Volume 29). The rate of changing the filters depends on the design of the wheel and the type of motor used to drive the filter. In practice, realistic time resolutions between two wavelengths are on the order of 5–100 milliseconds. However, the dwell time at each wavelength will depend on the integration time required to achieve an accepted signal-to-noise ratio with the camera used. Obviously, the time resolution is degraded for each wavelength required in a multiple spectral parameter experiment. At the present time, we are sacrificing temporal resolution for spatial signal-to-noise ratio of the image (DeBiasio *et al.*, 1987). Further improvements are expected with the use of acoustooptical tunable filters and other devices.

Rapid change of the emission wavelengths is a bigger challenge with fluorescence microscopes in the epiillumination configuration. Our earliest work with multiple spectral parameter imaging relied on manually changing the emission filter sets (DeBiasio *et al.*, 1987). However, motorized turrets containing multiple emission filter sets are now being developed and these should function with a time resolution of ~10–100 milliseconds. The successful development of dual control over excitation and emission wavelengths will be an important step toward the complete automation of the research fluorescence microscope. Rapid changes in emission wavelengths may soon be accomplished by imaging through a pair of acoustooptical tunable filters, one serving as the excitation filter and the other as the emission filter. In addition, improvements in the quality of interference filters and the promise of acousto optical tunable filters should permit the use of transmitted light optics for fluorescence. With this configuration it should be easier to implement the rapid methods for changing emission wavelengths.

Table II lists the filter sets that we are currently using for most of our five fluorescence parameter imaging experiments.

4. x, y, z Stage

A computer controlled stage with independent control of the *x*-, *y*-, and *z*-axes is essential for modern fluorescence imaging experiments. The *x*- and *y*-axis control is essential for identifying cells in a cell population that are going to be analyzed multiple times. The locations of selected cells are logged in and the scanning stage can return to specific cells automatically. The *z*-axis control is required for changing focus at different wavelengths and for through-focus image sets. There are now several sources for these stages, including most microscope manufacturers.

TABLE II

FILTER SETS[a]

Fluorescent probe	Excitation[b,c]	Dichroic[d]	Emission[b,c]
Hoechst 33342	G350/56 (Z)	395 (Z)	420/LP (Z)
Fluorescein actin	485/20 (Z)	510 (Z)	542/45 (Z)
Lissamine-rhodamine-dextran	546/12 (Z)	580 (Z)	590/40 (Ω)
$DiIC_1(5)$	615/30 (Ω)	650 (Ω)	695/70 (Ω)
Cy7-dextran	720/40 (Ω)	759 (Ω)	785/50 (Ω)

[a] All are interference filters, except G350/56 and 420 LP.
[b] Top number is center wavelength of the bandpass filter. Bottom number is total bandpass at half-maximum transmittance. If LP is on the bottom, the filter is a long-pass glass filter, in which case the top number is the wavelength of half-maximum transmittance.
[c] Z, Filter from Zeiss Optical; Ω, filter from Omega Optical, Inc., Brattleboro, VT.
[d] Wavelength of half-maximal transmittance for long-pass dichroic filters.

5. ENVIRONMENTAL CHAMBER

A cell environmental chamber for the microscope must maintain the health of the cells to the same level as that attainable in a tissue culture incubator. Therefore, temperature, gas, and pH must be maintained within a narrow physiological range. This is probably the aspect of quantitative fluorescence microscopy that is the most abused by investigators. Highly sophisticated physiological measurements on cells of questionable physiological condition are obviously not worth the effort. We have taken the position that mammalian cells should be investigated in the presence of bicarbonate and CO_2 since this is the most physiological condition (Bright *et al.*, 1987). Therefore, a closed environment is required for proper buffering. We have been using a closed chamber on the microscope since we did not want to enclose the whole microscope and turn it into an incubator at high temperature and humidity. However, whole microscope chambers have been constructed by others (see Wang, Volume 29).

We have used a modified Sykes-Moore chamber (51 mm in diameter; Bellco Glass Inc.). The rubber O-ring spacer was replaced by a stainless steel ring with three 18-gauge needles, silver soldered in place. A stainless steel baffle was also soldered across the entrance port to disperse the stream of perfusing fluid. Thin latex gaskets molded from dental dam were used to seal the chamber. A thermistor (10,000 Ω, 0.014-in. diameter; Thermometrics, Inc. Edison, NJ) was inserted into one of the needle ports to sense the chamber environment directly. The out-flow pH was monitored with a flow-through electrode (Markson Science, Inc., Phoenix,

AZ). The temperature control was provided by a combination of a thermo electrically heated stage mount that was designed to accept the Sykes-Moore chamber (Rainin Inst. Co., Inc., Boston, MA) and a modified air curtain (Sage Instrument Division Orion Research, Inc., Cambridge, MA) wired for feedback control by a microprocessor. This combination maintained the temperature of the culture fluid in the chamber at 37°C ± 0.2°.

We are not yet satisfied with the environmental chamber. The air curtain can alter the properties of lenses by causing strain, and the modified Sykes-Moore chamber in its current design is not suitable for high-resolution transmitted light contrast methods due to its thickness. This is a technical aspect that requires further development.

6. Multiple Ports for Different Detectors

Although the focus of this chapter is on multiple spectral parameter imaging, we believe that heretofore distinct modes of microscopy must be applied to the same cells to extract the most information. Therefore, future experiments will integrate the use of video-enhanced contrast microscopy, laser confocal scanning microscopy, quantitative measurements such as fluorescence redistribution after photobleaching, ratio imaging, and multiple spectral parameter imaging. In fact, different spectral regions can be used for various modes of microscopy. The microscope must have separate ports for each detector. Our most recent system takes advantage of the three separate ports available on the Zeiss IM35 microscope; one for an ISIT, another for a cooled CCD, and another for a confocal laser scanner.

C. Detectors

The various types of cameras available today are discussed in other chapters in this series (Spring and Lowy, and Aiken *et al.,* Volume 29). Multiple spectral parameter imaging has some particular requirements: the camera must be sensitive from ~400 to 900 nm in of order to take advantage of the new long-wavelength probes, as well as the standard probes (see Section II,A); the camera must have a high intrascene dynamic range to record fluorescence signals from cells with varying intensities; and the camera must be able to acquire the multiple spectral images with a time resolution faster than the rate of biological changes expressed by the cells. In fact, the investigator must integrate the issues of fluorescent probes, light sources, spectral selection, and detectors, since they form a continuum of problems. One solution for the future is the use of a cooled CCD with a pulsed light source in combination with the acoustooptical tunable filter described above. A large array CCD can be operated in the frame transfer mode, so that multiple images can be acquired rapidly (milliseconds) before having to read out the whole chip (Aikens *et al.*, Volume 29).

We have determined that a 576 × 384 Thompson-CFR CCD in a cooled housing (Photometrics Ltd., Tucson, AZ) has comparable sensitivity to an ISIT (DAGE/MTI, Michigan City, IN) at ~520 nm and much better sensitivity into the red region of the spectrum. Furthermore, the linear radiometric response, high dynamic range, and fixed geometric distortion of these devices adds to their value. The present readout time of seconds requires either a larger array to capture multiple parameters rapidly, or more rapiid readout electronics. Rapid advances in cooled CCD camera technology for microscopy are expected in the near future.

Most of our initial multiple spectral parameter imaging has employed a DAGE/MTI ISIT camera since we did not have a large array CCD chip and out time resolution was limited to many seconds. We accepted the low intrascene dynamic range, the decreased sensitivity at the longer wavelengths, and the requirements for extensive image restoration in order to perform the experiments (DeBiasio *et al.*, 1987). The gain and high voltage of the camera were adjusted for the "average cell" and were held constant for all images at all wavelengths. The light level at each wavelength was adjusted with neutral density filters and with the iris diaphragm. Higher irradiances at the specimen plane were required at the longer wavelengths since the camera was not very sensitive in the far red and near infrared.

D. Image Restoration

The general problem of characterizing the modulation transfer function of the imaging system and restoring the images to the "real" form is discussed by Young Chapter 1, and Jeričević *et al.*, Chapter 2, this volume. Therefore, the details of restoring images for shading and linearity and geometric distortions are not presented in this chapter, but they are important steps in the whole restoration process. The use of a wide range of wavelengths in performing multiple spectral parameter imaging experiments causes some specific problems that are discussed here.

E. Automating Acquisition at Multiple Wavelengths

The actual acquisition of the images at different wavelengths may be a time-consuming task if the microscope workstation does not facilitate part of the process. Switching the filter sets in excitation and emission path as well as the dichroic mirror may be done manually, but a computer-controlled electromechanical solution will allow automation of the procedure.

The stability of the position of the dichroic mirrors is of importance. Slight variation in this position causes translational shifts in the digitized images, calling for registration corrections afterward. In a normal microscope these translations are difficult to avoid, but if the registration error is systematic, than the time needed to compute the correction can be reduced significantly.

At each wavelength the focus needs to be readjusted and an autofocus system can easily make these small adjustments. The appropriate exposure time at each wavelength will be very different, considering the combined parameters of fluorophore efficiency, transmission coefficients in the optical path, and the wavelength dependence of the sensor characteristics. The exposure times may be estimated empirically by taking some images of a test cell. Other techniques can use a fast imprecise light measurement to estimate the intensity of the image, after which the actual image exposure can take place. This will reduce photobleaching and will, again, allow for automation.

Other techniques, such as screening the slide in phase-contrast or video-enhanced contrast (VEC) mode (to avoid photobleaching and to find the cells), may be facilitated by using a motorized stage with the possibility to record a list of stage positions which can be revisited in fluorescence mode.

If all of the suggested aspects are automated, then it is quite easy to just screen a slide to select the sites of interest in VEC. After that, the computer will acquire the multiple images at different wavelengths at those locations.

F. Image Processing

The set of images that we will have acquired consists typically of two to five monochromatic images per scene and will need several corrections before they can be analyzed or displayed. At this point, we will not discuss the normal corrections needed for acquiring digitized images, such as background subtraction and flat field correction. We assume that these operations have been applied to the images as part of a standard procedure (Jeričević *et al.*, Chapter 2, this volume; Galbraith *et al.*, 1988). Here we will introduce the special corrections needed to combine multiple parameter data. Image corrections can be divided into correcting: (1) registration, (2) magnification, and (3) histogram scaling.

1. Registration

Problems occur, as mentioned, due to variation in the positions of the dichroic mirrors. Images may be translated many pixels in any direction in the image plane. Translating images back over an integer number of pixels is a simple address operation in the image processor. Finding the exact translation vector yielding optimal registration is, however, a difficult problem. This translation vector may be found manually or automatically.

a. Manual Registration. The registration error can usually be seen clearly when two images are jointly displayed in an overlay mode where each image appears in a different color. Using a simple software solution, one can use a trial-and-error method to find the optimal registration. This is done by trying a translation vector and redisplaying the result to get a visual judgement from the images. However, this method is very cumbersome. Some imaging systems offer a way of moving two or three images over each other in real time, again in an overlay mode using three basic colors, red, green, and blue. With a mouse or trackball it is possible to move one of the images directly and to adjust quickly the registration based on visual criteria. This method is fast, but requires special features of the display hardware and some dedicated system software. The precision of the method can be as good as a single pixel, but depends on the images, in particular whether they offer good visual criteria such as common sharp edges.

b. Automatic Registration. For automatic implementations we first need to find alternative criteria to judge the registration property. Notice that a gray-value correlation between the images is probably not a good criterion, since we are not necessarily expecting the probes to be spatially correlated. Other criteria are necessary which may be depending on the sample and the choice of parameters that were stained. We have implemented two methods: one based on the fact that one parameter is a spatial subset of the other, and a method that involves the addition of fluorescent beads to the slide.

The registration of a "*spatial subset parameter*" image can be done by generating a binary mask of each of the images. Since we are expecting that the mask of the "subset parameter" should fit within the mask of the main parameter, we can use the logical exclusive OR of the two masks as a registration criterion. When the integration of the exclusive OR image is minimal, the registration is optimal. Fast iterative methods have been developed to find this optimum (Nederlof, 1987).

The "*bead method*" requires some extra work during the slide preparation. Multiple stained fluorescent beads with an appropriately small diameter are added to the sample at a moderate constant density. All

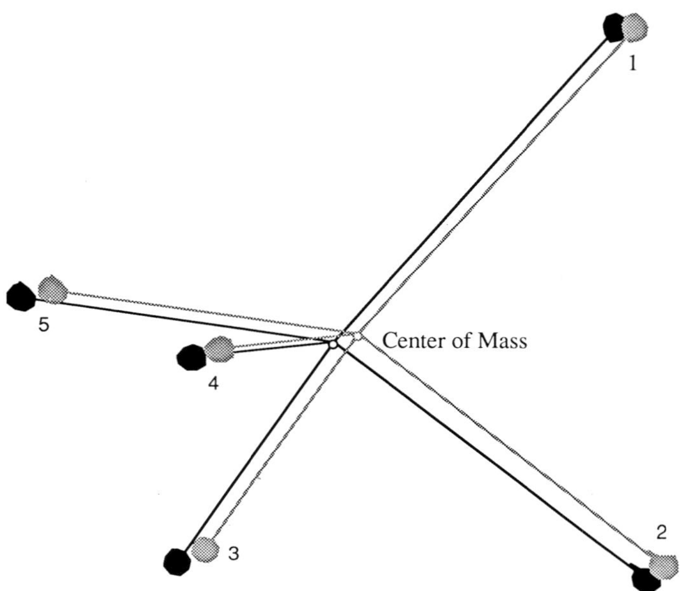

FIG. 2. The binary masks of multiple stained fluorescent beads (numbered 1–5) acquired at different wavelengths. The lines show the distances to the center of mass. It can be seen that there is an image shift as well as a difference in magnification. The calculation of the shift of the center of mass, and the comparison of the distances from the beads to their center of mass, can be used to estimate the corrections needed for image registration and magnification.

images acquired at multiple wavelengths will now contain a small number of fluorescent beads as bright marking spots. These spots can be used as markers for registration purposes. We have developed a method to detect rapidly the position of individual beads in an image. The center of mass of the beads is computed and is compared to that of the second image. The result is a translation vector specifying the displacement of the center of mass of the second image and thus of the second image as a whole. Figure 2 illustrates how the binary mask of the beads (numbered 1–5) and their center of mass shifts are used to compare images taken at two different wavelengths. The bead registration method was used with three of the images in Fig. 3 to create the three-color composite image in the frontispiece. The color composite image was generated by displaying the actin image in the red channel, the tubulin image in the green channel, and the nucleus image in the blue channel. A yellow color results when the red image and the green image contain overlapping regions of similar intensity. Nonmotile fibroblasts display prominent, linear actin-containing stress

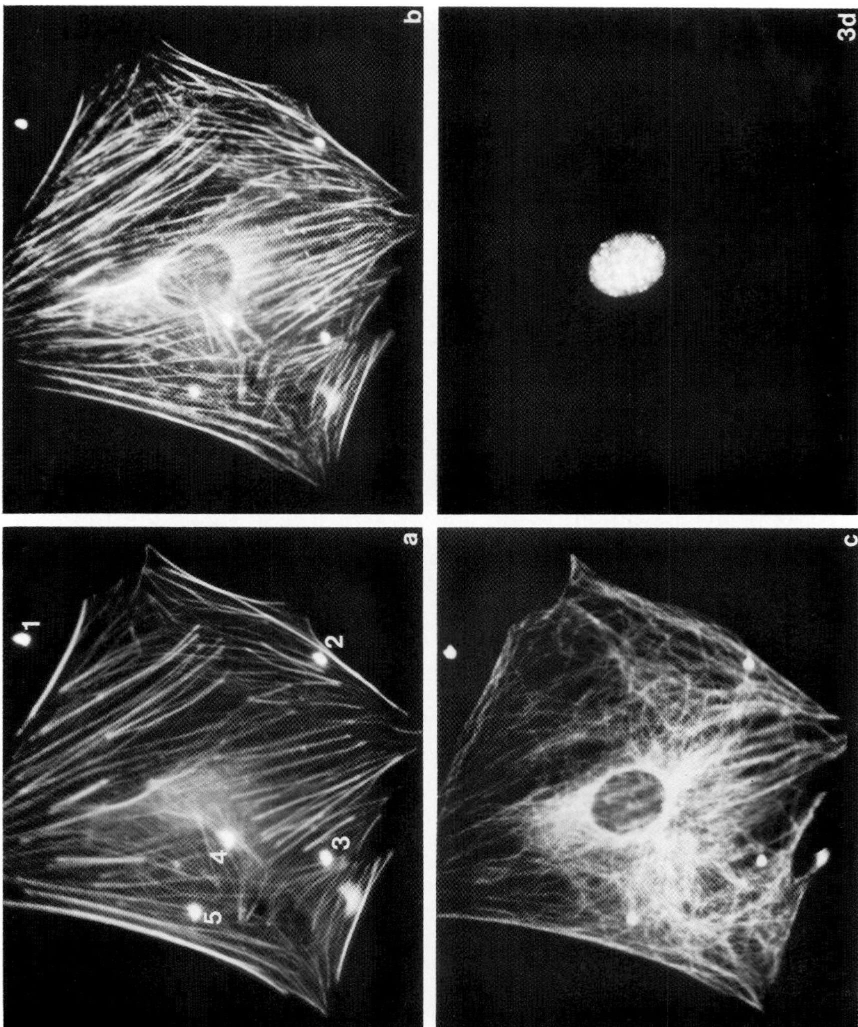

FIG. 3. Fluorescence images of (a) actin, (b) myosin, (c) tubulin, and (d) the nucleus in a single, well-spread, relatively nonmotile 3T3 fibroblast. Immunofluorescence processing methods were similar to those of Conrad *et al.* (1988) using rabbit antimyosin and mouse antitubulin as primary antibodies, and rhodamine-goat and anti-rabbit and Cy5-sheep anti-mouse as secondary antibodies. NBD-phallicidin and Hoechst 33342 were used to stain actin filaments and nuclei, respectively. Black and white images were background subtracted and enhanced for display purposes using a variety of image enhancement techniques, including histogram stretching, thresholding, and unsharp masking (Castleman, 1979). Image registration and magnification corrections were performed as outlined in the text (Section II,Fl,2) using the images of the multiple stained fluorescent beads (compare numbered beads here with those of Fig. 2). Bar = 20 μm.

fibers that span the length of the cell in a more-or-less parallel arrangement. Myosin (not shown) is found in its typical punctate periodic distribution along stress fibers as well as in diffuse arrays surrounding the nucleus. The myosin image is not displayed in the color composite image due to the inability to display clearly more than three parameters at the same time without obscurring detail, as discussed in Section II,G. Microtubules radiate outward toward the cell perimeter from a microtubule organizing center (MTOC) located near the cell center over the nucleus. The multiple fluorescent beads were also used to correct for magnification differences in the images (see below).

2. Magnification Correction

Even though modern optics have many corrections for wavelength-dependent magnification, it can still be shown that a difference in magnification of up to a few percent can occur at different wavelengths. A magnification difference prohibits proper image registration and should be corrected. We have incorporated the magnification correction in the 'bead method' in the following way. Since the beads are at a fixed distance from each other, they will be at a fixed distance to their center of mass. When the magnification changes, the center of mass can always be translated to the same position, but the distances of the beads toward the center of mass is proportional to the magnification. Thus, the analysis of the center of mass and the related distances can be used to estimate both the translation and the difference in magnification between the two images. Using an affine transformation we can then apply the correction for subpixel translation and magnification at the same time.

3. Scaling of Histograms

Ideally, all images will properly fill the dynamic range of the sensor and will thus have compatible histograms. However, it may be that due to the background subtraction, insufficient light, or a bad choice of the exposure time or camera settings, we have to deal with images that are unequally dark or bright. For display purposes it will cause a problem if one image is much darker than the other. It is therefore useful to do a linear scaling of each image so that the brightness levels match. Many applications will first require the extraction of intensity data, before scaling, in order to complete some quantitative analyses. One should save the original image for quantitative purposes.

G. Graphics Display of Multiple Parameter Data

After acquisition and processing, we now possess a properly registered stack of two to five monochromatic images with compatible brightness levels. The question remains how to display a data set like this to allow for the study of actual biological parameters.

A good solution is offered by the previously mentioned overlay method where three images can be displayed simultaneously, each in a different color (red, green, and blue). Using this technique, the different spectral images of separately labeled antibodies shown in Fig. 3 were simultaneously displayed and registered to produce the multiple parameter map of the frontispiece. An interactive application has proved to be very useful, since it offers the additional capability of switching each of the images on and off. It has become clear that, in trying to display all the parameters at the same time, one quickly surpasses the capacity of the medium and the observer. Interactive switching of the parameters offers an extra "dimension" needed to come to grips with the large amount of visual information. As an extension to this, an interactive way of changing the contrast and brightness of the individual images has been helpful in the same way.

The technique of overlaying three images in red, green, and blue may not be the perfect solution. Ideally, we should look at the images as a stack of "transparencies" which we want to observe simultaneously. Using this analog we can define some functions for the ideal interactive display application:

- Change the order of the slide stack (bring a slide to the front/send one to the back)
- Assign an arbitrary color (for instance the color of the probe) to each slide
- Modify contrast and brightness of each slide
- Assign an opacity function to each of the slides (this function may be directly related to the gray values, but may also be a step function or mask, or a function obtained from more advanced analysis)

The area of graphics display of multiple parameter images is in its infancy. Further work is required to be able to display multiple parameter images in both two and three dimensions. In addition, methods for feature extraction and statistical analyses of one parameter relative to other parameters have been initiated. A further problem, which has not been resolved, is how to produce high-quality multiple parameter hardcopy outputs.

H. Example of a Multiple Parameter Experiment

We have focused our attention on mammalian cells in culture. The goal of these experiments is to obtain information on multiple parameters within the same living cells. Cells grown on coverslips are labeled with multiple fluorescent probes in a specific sequence that minimally changes the morphology of the cells. There must be a balance between the optimal incorporation of each probe into the cell and the maintenance of the physiological health of the cells. Trauma due to the methods used to incorporate the probes and probe toxicity must be minimized. The sequence of the labeling steps and the optimal concentration of probe are determined experimentally with each new probe and every cell type investigated. A summary of some of the most recent methods for incorporating probes into living cells is presented in a chapter by McNeil in Volume 29 of this series. Issues of probe toxicity are discussed above in Section II,A.

1. Labeling Guidelines

The concentration of each probe must be determined separately, as well as together with the other fluorescent probes that will be used in the experiment. The optimal concentration of each probe will yield an acceptable signal-to-noise ratio, while minimizing any toxic effects. The use of a camera with a high intrascene dynamic range, such as the cooled CCD, permits the total fluorescence signal from each probe to vary over a large range. However, the use of one of the image intensifiers (see Spring and Lowy, Volume 29, this series) requires that the illumination and/or the camera gain and high voltage be varied for each probe or that the amount of labeling be adjusted.

A sequence of labeling multiple parameters must be chosen that both minimizes cell perturbations and preserves the optimal labeling of each parameter. Cell "recovery" periods involving the replacement of culture medium are important for allowing the cells to overcome the trauma of labeling and to permit the incorporation of the probe. Labeling methods that cause the most trauma are performed first to allow the cells to recover during other less perturbing labeling steps. Use one of the bulk cell loading methods, such as scrape loading or bead loading, early in the labeling sequence when well-spread cell morphologies are required. Fibroblasts usually require from 8 to 18 hours to reattach and to spread if scrape loading is used. Labeling endosomes through endocytosis with high concentrations of markers usually induces some transient cell trauma that requires 15–45 minutes of recovery. Direct microinjection of labeled

TABLE III

EXAMPLE OF A MULTIPLE PROBE LABELING SEQUENCE[a]

Step	Labeling	Parameter	Procedure	Recovery[b]
1	Scrape Loading	Cytoplasm (Cy7-dextran)[c]	Adherent nonconfluent 3T3 cells were scraped off plastic culture dishes in the presence of a 13 mg/ml solution of Cy-7-dextran (10K) in buffered saline. Dislodged cells were washed and replated onto glass coverslips	18 hours
2	Endocytosis	Endosomes (LRB-dextran)[d] (10K)	Coverslips were incubated for 2 hours in complete medium containing a 3.5 mg/ml concentration of LRB dextran (10K)	15 minutes
3	Microinjection	Actin (AF-actin)[e]	3T3 cells were microinjected with AF-actin (4 mg/ml, D:P = 0.8)	2 hours
4	Diffusion	Nucleus (Hoechst)[f]	Coverslips were incubated for 10 minutes in complete medium containing a final concentration of 7.5 μg/ml Hoechst	0 minutes
5	Diffusion	Mitochondria [diIC_1(5)][g]	Coverslips were incubated for a minimum of 20 minutes in complete medium containing a final concentration of 7.5×10^{-7} *M* diIC_1(5). This solution was not removed from the cells until immediately before mounting (membrane potential dye diffuses out of the cell and significantly reduces detectable fluorescence signal in 2.5–3.5 hours after removal)	0 minutes

[a] This procedure labels five cellular components of Swiss 3T3 fibroblasts with five spectrally distinct fluorescent probes (see DeBiasio *et al.*, 1987).

[b] Incubation took place in fresh medium with serum at 37°C, 5% CO_2 in all procedures.

[c] Cy7-isothiocyanate on aminoethylcarboxymidomethyl-dextran; dextran molecular weight 10,000 (Mujumdar *et al.*, 1988).

[d] LRB-dextran, lissamine-rhodamine-B-dextran; dextran molecular weight 10,000 (Molecular Probes Inc.).

[e] AF-actin, 5-iodoacetamidofluorescein-actin (Wang and Taylor, 1980).

[f] Hoechst 33342 (Sigma Chemical Co.).

[g] diIC_1(5), cationic potential-sensitive cyanine dye (Waggoner, 1985).

macromolecules also causes transient perturbation of cells and we usually allow 1–2 hours for recovery. Probes that are capable of diffusing across the cell membrane can be added to the cell culture during the recovery period of the preceding labeling step. Be careful with the diffusable probes, since toxic effects may be observed hours after labeling (Bright *et al.,* 1987).

The labeling sequence and the length of each recovery period depend on the type of probe used (i.e., membrane potential-sensitive dye, diffusable form of fura-2, or fluorescent analogs of proteins), and the labeling technique employed (i.e., scrape loading, bead loading, microinjection, diffusion). We have also found that the labeling sequence and concentration of probes used can actually vary depending on the stage of the cell cycle. An example of a five-parameter labeling experiment is presented in Table III.

References

Allen, G., and Lowe, G. (1973). *Biochem. J.* **133,** 679–686.

Barak, L. S., and Webb, W. W. (1981). *J. Cell Biol.* **90,** 595–604.

Barak, L. S., and Yocum, R. R. (1981). *Anal. Biochem.* **110,** 31–38.

Blasdel, G. G., and Salama, G. (1986). *Nature (London)* **321,** 579–586.

Bock, G., Hilchenbach, M., Schauenstein, K., and Wick, G., (1985). *J. Histochem. Cytochem.* **33,** 699–705.

Bratcher, S. C., Nitta, K., and Kronman, M. J. (1979). *Biochem. J.* **183,** 255–268.

Bock, G., Hilchenbach, M., Schauenstein, K., and Wick, G., (1985). *J. Histochem.* 1019–1033.

Castleman, K. R. (1979). "Digital Image Processing." Prentice Hall, New York.

Chen, R. F. (1968). *Anal. Biochem.* **25,** 412–416.

Cherry, R. J., Cogoli, A., Oppliger, M., Schneider, G., and Semenza, G. (1976). *Biochemistry* **15,** 3653–3656.

Conrad, P. A., Herman, I. H., and Taylor, D. L. (1988). In preparation.

Darzynkiewicz, Z., Kapuscinski, J., Traganos, F., and Crissman, H. A., (1987). *Cytometry* **8,** 138–145.

DeBiasio, R., Bright, G., Ernst, L. A., Waggoner, A. S., and Taylor, D. L. (1987). *J. Cell Biol.* **105,** 1613–1622.

Dolbeare, F. (1981). *In* "Modern Fluorescence Spectroscopy" (E. L. Wehry, ed.), pp. 251–293. Plenum, New York.

Dolbeare, F., (1983). *In* "Oncodevelopmental Markers. Biologic, Diagnostic, and Monitoring Aspects" (W. H. Fishman, ed.), pp. 207–217. Academic Press, New York.

Dolbeare, F., and Vanderlaan, M., (1979). *J. Histochem. Cytochem.* **27,** 1493–1495.

Duggan, J. X., DiCesare, J., and Williams, J. F. (1983). *In* "New Directions in Molecular Luminescence" (D. Eastwood, ed.), pp. 112–126. ASTM Special Technical Publication 822.

Ernst, L. A., Gupta, R. K., Mujumdar, R. B., and Waggoner, A. S. (1988). *Cytometry*, in press.

Freedman, J. C., and Laris, P. C. (1988). *In* "Spectroscopic Membrane Probes" (L. M. Loew, ed.), Vol. III, pp. 1–49. CRC Press, Boca Raton, Florida.

Galbraith, W., Ryan, K. W., Gliksman, N., Taylor, D. L., and Waggoner, A. S. (1988). *Compu. Med. Imaging Graph.* **12,** in press.

Garland, P. B., and Birmingham, J. J. (1986). *In* "Applications of Fluorescence in the Biomedical Sciences" (D. L. Taylor, A. S. Waggoner, R. F. Murphy, F. Lanni, and R. R. Birge, eds.) pp. 245–254. Liss, New York.

Garland, P. B., and Moore, C. H. (1979). *Biochem. J.* **183,** 561–572.

George, E. B., Nyirjesy, P., Basson, M., Pratap, P. R., Freedman, J. C., Ernst, L. A., and Waggoner, A. S. (1988). *J. Membr. Biol.,* in press.

Greenspan, P., and Fowler, S. D. (1985). *J. Lipid Res.* **26,** 81–789.

Grinvald, A., (1984). *Trends Neurosci.* **7,** 143–150.

Grinvald, A., Hildesheim, R., Farber, I. C., and Anglister, L. (1984). *Biophys. J.* **39,** 301–308.

Gross, D., Loew, L. M., and Webb, W. W. (1986). *Biophys. J.* **50,** 339–348.

Grynkiewicz, G., Poenie, M., and Tsien, R. Y. (1985). *J. Biol. Chem.* **260,** 3440–3450.

Hardy, R. R., Hayakawa, K., Parks, D. R., and Herzenberg, L. A. (1983). *Nature (London)* **306,** 270–272.

Haugland, R. P. (1983). *In* "Excited States of Biopolymers" (R. F. Steiner, ed), pp. 29–58.

Johnson, L. V., Walsh, M. L., Bockus, B. J., and Chen, L. B. (1981). *J. Cell. Biol.* **88,** 526–535.

Kapuscinski, J., and Darzynkiewicz, Z. (1987). *Cytometry* **8,** 129–137.

Kapuscinski, J., Darzynkiewicz, Z., and Melamed, M. (1982). *Cytometry* **2,** 201–211.

Kenner, R. A., and Aboderin, A. A. (1971). *Biochemistry* **10,** 4433–4440.

Kohen, E., Kohen, C., Hirschberg, J. G., Wouters, A. W., Thorell, B., Westerhoff, H. V., and Charyulu, K. K. N. (1983). *Cell Biochem. Function* **1,** 3–16.

Koller, E., and Wolfbeis, O. S. (1985). *Monatsh. Chem.* **116,** 65–75.

Kronick, M. N., and Grossman, P. D. (1983). *Clin. Chem.* **29,** 1582–1586.

Kubin, R. F., and Fletcher, A. N. (1983). *J. Luminescence* **27,** 455–462.

Kurtz, I., Dwelle, R., and Katzkaq, P. (1987). *Rev. Sci. Instrum.* **58,** 1996–2003.

Lee, L. G., Chen, C.-H., and Chiu, L. A. (1986). *Cytometry* **7,** 508–517.

Leytus, S. P., Patterson, W. L., and Mangel, W. F. (1983). *Biochem. J.* **215,** 253–260.

London, J., Zecevic, D., Loew, L. M., Orbach, H. S., and Cohen, L. B. (1986). *In* "Applications of Fluorescence in the Biomedical Sciences" (D. L. Taylor, A. S. Waggoner, R. F. Murphy, F. Lanni, and R. R. Birge, eds.). pp. 423–447. Liss, New York.

Luby-Phelps, K., and Taylor, D. L. (1988). *Cell Motil. Cytostruct.* **10,** 28–37.

Luby-Phelps, K., Taylor, D. L., and Lanni, F. (1986). *J. Cell Biol.* **102,** 2015–2022.

Mathies, R. A., and Stryer, L. (1986). *In* "Applications of Fluorescence in the Biomedical Sciences" (D. L. Taylor, A. S. Waggoner, R. F. Murphy, F. Lanni, and R. R. Birge, eds.), pp. 129–140. Liss, New York.

Mujumdar, R. B., Ernst, L. A., Mujumdar, S., Waggoner, A. S. (1988). *Cytometry*, in press.

Muirhead, K., Horan, P. K., and Poste, G., (1985). *Bio/Technology* **3,** 337–356.

Murphy, R. F., and Roederer, M. (1986). *In* "Applications of Fluorescence in the Biomedical Sciences" (D. L. Taylor, A. S. Waggoner, R. F. Murphy, F. Lanni, and R. R. Birge, eds.), pp. 545–566. Liss, New York.

Nederlof, M. A. (1987). Ingenieurs thesis, Delft University of Technology, Delft, The Netherlands.

Oi, V., Glazer, A. N., and Stryer, L. (1982). *J. Cell Biol.* **93,** 981–986.

Peters, R. (1981). *Cell Biol. Int. Rep.* **5,** 733–760.

Picciolo, G. L., and Kaplan, D. S. (1984). *Adv. Appl. Microbiol.* **30,** 197–234.

Ploem, J. S. (1986). *In* "Applications of Fluorescence in the Biomedical Sciences" (D. L. Taylor, A. S. Waggoner, R. F. Murphy, F. Lanni, and R. R. Birge, eds.), pp. 289–300. Liss, New York.

Pohl, F. M., Jovin, T. M., Baehr, W., and Holbrollk, J. J. (1972). *Proc. Natl. Acad. Sci. U.S.A.* **69,** 805–809.

Ransom, J. T., and Cambier, J. C. (1988). *In* "The Cell Physiology of Blood" (R. B. Gunn and J. C. Parker, eds). Rockefeller Univ. Press, New York, in press.

Sackett, D. L., and Wolff, J. (1987). *Anal. Biochem.* **167,** 228–234.

Salzberg, B. M., Obaid, A. L., Senseman, D. M., and Gainer, H. (1983). *Nature* (*London*) **306,** 36–40.

Seligmann, B., Haston, W. O., Wasvary, J. S., and Rediske, J. R. (1988). *In* "The Cell Physiology of Blood" (R. B. Gunn, and J. C. Parker, eds.). Rockefeller Univ. Press, New York, in press.

Shapiro, H. M. (1981). *Cytometry* **2,** 143–150.

Shapiro, H. (1985). "Practical Flow Cytometry." Liss, New York.

Simon, J. R., and Taylor, D. L. (1986). *In* "Methods in Enzymology" (R. B. Vallee, ed.), Vol. 134, pp. 487–507. Academic Press, Orlando, Florida.

Simon, J. R., Gough, A., Urbanik, E., Wang, F., Ware, B., and Taylor, D. L. (1988). *Biophys. J.,* in press.

Simons, E. R., Davies, T. A., Greengerg, S. M., Dunn, J. M., and Horn, W. C. (1988). *In* "The Cell Physiology of Blood" (R. B. Gunn and J. C. Parker, eds.). Rockefeller Univ. Press, New York, in press.

Sims, P. J., Waggoner, A. S., Wang, C. H., and Hoffman, J. F. (1974). *Biochemistry* **13,** 3315–3330.

Small, J. V., Zobeley, S., Rinnerthalen, G., and Faulstich, H. (1988). *J. Cell Sci.* **89,** 21–24.

Spring, K. R., and Smith, P. D. (1987). *J. Microsc.* **147,** 265–278.

Struck, D. K., Hoekstra, D., and Pagano, R. E. (1981). *Biochemistry* **20,** 4093–4099.

Taylor, D. L., Amato, P. A., McNiel, P. L., Luby-Phelps, K, and Tanasugarn, L. (1986). *In* "Applications of Fluorescence in the Biomedical Sciences" (D. L. Taylor, A. S. Waggoner, R. F. Murphy, F. Lanni, and R. R. Birge, eds.), pp. 347–376. Liss, New York.

Thorell, D. (1983). *Cytometry 4,* 61–65.

Titus, J. A., Haugland, R. P., Sharrow, S. O., and Segal, D. M. (1982). *J. Immunol. Methods* **50,** 93.

Valet, G., Raffael, A., Moroder, L., Wunsch, E., and Ruhenstroth-Bauer, G. (1981). *Naturwiss enschaften* **68,** 265–266.

Waggoner, A. S. (1985). *In* "The Enzymes of Biological Membranes" (A. Martonosi, ed.), Vol. I. pp. 313–331. Plenum, New York.

Waggoner, A. S. (1986). *In* "Applications of Fluorescence in the Biomedical Sciences" (D. L. Taylor, A. S. Waggoner, R. F. Murphy, F. Lanni, and R. R. Birge, eds.), pp. 3–28. Liss, New York.

Waggoner, A. S., and Stryer, L. (1970). *Proc. Natl. Acad. Sci. U.S.A.* **67,** 579–589.

Wang, Y.-L., and Taylor, D. L. (1980). *J. Histochem. Cytochem.* **28,** 11998–1206.

Wang, Y.-L., Heiple, J., and Taylor, D. L. (1982). *In* "Methods in Enzymology" (C. H. W. Hirs and S. N. Timasheff, eds.), Vol. 25, Part B, pp. 1–11. Academic Press, New York.

Webb, W. W., and Gross, D. (1986). *In* "Applications of Fluorescence in the Biomedical Sciences" (D. L. Taylor, A. S. Waggoner, R. F. Murphy, F. Lanni, and R. R. Birge, eds.), pp. 405–422. Liss, New York.

Weiss, M. J., and Chen, L. B. (1984). *Kodak Lab. Chem. Bull.* **55,** 1–4.

Wilderspin, A. F., and Green, N. M. (1983). *Anal. Biochem.* **132,** 449–455.

Wilson, H. A., Chused, T. M., Greenblatt, D., and Finkelman, F. D. (1988). *In* "The Cell Physiology of Blood" (R. B. Gunn and J. C. Parker, eds.). Rockefeller Univ. Press, New York, in press.

Wolf, B. M., and Waggoner, A. S. (1986). The Soc. Gen. Physiol. Ser. **40,** 101–113.

Wolfbeis, O. S., Furlinger, E., Kroneis, H., and Massoner, H. (1983). *Fresenius A. Anal. Chem.* **314,** 119–124.

Index

B

C

E

G

L

M

N

O

P

Q

T

CONTENTS OF RECENT VOLUMES

Volume 25

The Cytoskeleton
Part B. *Biological Systems and* in Vitro *Models*

Volume 26

Prenatal Diagnosis: Cell Biological Approaches

Volume 27

Echinoderm Gametes and Embryos

Volume 28

Dictyostelium Discoideum: Molecular Approaches to Cell Biology

Part I. *General Principles*

Part II. *Vesicular Traffic in Cells*

Part III. *Cell Motility*

Volume 29

Fluorescence Microscopy of Living Cells in Culture
Part A. *Fluorescent Analogs, Labeling Cells, and Basic Microscopy*